CONCEPTS AND CHALLENGES IN
Physical Science

THIRD EDITION

ANNOTATED TEACHER'S EDITION

Leonard Bernstein • Martin Schachter • Alan Winkler • Stanley Wolfe

STANLEY WOLFE
Project Coordinator

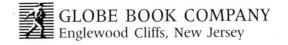 GLOBE BOOK COMPANY
Englewood Cliffs, New Jersey

Leonard Bernstein
Martin Schachter
Alan Winkler
Stanley Wolfe

 Globe Book Company
A Division of Simon and Schuster, Inc.
Englewood Cliffs, New Jersey 07632

PRINTED IN THE UNITED STATES OF AMERICA

10 9 8 7 6 5   95 94 93

ISBN: 1-55675-747-6

Contents

CONCEPTS AND CHALLENGES IN PHYSICAL SCIENCE PROGRAM

Student's Edition

Annotated Teacher's Edition

Teacher's Resource Book including

Lesson Review Worksheets
TechTerm Review Worksheets
Enrichment Worksheets
Evaluation Package
Foreign Language Supplement—Spanish

Laboratory Program, Student Edition

Laboratory Program, Annotated Teacher's Edition

CONCEPTS AND CHALLENGES
In Life, Earth, and Physical Science
Third Edition

Educators agree that teaching science successfully requires special tools. The essentials include a strategy for teaching that maximizes the potential for learning success and maintains a high level of interest and motivation in the classroom.

Concepts and Challenges in Life, Earth, and Physical Science translates this strategy into a practical teaching and learning tool. Experience *Concepts and Challenges in Science* and maximize your students' potential for learning success!

A wealth of resources to introduce your students to a wide range of science topics.

Life Science

Covering a broad spectrum of topics, this text provides a fascinating introduction to the myriad life forms that inhabit our world. Students will also learn about the many variables that *affect* life: The diseases that afflict different organisms...the effect of ecology on plants and animals...how various life forms adapt to changes.

Unit Topics: Scientific Methods and Skills • Needs of Living Things • Ecology •Cells and Organs • Classification • Simple Organisms • Plants • Plant Structure and Function • Animals Without Backbones • Animals With Backbones • Nutrition and Digestion • Support and Movement • Transport • Respiration and Excretion • Regulation and Behavior • Health and Disease • Reproduction and Development • Heredity and Genetics • Change Through Time

Earth Science

From our own world to the outer reaches of space, the text teaches all about the earth, and also explores other heavenly bodies of the universe. Students will gain insights into the various characteristics that make the earth unique in the solar system, and learn how the objects and phenomena of the universe that they study affect the way we live on our planet.

Unit Topics: Scientific Methods and Skills • Studying the Earth • Minerals • Rocks and How They Form • Wearing Down the Earth • Agents of Erosion • Building up the Earth • Plate Tectonics • The Rock Record • The Hydrosphere • The Atmosphere • Weather • Climate • Natural Resources • Exploring Space • The Solar System • Motion of the Earth • Stars

Physical Science

The principles, theories and assumptions that guide our understanding of the world and universe are explored in this text. Physics, chemistry, mathematics are just some of the major topics that help students comprehend the processes that shape every aspect of the universe. They'll learn what causes reflections in a mirror...the nature of matter...how machines work...and more.

Unit Topics: Scientific Methods and Skills • Force • Energy and Work • Motion •Machines • Heat • Waves • Sound • Light • Electricity • Magnetism • Matter • Density • Atoms • Compounds and Mixtures • Chemical Formulas • Chemical Reactions • Metals • Solutions • Suspensions • Acids, Bases, and Salts

THE STUDENT EDITION

The centerpiece of the *Concepts and Challenges in Science* approach to science instruction are student books specially designed to help students reading below grade-level grasp science concepts more readily, while motivating them to learn.

Each book introduces and reinforces the basis of all scientific activity—The Scientific Method. Manageable two-page lessons then present each topic with a "concept" on the left-hand page. The "challenge" on the facing, right-hand page then measures comprehension with engaging activities. Other features include:

- Learning objective and vocabulary lists to define goals and enhance learning for each lesson.
- One idea introduced in each paragraph to build student confidence by introducing concepts gradually.
- Frequent questions to reinforce understanding of concepts presented.
- Activities that develop interdisciplinary critical-thinking, problem-solving, and collaborative learning skills.
- Colorful illustrations and photographs that bring concepts to life.
- Unit Reviews that provide additional comprehension questions, study skills sharpeners and independent learning activities.

Equally important, students are introduced to scientists, science careers, and science's practical applications. This "real-world" emphasis enhances students' interest in the material—and motivates them to learn.

6-4 What are algae?

Objective ▶ Identify and describe different kinds of algae.

TechTerms

- **chlorophyll** (KLAWR-uh-fil): green pigment used by some organisms to make food
- **plankton** (PLANK-tun): organisms that float at the water's surface

Algae Algae are classified in the protist kingdom. All algae contain **chlorophyll** (KLAWR-uh-fil). Chlorophyll is a green pigment found in plant cells and algae. Plants and algae use chlorophyll to make food. The food-making process is called photosynthesis (foht-uh-SIN-thuh-sis).

▥ *Compare:* How are algae and plants alike?

One-Celled Algae Some kinds of algae are made up of only one cell. Most one-celled algae live in a watery environment. They float on the surface of the water. Algae make up **plankton** (PLANK-tun). Plankton are all the tiny plants and animals that float on the surface of the oceans and lakes. Although all algae have chlorophyll, all kinds of algae are not green. Fire algae are one-celled algae that are red in color. Golden-brown algae can have a color ranging from yellowish-green to golden-brown. These differences in color are due to other pigments, or colorings, in the algae.

▥ *Identify:* What is plankton?

Euglena The euglena is a unicellular, or one-celled, algae. It is both plantlike and animallike. Like all algae, the euglena contains chlorophyll. It uses the chlorophyll to make food from sunlight. When light is not available, the euglena also can take in food. A structure in the euglena called the eyespot can detect light. The euglena uses its eyespot to find light in the water. The euglena moves to the light by beating the flagellum that extends from its body.

▥ *Explain:* How is the euglena animallike?

Many-Celled Algae Some kinds of algae are made up of many cells. These are the multicellular algae. Pigments in multicellular algae may give them a green, red, or brown coloring. Most green algae live in fresh water.

Red and brown algae live in the ocean. Brown algae often are called seaweed or kelp. Brown algae are the largest and most complex kind of algae. If you have visited a beach, you may have seen brown algae washed up along the shoreline.

▶ *Infer:* Why are all types of algae not green?

UNIT 1

SCIENTIFIC METHODS AND SKILLS

CONTENTS

STUDY HINT As you read each lesson in Unit 1, write the lesson title and lesson objectives on a sheet of paper. After you complete each lesson, write the sentence or sentences that answer each objective.

---

used by

- **plankton** (PLANK-tun): organisms that float at the water's surface

Algae Algae are classified in the protist kingdom. All algae contain **chlorophyll** (KLAWR-uh-fil). Chlorophyll is a green pigment found in plant cells and algae. Plants and algae use chlorophyll to make food. The food-making process is called photosynthesis (foht-uh-SIN-thuh-sis).

▥ *Compare:* How are algae and plants alike?

One-Celled Algae Some kinds of algae are made up of only one cell. Most one-celled algae live in a watery environment. They float on the surface of the water. Algae make up **plankton** (PLANK-tun). Plankton are all the tiny plants and animals that float on the surface of the oceans and lakes. Although all algae have chlorophyll, all kinds...

LESSON SUMMARY

- All algae contain chlorophyll an own food by photosynthesis.
- Some kinds of algae are one-cell
- All algae are not green, but still phyll.
- The euglena is a single-celled both plantlike and animallike.
- Some algae are multicellular.
- Red and brown algae are multice

CHECK *Complete the following.*

make thei

The convenient Unit Opener page gives students a quick preview of concepts to be covered. The Study Hint provides helpful tips for more effective learning.

Only one idea is introduced in each short paragraph, an effective learning approach for the reluctant reader plus phonetic spellings of new or complex terms to aid in pronunciation.

Lesson Summaries, at the beginning of each "challenges" page reinforce concepts learned by providing an encapsulated review of key ideas.

LESSON SUMMARY

► All algae contain chlorophyll and make their own food by photosynthesis.

► Some kinds of algae are one-celled.

► All algae are not green, but still have chlorophyll.

► The euglena is a single-celled algae that is both plantlike and animallike.

► Some algae are multicellular.

► Red and brown algae are multicellular algae.

CHECK Complete the following.

1. Algae _____ make their own food.

2. Plants and algae both contain the green pigment _____ .

3. The euglena uses its _____ to detect light.

4. Fire algae are _____ in color.

5. The largest kind of algae is _____ .

APPLY Complete the following.

6. **Compare:** How is the Euglena similar to the trypanosome?

7. Suppose you needed to obtain a water sample that contained both algae and protozoans. Where would you go to obtain the sample?

Skill Builder...................

⚠ *Organizing* When you organize information, you put the information into some kind of order. A table is one way to organize information. Make a table with the following headings: "Type of Algae" and "Product." Use library references to determine the kinds of algae used to make the following products: toothpaste, marshmallows, ice cream, detergent, cheese, and silver polish.

◆○◆ SCIENCE CONNECTION ◆○◆○◆○◆○◆○◆○◆○◆○◆○◆○◆○◆○◆○◆○◆○◆
ALGAL BLOOMS

Algae make their own food by photosynthesis. For photosynthesis to occur, three things must be present -- chlorophyll, sunlight, and carbon dioxide. However, algae also must have water and other nutrients, such as nitrogen and phosphorous, to carry out their life processes. When all of these elements are present, algae can reproduce quickly and in large numbers. A sudden, large growth of algae is called an algal bloom.

Algal blooms often occur when wastes containing phosphates are dumped into ponds or lakes. Phosphates are commonly found in detergents. When an algal bloom occurs, algae can be seen living at the surface of a lake or pond in great numbers. However, an algal bloom can be a problem for organisms that live in the lake or pond. As the algae die, they are broken down by bacteria. The bacteria use oxygen from the water when they break down the algae. If the bacteria use too much oxygen, fishes and other organisms living in the water do not have enough oxygen to carry out their life processes. As a result, the organisms may die.

123

◆○◆ SCIENCE CONNECTION ◆○◆
ALGAL BLOOMS

Algae make their own food by photosynthesis. For photosynthesis to occur, three things must be present -- chlorophyll, sunlight, and carbon dioxide. However, algae also must have water and other nutrients, such as nitrogen and phosphorous, to carry out their life processes. When all of these elements are present, algae can reproduce quickly and in large numbers. A sudden, large growth of algae is called an algal bloom.

Algal blooms often occur when wastes containing phosphates are dumped into ponds or lakes. Phosphates are commonly found in detergents. When an algal bloom occurs, algae can be seen living at the surface of a lake or pond in great numbers. However, an algal bloom can be a problem for organisms that live in the lake or pond. As the algae die, they are broken down by bacteria. The bacteria use oxygen from the water when they break down the algae. If the bacteria use too much oxygen, fishes and other organisms living in the water do not have enough oxygen to carry out their life processes. As a result, the organisms may die.

Lesson Features include:
* *Science Connection Activities*
* *People in Science*
* *Leisure Activities*
* *Technology and Society*
* *Looking Back in Science*
* *Careers in Science*

Designed to motivate students to learn with thought-provoking facts, interesting insights and exercises that make science fun.

golden-br.... ....ese differences in color are ... to other pigments, or colorings, in the algae.

▐▶ *Identify:* What is plankton?

Euglena The euglena is a unicellular, or one-celled, algae. It is both plantlike and animallike. Like all algae, the euglena contains chlorophyll. It uses the chlorophyll to make food from sunlight. When light is not available, the euglena also can take in food. A structure in the euglena called the eyespot can detect light. The euglena uses its eyespot to find light in the water. The euglena moves to the light by beating the flagellum that extends from its body.

122

Red and brown algae live in the ocean. Brown algae often are called seaweed or kelp. Brown algae are the largest and most complex kind of algae. If you have visited a beach, you may have seen brown algae washed up along the shoreline.

▶ *Infer:* Why are all types of algae not green?

Frequent questions, interwoven throughout the lesson reinforce understanding and give students immediate insights into the relationship of scientific principles.

e their

hloro-

hat is

algae.

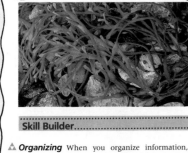

Skill Builder...................

⚠ *Organizing* When you organize information, you put the information into some kind of order. A table is one way to organize information. Make a table with the following headings: "Type of Algae" and "Product." Use library references to determine the kinds of algae used to make the following products: toothpaste, marshmallows, ice cream, detergent, cheese, and silver polish.

A wide variety of Challenge Features provide exercises designed to help students develop thinking and learning skills and helps to relate science to their daily lives: Skill Builders, Ideas in Action, Designing an Experiment, Infosearch, Health and Safety Tip, and State the Problem.

► Red and brown ....re multicellular algae.

CHECK *Complete the following.*

1. Algae _____ make their own food.

2. Plants and algae both contain the green pigment _____ .

3. The euglena uses its _____ to detect light.

4. Fire algae are _____ in color.

5. The largest kind of algae is _____ .

APPLY *Complete the following.*

6. **Compare:** How is the Euglena similar to the trypanosome?

The two-part lesson review encourages students to think about what they've learned, and helps them grasp abstract concepts by having them apply what they know to hypothetical situations.

STUDENT BOOK UNIT REVIEW

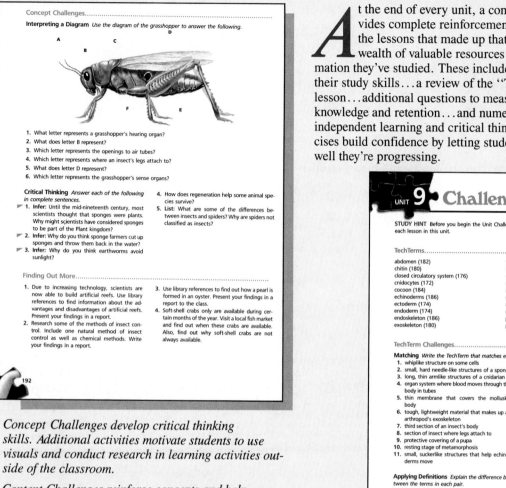

At the end of every unit, a comprehensive Unit Review provides complete reinforcement of all key concepts learned in the lessons that made up that unit. Students will find a wealth of valuable resources to help them retain the information they've studied. These include tips to help students sharpen their study skills...a review of the "Tech Terms" learned in each lesson...additional questions to measure concept and content knowledge and retention...and numerous activities that promote independent learning and critical thinking. In addition, these exercises build confidence by letting students see for themselves how well they're progressing.

Concept Challenges develop critical thinking skills. Additional activities motivate students to use visuals and conduct research in learning activities outside of the classroom.

Content Challenges reinforce concepts and help students practice science skills. A wide range and variety of questions hold student interest.

Reading Critically encourages the practice of interdisciplinary skills and measures concept understanding. Page numbers in parentheses help students find sections they may want to re-read.

TEACHER'S ANCILLARY MATERIALS

The **Teacher's Resource Book** is a complete package of teaching aids, conveniently organized in an easy-to-use binder. If you have ESL students, you'll appreciate the Spanish supplement. And all worksheets, reviews and testing material—more than 300 pages—are cross-referenced to Student Book lessons. All teaching aids are in reproducible Blackline Master format.

Vocabulary Practice *activities designed to make learning more fun.*

The Enrichment Worksheet *to help develop critical thinking skills, while motivating them to explore concepts further.*

A Lesson Review Worksheet *provides a variety of review exercises and reinforces key concepts.*

The Annotated Teacher's Edition provides complete Student Book text along with on-page annotations. It also contains other vaulable information, including:

- A correlation chart
- Teaching Tips
- Suggestions for cooperative learning activities

- A Materials List
- Supplies and address lists for equipment and materials

LABORATORY PROGRAM

Hands-on activities and experiments for students are central to the *Concepts and Challenges in Science* Laboratory Program. Fun and interesting, these formal laboratories enable students to see a wide range of basic scientific processes firsthand. Each Laboratory Program:

- Correlates to activities in the Student Edition to synthesize the lesson concept and help students see it in action.
- Gives students an opportunity to practice the Scientific Method and enhance their critical thinking skills.
- Encourages the use of science process skills as an aid to comprehension.
- Enhances comprehension with numerous drawings, tables, charts, and graphs.

With every experiment students perform, they'll gain not only understanding, but the *confidence* that builds learning success.

A question identifies the problem that's being explored.
Easily obtainable materials used in the experiment are clearly listed. The skills that students will use are also listed.
Step-by-step procedures, and diagrams are provided as a learning aid.
Ample space is provided for recording observations.
A series of simple questions to strengthen observation skills, and the ability to reach logical conclusions are included.

The Annotated Teacher's Edition of the Laboratory Program provides an overview of the laboratory activities and advice on teaching students reading below grade-level. Features include:

- A section on Reading Skills and laboratory procedures
- Using Laboratory Skills Worksheets
- Teaching Tips
- Suggestions for cooperative learning activities
- Guidelines for Safety
- Materials List
- A comprehensive Skills Matrix
- Supplies and address lists for equipment and materials

SKILLS DEVELOPMENT IN PHYSICAL SCIENCE

During the late 1980s, science teaching underwent some major shifts. Previous to the late 1980s, science teaching was based largely on recitation of vocabulary terms, facts, concepts, and principles for students to write down as notes. Mastery of the subject was evaluated upon each student's ability to memorize and recall isolated facts on short-answer-type exams. Science curricula and pedagogical strategies have shifted in philosophy and in practice. Science curricula stress relationships among the facts, concepts, and principles of each discrete science and among the different sciences. The infusion of reading, writing, and study skills also has been noted. As well, science instruction has begun to focus on teaching the methods, processes, and skills of science and on promoting the development of critical thinking by students. CONCEPTS AND CHALLENGES IN PHYSICAL SCIENCE has been developed with these new directions in mind. The elements have been incorporated into the text in a readable, relevant, and lively textbook. The writing style is highly motivating, interesting, and challenging without being overwhelming.

Study Skills

Good study skills are important to all subject areas. To help students develop their study skills, *Concepts and Challenges* employs the following features or techniques:

- **Study Hints** Each unit opening page presents a Study Hint for the students. Six different strategies for approaching the units in the text are used. Each Study Hint is used three times throughout the text.

- **Unit Challenge Study Hints** Each unit review (called the Unit Challenge) presents a Study Hint as its first element. The Study Hint is uniform throughout the text. The Study Hint provides students with a useful tip for reviewing the unit *before* beginning the exercises.

To help students develop good study skills, you may want to suggest that students follow one or more of these procedures as they study physical science:

1. Previewing Read the title and objective of each lesson. Review the TechTerms, their phonetic respellings, where applicable, and their definitions.

2. Note Taking Each lesson is divided into several short sections made up of one or two paragraphs. Each paragraph has a topic sentence. Encourage students to write the topic sentence for each paragraph.

3. Re-reading Encourage students to re-read short sections if they cannot answer the question at the end of the section.

4. Observing and Analyzing Encourage students to make use of the graphics on each page to help them understand a concept.

5. Questioning Encourage students to write questions about the content as they read the lesson.

6. Answering Questions Writing often reinforces learning. Have students write answers to in-text questions and assessment questions. Have students discuss their written responses in small groups or as a class.

Communication Skills

Communication skills, such as reading, writing, speaking, and listening, are the core of subject mastery. In fact, good communication skills are needed to acquire and to apply knowledge in all disciplines.

CONCEPTS AND CHALLENGES IN PHYSICAL SCIENCE has been developed specifically to incorporate these skills and to make these skills an integral part of the program. These skills have been used through the textbook in many forms.

Comprehension and Reading

The content, style, format, and overall appearance of a text play a major role in reading comprehension. *Concepts and Challenges* has been developed with student comprehension in mind.

1. Each lesson is two pages—a self-contained lesson. While reading or studying, students do not have to move from page-to-page. They can concentrate on the complete concept before them.

2. Each lesson is visually attractive and non-threatening.

3. Each lesson has a consistent format.

   a. **Lesson Title** Each title is presented as a question. The questions provide students with a clear aim of the day's lesson.

   b. **Objective** One or two clearly stated objectives stated in behavioral terms provide students with learning goals.

   c. **TechTerms** Students are introduced to new terms *prior* to beginning the lesson.

   d. **One-idea paragraphs** Each paragraph stresses one idea, building to the lesson concept.

Frequent Review

Each short section in a lesson is followed by a question. Questions stress reading comprehension and skills development. Reading comprehension questions have answers explicitly stated in the reading. If the student is unable to answer the question, the student should be able to find the answer by re-reading the passage. Skills questions are more challenging but can be answered by re-reading a passage or using the graphics on the page.

The gratification of answering these questions successfully reinforces learning and acts as motivation to read each short section carefully. The format encourages students to re-read a short section when necessary until the content or the concept is understood.

Writing

Writing is important to all academic disciplines. Writing involves a number of skills including the ability to collect information, to analyze information, and to organize the information. Writing involves formulating ideas and sequencing relationships, drafting, ordering, and articulating paragraphs, and revising and polishing written materials to reach a final draft.

CONCEPTS AND CHALLENGES IN PHYSICAL SCIENCE enhances writing competence by correlating writing activities directly to the text. In this way, reading and writing complement each other. Challenge features, such as Skill Builders and InfoSearch, encourage students to read and write about various topics. Skill Builders also stress the application of science and language arts skills. Other features, such as Design an Experiment, reinforce scientific skills and processes while providing students with practice that combines reading, writing, and communicating information to another person in a clear, concise way.

Vocabulary

The study of physical science requires a working knowledge of words and terms that may be unfamiliar to students. *Concepts and Challenges* introduces vocabulary terms only when they are essential to the key concept or principle in each lesson. Lessons generally have between one and four new terms. In this way, students are introduced to important terms and develop a working physical science vocabulary without the burden of memorizing too many new terms.

Each term is introduced before the lesson under the heading TechTerms. The terms are phonetically respelled and defined. Within the lesson, the science terms are boldfaced, phonetically respelled, and defined in context. Students are thus provided with a preview of terms and a reinforcement of the same terms.

Recognizing that vocabulary skills development is important, vocabulary skills, such as using prefixes, suffixes, and root words, identifying word origins, and so on, are further highlighted and developed in many Skill Builder features. Vocabulary development also is reinforced in the Unit Challenge exercises. Besides evaluating knowledge of vocabulary, these exercises also stress word relationships and applying definitions.

Science Process Skills and Methods

The emphasis on science process skills and their relationship to scientific methods has become an integral part of science teaching pedagogy. The implementation of science process skills and methods is a unique and important feature of *Concepts and Challenges in Physical Science.* Throughout the text, science process skills are highlighted and integrated with the content and assessments. The nine skills are organizing, classifying, measuring, observing, inferring, predicting, analyzing, modeling, and hypothesizing. Each skill is identified by name, a symbol, and a definition. Methods also are highlighted in various features of the text.

- Lesson 1-2 is devoted to explaining the nine highlighted skills in the text in an interesting, readable format. Two additional skills—communicating and researching—are introduced.

- Lesson 1-3 is devoted to explaining scientific methods.

- Lessons 1-4 through 1-7 stress measuring while using several other skills throughout each lesson.

Concepts and Challenges also stresses science process skills and methods by means of the following techniques or features:

- highlighting with symbols those questions that require the use of a particular skill.

- incorporating a wide range of science process skills in Lesson Activities and Skill Builders

- encouraging application of science process skills in such lesson features as Activities, State the Problem, and Design an Experiment, and in the Critical Thinking and Interpreting a Diagram sections of the Unit Challenges.

Mathematics Skills

The language of the scientists includes mathematics. Throughout *Concepts and Challenges in Physical Science*, students are encouraged to apply basic arithmetic skills. The program also illustrates the close relationship between science and mathematics.

In *Concepts and Challenges*, students are provided opportunities to add, subtract, multiply, and divide using whole numbers and decimals. Students also are encouraged to use mathematical formulas to solve problems. Most formulas are presented in colorful display type. All mathematics questions are prefaced with the skill term Calculate.

Mathematical skills are necessary to understand measuring tools and to use these tools accurately. *Concepts and Challenges in Physical Science* encourages students to use a variety of measuring tools. For example, students are provided with opportunities to use metric rulers, calibrated glassware, and balances.

Skills Matrix

A comprehensive skills and features matrix is provided for *Concepts and Challenges in Physical Science* on pages T-xii–T-xvi. The scope and sequence of the skills used throughout the program can be used as a teaching tool to help develop and reinforce student application of skills.

COMPREHENSIVE SKILLS AND FEATURES MATRIX BY LESSON

	1-1	1-2	1-3	1-4	1-5	1-6	1-7	2-1	2-2	2-3	2-4	2-5	2-6	2-7	2-8	2-9	3-1	3-2	3-3	3-4	3-5	3-6	3-7	4-1	4-2	4-3	4-4	4-5	4-6	5-1
SCIENCE SKILLS																														
Observe		●					●					●																		
Measure		●												●			●													
Infer	●	●	●		●			●	●	●		●													●				●	●
Organize																														
Classify		●										●							●						●				●	
Predict		●									●														●	●	●			
Hypothesize		●								●		●	●	●		●	●												●	
Model		●																												
Analyze	●	●	●	●	●		●	●	●	●		●		●					●	●				●		●	●			
Research		●																												
Communicate		●																												
INTERDISCIPLINARY SKILLS																														
Define			●	●		●			●	●		●		●	●		●					●						●		●
Build Vocabulary																										●				
Finding Main Ideas																														
Identify	●	●	●		●			●		●		●		●	●	●		●	●		●	●		●	●	●				
Locate																														
Compare			●		●						●			●			●				●			●						
Contrast																						●					●			
Sequence																														
Calculate			●	●	●		●					●	●								●	●	●					●		●
CHALLENGE FEATURES																														
Skill Builder	●		●		●								●		●									●	●	●		●	●	
InfoSearch									●	●												●								
Ideas in Action		●		●				●				●		●			●									●				●
Health & Safety					●										●					●									●	
State the Problem								●			●									●										
Design an Experiment				●							●							●				●								
LESSON FEATURES																														
Activity		●						●	●	●		●					●							●		●				
Career	●							●							●							●				●				
Technology																		●											●	●
Science Connection			●												●	●		●										●		
Leisure Activity												●					●										●			
People in Science		●									●									●									●	
Looking Back in Science				●																										

5-2	5-3	5-4	5-5	5-6	6-1	6-2	6-3	6-4	6-5	6-6	6-7	6-8	6-9	7-1	7-2	7-3	7-4	7-5	7-6	8-1	8-2	8-3	8-4	8-5	8-6	8-7	8-8	8-9	9-1	9-2	9-3	9-4	9-5	9-6	9-7	9-8	9-9	9-10	10-1	10-2	10-3

Category	Skill	10-4	10-5	10-6	10-7	10-8	10-9	11-1	11-2	11-3	11-4	11-5	11-6	11-7	11-8	11-9	12-1	12-2	12-3	12-4	12-5	12-6	12-7	12-8	12-9	13-1	13-2	13-3	13-4	13-5	14-1
SCIENCE SKILLS	Observe								•		•						•			•		•		•					•		
	Measure												•																		
	Infer	•		•			•	•		•	•	•	•	•	•	•		•		•			•								
	Organize																														
	Classify									•	•		•				•	•	•	•			•								
	Predict								•	•	•			•	•	•													•		
	Hypothesize	•		•		•					•	•		•									•	•							•
	Model		•							•				•														•			
	Analyze				•					•										•			•	•		•		•			
	Research											•																•			
	Communicate																														
INTERDISCIPLINARY SKILLS	Define		•	•	•				•	•	•	•					•	•	•	•		•	•	•	•		•		•	•	•
	Build Vocabulary																														
	Finding Main Ideas																														
	Identify	•		•		•	•										•			•		•	•	•	•						•
	Locate																														
	Compare												•				•						•								
	Contrast	•								•				•	•	•				•											
	Sequence																									•					
	Calculate				•												•									•	•	•			
CHALLENGE FEATURES	Skill Builder		•		•				•		•			•		•		•		•							•	•			
	InfoSearch			•				•		•			•		•		•					•		•							•
	Ideas in Action		•											•			•		•									•			
	Health & Safety																				•									•	
	State the Problem					•																									
	Design an Experiment	•					•																•		•						
LESSON FEATURES	Activity	•						•		•		•					•			•							•		•		
	Career			•	•										•				•							•		•			
	Technology			•					•																	•					•
	Science Connection					•										•		•								•					
	Leisure Activity																					•								•	
	People in Science													•								•	•								
	Looking Back in Science		•					•				•																			

14-2	14-3	14-4	14-5	14-6	15-1	15-2	15-3	15-4	15-5	15-6	15-7	15-8	16-1	16-2	16-3	16-4	16-5	16-6	17-1	17-2	17-3	17-4	17-5	17-6	17-7	18-1	18-2	18-3	18-4	18-5	18-6	19-1	19-2	19-3	19-4	19-5	19-6	19-7	19-8	19-9	20-1

		20-2	20-3	20-4	20-5	21-1	21-2	21-3	21-4	21-5	21-6
SCIENCE SKILLS	Observe		●		●			●			
	Measure										
	Infer		●	●	●	●				●	
	Organize										
	Classify	●		●	●		●		●		
	Predict	●	●					●			
	Hypothesize		●		●		●				
	Model	●	●								
	Analyze				●						
	Research										
	Communicate										
INTERDISCIPLINARY SKILLS	Define	●		●	●				●	●	●
	Build Vocabulary										
	Finding Main Ideas										
	Identify	●	●		●	●	●		●	●	
	Locate										
	Compare			●							
	Contrast		●								
	Sequence										
	Calculate										
CHALLENGE FEATURES	Skill Builder	●	●				●		●		
	InfoSearch			●							
	Ideas in Action								●		
	Health & Safety				●	●					
	State the Problem										
	Design an Experiment							●			●
LESSON FEATURES	Activity		●		●			●	●		●
	Career										
	Technology										
	Science Connection	●		●		●	●				
	Leisure Activity										
	People in Science										
	Looking Back in Science								●		

MATERIALS LIST
(per 30 students)

Item	Quantity	Lesson
Alcohol, rubbing	1 bottle	9-6
Aluminum foil	1 roll	17-1
Ammonium chloride	1 bottle	19-5
Balances	30	13-4, 17-1, 17-7
Bar magnets	60	11-2, 11-4
Batteries, 1.5-volt	60	11-6
Beakers	120	1-3, 1-4, 1-5, 6-7, 9-6, 18-6, 19-4, 19-5, 21-3
Beakers, 100-mL	60	17-7
Beakers, 150-mL	30	15-1
Blocks, wood	30	5-3
Boards, wood	30	5-5
Books	90	3-6, 4-1, 5-4, 5-5, 9-7
Bottles or jars, narrow-necked	30	12-4
Bowls	30	7-5, 20-3
Boxes, cardboard	60	6-8
Buttons	30	19-9
Calcium chloride	1 bottle	17-7
Cardboard	30 pieces	4-1, 9-8
Combs, plastic or rubber	30	10-1
Containers	30	11-4
Copper strips	30	10-4
Corks	30	11-4
Corn syrup	1 bottle	13-2
Cups, styrofoam	30	8-2
Drug labels	90	20-1
Eggbeaters	30	20-3
Eggs, hard-boiled	30	18-5
Eggs, raw	30	20-3
Eyedroppers	30	6-7, 21-6
Filter paper	1 box	9-6
Flashlights	30	7-4, 9-1
Food coloring, dark	1 bottle	6-7
Glasses, drinking	30	8-1, 12-1, 19-9
Glycerine	1 bottle	13-2
Golf balls	30	13-4
Graduated cylinders	30	1-4, 13-2, 13-4, 17-7, 19-4, 21-3
Hand lenses	30	20-5

Item	Quantity	Lesson
Heat sources	30	1-5, 6-7, 19-5, 19-7, 21-3
Hydrogen peroxide	1 bottle	15-1
Ice cubes	1 bag/ lesson	1-3, 1-5, 12-4
Index cards	90	9-1
Iron filings	1 box	11-2
Knives	30	9-6, 15-1
Leaves	30	9-6
Lemons	30	10-4
Lenses, convex	15	9-8
Litmus paper, blue	1 package	21-6
Litmus paper, red	1 package	21-6
Marbles	30	4-1
Marking pens	30	7-4, 12-1
Masses, small	30	2-3
Matches	1 box	17-1
Measuring cups	30	20-3
Metersticks	30	3-6, 5-3
Metric rulers	30	1-5, 2-1, 4-1, 5-5, 7-4, 8-6, 9-7, 9-8
Mirrors, flat	30	7-4, 9-7
Modeling clay, red	1 package	15-6
Modeling clay, white	1 package	15-6
Nail polish, clear	1 bottle	18-6
Nails or thumbtacks	30	5-4
Nails, iron	90	11-6, 18-6
Needles, sewing	30	11-4
Objects, small	60	2-3, 2-6
Pails	30	12-1
Pans, shallow	90	6-3
Paper	30 sheets/ lesson	7-4, 9-8, 10-1, 16-2, 16-6, 17-1, 17-2, 20-1
Paper clips	1 box	11-6
Paper towels	1 roll	6-3
Paper, construction	30 sheets	11-2
Paper, graph	30 sheets	6-5
Pencils	30	6-5, 16-2, 16-6, 17-2, 19-9, 20-1
Pennies	30	7-5
Petroleum jelly	1 large jar	18-6, 20-5
Plastic wrap	1 roll	6-8

Item	Quantity	Lesson
Potassium carbonate	1 bottle	17-7
Potatoes, raw	30	15-1
Protractors	30	7-4
Pulleys, movable	30	5-4
Red cabbage leaves	120	21-3
Ribbons	30	7-3
Ring stands	30	6-7
Rocks, small	30	13-4
Rope	90 meters	7-3
Safety goggles	30 pairs	11-2
Salt	1 box	1-3, 19-7
Samples, assorted		21-3, 21-6
Sandpaper	30 sheets	2-6
Scissors	30	6-8
Slides, glass	120	20-5
Spoons	30	19-4, 19-7, 19-9, 21-6
Spoons, silver	30	18-5
Spring scales	30	2-3, 2-6, 3-6, 5-4, 5-5
Stirring rods	30	1-3, 17-7, 19-5
Strainers	30	21-3
String	1 large ball	3-6, 5-4, 5-5, 8-2, 19-9
Sugar cubes	1 box	19-4
Sugar, granulated	1 box	19-9
Tablespoons	30	1-3
Tape	1 roll	2-6, 6-8

Item	Quantity	Lesson
Test tube racks	30	8-6, 1-4
Test tubes	150	1-4, 8-6
Thermometers, Celsius	60	1-5, 6-8, 19-5
Thermometers, unmarked	30	1-5
Thumbtacks	30	11-4
Tissues	1 box	12-1
Toothpicks	1 box	15-6
Tuning forks	30	8-1
Vegetable oil	1 bottle	13-2, 20-3
Vinegar	1 bottle	18-6, 20-3
Voltmeters	30	10-4
Watch or clock with second hand	1	4-1
Water		1-3, 1-4, 6-3, 6-7, 7-5, 8-1, 8-6, 11-4, 12-1, 12-4, 13-2, 13-4, 17-1, 18-6, 19-4, 19-5, 19-7, 19-9, 21-3, 21-6
Water, distilled	3 liters	
Wax pencils	30	1-3, 1-5
Wire gauze	30 pieces	6-7
Wire, insulated	2 rolls	10-4, 11-6
Zinc strips	30	10-4

UNIT AND LESSON TEACHING STRATEGIES

UNIT 1
SCIENTIFIC METHODS AND SKILLS (p. 1)

Previewing the Unit

Read the titles of the lessons in this unit aloud. Identify terms in the titles with which students are unfamiliar. Write these terms on the chalkboard. Using the Glossary of the text or a dictionary, define the terms as a class. Have students write the terms and their definitions in their notebooks.

Bulletin Board Suggestions

1. Assemble a bulletin board entitled "Branches of Physical Science." Have students look for photographs and articles illustrating or explaining the two main branches of physical science. Discuss with students what each branch covers.

2. Assemble a bulletin board showing the metric units for length, mass, and volume, and the prefixes that are used to change the size of metric units. Have students practice converting from one metric unit to another using the bulletin board as a guide.

LESSON 1-1
What is physical science? (p. 2)

Teaching Strategies

PREVIEW Before beginning this lesson, have students scan the lesson looking for words with which they are unfamiliar. Have students work in pairs or small groups to define each of the words on their lists.

CLASS ACTIVITY Have students review the titles of the units and lessons in the Table of Contents and classify each unit of the text into its appropriate branch of physical science.

REINFORCEMENT Randomly read out the names of the specialized fields in physical science listed in Table 1 on page 2. Have students describe what is studied in each field.

Answers to Challenges (p. 3)

Apply

7. Answers will vary. Accept all logical responses.

8. Answers will vary. Accept all logical responses.

9. biology and chemistry

Skill Builder: Building Vocabulary

Thermo-: pertaining to heat

Teaching Tips for Career in Physical Science
Scientific Illustrator

CLASS ACTIVITY Have students flip through the pages of their textbooks to see the work of scientific illustrators. Ask students to identify illustrations showing a process, a structure, a cycle, and an equation.

Question

RELATE Why would a background in science be helpful to a scientific illustrator? (Answers will vary. Accept all logical responses.)

LESSON 1-2
What are science skills? (p. 4)

Teaching Strategies

PREVIEW Have students write the lesson title and objective in their notebooks. As students read the lesson, tell them to write the sentence or sentences that provide the information that meets the objective.

DISCUSSION Describe each science skill and give students an example of how people use these skills every day. Make sure students understand the difference between observations and inferences, and hypotheses and predictions.

REINFORCEMENT Emphasize that people make observations using all of their senses, not just with the sense of sight. Have students describe the classroom. Write their observations on the chalkboard.

DEMONSTRATION Show students articles from scientific journals, newspapers, and so on, to demonstrate one way scientists communicate.

CLASS ACTIVITY Have students make charts in which they classify the following items: bread, shirts, cats, nickels, dimes, milk, pants, socks, quarters, orange juice, people, cookies, pennies, shoes, fishes, dogs, dollar bills, eggs.

COOPERATIVE/COLLABORATIVE LEARNING Have students work in pairs to answer the Check and Apply questions.

Answers to Challenges (p. 5)
Apply

7. Answers will vary. Accept all logical responses.

8. Answers will vary. Accept all logical responses.

9. Answers will vary. Accept all logical responses.

Ideas in Action

Answers will vary. Accept all logical responses.

Teaching Tips for People in Science
Stephen Hawking

INTRASCIENCE CONNECTION Black holes form from very massive stars. Towards the end of its life cycle, a very massive star explodes and forms a supernova. The core left behind after a supernova is so dense, that it is swallowed up by its own gravity.

Questions

1. DEFINE What is cosmology? (study of how universe began and how it may develop in the future)

2. How do you think scientists discover black holes? (Answers will vary. Accept all logical responses.)

LESSON 1-3
What is scientific method?

(p. 6)

Teaching Strategies

PREVIEW Before beginning this lesson, read the list of Tech-Terms aloud so students can hear their pronunciations.

DISCUSSION List the steps of scientific method on the chalkboard and describe each step. Emphasize that because each problem is different, scientists may use all or part of scientific method in a way that is suited to each particular problem.

EXTENSION When designing an experiment, scientists look at variables. A variable is anything that can affect the results of an experiment. Discuss with students why a well-designed experiment tests only one variable at a time.

Answers to Challenges (p. 7)
Check

1. information

2. to test a hypothesis

3. model, or guide, used to gather information and solve problems

4. They use tables, charts, graphs, and diagrams.

5. They read and communicate with one another.

Apply

6. Answers will vary. Accept all logical responses.

7. No, because every problem is unique.

8. Answers will vary. Accept all logical responses.

Skill Builder: Writing a Laboratory Report

Check students' reports for a logical use of the steps of scientific method.

Skill Builder: Organizing Data

Check for a logical organization of students' data.

Teaching Tips for Activity
Controlling Variables

Skills: *observing, comparing*

COOPERATIVE/COLLABORATIVE LEARNING Have students work in pairs.

Questions

1. beaker with salt

2. salt

3. Salt makes ice melt faster.

LESSON 1-4
How are length and area measured? (p. 8)

Teaching Strategies

PREVIEW Before students begin the lesson, have them read the Lesson Summary.

REINFORCEMENT Before discussing SI, review the rules for multiplying and dividing by 10, 100, and 1000. This will help students convert from one metric unit to another.

DEMONSTRATION Show students a meterstick and a metric ruler. Point out centimeter and millimeter divisions. Demonstrate the proper way to measure using a metric ruler.

CLASS ACTIVITY Using a meterstick, measure the length and width of your classroom. Then have students calculate the area of the classroom.

Answers to Challenges (p. 9)
Apply

8. 120 square meters

9. **1.** e; **2.** d; **3.** f; **4.** a; **5.** c; **6.** b

10. Answers will vary. Accept all logical responses.

Ideas in Action

Answers will vary. Accept all logical responses.

Teaching Tips for Science Connection
Natural Institute of Standards and Technology

EXTENSION Have students write to the National Institute of Standards and Technology to find out the standards of measurement for light, color, electric energy, radioactivity, X-ray intensity, and sound.

Questions

1. Why are standard measurements important? (Answers will vary. Accept all logical responses.)

2. LIST What are three institutes of the National Institute of Standards and Technology? (Institute for Basic Standards; Institute for Materials Research; Institute for Computer Sciences and Technology)

LESSON 1-5
How are mass and weight measured? (p. 10)

Teaching Strategies

PREVIEW Have students write the lesson title and objectives in their notebooks. As students read the lesson, tell them to write down the sentence or sentences that meet the objectives.

CLASS ACTIVITY If possible, have students find the mass of different objects using a triple-beam balance.

REINFORCEMENT Be sure students understand the difference between mass and weight. Explain that on Earth, mass and weight often are used interchangeably, but that they are not the same. Emphasize this point by calculating the weight of a person on Earth and on the moon. (A person's weight on the moon would be about one-sixth of the person's weight on the earth.)

COOPERATIVE/COLLABORATIVE LEARNING Answer the Check and Apply questions as a class.

Answers to Challenges (p. 11)
Apply

7. Mass is a measure of how much matter an object contains. Weight is a measure of the pull of gravity on an object.

8. Answers will vary. Accept all logical responses.

9. On Earth; gravity on Earth is greater than on Mars because gravity increases with mass.

10. gram

Designing an Experiment

Have volunteers describe their experiments to the rest of the class.

Teaching Tips for Looking Back in Science
Weight Measurements of the Past

CLASS ACTIVITY To emphasize the need for standard measurements, have three students of varying sizes come to the front of the classroom. Compare the size of their hands. Find out which student can hold the most in one hand. Have students discuss the problems that would arise if scientists did not use standard measurements.

Question

What problems could result from using a unit of weight such as the talent? (Answers will vary. Accept all logical responses.)

LESSON 1-6
How is volume measured?
(p. 12)

Teaching Strategies

PREVIEW Before beginning this lesson, read the list of Tech-Terms aloud so students can hear their pronunciations.

CLASS ACTIVITY Bring empty 1-liter and 2-liter bottles to class. Ask students how much liquid each bottle holds. Most students are familiar with these bottles. Point out that the liter is the basic unit of volume in the metric system.

DEMONSTRATION Pour water into a graduated cylinder and show students the correct way to read the volume of a liquid.

REINFORCEMENT Some students may be confused by the relationship between cubic centimeters and milliliters. Show students that 1 cm^3 is equal to 1 mL by calculating the volume of a rectangular plastic container. Then use a graduated cylinder to measure out how much water the container will hold in milliliters. Students will see that the values are equal.

Answers to Challenges (p. 13)
Apply

6. 8 cm^3

7. 90 mL

8. displacement

Skill Builder: Calculating

Check students' tables.

Teaching Tips for Activity
Using a Graduated Cylinder

Skills: *measuring, analyzing*

COOPERATIVE/COLLABORATIVE LEARNING Have students work in pairs.

Questions

1. Answers will vary. Accept all logical responses.

2. a. Answers will vary. Accept all logical responses. **b.** Answers will vary. Accept all logical responses.

LESSON 1-7
How is temperature measured? (p. 14)

Teaching Strategies

PREVIEW Have students write the lesson title and objective in their notebooks. As students read the lesson, tell them to write down the sentence or sentences that meet the objective.

DISCUSSION Bring several different thermometers to class. Point out the bulb and the liquid on each thermometer. Show students how to read a thermometer. Ask students what will happen to the liquid in the thermometer if it is heated. (It will rise.) Make sure students understand the relationship between temperature and heat energy.

CLASS ACTIVITY Tell students to record the daily high and low temperatures in degrees Celsius for the next week. Point out that while the United States uses the Fahrenheit scale, most countries use the Celsius scale. In the U.S., temperature readings usually are given in both degrees Fahrenheit and degrees Celsius.

EXTENSION Have students find out the equations used to convert from Fahrenheit to Celsius and Celsius to Fahrenheit. Have students practice converting from one temperature scale to another.

CLASS ACTIVITY Using Table 1, read out the temperatures of things listed on one of the scales. Have students identify the temperature readings on the other two scales.

Answers to Challenges (p. 15)
Check
1. Fahrenheit
2. by looking at the mark on the tube where the liquid stops
3. how hot or cold something is
4. liquid rises
5. Kelvin
6. absolute zero

Apply
7. 37 °C; 98.6 °F; 310 K
8. 100 °C; 212 °F
9. 22 °C; 295 K
10. add 273 degrees

Health and Safety Tip
A person's temperature may rise when germs enter the body. Body temperature rises to fight off the germs, which will die at high temperatures. However, body cells also are destroyed at high temperatures, which is why high fevers can be very dangerous. Any person with a high fever should see a doctor immediately.

Teaching Tips for Activity
Calibrating a Thermometer
Skills: *analyzing, observing*

COOPERATIVE/COLLABORATIVE LEARNING Answer the questions as a class.

Questions
1. **a.** 100 °C **b.** 0 °C
2. 10

Answers to Unit Challenges (pp. 16–18)
Understanding the Features: Reading Critically
1. Answers will vary. Accept all logical responses.
2. Institute for Basic Standards; Institute for Materials Research; Institute for Computer Sciences and Technology
3. make drawings, diagrams, and sketches to illustrate scientific concepts
4. study of how the universe began and how it may develop in the future

Interpreting a Table
1. kilometer
2. 2013 km
3. meter
4. 645,000
5. The metric system is not commonly used for everyday measurements in the United States.

Critical Thinking
1. Answers will vary. Accept all logical responses.
2. Answers will vary. Accept all logical responses.
3. 2 liters
4. Every problem is unique.
5. Every substance has specific melting and boiling points.

FORCE (p. 19)

Previewing the Unit

Read the titles of the lessons in this unit aloud. Identify terms in the titles with which students are unfamiliar. Write these terms on the chalkboard. Using the Glossary of the text or a dictionary, define the terms as a class. Have students write the terms and their definitions in their notebooks. Then, have students work in small groups to carry out the task in the Study Hint.

Bulletin Board Suggestion

Assemble a bulletin board showing different kinds of forces. Tell students to find and collect pictures showing forces in nature and people applying forces. Attach the pictures to the bulletin board. Identify whether the force in each picture is a push or a pull.

LESSON 2-1
What is a force? (p. 20)

Teaching Strategies

PREVIEW Before beginning this lesson, have students scan the lesson for the science process skill symbols. Have students identify each skill used in the lesson. Students should then review the definition given for each skill on page 4 of their text.

DISCUSSION Ask a volunteer to open the classroom door. Point out that the student had to either push or pull the door to open it. Define force as a push or a pull. Discuss the different kinds of forces in nature, such as electrical force, magnetic force, and gravity.

DEMONSTRATION To show students a magnetic force, pick up a metal paper clip with a magnet.

INTERDISCIPLINARY CONNECTION Emphasize that the scientific definition of a force is a push or a pull. The scientific definition of force is different from its common definitions, such as its meaning in the terms "police force" and "forces of good and evil."

RETEACHING OPTION Bring in an anti-static product. Most students are familiar with static electricity and its effect on clothing. Tell students that static electricity is an electrical force. Define electrical force and then describe other kinds of forces.

Answers to Challenges (p. 21)
Apply

8. in the direction of the pull on the string

9. the textbook, because it weighs more

10. The force of the push starts the book moving; the force of gravity pulls it to the floor.

State the Problem

How is a compass needle affected by a magnet?

Ideas in Action

On a seesaw, if the heavier person is on the lower end, the lighter person must adjust his or her position.

Teaching Tips for Activity
Interpreting Force Diagrams

Skills: *measuring, interpreting*

COOPERATIVE/COLLABORATIVE LEARNING Answer the questions as a class.

Questions

1. 3 N

2. 5 N

3. a. 3 N **b.** 2 N

LESSON 2-2
What is gravity? (p. 22)

Teaching Strategies

PREVIEW Have students read the lesson feature before beginning the lesson. Discuss the concepts presented in the feature. Ask students how they think this information relates to topics they have already studied or to the lesson they are about to study.

DISCUSSION Write the term "gravity" on the chalkboard. Ask students to describe what they think of when they hear the term "gravity." List student responses on the chalkboard. Use any student knowledge of gravity as a springboard for a discussion of universal gravitation.

REINFORCEMENT Emphasize that on the earth, all objects fall in the direction of the earth's center. Toss an object in the air, and elicit the word "up" as being away from the earth's center and "down" as being toward the earth's center.

EXTENSION You may wish to show students the equation describing the force of gravity between two objects: $F = G \times M_1 \times M_2/R_2$, where F = the force of gravity, M_1 = the mass of one object, M_2 = the mass of the second object, R = distance between the two objects, and G = a constant value (the gravitational constant). By substituting numbers into the equation, you can demonstrate how mass and distance affect gravitational force.

Answers to Challenges (p. 23)
Apply

7. The earth has more mass than the apple.

8. The sun has a stronger force of gravity.

InfoSearch

You may wish to discuss students' questions and answers as a class.

Teaching Tips for Career in Physical Science
Astronaut

INTRASCIENCE CONNECTION There are two main kinds of astronauts, pilots and mission specialists. Pilots fly the spacecraft. Mission specialists usually have advanced degrees in a field of science, mathematics, or engineering. During a flight, mission specialists carry out many important scientific experiments.

Questions

1. ANALYZE Why do you think astronauts need psychological training? (Answers will vary. Accept all logical responses.)

2. DESCRIBE How would low gravity affect your daily activities? (Answers will vary. Accept all logical responses.)

LESSON 2-3
How does a spring scale work? (p. 24)

Teaching Strategies

PREVIEW Before beginning this lesson, write the title of each section in the lesson on the chalkboard. Have students copy these titles in their notebooks using an outline format. As students read each section, they should write the topic sentence for each paragraph in the section in their notebooks.

REINFORCEMENT To begin this lesson, review some of the metric units introduced in Unit 1. Ask students to name the basic units of distance, volume, and mass in the metric system (meter, liter, and gram, respectively). Then introduce the newton, which is the basic unit of force in the metric system. Tell students that this unit is named in honor of Isaac Newton.

DISCUSSION Refer students to the diagram of a spring scale on p. 24. Have students locate the main parts of the scale as you describe how a spring scale works.

CLASS ACTIVITY Bring a spring scale to class and demonstrate how it works. Have students take turns weighing common objects on the spring scale. List the weights on the chalkboard. After each student has had a turn, tell students to list all the objects in order of increasing weight.

COOPERATIVE/COLLABORATIVE LEARNING Have students work in pairs to answer the Check and Apply questions.

Answers to Challenges (p. 25)
Apply

7. 10 N; because weight is a force

8. force of the earth's gravity

9. It is 1/6 the force of the earth's gravity.

InfoSearch

You may wish to discuss students' questions and answers as a class.

Teaching Tips for Activity
Using a Spring Scale

Skills: *measuring, relating*

COOPERATIVE/COLLABORATIVE LEARNING Have students work in pairs.

Questions

1. Answers will vary.

2. gravity

3. 9.8 N

LESSON 2-4
What is air resistance? (p. 26)

Teaching Strategies

PREVIEW Before beginning this lesson, read the list of Tech-Terms aloud so students can hear their pronunciations.

DEMONSTRATION Demonstrate the relationship between air resistance and surface area. Drop a flat sheet of paper and a crumpled-up sheet of paper to the floor. Have students observe that the crumpled piece of paper hits the ground first. Point out that the sheet of paper has a larger surface area than the crumpled paper, and therefore is affected by a greater air resistance.

DEMONSTRATION Use a "Guine and Feather" tube to compare the falling rates in air and in a vacuum.

REINFORCEMENT To help students remember the meaning of terminal velocity, point out that "terminal" means "end or final." Terminal velocity is the final velocity a falling object attains.

Answers to Challenges (p. 27)
Apply

6. No. The hammer will hit the ground at a faster speed.

7. to increase the air resistance and slow their landing speed

8. On the moon, a hammer and feather would hit the ground at the same time.

9. Opening a parachute would slow a skydiver's fall.

Designing an Experiment

Have volunteers describe their experiments to the rest of the class.

Teaching Tips for People in Science
Galileo Galilei

INTRASCIENCE CONNECTION A space probe named *Galileo* in honor of Galileo will explore more of Jupiter. It was launched in 1989 and will reach Jupiter in 1995. Scientists hope to learn more about Jupiter and its moons from information gathered by the space probe.

EXTENSION Have interested students find out more about Galileo. Tell students to write a brief biography on his life.

Questions

1. DEFINE What is acceleration? (rate of change in speed)

2. LIST What were three of Galileo's contributions to science? (Students should list three contributions given in the feature.)

LESSON 2-5
What is friction? (p. 28)

Teaching Strategies

PREVIEW Have students write the lesson title and objective in their notebooks. As students read the lesson, tell them to write down the sentence or sentences that provide the information needed in the objective.

DEMONSTRATION To introduce this lesson, push a book across your desk or across the floor. Ask students why the book eventually slows down and stops moving. Discuss all student responses. Describe the forces acting on the book. Identify friction as the force that causes the book to stop moving.

REINFORCEMENT Write the following head on the chalkboard: *Types of Friction*. From this head draw arrows to each of the following heads: *Sliding friction, Air resistance, Rolling friction*. Describe an example of each of these types of friction. Have students copy the information in their notebooks under the proper heads.

COOPERATIVE/COLLABORATIVE LEARNING Have a student describe the uses and disadvantages of friction.

Answers to Challenges (p. 29)
Apply

5. more; so that the car can move

6. to increase the friction between the snow and a car's tires

7. a. air resistance **b.** rolling friction **c.** sliding friction **d.** air resistance

8. Check students' diagrams.

State the Problem

How much force is needed to pull the box up the ramp?

Teaching Tips for Leisure Activity
Ice Skating

DISCUSSION Have any students who have gone ice skating or seen an ice show describe their experiences to the rest of the class.

INTERDISCIPLINARY CONNECTION Figure skating is one of the most popular events in the Winter Olympics. Bring in action pictures showing champion figure skaters such as Jill Trenary, Brian Boitano, Dorothy Hamill, Scott Hamilton, or Katerina Witt. Point out that sliding friction enables these figure skaters to perform.

Questions

1. INFER What kind of friction allows you to roller skate? (rolling friction)

2. EXPLAIN Why should you not skate on a pond unless you know the ice can support your weight? (It is unsafe.)

LESSON 2-6
How can friction be reduced? (p. 30)

Teaching Strategies

PREVIEW Before beginning this lesson, have students scan the lesson looking for words with which they are unfamiliar. Have students work in pairs or small groups to define each of the words on their lists.

DEMONSTRATION Demonstrate how friction can increase the amount of work needed to move an object. Using a spring scale, drag a 100-g mass a distance of 10 cm across some sandpaper. Calculate the amount of work done. Repeat the measurement using smooth paper in place of sandpaper. Compare the amount of work done in each case.

DEMONSTRATION Demonstrate how lubrication can make the driving of screws into wood easier. Drive a screw into a block of wood. Coat an identical screw with soap and repeat the exercise. Compare the effort needed in each case.

RETEACHING OPTION Display a can of motor oil in front of the classroom. Most students know that cars need oil. Explain how motor oil acts as a lubricant and reduces friction between the metal parts of a car.

Answers to Challenges (p. 31)
Apply

5. wooden floor, because there is less friction between the chair and the floor

6. a. 25 N **b.** 22 N

7. It makes it easier for a car to skid.

8. Reducing the amount of contact with the water reduces the force of friction between the boat and the water.

Ideas in Action

People walk more carefully on wet or icy streets because friction is reduced and it is easier to slip.

Teaching Tips for Activity
Measuring Friction

Skills: *calculating, comparing, measuring*

COOPERATIVE/COLLABORATIVE LEARNING Have students work in pairs.

Questions

1. Answers will vary.

2. Answers will vary.

3. A smooth surface has less friction than a rough surface.

LESSON 2-7
What is pressure? (p. 32)

Teaching Strategies

PREVIEW Before beginning this lesson, have students scan the lesson for the science process skill symbols. Have students identify each skill used in the lesson. Students should then review the definition given for each skill on page 4 of their text.

DISCUSSION Define pressure as the amount of force acting on a surface. Refer students to the equation for calculating pressure on p. 32 and work through the problem. Then show students how pressure can be changed by changing the force or area.

REINFORCEMENT Emphasize that pressure is equal to force divided by area. Students often confuse force and pressure, and use the terms interchangeably.

DEMONSTRATION Show how the same force may produce different pressures. Use a block of pine wood, some common nails, and a hammer. Position a nail point down on the wood and strike it one blow with the hammer. Repeat this with a nail pointed head down. Compare the different degrees of penetration of the nails. Compare the area of nail that is in contact with the wood. Point out that the same amount of force produced different pressures.

INTRASCIENCE CONNECTION Blood is a liquid tissue. Have students feel their pulses. Tell students that when they feel their pulses, they are feeling blood pressure against the walls of an artery. The blood pressure in an artery rises and falls with each beat of the heart.

DEMONSTRATION Demonstrate the effect of air pressure on an empty can by placing a small amount of water into a screw-top can. With the top of the can removed, place the can over a heat source and boil the water. The water vapor will drive much of the air out of the can. While the water is boiling, carefully replace the cap and remove the can from the heat. As the can cools, the water vapor will condense and create a partial vacuum in the can. The air pressure outside the can will crush the can.

Answers to Challenges (p. 33)
Apply

6. If you make the area where the water comes out smaller, the pressure will increase.

7. a. 5 N/m^2 **b.** The pressure is increased to 6 N/m^2.
c. Increase the area to 20 m^2 or decrease the force to 30 N.

Skill Builder: Measuring

1. 120 N

2. 3 N

3. 8 N

Teaching Tips for Science Connection
Hydraulics

DISCUSSION Refer students to the picture of the barber's chair, and describe how a hydraulic lift works. Make sure students understand why the amount of force is greater in the wide container than in the narrow container. To help students understand hydraulics, you may wish to plug sample numbers in the equation, force = pressure × area.

Questions

1. What are some uses for hydraulic lifts? (Answers will vary. Accept all logical responses.)

2. EXPLAIN How does a hydraulic lift make work easier? (It increases effort forces.)

LESSON 2-8
How is fluid pressure measured? (p. 34)

Teaching Strategies

PREVIEW Before beginning this lesson, read the list of Tech-Terms aloud so students can hear their pronunciations.

DEMONSTRATION Demonstrate that air exerts pressure. Place a piece of cardboard on top of a glass filled with water. Hold the glass sideways. Ask students why the cardboard and water remain in place. (Air pressure is pushing sideways against the cardboard.) Then tilt the glass in other directions. Have students note that air pressure pushes in all directions.

DISCUSSION Ask students if they ever noticed their ears pop when taking a trip to the mountains or an airplane trip. Ask students why their ears pop. Discuss student responses. Then describe how elevation affects air pressure. Point out that as elevation increases, air pressure decreases. When a person travels to high elevations, the inside of the person's ears has a higher pressure than the surrounding air. When pressure equalizes inside and outside the ears, a popping sound is heard.

DEMONSTRATION Show students an aneroid or mercury barometer and a manometer. Point out that an aneroid barometer does not contain any liquid.

EXTENSION Have interested students research the development of submarines. Tell students to make a time line of important dates and inventions to summarize their findings.

Answers to Challenges (p. 35)
Apply

6. It decreases.

7. The pressure on it increases.

8. Pressure inside the body is equal to the pressure outside the body.

Health and Safety Tip

To avoid discomfort in the ears while taking an airplane trip, a person should chew gum, or swallow.

Teaching Tips for Science Connection
The "Bends"

DISCUSSION Before reading this feature, ask students if they have ever gone, or would ever like to go, scuba diving. Have any students who have been scuba diving describe their experiences to the rest of the class. Ask students why they think scuba diving would be fun, what they would expect to see, and what might be frightening about scuba diving.

Questions

1. Why must divers return to the surface slowly? (to avoid the bends)

2. HYPOTHESIZE What do you think could happen if gas bubbles got trapped in the bloodstream? (injury or death)

LESSON 2-9
What is Bernoulli's principle?
(p. 36)

Teaching Strategies

PREVIEW Before beginning this lesson, write the title of each section in the lesson on the chalkboard. Have students copy these titles in their notebooks using an outline format. As students read each section, they should write the topic sentence for each paragraph in the section in their notebooks.

DEMONSTRATION Introduce this lesson by performing the simple activity described on p. 36. Blow over the top of a sheet of paper. Have students observe how the sheet of paper rises. Point out that this experiment demonstrates Bernoulli's principle. Explain that Bernoulli's principle states that as the speed of a fluid increases, its pressure decreases. Be sure students understand that both gases and liquids are fluids. Students tend to think of only liquids as fluids.

DISCUSSION Refer students to the pictures of the airplane wing and the whole airplane on p. 36. Describe how airplane wings are designed to use Bernoulli's principle. Then have students identify the three forces on an airplane.

RETEACHING OPTION Bring a model airplane to class. Using the model, describe the forces that combine to make an airplane fly.

COOPERATIVE/COLLABORATIVE LEARNING Answer the Check and Apply questions as a class.

Answers to Challenges (p. 37)
Apply
6. There is more air resistance in a stiff wind than in calm air.

7. It gives the plane more thrust.

8. to slow the speed of the plane

Ideas in Action
When several faucets or appliances that use water are on at the same time, water pressure decreases.

Teaching Tips for Career in Physical Science
Automobile Designer
INTERDISCIPLINARY CONNECTION Have a contest to see who can come up with the best automobile design. Have students vote on each others' designs.

Questions
1. **INFER** What is the "form and function" of a car? (form is the design and look; function is how a car runs, its safety, and so on)

2. Why must an automobile designer know about modern technology? (A car is a technological tool.)

Answers to Unit Challenges (pp. 38–40)
Understanding the Features: Reading Critically
1. An astronaut has to learn how to move around in low gravity.

2. telescope

3. sliding

4. In shock absorbers, a large force on one end is changed to a small force on the other end.

5. to avoid the bends

6. to make a car more fuel efficient

Critical Thinking
1. There is no air on the moon.

2. Air pressure decreases because there are fewer air particles.

3. Movement would not be possible.

4. toes; force per unit area is greater

Interpreting a Diagram
1. F_2

2. F_1

3. F_3

4. 50 N

5. 10 N

6. 20 N

UNIT 3
ENERGY AND WORK
(p. 41)

Previewing the Unit

Have students carry out the task in the Study Hint. Then have a representative from each pair write their list of words on the chalkboard. Compare students' lists and the definitions of each word on their lists.

Bulletin Board Suggestion

Using the diagrams on p. 44 as a guide, assemble a bulletin board showing the different forms of energy. At the top of the bulletin board, write the following head: *Five Main Forms of Energy*. Divide the bulletin board into five sections. Label each section with one of the following heads: *Mechanical Energy, Electromagnetic Energy, Heat Energy, Chemical Energy, Nuclear Energy*. Attach pictures illustrating each of these forms of energy in the appropriate sections.

LESSON 3-1
What are the two basic kinds of energy? (p. 42)

Teaching Strategies

PREVIEW Before beginning this lesson, read the list of Tech-Terms aloud so students can hear their pronunciations.

DEMONSTRATION Make a pendulum by tying a weight to the end of a string. Attach the other end of the string to a ring stand. Raise the pendulum. Explain that when the pendulum is raised, it has potential energy. Then swing the pendulum. Point out that when the pendulum is moving, it has kinetic energy.

DEMONSTRATION Demonstrate how potential energy can be stored in a spring. Tie a spring with thread to compress the spring. Place the spring between two small objects, such as two small carts. Release the spring by cutting the thread. Have students observe what happens. (The objects will be pushed apart.) Point out that the potential energy of the compressed spring changed to kinetic energy when the spring was released.

REINFORCEMENT Be sure students understand the differences between potential energy and kinetic energy. To help students understand the meanings of these two basic kinds of energy, have students look up the definitions of the terms "potential" and "kinetic" in a dictionary.

COOPERATIVE/COLLABORATIVE LEARNING Have students work in pairs to answer the Check and Apply questions.

Answers to Challenges (p. 43)
Apply

7. Potential energy is stored energy. Kinetic energy is the energy of motion.

8. Figure A shows potential energy; Figure B shows both potential and kinetic energy.

Teaching Tips for Leisure Activity
Archery

DISCUSSION If possible, bring in a bow to show the class as you discuss the sport of archery.

CLASS ACTIVITY Have students work in small groups to make posters illustrating the safety rules for archery.

Question

RELATE What is the relationship between the potential energy of the bow and the distance the arrow travels? (The greater the potential energy, the farther the arrow will travel.)

LESSON 3-2
What are different forms of energy? (p. 44)

Teaching Strategies

PREVIEW Have students read the lesson feature before beginning the lesson. Discuss the concepts presented in the feature. Ask students how they think this information relates to topics they have already studied or to the lesson they are about to study.

CLASS ACTIVITY Write the following head on the chalkboard: *Main Forms of Energy*. From this head, draw arrows to each of the following heads: *Mechanical, Electromagnetic, Heat, Chemical, Nuclear*. Describe each of these main forms of energy and give students examples. Tell students to copy the information in their notebooks under the proper heads.

REINFORCEMENT Refer students to the picture on p. 44 showing the different forms of energy. Have students identify each form shown.

REINFORCEMENT Bring in pictures representing the five main forms of energy. Display the pictures in front of the classroom in random order. Have students classify each picture according to the energy forms shown.

Answers to Challenges (p. 45)
Check

1. food

2. five

3. energy in moving things

4. electromagnetic

5. heat

6. chemical

7. energy stored in the nuclei of atoms

Apply

8. Chemical and nuclear energy are potential energy. Mechanical, electromagnetic, and heat energy are kinetic energy.

9. a. chemical, heat **b.** chemical, heat **c.** electromagnetic, heat **d.** chemical, heat **e.** mechanical **f.** nuclear, heat

Designing an Experiment

You may wish to have volunteers describe their experiments to the rest of the class.

Teaching tips for Technology and Society

Nuclear Reactors

DISCUSSION Have a class debate about the use of nuclear energy.

EXTENSION Have students collect and bring in newspaper and magazine articles pertaining to nuclear energy. Read and discuss the articles as a class.

Questions

1. **COMPARE** What is the difference between nuclear fission and nuclear fusion? (In fission, an atomic nucleus is split into smaller nuclei. In fusion, two small nuclei are joined to form a larger nucleus.)

2. **IDENTIFY** What is the most common fuel in nuclear reactors? (uranium)

LESSON 3-3
How does energy change form? (p. 46)

Teaching Strategies

PREVIEW Have students write the lesson title and objective in their notebooks. As students read the lesson, tell them to write down the sentence or sentences that provide the information needed in the objective.

DEMONSTRATION Show students simple examples of energy changing forms. Turn off and on the classroom lights. Point out that turning on a light changes electrical energy into heat energy and light energy. Strike a match to change chemical energy into heat energy and light energy.

CLASS ACTIVITY Tell students to rub their hands together. Ask students why their hands get warm. (Rubbing the hands together changes mechanical energy into heat energy.)

DISCUSSION Ask students what they think of when they hear the word "pollution." List students' responses on the chalkboard. Students probably are aware that water, land, and air pollution are major problems. Point out that thermal pollution also is a kind of pollution. Describe how thermal pollution occurs and how it may harm living things.

REINFORCEMENT To help students remember that thermal pollution is caused by waste heat, have them look up the definition of the prefix "thermo-" in a dictionary. (thermo-: pertaining to heat) Then ask students to think of familiar terms including the prefix thermo-. (thermos bottle, thermal underwear, thermometer, and so on)

Answers to Challenges (p. 47)
Apply

8. **a.** Chemical energy is changed in heat energy. **b.** Electrical energy is changed into heat energy. **c.** Electrical energy is changed into light energy.

9. Answers will vary. Accept all logical responses.

Ideas in Action

Discuss students' answers as a class.

Ideas in Action

Answers will vary. Accept all logical responses.

Teaching tips for Science Connection

Thermography

EXTENSION The heat energy detected by a thermograph is infrared energy. All objects radiate infrared energy naturally.

Questions

1. **DESCRIBE** What are three uses of thermography? (Students should restate three uses given in the feature).

2. What energy changes are involved in thermography? (Heat energy is changed into electrical energy.)

LESSON 3-4
What is conservation of energy? (p. 48)

Teaching Strategies

PREVIEW Before beginning this lesson, write the title of each section in the lesson on the chalkboard. Have students copy these titles in their notebooks using an outline format. As students read each section, they should write the topic sentence for each paragraph in the section in their notebooks.

DISCUSSION Describe the law of conservation of energy. Emphasize that energy can never be made or destroyed.

DEMONSTRATION Demonstrate the law of conservation of energy with a simple activity. Bounce a rubber ball in front of the class. Have students observe that each bounce of the ball is lower than the preceding one. Point out that some of the mechanical energy of the bouncing ball is converted to heat and some of the mechanical energy is converted to sound.

REINFORCEMENT Write Einstein's equation on the chalkboard ($E = mc^2$). Point to each letter in the equation and have students identify what the letter represents. (E = energy; m = mass; c = the speed of light) Explain in terms of the equation how mass can be converted to energy. You may wish to tell students that the speed of light is about 3×10^8 m/s. Point out that this large number is squared in the equation. By plugging a 1 for m into the equation, students can observe how a small amount of matter can be changed into a huge amount of energy.

COOPERATIVE/COLLABORATIVE LEARNING Answer the Check and Apply questions as a class.

Answers to Challenges (p. 49)
Apply

7. In a player's hand, the basketball has potential energy. When it is thrown through the hoop, it has kinetic energy. When it is caught again, it has potential energy.

8. Mass is changed into heat and light.

State the Problem

At which point in the pendulum's swing is kinetic energy greatest? At which point is potential energy greatest?

Teaching Tips for People in Science
Albert Einstein

DISCUSSION Some of your students may think of the name "Einstein" as a synonym for "genius." Have students read the feature and discuss Einstein's accomplishments. Then ask students why Einstein is considered a genius.

EXTENSION The photoelectric cell, or electric eye, resulted from Einstein's work concerning the nature of light. The electric eye made possible sound motion pictures, television, and many other inventions. Have interested students find out about the photoelectric cell and how it works. Tell students to write their findings in a report.

Questions

1. RELATE What is the theory of relativity? (Matter and energy are two forms of the same thing.)

2. Why do you think Einstein is sometimes called the Father of the Atomic Age? (Answers will vary. Accept all logical responses).

Teaching Tips for Career in Physical Science
Construction Worker

DISCUSSION The sight of construction workers at work is probably familiar to all students. Before students read the feature, have them discuss the different kinds of jobs they have observed construction workers performing.

Questions

1. INFER Why must construction workers be aware of safety at all times? (Answers will vary. Accept all logical responses.)

2. What do contractors do? (They plan buildings to be safe and practical for their intended use.)

LESSON 3-6
How can work be measured?
(p. 52)

Teaching Strategies

PREVIEW Have students write the lesson title and objective in their notebooks. As students read the lesson, tell them to write the sentence or sentences that provide the information needed in the objective.

REINFORCEMENT This lesson is a straightforward presentation of how to calculate work. Provide students with many examples to reinforce the concept. Some students may be confused by the example on p. 52 which involves a spring scale pulling a 10-N box. Point out that you always must measure the force applied in the direction of motion. Use the diagrams on p. 52 to illustrate your explanation.

CLASS ACTIVITY Tell students to calculate how much work they do in climbing up a flight of stairs. First tell students to measure the height of each stair and their own weights in newtons. (One pound is equal to 4.5 newtons.) Then have students calculate how much work they do in climbing one stair (weight in newtons time height of stair). Finally, have students calculate how much work is done in climbing the whole flight of stairs. (Work done in climbing one stair times total number of stairs.)

Answers to Challenges (p. 53)
Apply

7. a. 8 joules **b.** 90 joules

8. 80 joules is greater than 20 joules

InfoSearch

You may wish to discuss students' questions and answers as a class.

Teaching Tips for Activity
Measuring Work

Skills: *measuring, comparing*

COOPERATIVE/COLLABORATIVE LEARNING Have students work in pairs.

LESSON 3-5
What is work? (p. 50)

Teaching Strategies

PREVIEW Before beginning this lesson, have students scan the lesson for the science process skill symbols. Have students identify each skill used in the lesson. Students should then review the definition given for each skill on page 4 of their text.

DISCUSSION Ask students what they think of when they hear the word "work." Be sure students understand that a word used in a scientific context has a definite meaning. Explain that work is equal to force times distance.

CLASS ACTIVITY Introduce this lesson by assigning an impossible task. Ask a volunteer to try to push down the classroom wall. Ask another volunteer to push a pencil across your desk. Point out that the first student has done no work because the wall did not move. Explain that the second student did do work because the pencil moved. Use this simple activity to introduce the definition of work as force times distance.

REINFORCEMENT Refer students to the diagram on p. 50 comparing the direction of applied force and the direction of motion. Be sure students understand that in order for work to be done, the direction of the applied force and the direction of motion must be the same.

Answers to Challenges (p. 51)
Apply

6. a. no **b.** no **c.** yes **d.** no **e.** yes

Health and Safety Tip

You may wish to demonstrate the correct way to lift a heavy object from the floor.

Questions

1. Answers will vary. Accept all logical responses.

2. Answers will vary. Accept all logical responses.

3. No; more work was done lifting the book.

LESSON 3-7
What is power? (p. 54)

Teaching Strategies

PREVIEW Before beginning this section, have students scan the lesson looking for words with which they are unfamiliar. Have students work in pairs or small groups to define each of the words on their lists.

DISCUSSION Bring a variety of light bulbs to class and display them in front of the class. Hold up each light bulb and point out its wattage. Most students are probably familiar with watts as units of electric power. Use students' knowledge as a springboard for a discussion of power. Be sure students understand the difference between work and power.

CLASS ACTIVITY Have students calculate the power it takes to run up a flight of stairs by measuring how much work is done in climbing the stairs and how much time it takes to climb the stairs.

REINFORCEMENT Provide students with many examples using the formula for calculating power.

COLLABORATIVE/COOPERATIVE LEARNING Have students quiz each other on the material in this lesson.

Answers to Challenges (p. 55)

Apply

9. a. 100 watts **b.** 900 watts **c.** 2000 watts **d.** 150 watts

Designing an Experiment

Have volunteers describe their experiments to the rest of the class.

Teaching Tips for Looking Back in Science

Horsepower

RETEACHING OPTION Bring in pictures showing people using horses to do work. Point out that before the invention of automobiles and other machines, horses were used to plow soil, pull wagons, and so on. Then have students read the feature on horsepower.

Questions

1. **CALCULATE** How many watts are equal to 2 horsepower? (1500 W)

2. Why do you think James Watt used the power of a horse as the standard unit of power for the steam engine? (Horses were commonly used to do work.)

Answers to Unit Challenges (pp. 56–58)
Understanding the Features: Reading Critically

1. arm guard and a special glove

2. Two small nuclei are joined to form a larger nucleus.

3. A device that turns invisible heat energy into a visible picture.

4. 1921

5. Answers will vary. Accept all logical responses.

6. Answers will vary. Accept all logical responses

Critical Thinking

1. Potential energy is stored energy. Kinetic energy is energy of motion.

2. When a pendulum is at the top of its swing, it has potential energy. When the pendulum swings, its potential energy changes to kinetic energy.

3. Answers will vary. Accept all logical responses.

4. A person can use energy trying to move an object, but not be successful in moving the object.

5. Power is equal to the amount of time it takes to do work.

6. Answers will vary. Accept all logical responses.

Interpreting a Diagram

1. 10 N

2. 2 meters

3. 10 N

4. 20 J

5. 10 N

6. 2 meters

7. 4 N

8. 8 J

9. different; more work is being done to lift the object than to pull it

Previewing the Unit

Read the titles of the lessons in this unit aloud. Identify terms in the titles with which students are unfamiliar. Write these terms on the chalkboard. Using the Glossary of the text or a dictionary, define the terms as a class. Have students write the terms and their definitions in their notebooks.

Bulletin Board Suggestion

Assemble a bulletin board entitled "Newton's Laws of Motion." Attach pictures illustrating each of Newton's three laws of motion. Use the diagrams on pp. 66–70 as guides. Label each picture with a 1, 2, or 3 according to the law of motion it illustrates.

LESSON 4-1
What are speed and velocity?

(p. 60)

Teaching Strategies

PREVIEW Have students write the lesson title and objective in their notebooks. As students read the lesson, tell them to write down the sentence or sentences that provide the information needed in the objective.

CLASS ACTIVITY To introduce this lesson, measure the length of the classroom in meters. Record the measurement on the chalkboard. Ask a volunteer to walk the length of the classroom. Using a stopwatch, measure how long it takes the student to walk the length of the classroom. Then calculate the student's average speed.

CLASS ACTIVITY Have two volunteers come up to the front of the classroom. Tell the two students to walk around the classroom in opposite directions. Point out that even if the two students were traveling at the same speed they would have different velocities. Define velocity as speed and direction.

REINFORCEMENT Give students several practice problems on calculating average speed.

EXTENSION Have students calculate their approximate speeds in getting to school. Tell students to find out how far they live from the school and to time how long it takes them to get to school. Using these measurements, students can figure out their average speeds.

RETEACHING OPTION Show students any filmstrip in which cars, or other vehicles, are moving. Use the filmstrip to discuss motion, average speed, instantaneous speed, and velocity.

Answers to Challenges (p. 61)
Check

1. change in position

2. speed = distance/time

3. total speed divided by total time

4. speed at any one instant

5. its speed and direction

Apply

6. 80 km/hr; 80 km/hr east

7. speed; at no instant should a person exceed the speed limit

Skill Builder: Analyzing a Graph

40 km/hr; 0; 75 km/hr

Teaching Tips for Activity
Measuring Average Speed

Skills: *measuring, calculating*

COOPERATIVE/COLLABORATIVE LEARNING Have students work in pairs.

Questions

1. Answers will vary. Accept all logical responses.

2. Answers will vary. Accept all logical responses.

LESSON 4-2
What is acceleration? (p. 62)

Teaching Strategies

PREVIEW Before beginning this lesson, write the title of each section in the lesson on the chalkboard. Have students copy these titles in their notebooks using an outline format. As students read each section, they should write the topic sentence for each paragraph in the section in their notebooks.

DISCUSSION Most students probably are familiar with acceleration in relation to cars. Acceleration often is used as a selling feature in advertisements for cars. Ask students what they think acceleration means. Discuss student responses. Define acceleration as a change in speed or direction. Be sure students understand that deceleration is a kind of acceleration. Then describe how to measure acceleration, using the problem on p. 62 as an example.

REINFORCEMENT Give students several practice problems on calculating acceleration.

INTERDISCIPLINARY CONNECTION A car's ability to accelerate quickly is an important feature for safety. The quicker a car can accelerate, the less chance there is of an accident when entering highways, or if there is a need to stop suddenly.

COOPERATIVE/COLLABORATIVE LEARNING Answer the Check and Apply questions as a class.

Answers to Challenges (p. 63)
Apply

7. b, d, e

8. Answers will vary. Accept all logical responses.

Skill Builder: Interpreting Tables

9.8 m/sec/sec; 9/8 m/sec/sec; 9.8 m/sec/sec. It is constant.

Teaching Tips for Career in Physical Science
Airplane Pilot

INTERDISCIPLINARY CONNECTION The majority of airplane pilots begin their careers as officers in the Air Force.

Questions

1. Why do you think special training is necessary to become an instructor pilot? (Answers will vary. Accept all logical responses.)

2. DEFINE What is a test pilot? (Pilot who flies experimental airplanes.)

Questions

1. Why do you think marathon runners must train each day? (to maintain their level of fitness)

2. RELATE Why do you think running is a very popular sport? (Answers will vary. Accept all logical responses.)

LESSON 4-3
What are balanced and unbalanced forces? (p. 64)

Teaching Strategies

PREVIEW Before beginning this lesson, have students scan the lesson for the science process skill symbols. Have students identify each skill used in the lesson. Students should then review the definition given for each skill on page 4 of their text.

REINFORCEMENT Before beginning this lesson, review some of the concepts covered in Unit 2. Ask students to define force (push or pull) and to name some forces (gravity, weight, magnetic forces).

DISCUSSION Refer students to the diagrams showing tug-of-war games on p. 64. Point out that balanced forces do not cause a change in motion. Then describe the two ways unbalanced forces can change the motion of an object.

DEMONSTRATION Show students unbalanced forces using a rubber band. Tie a weight to the end of the rubber band. Hold the other end of the rubber band and let go of the weight. Tell students to note how the weight bobs up and down. Explain the motion of the weight in terms of unbalanced forces. Point out the two forces acting on the weight, the downward force of gravity and the upward tension in the rubber band. Tell students that when the weight is first let go, the force of gravity is greater than the tension. Thus, the weight moves downward. As the rubber band stretches, the upward tension increases until it exceeds gravity. Thus, the weight moves upward. As the weight moves upward, tension decreases, and once again the force of gravity is greater. Thus, the weight moves downward. This pattern of motion repeats itself over and over, causing the weight to bob.

Answers to Challenge (p. 65)
Apply

7. unbalanced; speed of skater changed

8. left to right; skater continues moving in the same direction

Ideas in Action

Answers will vary. Accept all logical responses.

Teaching Tips for Leisure Activity
Running

EXTENSION Ask the physical education teacher in your school to talk to your class about the health benefits of running, the proper way to run, the best equipment to use, and safety precautions that should be followed.

LESSON 4-4
What is Newton's first law of motion? (p. 66)

Teaching Strategies

PREVIEW Before beginning this section, have students scan the lesson looking for words with which they are unfamiliar. Have students work in pairs or groups to define each of the words on their lists.

DEMONSTRATION Point to a book on your desk. Ask students what would cause the book to move. Elicit the response that the book will stay at rest unless it is acted upon by an unbalanced force. Push the book with your hand. Have students observe how your applied force causes the book to move.

EXTENSION Some students may be confused by Newton's first law of motion because everyday experiences seem to contradict the law. For example, a constant force must be applied to a car to keep the car moving forward. Be sure students understand that the force of friction is working against the car. In outer space there is no friction. Unless an outside force acts or a spacecraft, it will move at a constant speed forever.

INTERDISCIPLINARY CONNECTION Use a discussion of the effects of inertia to emphasize the importance of using seat belts. Each year the use of seat belts saves thousands of lives.

Answers to Challenges (p. 67)
Apply

7. The quarter falls into the glass; it tends to stay at rest when the card is flicked away.

8. Yes, because an object in motion tends to keep moving.

Skill Builder: Building Vocabulary

The word *iners* means inactive. Inert materials are not active. An inert gas is a gas that does not readily combine with other substances in a chemical reaction.

Teaching Tips for Science Connection
Seat Belts

RETEACHING OPTION Obtain a film showing the benefits of seat belt use. You might be able to obtain one from a local automobile safety school, police or fire department, or Department of Motor Vehicles.

CLASS ACTIVITY In some states, wearing seat belts is required by law. Have a class debate on whether seat belt use should be mandatory.

Question

RELATE Why could seat belts be called "anti-inertia" belts? (A seat belt keeps a person from moving forward when a car stops. Without seat belts, a person would continue to move forward because of inertia.)

LESSON 4-5
What is Newton's second law of motion? (p. 68)

Teaching Strategies

PREVIEW Have students read the lesson feature before beginning the lesson. Discuss the concepts presented in the feature. Ask students how they think this information relates to topics they have already studied or to the lesson they are about to study.

DISCUSSION Write the equation for Newton's second law of motion on the chalkboard: $F = ma$. Identify each variable in the equation and describe the relationship among force, mass, and acceleration.

REINFORCEMENT Review the units for mass and acceleration. Then introduce the newton as a unit of force.

COOPERATIVE/COLLABORATIVE LEARNING Have a volunteer describe Newton's second law of motion to the rest of the class.

Answers to Challenges (p. 69)
Apply
8. 2 newtons

9. 100 newtons

10. 2 kg

Skill Builder: Interpreting a Diagram

B; There is a greater applied force.

Teaching Tips for People in Science
Sir Isaac Newton

EXTENSION If possible, obtain a copy of Newton's book *Principia* to show the class.

DISCUSSION Use the question that ends this feature to motivate a discussion of how scientists build upon the work of other scientists.

Questions

1. Why do you think Isaac Newton was knighted? (for his many accomplishments in science)

2. **NAME** What were three fields in which Newton studied? (physics, astronomy, mathematics)

LESSON 4-6
What is Newton's third law of motion? (p. 70)

Teaching Strategies

PREVIEW Before beginning this lesson, have students review the TechTerms and their definitions and read the Lesson Summary.

DEMONSTRATION To introduce this lesson, blow up a balloon and let it go. Point out the forces that propel the balloon. The sides of the balloon exert a force that pushes air out of the balloon. This is the action force. The air exerts an equal and opposite force on the balloon. This is the reaction force.

DISCUSSION Compare the balloon to a rocket. Describe the action and reaction forces that propel a rocket upward.

COOPERATIVE/COLLABORATIVE LEARNING To summarize this unit, ask three volunteers to each describe one of Newton's three laws of motion.

Answers to Challenges (p. 71)
Apply
7. 100 N; 100 N

8. feet pushing against the ground; ground pushing against your feet

9. Answers will vary. Accept all logical responses.

Health and Safety Tip

Discuss students' findings as a class.

Teaching Tips for Technology and Society
National Aerospace Plane

DISCUSSION After students read the feature, have them answer the question that ends the feature. Discuss the advantages and disadvantages of an aerospace plane.

Questions

1. **DEFINE** What does the word "aerospace" mean? (The earth's atmosphere and the space outside it.)

2. **APPLY** What is Mach 3? (more than three times the speed of sound)

Answers to Unit Challenges (pp. 72–74)
Understanding the Features: Reading Critically

1. anti-inertia belts

2. 42.2 km

3. It would take off from a regular airport.

4. calculus

5. Answers will vary. Accept all logical responses

Understanding a Diagram

1. velocity; It shows both speed and direction.

2. West

3. East

4. 90 km/hr; instantaneous speed

5. 180 km

Critical Thinking

1. Average speed is equal to the total distance traveled divided by the total time for the trip. Instantaneous speed is speed at any one instant.

2. when it is not changing speed or direction

3. the first car, because it takes less time

4. in a direction opposite the direction the object is moving; the object slows down.

Previewing the Unit

Read the titles of the lessons in this unit aloud. Identify the terms in the titles with which students are unfamiliar. Using the glossary of the text or a dictionary, define the terms as a class. Have students write the terms and their definitions in their notebooks.

Bulletin Board Suggestions

1. Make a bulletin board entitled *Simple Machines*. Divide the bulletin board into six sections labeled *Lever, Pulley, Wheel and Axle, Inclined Plane, Wedge*, and *Screw*. Attach pictures showing these six simple machines under the appropriate headings.

2. Assemble a bulletin board showing the equations introduced in this lesson. Write the equations for finding the mechanical advantage of a machine, work input, work output, and efficiency. Tell students to use the bulletin board if they need help solving problems related to these equations.

LESSON 5-1
What is a simple machine?
(p. 76)

Teaching Strategies

PREVIEW Before beginning this lesson, have students scan the lesson for the science process skill symbols. Have students identify each skill used in the lesson. Students should then review the definition given for each skill on page 4 of their text.

DEMONSTRATION Bring a variety of machines to class, such as a jar opener, a hammer, a portable typewriter, and an eggbeater. Demonstrate how each machine makes work easier by changing the size of an effort force, the direction of a force, or the speed of a force.

REINFORCEMENT Give students several practice problems on finding the mechanical advantage of a machine.

RETEACHING OPTION Write the following head on the chalkboard: *Simple Machines*. From this head draw arrows to each of the following heads: *Lever, Pulley, Inclined Plane, Wedge, Screw, Wheel and Axle*. Identify these heads as the names of the six simple machines. Have students copy the chart in their notebooks.

COOPERATIVE/COLLABORATIVE LEARNING Have each student describe how they have used a machine to make work easier.

Answers to Challenges (p. 77)
Check

1. resistance force divided by effort force

2. force that opposes the effort force

3. changing size, direction, or speed of a force

4. effort force

5. number of times a machine multiplies the effort force

Apply

6. a. 10 **b.** 5 **c.** 10 **d.** 8 **e.** 5

7. increases the speed of a force

Ideas in Action

Answers will vary. Accept all logical responses.

Teaching Tips for Technology and Society
Robots

DISCUSSION Before students read this feature, ask them to describe a robot. Many students probably will describe the human-like robots depicted in science fiction books and movies. Point out that most robots do not look like humans. Discuss how robots help people do work now and how robots might change people's lives in the future.

Questions

1. STATE What is one advantage robots have over human workers? (Students should restate one of the advantages given in the feature.)

2. HYPOTHESIZE How do you think robots are used in manufacturing? (Answers will vary. Accept all logical responses.)

LESSON 5-2
What is efficiency? (p. 78)

Teaching Strategies

PREVIEW Before beginning this lesson, write the title of each section in the lesson on the chalkboard. Have students copy these titles in their notebooks using an outline format. As students read each section, they should write the topic sentence for each paragraph in the section in their notebooks.

REINFORCEMENT This lesson lays the foundation for an understanding of how machines can make work easier. Be sure students understand the equations for work output, work input, and efficiency. Write the equations on the chalkboard. Give students many practice problems on solving for work input, work output, and efficiency.

Answers to Challenges (p. 79)
Apply

7. Answer will vary. Accept all logical responses.

8. 50%

9. Answer will vary. Accept all logical responses.

10. 50%

Skill Builder: Researching

Check and discuss students' time lines.

Teaching Tips for Leisure Activity
Bicycling

EXTENSION Bring in pictures showing the development of the bicycle. Have students compare early models with modern bicycles.

Questions

1. **ANALYZE** Why do you think a bicycle is a conservation-wise means of transportation? (It does not use fuel.)

2. **LIST** What are two bicycle safety rules you should follow? (Answers will vary. Accept all logical responses.)

LESSON 5-3
How does a lever work? (p. 80)

Teaching Strategies

PREVIEW Before beginning this lesson, have students review the TechTerms and their definitions and read the Lesson Summary.

DISCUSSION Copy the diagram of the lever on p. 80 onto the chalkboard. Point out and describe the effort arm, the resistance arm, and the fulcrum. Then discuss how to find the mechanical advantage of a lever and compare the three classes of levers.

CLASS ACTIVITY Display a variety of levers in front of the classroom, for example, a crowbar, a hammer, scissors (first-class levers), a bottle opener, a nutcracker (second-class levers), a fishing rod, a baseball bat, and a rake (third-class levers). Have students classify each lever as a first-, second-, or third-class lever.

DEMONSTRATION Bring a car jack to class. Show the various parts of the jack. Be sure students understand that the handle of the jack acts as a lever. Then measure the effort arm and the resistance arm of the car jack to find its ideal MA.

INTERDISCIPLINARY CONNECTION Many large buildings, such as the Pyramids of Egypt, were built before the invention of modern machines. The buildings were made of heavy stone. Ancient people were able to move heavy stones using long sticks as levers.

Answers to Challenges (p. 81)
Apply

6. 4

7. increase effort arm length; decrease resistance arm length

8. first-class:scissors, hammer; second-class:nutcracker, bottle opener; third-class: hockey stick

InfoSearch

You may wish to discuss students' questions and answers as a class.

Teaching Tips for Activity
Using a Lever

Skills: *comparing, analyzing*

COOPERATIVE/COLLABORATIVE LEARNING Have students work in pairs.

Questions

1. Answers will vary.

2. closer; increases length of effort arm

LESSON 5-4
How do pulleys work? (p. 82)

Teaching Strategies

PREVIEW Before beginning this lesson, have students scan the lesson looking for words with which they are unfamiliar. Have students work in pairs or small groups to define each of the words on their lists.

DISCUSSION Refer students to the diagrams of pulleys on p. 82. Explain the difference between a fixed pulley and a movable pulley. Be sure students understand why the mechanical advantage of a fixed pulley is 1.

DEMONSTRATION Bring a variety of pulleys to class. Set up and show the operation of a single fixed pulley and a simple movable pulley. Use a spring scale to show the mechanical advantage of each pulley.

RETEACHING OPTION Bring in pictures showing everyday uses of pulleys, such as a person hanging clothes on a clothesline or a person raising a flag on a flagpole. Point out how pulleys help people do work.

COOPERATIVE/COLLABORATIVE LEARNING Have a volunteer compare fixed and movable pulleys.

Answers to Challenges (p. 83)
Apply

6. fixed; It changes direction of the effort force and its wheel does not move.

7. 2; 2; 4

Skill Builder: Modeling

Have volunteers show and describe their models to the rest of the class.

Teaching Tips for Activity
Using a Movable Pulley

Skills: *comparing, measuring*

COOPERATIVE/COLLABORATIVE LEARNING Have students work in pairs.

Questions

1. Answers will vary.

2. Answers will vary.

3. 3

LESSON 5-5
How does an inclined plane work? (p. 84)

Teaching Strategies

PREVIEW Have students write the lesson title and objective in their notebooks. As students read the lesson, tell them to write down the sentence or sentences that provide the information needed in the objective.

CLASS ACTIVITY To introduce the lesson, have students look up the meanings of the words "inclined" and "plane" in a dictionary (inclined: slanted; plane: flat surface). Tell students to use the definitions to write the definition of "inclined plane" in their own words.

DEMONSTRATION Using a wooden plank, set up a ramp. Show the mechanical advantage of the ramp by loading masses into a small wheeled cart. Using a spring scale, lift the cart straight up. Then pull the cart up the ramp using the spring scale. Students will see that less force is needed to move the cart up the ramp.

REINFORCEMENT Set up a wooden plank to make inclined planes with different MAs. Have students measure the length and height of the planks to calculate MA.

DISCUSSION Be sure students understand that a person uses less force when using an inclined plane, but must move a longer distance. Relate this to the equation for finding the MA of an inclined plane.

Answers to Challenges (p. 85)
Apply
7. increase its length; lower its height

8. 200 N

Skill Builder: Interpreting a Diagram
B; Its length divided by its height results in the lowest of the three MAs.

Teaching Tips for Activity
Finding the MA of an Inclined Plane
Skills: *measuring, analyzing*

COOPERATIVE/COLLABORATIVE LEARNING Answer the questions as a class.

Questions
1. Answers will vary. Accept all logical responses.

2. Answers will vary. Accept all logical responses.

3. Answers will vary. Accept all logical responses.

LESSON 5-6
What is a compound machine? (p. 86)

Teaching Strategies

PREVIEW Have students read the lesson feature before beginning the lesson. Discuss the concepts presented in the feature. Ask students how they think this information relates to topics they have already studied or to the lesson they are about to study.

DISCUSSION Point out that most machines are compound machines. Define a compound machine as a machine that combines two or more simple machines. Refer students to the diagram of a bicycle on p. 86 to observe the simple machines that make up a bicycle.

DEMONSTRATION Hold up a pair of scissors in front of the class and point out the three simple machines that make up a pair of scissors.

EXTENSION Have interested students find out how to calculate the mechanical advantage of a compound machine. Tell students to present their findings to the class in an oral report.

Answers to Challenges (p. 87)
Check
1. lever, screw, wedge

2. machine that combines two or more simple machines

3. wheel and axle

4. the mechanical advantages of all its simple machines

Apply
5. It is greater.

6. It increases the speed of a force.

7. Check students' diagrams.

Health and Safety Tip
You may wish to display students' posters in the classroom.

Teaching Tips for Career in Physical Science
Machinist
EXTENSION If possible, arrange for a machinist to visit your class. Ask the machinist to describe a typical workday and to show students some of the tools that are used to make machine parts.

Questions
1. INFER Why do you think many machinist jobs are available in the aerospace industry? (Answers will vary. Accept all logical responses.)

2. What machine parts do you think are standardized? (Answers will vary. Accept all logical responses.)

Answers to Unit Challenges (pp. 88–90)
Understanding the Features: Reading Critically
1. human-like

2. Computers are now used to operate many machine tools.

3. to move faster

Critical Thinking
1. Work is equal to a force times a distance.

2. Answers will vary. Accept all logical responses.

3. Answers will vary. Accept all logical responses.

4. block and tackle; has the greater MA

Interpreting a Diagram
1. 2 meters

2. 0.5 meter

3. 4

5. 150 N

6. first-class lever; The fulcrum is between the effort and the resistance.

Previewing the Unit

Read the titles of the lessons in this unit aloud. Identify terms in the titles with which students are unfamiliar. Write these terms on the chalkboard. Using the Glossary of the text or a dictionary, define the terms as a class. Have students write the terms and their definitions in their notebooks. Then, have students work in small groups to carry out the task in the Study Hint.

Bulletin Board Suggestions

1. Assemble a bulletin board showing how heat energy can be transferred. Attach pictures illustrating conduction, convection, and radiation. Label each picture. Use the bulletin board when describing how heat is transferred through solids, liquids, gases, and empty space.

2. Assemble a bulletin board showing how solar radiation affects the earth. At the top of the bulletin board, attach a picture of the sun. Beneath the picture of the sun, attach pictures illustrating passive and active solar heating systems, fossil fuels, photosynthesis, the water cycle, and so on.

LESSON 6-1
What is heat? (p. 92)

Teaching Strategies

PREVIEW Before beginning this lesson, have students scan the lesson for the science process skill symbols. Have students identify each skill used in the lesson. Students should then review the definition given for each skill on page 4 of their text.

EXTENSION Benjamin Thompson also disproved the caloric theory by weighing heated objects. Thompson believed that if heat was a fluid, it should have weight. However, he never detected a change in the weight of an object that was heated.

DISCUSSION Have students name the five main forms of energy (mechanical, electromagnetic, heat, chemical, nuclear). Then discuss heat as a form of energy. Develop the concept that heat energy is related to the motion of molecules. Students will learn more about the motion of particles in solids, liquids, and gases in Unit 12.

DEMONSTRATION Pour equal amounts of cold water and hot water into separate beakers. Add a drop of food coloring to each beaker. Tell students to note in which beaker the food coloring spreads faster (beaker containing hot water). Point out that the particles of water are constantly moving, so the food coloring is being spread throughout the water. The particles in the hot water are moving faster, causing the food coloring in hot water to spread out faster.

CLASS ACTIVITY Have students rub their hands together to show that friction produces heat.

DEMONSTRATION To show students that heat can do work, hold a pinwheel over a hot light bulb. Explain that heat from the light bulb makes the air around the pinwheel move. The moving air turns the pinwheel.

Answers to Challenges (p. 93)
Apply

7. kinetic energy; When heat is added to matter, the particles of matter move faster and faster.

8. The particles of water move more quickly and spread out.

9. Mechanical energy is changed into heat energy.

Health and Safety Tip

If you spill boiling water on yourself, you should cover the burn with a sterile dressing and get medical help immediately.

Teaching Tips for People in Science
Benjamin Thompson (Count Rumford)

INTERDISCIPLINARY CONNECTION During the Revolutionary War, Benjamin Thompson was charged with being a Tory. You may wish to tell students that Tories were people who remained loyal to England during the war.

Question

LIST What were three of Rumford's inventions? (kitchen range, double boiler, drip coffee pot)

LESSON 6-2
How is heat measured? (p. 94)

Teaching Strategies

PREVIEW Before beginning this lesson, write the title of each section in the lesson on the chalkboard. Have students copy these titles in their notebooks using an outline format. As students read each section, they should write the topic sentence for each paragraph in the section in their notebooks.

DEMONSTRATION Pour 100 grams, or 100 cm^3, of water into a beaker. Place a thermometer in the beaker. Record the temperature on the chalkboard. Then heat the water for a short time. Measure the temperature of the heated water. Calculate the amount of heat, or calories, that was added to the water.

DISCUSSION All students are probably familiar with the term "Calorie." Many students probably associate the term only with dieting. Ask students that they think of when they hear the word "Calorie." Discuss student responses. Then explain that Calories are used to measure the amount of energy in food.

REINFORCEMENT Emphasize that a kilocalorie and a Calorie are the same, and that a calorie is 0.001 kilocalorie. Show students some food labels listing Calories. Be sure students understand that the number of Calories is listed on food labels.

COOPERATIVE/COLLABORATIVE LEARNING Have students work in pairs to answer the Check and Apply questions.

Answers to Challenges (p. 95)
Apply

8. One calorie is only one thousandth of a kilocalorie (Calorie).

9. Heat energy is released when food is broken down.

Skill Builder: Researching and Organizing

Check students' menus.

Teaching Tips for Technology and Society

Cryogenics

REINFORCEMENT Read the term "cryogenics" aloud so that students can hear its pronunciation.

EXTENSION The supercold probe used in cryogenic surgery is cooled by liquid nitrogen. The probe can be used to cut away or destroy diseased tissue.

Questions

1. DEFINE What is absolute zero? (the lowest possible temperature that can be reached)

2. What happens to hydrogen and oxygen at very low temperatures? (They become liquids.)

LESSON 6-3

What is temperature? (p. 96)

Teaching Strategies

PREVIEW Before beginning this lesson, have students scan the lesson looking for words with which they are unfamiliar. Have students work in pairs or small groups to define each of the words on their lists.

REINFORCEMENT Before beginning this lesson, remind students that temperature is a measure of how much heat energy an object contains and that temperature is measured with a thermometer.

DEMONSTRATION To show students the difference between heat and temperature, fill two beakers, one large and one small, with hot water. Use a thermometer to make sure the temperature of the water is the same in each beaker. Then place several ice cubes into each beaker. Students will observe that when the ice cubes melt, the temperature change is less in the large beaker than in the small beaker. Point out that there was less change in temperature in the large beaker because the large beaker has more heat energy.

DEMONSTRATION Show how heat moves from a place with a high temperature to a place with a lower temperature. Obtain two polystyrene cups. Fill one with boiling water and the other with cold water. Place a heavy U-shaped bar with one end in each cup. Place a thermometer into each cup. Students will be able to observe that the hot cup transfers heat to the cold cup.

Answers to Challenges (p. 97)

Apply

8. Heat is moving from the room to the ice because heat moves from a place with a high temperature to a place with a low temperature.

Skill Builder: Researching

1. 57.7 °C

2. −88.3 °C

3. 10^{-5}K; 100,000,000°

4. Mercury; Pluto

Teaching Tips for Activity

Observing Temperature Differences

Skills: *observing, hypothesizing*

COOPERATIVE/COLLABORATIVE LEARNING Answer the questions as a class.

SAFETY TIP Caution students not to use boiling water in Step 1.

Questions

1. yes

2. Answers will vary. Accept all logical responses.

LESSON 6-4

What is freezing point? (p. 98)

Teaching Strategies

PREVIEW Have students write the lesson title and objective in their notebooks. As students read the lesson, tell them to write down the sentence or sentences that provide the information needed in the objective.

DEMONSTRATION To introduce this lesson bring three glasses of water to class, one containing water right from the tap, one containing water that has been in the freezer for 5 minutes, and one containing water that has been in the freezer for 15 minutes. Measure the temperature of the water in each glass. Write the temperatures on the chalkboard. Have students note that as water freezes, the temperature drops until it reaches the freezing point of water.

CLASS ACTIVITY Have students work in small groups to make a line graph showing the information in Table 1. Tell students to plot time along the x-axis and temperature along the y-axis.

REINFORCEMENT Be sure students understand that the freezing point and melting point of a substance are the same. Ask students what would be the melting point of solid forms of the substances listed in Table 2. (same as their freezing points)

Answers to Challenges (p. 99)

Check

1. It changes to ice.

2. 0 °C

3. freezing point

4. 0 °C

5. No; every liquid has its own freezing point.

6. They are the same.

7. It changes to a liquid.

8. 0 °C

Apply

9. The substance has reached its melting point.

Skill Builder: Calculating

800 calories; 20,000 calories; 160,000 calories

Teaching Tips for Science Connection
Sublimation

DISCUSSION Bring in some mothballs or dry ice to show the class. Explain that both these substance undergo sublimation. Describe sublimation and some of the uses of materials that sublime.

Questions

1. RELATE What is dry ice? (solid carbon dioxide)

2. DEFINE What is sublimation? (process of changing directly from a solid to a gas)

LESSON 6-5
What is boiling point? (p. 100)

Teaching Strategies

PREVIEW Before beginning this lesson, have students review the TechTerms and their definitions and read the Lesson Summary.

DEMONSTRATION Show students the boiling point of water. Heat a beaker of water over a Bunsen burner. Record the temperature every 20 seconds. Have students observe that once the water begins to boil, the temperature stays at 100 °C. Point out the small bubbles that appear in the water, which show that a gas is being formed.

DEMONSTRATION The evaporation of liquids can be demonstrated using ammonia or perfume. If allowed to stand in an open container in front of the classroom, the liquid will evaporate. Diffusion will carry the gas molecules through the room.

INTERDISCIPLINARY CONNECTION Students may be aware that people often add salt when boiling water to cook pasta and other foods. Ask students how they think salt affects the boiling point of water. (Salt raises the boiling point of the water. Thus, the food cooks faster.)

Answers to Challenges (p. 101)
Apply

8. 357 °C

9. 118 °C

10. acetone

11. Answers will vary. Accept all logical responses.

InfoSearch

You may wish to discuss students' questions and answers as a class.

Teaching Tips for Activity
Relating Boiling Point and Elevation

Skills: *graphing, analyzing*

COOPERATIVE/COLLABORATIVE LEARNING Answer the questions as a class

Questions

1. It decreases

2. a. 100 °C **b.** 93.4 °C **c.** 96.6 °C **d.** 96.6 °C **e.** 100 °C
f. 100 °C

LESSON 6-6
What is conduction? (p. 102)

Teaching Strategies

PREVIEW Before beginning this lesson, read the list of Tech-Terms aloud so students can hear their pronunciations.

DISCUSSION Hold up a metal cooking pot in front of the classroom and discuss why metals are used as cooking utensils. Point out that metals are good conductors of heat. Define a conductor as a substance that allows heat to move through it easily. Then compare conductors and insulators.

DEMONSTRATION Use marbles, a glass tray, and an overhead projector to show a model of conduction. Cover about 3/4 of the bottom of the tray with a single layer of marbles. Place the tray on the overhead projector. Agitate the marbles at one end of the tray. Students will note that the disturbance is "conducted" to the marbles at the other end.

DISCUSSION Display examples of insulators in front of the classroom. Discuss the uses of insulators in homes, thermos bottles, ice chests, and so on.

DEMONSTRATION To show students that wood and wax are two examples of insulators, hold up a lighted candle and a lighted match. Ask students why your fingers are not burned. (Heat is not conducted.)

RETEACHING OPTION Obtain two containers, one made of metal, the other of expanded polystyrene. Fill both containers with boiling water. Periodically, place thermometers in each container and compare the temperature. Students will note that the temperature of the water in the metal container drops faster than the temperature of the water in the polystyrene container. Identify metals as good conductors and polystyrene as a good insulator.

Answers to Challenges (p. 103)
Check

1. conductors

2. Particles of matter bump into one another.

3. a substance that conducts heat easily

4. copper and silver

5. a substance that does not conduct heat easily

6. wood, paper, wax, or air

7. to keep them warm in winter and cool in summer

Apply

8. Copper is a good conductor of heat.

9. Plastic and wood are good insulators.

Ideas in Action

Answers will vary. Accept all logical responses.

Skill Builder: Researching

Fiberglass is often used to insulate homes. Goose down is used in clothing.

Teaching Tips for Science Connection
Home Insulation

CLASS ACTIVITY The fuel used to heat a home may be gas, oil, coal, electricity, solar, or wood. Have students find out what type of fuels are used to heat their homes, and what kinds of insulation are used in their homes to prevent heat loss. Compile a class list of fuels and insulators.

Questions

1. RELATE Why does good insulation cut fuel use? (Insulation prevents heat loss. Fuel is used to heat a home.)

2. LIST What are two insulators that can be used in a home? (fiberglass and liquid plastic foam)

LESSON 6-7
What is convection? (p. 104)

Teaching Strategies

PREVIEW Before beginning this lesson, write the title of each section in the lesson on the chalkboard. Have students copy these titles in their notebooks using an outline format. As students read each section, they should write the topic sentence for each paragraph in the section in their notebooks.

DISCUSSION Discuss how some room heaters make use of convection. Students may be familiar with these heating systems. Steam or hot water moves through a heater. It heats some of the air in the room. The warm air then rises. Cool air sinks and moves toward the heater. This occurs in an endless cycle of rising and sinking air.

REINFORCEMENT Be sure that students understand that convection currents are due to differences in density. Point out that warm air is less dense than cold air because the particles in warm air are farther apart.

INTRASCIENCE CONNECTION Relate convection currents to large movements of air in the atmosphere. They are a cause of winds and other weather conditions. When a warm air mass meets a cold air mass, the warm air mass rises over the cold air mass.

DEMONSTRATION Show that the heat from a candle is carried upward by a convection current. A strip of paper held close to the side of candle will not catch fire. Hold it above and it quickly bursts into flame. The heat is carried to the paper above by a convection current of hot gases.

DEMONSTRATION Demonstrate convection in liquids. Place some sawdust into a beaker. Fill the beaker with water. Allow the sawdust to become soaked with water and to settle to the bottom. Heat the beaker and have the class note the movement of the sawdust.

Answers to Challenges (p. 105)
Apply

6. Convection currents carry smoke up a chimney.

7. The girl on the ladder because warm air rises.

Ideas in Action

Discuss students' results. To get fresh air into a room, a window should be opened from both the top and bottom.

Teaching Tips for Activity
Observing Convection Currents in Water

Skill: *observing*

COOPERATIVE/COLLABORATIVE LEARNING Have students work in pairs.

Questions

1. It moved in a circle through the water.

2. convection current

3. Warm water rises.

LESSON 6-8
What is radiation? (p. 106)

Teaching Strategies

PREVIEW Before beginning this lesson, have students scan the lesson for the science process skill symbols. Have students identify each skill used in the lesson. Students should then review the definition given for each skill on page 4 of their text.

CLASS ACTIVITY Allow students to examine a radiometer. Show how it operates in sunlight, but stops when the sunlight is blocked. Point out that heat and light energy from the sun cause the movement of the radiometer.

EXTENSION If possible, arrange for the custodian of the school to describe the school's heating system to your class. Tell students to prepare questions they would like to ask beforehand.

DEMONSTRATION Demonstrate the conversion of radiant heat and light energy into electricity. Set up a bank of solar cells with a miniature motor attached. Set up a single solar cell in series with a galvanometer or ammeter. Students will note the current produced in sunlight.

DISCUSSION Point out to the class that on Earth, heat most often travels by conduction, convection, and radiation. Use an electric heater as an example. The case of the heater is warmed by the convection current of hot air rising from the glowing heater coil. Heat travels through the case by conduction. Heat is transmitted to the room by convection and radiation. Heat from convection can be felt above the heater. Heat from radiation is projected from the front of the heater. Most electric heaters have a reflective surface behind the heating coils. This reflects radiant energy out into the room.

Answers to Challenges (p. 107)
Apply

8. radiation

9. a. radiation **b.** conduction **c.** convection

InfoSearch

You may wish to discuss students' questions and answers as a class.

Teaching Tips for Activity
Measuring the Effect of Passive Solar Heating

Skills: *observing, comparing*

SAFETY TIP Caution students to be careful when using scissors.

COOPERATIVE/COLLABORATIVE LEARNING Have students work in pairs.

Questions

1. box with window facing the sun
2. toward the sun

What is thermal expansion?
(p. 108)

Teaching Strategies

PREVIEW Before beginning this lesson, have students scan the lesson looking for word with which they are unfamiliar. Have students work in pairs or small groups to define each of the words on their lists.

DISCUSSION Discuss thermal expansion. Relate thermal expansion to the movement of molecules. Point out that water expands when it freezes, whereas other substances contract.

DISCUSSION Relate the thermal expansion of liquids to the movement of mercury in a thermometer.

DEMONSTRATION To show students the thermal expansion of gases, stretch a balloon over the mouth of a flask. Gently heat the flask. Students will observe that the balloon gets larger. Point out that the balloon gets larger as heated air expands.

DEMONSTRATION To show that water expands when it freezes, fill a container with water to the very top. Cover the container and place it in a freezer. Have the students note that when the water froze, it cracked the container.

Answers to Challenges (p. 109)
Check

1. It would break.
2. thermal expansion
3. They contract.
4. They expand.
5. It expands.
6. It is less dense than water.
7. They expand.
8. They contract.

Apply

9. The particles spread out; the volume increases.
10. Answers will vary. Accept all logical responses.

State the Problem

The ball enlarged when it was heated so it will no longer fit through the ring.

Teaching Tips for Science Connection
Bimetallic Thermostats

DEMONSTRATION Demonstrate a bimetallic strip. Point out the two metals that make up the bimetallic strip.

Questions

1. **EXPLAIN** What happens to the two metals in a thermostat when they are heated? (They expand.)
2. Why does a bimetallic strip bend when it is heated? (One metal expands and contracts faster than the other.)

Answers to Unit Challenges (pp. 110–112)
Understanding the Features: Reading Critically

1. There is less bleeding and less pain.
2. They have a metal strip made up of two metals.
3. high; An insulating material with a high R-value is best at preventing heat loss.
4. The water began to boil.
5. It keeps foods frozen without making a mess by melting.

Critical Thinking

1. Answers will vary. Accept all logical responses.
2. Mercury rises when it is heated and falls when it cools.
3. Answers will vary. Accept all logical responses.
4. Conduction is heat transfer in solids. Convection is heat transfer in gases and liquids. Radiation is the transfer of heat through space.
5. It showed that heat did not flow, but was a form of energy.

Interpreting a Table

1. the freezing of water
2. degrees Celsius
3. in a graph
4. 25 min
5. measuring, observing
6. It took 10 min for the water to freeze completely.

WAVES (p. 113)

Previewing the Unit

Have students complete the task in the Study Hint. After students have finished their lists, have a student from each pair write their word list on the chalkboard. Compare students' word lists and definitions.

Bulletin Board Suggestions

1. Assemble a bulletin board illustrating the features of a wave. Attach a diagram of a wave to the bulletin board. Use the diagram on p. 118 as a guide. Label the crest, the trough, the wavelength, and the amplitude. Use the bulletin board to review wave features.

2. Assemble a bulletin board showing waves and their effects. Attach pictures showing ocean waves, earthquake damage, and so on, to the bulletin board.

Answers to Challenges (p. 115)

Apply

7. The boat moves up and down as the waves pass by.

8. the wheat

9. the kinetic energy of the quake

Health and Safety Tip

Before a hurricane, people should evacuate low-lying coastal areas, and put tape over their windows. During a hurricane, people should remain indoors away from windows. People should not use the telephone during a hurricane.

Teaching Tips for Leisure Activity

Surfing

DISCUSSION Allow any students who have ever been surfing to describe their experiences to the rest of the class.

INTERDISCIPLINARY CONNECTION The sport of surfing has greatly influenced the songwriting of various musical groups. Dozens of songs have been written about the sport, which often is associated with a relaxed way of life.

Questions

1. **RELATE** Why must surfers be good swimmers? (Any water sport requires being able to swim well for safety reasons.)

2. **IDENTIFY** Where are the best places to go surfing in the United States? (Hawaii and Southern California)

LESSON 7-1
What is a wave? (p. 114)

Teaching Strategies

PREVIEW Have students read the lesson feature before beginning the lesson. Discuss the concepts presented in the feature. Ask students how they think this information relates to topics they have already studied or to the lesson they are about to study.

DEMONSTRATION To introduce this lesson, drop a small stone into a pan of water. Tell students to observe the small circular waves that move outward. Define waves as disturbances that transfer energy from place to place. Be sure students understand that in addition to ocean waves, there are other types of waves, such as sound waves and light waves.

DISCUSSION The movement of particles in a wave is difficult for students to grasp. Many students think that water travels forward with the wave. Emphasize that only the energy in a wave moves forward. The particles in a medium only move in small circles. Refer students to the diagram on page 114 showing the movement of water particles in a wave to illustrate your description.

DEMONSTRATION Float a cork, or other small object, in a large tub of water. Make waves by tapping a spoon on the water surface. Have students note how the object moves. Point out that water movement is mainly up and down.

CLASS ACTIVITY To show students that matter is not transferred when a wave passes, perform this simple activity. Tie knots of colored strings to a rope. Have two students each hold one end of the rope. Tell one student to move the rope up and down to produce waves. Students will observe that the colored strings move up and down as a wave passes, but they do not move forward.

LESSON 7-2
What are two kinds of waves? (p. 116)

Teaching Strategies

PREVIEW Before beginning this lesson, read the list of Tech-Terms aloud so students can hear their pronunciations.

INTERDISCIPLINARY CONNECTION Before beginning this lesson, draw a right angle on the chalkboard. This drawing will help students understand motion in a transverse wave.

DEMONSTRATION To show students transverse waves, tie a rope to a doorknob, or other structure. Move the rope up and down sharply. Point out that the direction of the wave is towards the door, but the rope moves up and down.

DEMONSTRATION Demonstrate a longitudinal wave with a coiled spring. Stretch the spring and then squeeze some coils together. Then release the coils. Point out the compression and rarefaction of the waves. Show students that each compression is followed by a rarefaction.

INTRASCIENCE CONNECTION Thunder is caused when giant storm clouds give off electricity. The electricity causes the air to warm and expand quickly. The thunder people hear is caused by this expanding and contracting of the air. People first see lightning and then hear thunder. The lapse in time is due to the fact that light waves move faster than sound waves.

RETEACHING OPTION Write the following head on the chalkboard: *Waves.* From this head draw arrows to each of the following heads: *Transverse, Longitudinal.* Describe the way the particles of the medium move in transverse and longitudinal waves. Tell students to copy the information in their notebooks under the proper heads.

Answers to Challenges (p. 117)

Apply

7. transverse; The water moves up and down.

8. longitudinal

9. transverse

Skill Builder: Building Vocabulary

transverse: crosswise; longitudinal: placed lengthwise; compression: a compressing or being compressed or pressed together; rarefaction: not dense

Teaching Tips for Science Connection

Earthquake Waves

EXTENSION To find the epicenter of an earthquake, scientists need seismograms from three stations. A circle is drawn on a map around each station. Each station is at the center of its circle. There is only one point where all three circles cross. The epicenter is near this point.

DEMONSTRATION If possible, obtain an actual seismogram to show students. Point out the seismic waves, which are recorded on a seismogram.

Questions

1. IDENTIFY What type of seismic waves cause the greatest damage? (L-waves)

2. DEFINE What is the Richter scale? (a measure of how much energy an earthquake releases)

LESSON 7-3
What are the features of a wave? (p. 118)

Teaching Strategies

PREVIEW Have students write the lesson title and objectives in their notebooks. As students read the lesson, tell them to write down the sentence or sentences that provide the information needed in the objective.

REINFORCEMENT Remind students that the crest is the high point of a wave. Then ask students what the low point of a wave is called. (trough)

RETEACHING OPTION Use an overhead projector and transparency to show the features of a wave. Redraw the diagram on p. 118 showing the features of a wave onto the transparency. Point out and describe the amplitude, wavelength, and frequency of the wave shown. Have students copy the diagram in their notebooks.

REINFORCEMENT Be sure students understand the relationship among frequency, wavelength, and speed of a wave by giving students several practice problems. For example, ask students to find the speed of a wave whose frequency is 3 Hz and wavelength is 2 m. (6 m/s; 3 waves/second × 2 meters/wave)

COOPERATIVE/COLLABORATIVE LEARNING Have students work in pairs to answer the Check and Apply questions.

Answers to Challenges (p. 119)

Apply

8. 500 m/sec

9. 2.5 Hz

10. 5 m

InfoSearch

You may wish to discuss students' questions and answers as a class.

Teaching Tips for Activity

Observing Waves in a Rope

Skills: *observing, analyzing*

COOPERATIVE/COLLABORATIVE LEARNING Have students work in pairs.

Questions

1. It moved up and down.

2. transverse

3. a. Frequency increased. **b.** Wavelength decreased.

LESSON 7-4
How are waves reflected?
(p. 120)

Teaching Strategies

PREVIEW Before beginning this lesson, write the title of each section in the lesson on the chalkboard. Have students copy these titles in their notebooks using an outline format. As students read each section, they should write the topic sentence for each paragraph in the section in their notebooks.

CLASS ACTIVITY If possible, arrange for your class to go on a field trip to observe the action of ocean waves. Point out some of the shoreline features caused by wave erosion.

DISCUSSION All students are familiar with seeing their own reflections in a mirror. Use mirror reflections as a spring board for a discussion of the reflection of waves. Tell students that reflection is the bouncing back of a wave after striking a barrier, such as a mirror. Point out that light waves reflect from a mirror to produce images.

DEMONSTRATION Tie a rope to a doorknob. Move the rope up and down to make waves. Tell students to observe that when a wave reaches the doorknob, it is reflected.

DEMONSTRATION If possible, use a ripple tank to show students the law of reflection. If a ripple tank is not available, simply make waves in a pan of water. Have students note that the waves reflect off the sides of the pan.

Answers to Challenges (p. 121)

Check

1. It bounces back.

2. the bouncing back of a wave from a barrier

3. incident wave

4. reflected wave

5. line at right angles to a barrier

6. 90°

Apply

7. The angle of incidence is the angle at which an incident wave strikes a barrier. The angle of reflection is the angle at which the wave is reflected. The angle of incidence is equal to the angle of reflection.

8. c

9. d

10. a

11. b

Skill Builder: Researching

Check students' reports.

Teaching Tips for Activity

Measuring the Angle of Incidence and Angle of Reflection

Skills: *measuring, observing*

COOPERATIVE/COLLABORATIVE LEARNING Have students work in pairs.

Questions

1. Answers will vary.

2. Answers will vary.

LESSON 7-5

How are waves refracted?

(p. 122)

Teaching Strategies

PREVIEW Before beginning this section, have students scan the lesson looking for words with which they are unfamiliar. Have students work in pairs or small groups to define each of the words on their lists.

DEMONSTRATION To introduce this activity, perform a simple experiment. Place a pencil into a glass of water at an angle. Ask students to describe how the pencil appears. (Elicit from students that the pencil appears to be broken where it enters the water.) Explain that the pencil looks broken because light waves are refracted, or bent, when they move from air into water.

DISCUSSION Describe the three laws of refraction. Refer students to the diagrams on p. 122 showing the three laws of refraction.

REINFORCEMENT Write the densities of some common substances on the chalkboard. You may wish to use some of the substances listed in Table 1 on page 244 and Table 1 on page 248 (in Unit 13). Give students situations of waves moving from one of the listed substances into another. Have students use the three laws of refraction to explain how the wave is bent.

COOPERATIVE/COLLABORATIVE LEARNING Have a volunteer describe the three laws of refraction to the rest of the class.

Answers to Challenges (p. 123)

Apply

7. **a.** toward the normal **b.** away from the normal **c.** directly back

8. refraction of light waves

InfoSearch

You may wish to discuss students' questions and answers as a class.

Teaching Tips for Activity

Observing the Effects of Refraction

Skills: *observing, analyzing*

COOPERATIVE/COLLABORATIVE LEARNING Students must work in pairs.

Questions

1. It was blocked by the side of the bowl.

2. Refraction of light waves makes the penny appear closer to the surface of the water than it is.

3. Check students' diagrams. Diagrams should show light waves being bent away from the normal.

LESSON 7-6

What is the Doppler Effect?

(p. 124)

Teaching Strategies

PREVIEW Before beginning this lesson, have students scan the lesson for the science process skill symbols. Have students identify each skill used in the lesson. Students should then review the definition given for each skill on page 4 of their text.

REINFORCEMENT Before discussing the Doppler effect, remind students that frequency is the number of complete waves that pass a point each second.

DISCUSSION Describe the Doppler effect as it pertains to sound waves, but make sure students understand that the Doppler effect is a property of all waves.

RETEACHING OPTION Bring a portable radio to class. Turn on the radio, at a very low volume. Stand in the middle of the classroom. Tell students to quietly listen to the radio. Then start walking toward the front of the classroom. For students in the back of the classroom, the sound waves will appear to have a lower frequency. For students in the front of the classroom, the sound waves will appear to have a greater frequency. Point out that this apparent change in the frequency of waves is the Doppler effect.

Answers to Challenges (p. 125)

Apply

7. As the car approaches, the sound of the horn is high pitched. As the car passes you and moves away, the sound of the horn becomes lower pitched.

8. toward the car; The frequency is apparently becoming greater.

Ideas in Action

Discuss students' findings as a class.

Teaching Tips for Science Connection

Red Shift

DISCUSSION To help students understand red shift, describe the relationship between frequency and wavelength. Frequency increases as wavelength decreases. This accounts for the red shift. As frequency apparently becomes lower, the wavelength appears to change and become longer.

Questions

1. EXPLAIN What determines the color of light? (wavelength)

2. What did astronomers find out by studying the red shifts of galaxies? (All of the galaxies are moving away from the earth and from each other.)

Answers to Unit Challenges (pp. 126–128)

Understanding the Features: Reading Critically

1. for their own safety

2. The universe is expanding.

3. three; primary, or P-waves, secondary, or S-waves, long, or L-waves.

Critical Thinking

1. They are reflected.

2. Answers will vary. Accept all logical responses.

3. Answers will vary. Accept all logical responses.

Interpreting a Diagram

1. c

2. a

3. b

4. speed = frequency × wavelength

5. It decreases.

6. It decreases.

Previewing the Unit

Read the titles of the lessons in this unit aloud. Identify terms in the titles with which students are unfamiliar. Write these terms on the chalkboard. Using the Glossary of the text or a dictionary, define the terms as a class. Have students write the terms and their definitions in their notebooks.

Bulletin Board Suggestions

1. Assemble a bulletin board showing musical instruments. At the top of the bulletin board write the head *Musical Instruments.* Divide the bulletin board into three sections. Label each section with one of the following heads: *Percussion Instruments, String Instruments, Wind Instruments.* Have students find and collect pictures of a wide variety of musical instruments. Tell students to label each picture with the name of the musical instrument. As a class, classify each instrument as a percussion, string, or wind instrument. Then attach all the pictures to the bulletin board in the appropriate sections.

2. Make a bulletin board showing the path of sound waves through the human ear. Attach a diagram of the human ear to the bulletin board. Label each part of the outer, middle, and inner ear. Draw arrows on the diagram showing how vibrations are transferred from one part of the ear to another.

LESSON 8-1
What is sound? (p. 130)

Teaching Strategies

PREVIEW Before beginning this lesson, have students scan the lesson for the science process skill symbols. Have students identify each skill used in the lesson. Students should then review the definition given for each skill on page 4 of their text.

DISCUSSION To introduce this lesson, ask students to describe all the sounds that can be heard in the classroom. List student responses on the chalkboard. Then explain that sound is a form of energy and that all sounds are caused by vibrations.

CLASS ACTIVITY Tell students to place their fingers on their windpipes and speak. Students will feel the vibrations from the sounds of their voices.

REINFORCEMENT When discussing sound waves, remind students that a rarefaction is part of a medium where the particles are far apart and a compression is part of a medium where the particles are close together.

Answers to Challenges (p. 131)
Apply

8. The vibration from the speaker will cause the water to vibrate.

9. No; all sounds are caused by vibrations.

10. Answers will vary. Accept all logical responses.

Ideas in Action
Answers will vary. Accept all logical responses.

Ideas in Action
Students should be able to hear sounds in the next room.

Teaching Tips for Activity
Observing Vibrations in a Tuning Fork
Skill: *observing*

COOPERATIVE/COLLABORATIVE LEARNING Have students work in pairs.

Questions

1. The tuning fork vibrated. Yes.

2. The vibrating tuning fork caused water to splash out of the glass.

LESSON 8-2
How do sound waves travel?
(p. 132)

Teaching Strategies

PREVIEW Have students write the lesson title and objective in their notebooks. As students read the lesson, tell them to write down the sentence or sentences that provide the information needed in the objective.

REINFORCEMENT Have students recall that a rarefaction and a compression are parts of a longitudinal wave.

DISCUSSION Identify sound waves as longitudinal waves. Describe how particles of a medium travel in a longitudinal wave. Refer students to the diagram on p. 132 to observe the movements of sound waves through air.

DEMONSTRATION Use a coiled spring to show students how particles travel in a longitudinal wave.

CLASS ACTIVITY Have a few students put their ears against the wall of the classroom. Ask the students to describe what sounds they hear in the next room. Use this activity to show students that sound waves can travel through solids.

DEMONSTRATION To demonstrate that sound waves travel through liquids, hit two rocks together while holding them under water in a large jar. Students will be able to hear the sound made when the rocks hit each other.

INTRASCIENCE CONNECTION Have interested students research the communication of either dolphins or whales. Tell students to present their findings to the class in an oral report.

Answers to Challenges (p. 133)
Apply

8. No, because sound cannot travel in a vacuum.

9. a

State the Problem

Sound waves cannot travel in a vacuum.

Teaching Tips for Activity
Making a String Telephone

Skills: *modeling, analyzing*

COOPERATIVE/COLLABORATIVE LEARNING Have students work in pairs.

Questions

1. Yes, the sounds should be fairly clear.

2. through the string

3. At some point, the distance will be too great for sounds to be transmitted clearly.

LESSON 8-3
What is the speed of sound?

(p. 134)

Teaching Strategies

PREVIEW Have students read the lesson feature before beginning the lesson. Discuss the concepts presented in the feature. Ask students how they think this information relates to the topics they have already studied or to the lesson they are about to study.

DISCUSSION Provide motivation for this lesson by having students recall their experiences with firework displays. Students may recall that the burst of color from a skyrocket can be seen before the sound of the explosion can be heard. Ask students to explain this occurrence. Discuss all student responses. Then point out that differences in the speed of light and the speed of sound account for the time lapse. Follow up the discussion of fireworks with a discussion of thunder and lightning, which are described on p. 134.

REINFORCEMENT To familiarize students with the speed of sound in different materials, have students organize the materials listed in Table 1 on p. 134 in order of increasing speed of sound.

EXTENSION Have interested students research the development of airplanes. Tell students to make a time line to summarize their findings.

COOPERATIVE/COLLABORATIVE LEARNING Answer the Check and Apply questions as a class.

Answers to Challenges (p. 135)
Apply

5. The sound travels faster through the metal tracks than through the air.

6. **a.** The speed of sound in cork is not as fast as in other solids. **b.** rubber

Designing an Experiment

Have volunteers describe their experiments to the rest of the class.

Teaching Tips for Technology and Society
Supersonic Airplanes

DISCUSSION Use this feature to discuss the advantages and disadvantages of technological progress.

EXTENSION Have students find out more about the life of Chuck Yeager and write a brief biography on his life.

Questions

1. **INFER** What is Mach 3? (three times the speed of sound)

2. **DEFINE** What is a sonic boom? (loud noise caused when a plane breaks the sound barrier)

LESSON 8-4
What is an echo? (p. 136)

Teaching Strategies

PREVIEW Before beginning this section, have students scan the lesson looking for words with which they are unfamiliar. Have students work in pairs or small groups to define each of the words on their lists.

CLASS ACTIVITY Echoes can frequently be demonstrated in a schoolyard that has a walled-in area. Take your class to an echo-producing area and have students listen to their own echoes. Be sure students understand that echoes are caused when sound waves are reflected.

INTRASCIENCE CONNECTION Show students a map of the ocean floor. Point out features on the ocean floor, such as valleys, mountains, and guyots. Then describe how SONAR is used to map the ocean floor.

REINFORCEMENT Provide students with examples for calculating ocean depths. Tell students to make sure they divide the time for the total trip by 2 before multiplying by the speed of sound.

Answers to Challenges (p. 137)
Apply

6. The material does not absorb sound.

7. Answers will vary. Students should recognize that carpeting and other materials that absorb sound energy would probably be used.

8. 4500 meters

InfoSearch

You may wish to discuss students' questions and answers as a class.

Teaching Tips for Career in Physical Science
Acoustics Engineer

EXTENSION If possible, arrange for your class to visit a concert hall. Ask a tour guide to describe some of the materials that are used in the hall to ensure good sound quality.

Questions

1. **DEFINE** What is acoustics? (science of sound)

2. What is a "dead spot" in a concert hall? (place where it is hard to hear music clearly)

LESSON 8-5
What is intensity? (p. 138)

Teaching Strategies

PREVIEW Before beginning this lesson, write the title of each section in the lesson on the chalkboard. Have students copy these titles in their notebooks using an outline format. As students read each section, they should write the topic sentence for each paragraph in the section in their notebooks.

DISCUSSION To provide motivation for this lesson, begin the class by whispering. Gradually increase loudness until you are speaking very loudly. Point out that increasing the amount of energy increases the loudness of sound. Then describe the relationship between amplitude and loudness.

DEMONSTRATION To demonstrate that loud sounds have more energy than soft sounds, slap a meter stick on your desk using varying amounts of energy.

REINFORCEMENT The diagrams on p. 138 illustrate the difference in amplitude of waves for loud and soft sounds. Be sure students understand that the transverse waves are only a means of comparing loudness, or amplitude, of different sounds. Emphasize that sound waves are longitudinal waves.

CLASS ACTIVITY Have students study the intensity of different sounds listed in Table 1 on p. 138. Ask students which sounds listed in Table 1 are dangerous. (jet engine, rocket engine)

Answers to Challenges (p. 139)
Apply

7. 60 dB

8. Intensity decreases as distance increases.

9. a rock concern, because the decibel level is higher

Health and Safety Tip

Discuss students' findings as a class.

Teaching Tips for Science Connection
Noise Pollution

REINFORCEMENT Many young people play their stereos and radios at very high volumes. Emphasize that loud noise can cause permanent hearing loss and that people should not play music at very loud volumes, especially when they are wearing earphones.

Questions

1. **HYPOTHESIZE** How can you help prevent noise pollution? (Answers will vary. Accept all logical responses.)

2. **INFER** Why do you think noise pollution is more of a problem in big cities than in small towns? (There is more traffic, construction, and so on in a big city.)

LESSON 8-6
What are frequency and pitch? (p. 140)

Teaching Strategies

PREVIEW Before beginning this lesson, read the list of Tech-Terms aloud so students can hear their pronunciations.

REINFORCEMENT Remind students that the frequency of a wave is the number of complete waves per second. Ask students what unit is used to measure frequency. (hertz)

DEMONSTRATION Review the fact that sound may differ in loudness. Demonstrate this fact, using two tuning forks of the same pitch. Strike one tuning fork gently to produce a soft sound. Strike the other more forcefully to produce a louder sound. Then show tuning forks with different pitches. Call attention to the frequencies that are marked on each fork. Explain that the number represents the number of times the tuning fork vibrates each second when struck.

CLASS ACTIVITY Have students hold plastic rulers, so that the end of the rulers extend over the edge of their desks. Tell students to pluck their rulers to make the rulers vibrate. Have students note the pitch and vibration of the ruler. Then tell students to move the rulers so that less of the rulers extend over the desks. Again, tell students to pluck the rulers. Have students note that when a shorter length of ruler vibrates, it vibrates faster and produces a higher pitch.

RETEACHING OPTION Bring a dog whistle to class and blow into it. Many students probably know that humans cannot hear a dog whistle. Use this knowledge as a springboard for a discussion of ultrasonic frequencies, and the relationship between frequency and pitch.

Answers to Challenges (p. 141)
Apply

8. the 20-Hz sound; because it has a lower frequency

9. Its frequency increases as it moves toward you.

10. Answers will vary. Accept all logical responses.

11. Answers will vary. Accept all logical responses.

InfoSearch

You may wish to discuss students' questions and answers as a class.

Teaching Tips for Activity
Observing Change in Pitch

Skills: *observing, comparing*

COOPERATIVE/COLLABORATIVE LEARNING Answer the questions as a class.

Questions

1. The pitch changes as the height of the water column changes.

2. the test tube with the 10-cm column of water

3. the test tube with the 2-cm column of water

LESSON 8-7
What is sound quality? (p. 142)

Teaching Strategies

PREVIEW Before beginning this lesson, have students review the TechTerms and their definitions and read the Lesson Summary.

DEMONSTRATION If possible, obtain an oscilloscope to show the difference between a pure tone and overtones.

CLASS ACTIVITY Allow students to speak into the microphone of an oscilloscope to observe the patterns their voices make. Students will observe that each person's voice pattern is unique.

EXTENSION If any students play stringed instruments, ask them to demonstrate a fundamental tone and overtones. Emphasize that musical instruments have different sound qualities.

COOPERATIVE/COLLABORATIVE LEARNING Have students work in pairs to answer the Check and Apply questions.

Answers to Challenges (p. 143)
Apply

7. All voices would sound the same.

8. because they each have a different timbre

Ideas In Action

You may wish to play some instrumental music and identify musical instruments as a class.

InfoSearch

You may wish to discuss students' questions and answers as a class.

Teaching Tips for People in Science
Thomas Alva Edison

EXTENSION Thomas Edison defined genius as "one percent inspiration and ninety percent perspiration." Ask students to describe in their own words what Edison meant by this statement.

Questions

1. LIST What were three of Thomas Edison's inventions? (light bulb, motion picture camera, phonograph)

2. What was the earliest phonograph called? (gramophone)

LESSON 8-8
What is music? (p. 144)

Teaching Strategies

PREVIEW Before beginning this lesson, have students scan the lesson for the science process skill symbols. Have students identify each skill used in the lesson. Students should then review the definition given for each skill on page 4 of their text.

DISCUSSION Tell the students that the three parts of music are rhythm, melody, and harmony. During this lesson, an inter-disciplinary approach should be taken. The point should be made that, although music, science, and mathematics are different areas of study, and that even though aptitudes, or talents, in these areas are usually considered to be quite different and unrelated, music and the sciences have much in common.

RETEACHING OPTION Make a chart showing the three main groups of musical instruments. Write the following head on the chalkboard: *Musical Instruments*. From this head draw arrows to each of the following heads: *Percussion Instruments, String Instruments, Wind Instruments*. From the head Percussion Instruments, draw an arrow to the following example: *Drum*. From the head *String Instruments*, draw arrows to the following examples: *Guitars, Violins, Cellos*. From the head *Wind Instruments*, draw arrows to the following examples: *Trumpets, Clarinets, Flutes*. Have students copy the chart in their notebooks.

Answers to Challenges (p. 145)
Apply

7. Music has rhythm, melody, and harmony. Noise does not.

8. a. wind **b.** percussion **c.** string

InfoSearch

You may wish to discuss students' questions and answers as a class.

Teaching Tips for Leisure Activity
School Band

DISCUSSION Have any students who are members of the school band describe their practice sessions, performances, and so on, to the rest of the class.

EXTENSION In addition to practicing their musical instruments, most band members also learn and practice marching skills.

Question

How is being a member of a band different from playing a musical instrument alone? (Answers will vary. Accept all logical responses.)

LESSON 8-9
How do you hear? (p. 146)

Teaching Strategies

PREVIEW Have students write the lesson title and objective in their notebooks. As students read the lesson, tell them to write down the sentence or sentences that provide the information needed in the objective.

REINFORCEMENT Refer students to the diagram of the human ear on p. 146. Have students trace the path of sound waves through the ear.

DEMONSTRATION If possible, obtain a three-dimensional model of the human ear to show your class.

Answers to Challenges (p. 147)

Apply

4. e, g, h, b, a, c, f, d

InfoSearch

You may wish to discuss students' questions and answers as a class.

Teaching Tips for Looking Back in Science

The Invention of the Telephone

INTERDISCIPLINARY CONNECTION In addition to inventing the telephone, Alexander Graham Bell was also a founding member of the National Geographic Society in 1888. He served as its president from 1898 to 1903.

Question

How would your life be different if there were no telephone communication? (Answers will vary. Accept all logical responses.)

Answers to Unit Challenges (pp. 148–150)

Understanding the Features: Reading Critically

1. reduce traffic; pass laws

2. sign language for the deaf

3. twice the speed of sound

4. phonograph

5. how well all the different musical sounds come together as one sound

6. too many echoes would distort the music

Critical Thinking

1. Particles are closer together in a solid.

2. It is transferred to the bones of the middle ear.

3. It pushes on the air in front of it.

4. 50 dB; loudness increases with amplitude

5. loss of hearing

Interpreting a Diagram

1. eardrum

2. cochlea and auditory nerve

3. auditory nerve

4. because they are shaped like those objects

5. It helps funnel sound waves into the ear canal.

Previewing the Unit

Read the titles of the lessons in this unit aloud. Identify terms in the titles with which students are unfamiliar. Write these terms on the chalkboard. Using the Glossary of the text or a dictionary, define the terms as a class. Have students write the terms and their definitions in their notebooks. Then have students work in small groups to carry out the task in the Study Hint.

Bulletin Board Suggestions

1. Assemble a bulletin board showing the electromagnetic spectrum. Use the diagram on p. 168 as a guide. Label the various types of electromagnetic waves that make up the electromagnetic spectrum. Include arrows at the top of the spectrum showing the relationship between wavelength and frequency. Use the bulletin board to review the electromagnetic spectrum and the properties of light.

2. Assemble a bulletin board entitled "Uses of Light." Attach pictures that illustrate the countless uses of light in people's daily lives.

LESSON 9-1
What is light? (p. 152)

Teaching Strategies

PREVIEW Have students write the lesson title and objective in their notebooks. As students read the lesson, tell them to write down the sentence or sentences that provide the information needed in the objective.

DEMONSTRATION Use a radiometer to demonstrate that light is a form of energy. Show that the vanes of the instrument turn when exposed to light. When the light source is removed, they stop. Review the concept that energy is the ability to do work and that since light can do work, it is a form of energy.

CLASS ACTIVITY Obtain a piece of rubber hose. Bend it in the middle and point one end toward a light source. Have students take turns looking through the other end. Students will not be able to see the light source since light travels in straight lines. Then straighten out the hose. Again have students take turns looking through the hose. This time students will be able to see the light.

DEMONSTRATION Show students what is meant by a ray of light. Tape a flashlight so that only a narrow beam of light comes from it. In a darkened room, direct the beam of light parallel to the chalkboard. Students will be able to observe a ray of light.

Answers to Challenges (p. 153)
Apply

6. Yes; light coming from the light bulb

7. Answers will vary. Accept all logical responses.

InfoSearch

You may wish to discuss students' questions and answers as a class.

Teaching Tips for Activity
Observing that Light Travels in Straight Lines

Skills: *observing, analyzing*

COOPERATIVE/COLLABORATIVE LEARNING Answer the questions as a class.

Questions

1. Yes.

2. No; light is blocked.

3. Light travels in straight lines.

LESSON 9-2
How do light waves travel?
(p. 154)

Teaching Strategies

PREVIEW Have students read the lesson feature before beginning the lesson. Discuss the concepts presented in the feature. Ask students how they think this information relates to topics they have already studied or to the lesson they are about to study.

REINFORCEMENT Redraw the diagram on p. 154 showing the properties of a light wave on the chalkboard. Label the crest, the trough, the amplitude, and the wavelength.

RETEACHING OPTION Using a piece of wire, make a rigid model of several wavelengths of a transverse wave. Use the model to demonstrate amplitude and wavelength. Move the model past a fixed point to illustrate frequency.

COOPERATIVE/COLLABORATIVE LEARNING Have students work in pairs to answer the Check and Apply questions.

Answers to Challenges (p. 155)
Apply

6. a. sound b. sound c. light d. light e. light

7. Ultrasonic sound cannot be heard by the human ear.

Skill Builder: Predicting

Answers will vary. Accept all logical responses.

Teaching Tips for Leisure Activity
Photography

DEMONSTRATION You may wish to show students the different parts of a camera described in the feature.

EXTENSION Have students bring in some of their own photographs to display and discuss.

Questions

1. **IDENTIFY** In a camera, what is the opening for light to enter called? (aperture)

2. **INFER** What is the function of a camera's lens? (to focus light)

LESSON 9-3
What are sources of light?
(p. 156)

Teaching Strategies

PREVIEW Before beginning this lesson, have students scan the lesson looking for words with which they are unfamiliar. Have students work in pairs or small groups to define each of the words on their lists.

DEMONSTRATION Show the difference between a luminous object and an illuminated object. Use the sun, a classroom light, or other light source as an example of a luminous object. Point out that any object that is visible and is not luminous must be illuminated.

REINFORCEMENT Emphasize that most objects are illuminated objects. They do not give off their own light. They simply reflect light.

DISCUSSION Explain the differences among opaque, translucent, and transparent materials. For demonstration purposes, three identically shaped pieces of glass are useful. One piece should be clear, one frosted, and one painted.

Answers to Challenges (p. 157)
Apply

8. translucent

9. a. luminous **b.** illuminated **c.** luminous **d.** luminous **e.** illuminated

Designing an Experiment

Have volunteers describe their experiments to the rest of the class.

Teaching Tips for Science Connection
Shadows

INTRASCIENCE CONNECTION During a lunar eclipse, only regions within the umbra have a total eclipse. Regions within the penumbra experience a partial eclipse.

DEMONSTRATION Place an object on a piece of white paper and hold it in front of a lamp. Have students observe the shadow that is formed. Point out the umbra and penumbra of the shadow.

Questions

1. DEFINE What is the penumbra? (gray, outer part of a shadow)

2. EXPLAIN Why do most shadows not have umbras? (Most light comes from extended light sources. Umbras only form when a point source of light shines on an object.)

LESSON 9-4
How do you see? (p. 158)

Teaching Strategies

PREVIEW Before beginning this lesson, write the title of each section in the lesson on the chalkboard. Have students copy these titles in their notebooks using an outline format. As students read each section, they should write the topic sentence for each paragraph in the section in their notebooks.

DEMONSTRATION If possible, show students a model of the eye.

REINFORCEMENT Refer students to the diagram of the eye on p. 159. Have students locate the parts of the eye as you describe their functions.

CLASS ACTIVITY Have students work with partners to observe how light affects the pupils of the eyes. Tell students to take turns observing each other's eyes in normal light and after their eyes have been closed for one minute.

Answers to Challenges (p. 159)
Apply

7. Light, or electromagnetic energy, is converted to electrical energy.

8. The iris expanded to let more light into your eye.

Health and Safety Tip

The sun's rays and rays from a laser beam can destroy the photoreceptor cells in the retina.

Teaching Tips for Science Connection
Corrective Lenses

EXTENSION You may wish to display a convex lens and a concave lens. Convex lenses are used to correct farsightedness. Concave lenses are used to correct nearsightedness.

Questions

1. LIST What are two corrective lenses? (eyeglasses; contact lenses)

2. How do objects that are nearby appear to a person who is farsighted? (blurry)

LESSON 9-5
What is color? (p. 160)

Teaching Strategies

PREVIEW Before beginning this lesson, have students review the TechTerms and their definitions and read the Lesson Summary.

REINFORCEMENT Be sure students understand that white light is composed of the seven basic colors of light. Bring in a picture showing a rainbow. Explain that rainbows are formed when raindrops act as natural prisms. Have students identify the seven basic colors that make up a rainbow.

DEMONSTRATION Use a prism to disperse white light into the seven basic colors of light.

CLASS ACTIVITY Hold up different objects that are red, orange, yellow, green, blue, indigo, violet, white, or black. For each object, ask what color(s) of light are observed and what color(s) are reflected.

Answers to Challenges (p. 161)
Apply

5. violet; red

6. In a rainbow, sunlight is separated into the colors of the spectrum.

InfoSearch

You may wish to discuss students' questions and answers as a class.

Teaching Tips for Science Connection
Colorblindness

INTRASCIENCE CONNECTION Colorblindness is a sex-linked trait. The sex chromosomes (X and Y chromosomes) carry genes for certain traits, such as colorblindness. The gene for colorblindness is a recessive gene that is carried on the X chromosome. Since a woman receives two X chromosomes, she would have to inherit two recessive genes for colorblindness to show the trait. A man only has to inherit one X chromosome carrying the gene for colorblindness.

Question

LIST What are three kinds of colorblindness? (red-blind; blue-blind; green-blind)

LESSON 9-6
What is photosynthesis? (p. 162)

Teaching Strategies

PREVIEW Before beginning this lesson, read the list of Tech-Terms aloud so students can hear their pronunciations.

CLASS ACTIVITY Place a plant in the classroom with its leaves facing away from the sun. Tell students to observe the plant for the next week. Ask students how the leaves responded to sunlight. (The leaves turned toward the sun.)

EXTENSION Have the students observe prepared slides of plant cells under a microscope. Tell students to sketch what they see. Identify the parts of a plant cell, including the chloroplasts. Chloroplasts are small, green structures in a plant cell that contain chlorophyll.

REINFORCEMENT The importance of photosynthesis cannot be overemphasized. Photosynthesis is a source of food for organisms. It is also the main source of oxygen on the earth.

EXTENSION Write the equation summarizing photosynthesis on the chalkboard:

$$6CO_2 + 6H_2O + energy \longrightarrow C_6H_{12}O_6 + 6O_2$$

Point out that during photosynthesis, plants use carbon dioxide, water, and the sun's energy to make glucose; oxygen is given off as a waste product.

EXTENSION In addition to chlorophyll, plants have other pigments that help absorb sunlight. One group of accessory pigments is the carotenoids. Carotenoids are yellow, orange, or brown. In the fall, plants break down the chlorophyll in their leaves and the colors of the carotenoids can be seen.

COOPERATIVE/COLLABORATIVE LEARNING Have students quiz each other on the material in this lesson.

Answers to Challenges (p. 163)
Apply

7. The stored energy in the plant is used by the animal.

8. The plants use the energy to grow and reproduce.

InfoSearch

You may wish to discuss students' questions and answers as a class.

Teaching Tips for Activity
Removing Chlorophyll from Green Leaves

Skills: *observing, predicting*

SAFETY TIP Do this activity in a well-ventilated room. Caution students to be careful when using a knife.

COOPERATIVE/COLLABORATIVE LEARNING Have students work in pairs.

Questions

1. green

2. red, orange, brown, yellow

3. in the fall

LESSON 9-7
How do mirrors reflect light?
(p. 164)

Teaching Strategies

PREVIEW Before beginning this lesson, have students scan the lesson for the science process skill symbols. Have students identify each skill used in the lesson. Students should then review the definition given for each skill on page 4 of their text.

DEMONSTRATION A bouncing ball is a useful analog for reflecting light. However, because it is difficult to make any measurements, it should remain a qualitative demonstration. The spin of the ball is one of the things that make a measurement difficult. If an air table is available, you will be able to show very clearly how objects without spin obey the law of reflection.

RETEACHING OPTION Write the following head on the chalkboard: *Reflection.* From this head, draw arrows to each of the following heads: *Regular Reflection, Diffuse Reflection.* Describe and compare these two types of reflection. Have students copy the information in their notebooks under the proper heads.

DEMONSTRATION Use a smooth and crumpled piece of aluminum foil to show students regular and diffuse reflection.

Answers to Challenges (p. 165)
Apply

7. The light is completely absorbed.

8. Diffuse; water droplets make the mirror a rough surface.

Skill Builder: Measuring

The distance from the object to the mirror is the same as the distance from the image to the mirror. The heights of the object and the image are the same. The image is formed by a flat mirror is the same distance behind the mirror as the object is in front of the mirror, and is the same size as the object.

Teaching Tips for Activity
Observing Mirror Images

Skills: *measuring, observing, analyzing*

T36

COOPERATIVE/COLLABORATIVE LEARNING   Answer the questions as a class.

Questions

1. 3-cm mark; 6-cm mark

2. No; it was backwards and upside down.

3. Answers will vary. Accept all logical responses.

LESSON 9-8
How do lenses refract light?

(p. 166)

Teaching Strategies

PREVIEW   Before beginning this lesson, have students review the TechTerms and their definitions and read the Lesson Summary.

REINFORCEMENT   Remind students that when light moves from one medium to another, it is bent, or refracted.

DEMONSTRATION   Show students how light is refracted as it passes into water. Fill a small tank with water. Add some eosin to the water to make the effect more visible. Shine a ray of light into the water. Students will see that the ray of light is bent. Stand a ruler in a vertical position in the water to serve as a normal line. Students will note that as light passes from air to water, it is bent toward the normal. As the light ray leaves the tank, it is bent away from the normal.

CLASS ACTIVITY   Bring a variety of double convex, double concave, and convex-concave lenses to class for students to observe. Students will be best able to see the differences between the shapes of these lenses if cross sections are used for illustration.

REINFORCEMENT   Students may be helped to remember the difference between convex and concave by thinking of "caved in" for concave.

DEMONSTRATION   Shine a light source through a concave lens and a convex lens to show students the actual effect these lenses have on light. A blackboard optics kit is excellent for these demonstrations.

Answers to Challenges (p. 167)
Apply

7. concave

8. convex

Ideas in Action

Answers will vary. Accept all logical responses.

Teaching Tips for Activity
Forming a Real Image with a Convex Lens

Skills: *observing, relating*

COOPERATIVE/COLLABORATIVE LEARNING   Students must work in pairs.

Questions

1. Answers will vary. Accept all logical responses.

2. Answers will vary. The distance at which a clear image is formed is the focal length of the lens.

LESSON 9-9
What is the electromagnetic spectrum? (p. 168)

Teaching Strategies

PREVIEW   Before beginning this lesson, write the title of each section in the lesson the chalkboard. Have students copy these titles in their notebooks using an outline format. As students read each section, they should write the topic sentence for each paragraph in the section in their notebooks.

DISCUSSION   Refer students to the diagram of the electromagnetic spectrum on page 168 and discuss the different types of electromagnetic waves. Be sure students understand that wavelength and frequency are inversely proportional. Ask students which type of waves have a longer wavelength, gamma rays or radio waves. (radio) Then ask which waves have a higher frequency, gamma rays or radio waves. (gamma)

REINFORCEMENT   Be sure students understand that radio waves are used for television as well as radio broadcasting.

DEMONSTRATION   If possible, bring in an X-ray image to show the class.

Answers to Challenges (p. 169)
Apply

8. The microwaves are not absorbed by the glass.

Health and Safety Tip

X-ray technicians wear lead aprons because lead blocks X-rays from entering the body. Pregnant women are advised not to have X-rays taken because X-rays may possibly cause birth defects in their babies.

Teaching Tips for Career in Physical Science
X-ray Technician

EXTENSION   If possible, arrange for an X-ray technician to visit your class. Ask the X-ray technician to describe a typical workday and the precautions that are taken to prevent overexposure to X-rays.

Questions

1. Why do dentists use X-rays? (to find cavities)

2. What do X-ray technicians do? (operate X-ray machines)

LESSON 9-10
What are lasers? (p. 170)

Teaching Strategies

PREVIEW   Before beginning this section, have students scan the lesson looking for words with which they are unfamiliar. Have students work in pairs or small groups to define each of the words on their lists.

DISCUSSION   To motivate this lesson, bring in a compact disk, or CD, and display it in front of the classroom, or show students the bar codes on supermarket products. Tell students that both items illustrate a use of lasers. Then have any students who have seen a laser show describe it to the rest of the class.

DEMONSTRATION If you are fortunate enough to have a laser in your school, demonstrate it as part of this lesson. Show the interference effects and the high energy of the laser beam.

Answers to Challenges (p. 171)
Apply

8. White light is made up of many different wavelengths. Laser light is made up of only one wavelength.

9. No; white light spreads out.

10. red

Skill Builder: Building Vocabulary

"Mono-" means "one," "chromatic" means "of or relating to color." Therefore, "monochromatic" means "having one color." Laser light has only one color.

Teaching Tips for Science Connection
Holography

EXTENSION Students will be fascinated with holograms. These are inexpensive and are available from many scientific supply houses. Many holograms can be viewed with monochromatic light. Pupils will note the three-dimensional feature of the image, permitting the viewer to get a different perspective by changing his or her angle of view. Point out that any portion of the hologram contains information about the entire image. If half of the hologram is taped over, the full image is still available but with diminished quality. The appearance of the hologram may be offered as additional evidence of the wave nature of light.

Question

DEFINE What is holography? (photography that uses laser light)

Answers to Unit Challenges (pp. 172–174)
Understanding the Features: Reading Critically

1. aperture

2. Corrective lenses help nearsighted people see things that are far away.

3. umbra

4. It is inherited from a person's parents.

5. to show three-dimensional views of building

6. The high energy of X-rays may damage living tissue.

Critical Thinking

1. Answers will vary. Accept all logical responses.

2. It bends, or refracts, white light.

3. There is no interference.

4. It is made up of only one wavelength of color.

5. Answers will vary. Accept all logical responses.

6. pupil; retina; lens of the eye

Interpreting a Diagram

1. convex

2. A

3. B

4. Check students' diagrams.

5. brought together

Previewing the Unit

Have students complete the task in the Study Hint. Then compare students' word lists and definitions as a class.

Bulletin Board Suggestions

1. Make a bulletin board comparing a series circuit and a parallel circuit. Divide the bulletin board into two sections. Label one section *Series Circuit* and the other *Parallel Circuit*. Draw diagrams showing these two types of circuits in the appropriate section. Use the diagrams on pages 184 and 186 as guides. Label the source of electric current, the wire, the switches, and the load on each diagram. You also may wish to include symbols and values for voltage, resistance, and electric current to familiarize students with these symbols and to illustrate Ohm's law.

2. Assemble a bulletin board showing the countless uses of electricity. Have students find and collect pictures of appliances and other electric devices. Make a collage with all the pictures.

LESSON 10-1
What is electricity? (p. 176)

Teaching Strategies

PREVIEW Before beginning this lesson, read the list of Tech-Terms aloud so students can hear their pronunciations.

DEMONSTRATION Use styrofoam balls or paper circles to represent atoms.

DISCUSSION All students probably have had some direct experience with the effects of a static electric charge, although they may not have recognized it as such. The plastics and synthetic fabrics in wide use today are particularly susceptible to acquiring a charge. This accounts for their clinging quality. Crackling sounds and even visible discharges can often be noticed when handling these fabrics in dry air. The self-adhering behavior of plastic food wrap is also a result of static charge. Discuss experiences students have had with static electricity to introduce this lesson.

DEMONSTRATION Demonstrate the law of electrostatic charge. Rub rubber rods with fur to add electrons. Show that these rods repel each other. Rub glass rods with silk or plastic to remove electrons. Show that these rods also repel each other. Rub a rubber rod with fur and a glass rod with silk. Show that they attract each other. From these demonstrations, draw the inference that there must be two different types of charges, one on the rubber rods and one on the glass rods. Elicit from students that like charges repel each other and unlike charges attract each other.

Answers to Challenges (p. 177)
Apply

6. The cloud has a negative charge.

7. It will have a positive charge.

8. Answers will vary. Accept all logical responses.

Health and Safety Tip

Someone who receives an electric shock needs immediate medical attention. An ambulance should be called immediately. The vital signs of the victim should be checked. Artificial respiration or cardiopulmonary resuscitation may be necessary.

Teaching Tips for Activity
Observing Electric Charges

Skills: *observing, analyzing*

COOPERATIVE/COLLABORATIVE LEARNING Have a volunteer demonstrate this activity for the rest of the class.

Questions

1. They were attracted to the comb.

2. Running the comb through your hair caused an electric charge to build up on the comb. The electric charge attracted the pieces of paper.

LESSON 10-2
What are insulators and conductors? (p. 178)

Teaching Strategies

PREVIEW Before beginning this lesson, have students scan the lesson for the science process skill symbols. Have students identify each skill used in the lesson. Students should then review the definition given for each skill on page 4 of their text.

DISCUSSION Provide motivation for this lesson with a discussion of electrical wiring. Many students probably are aware that metals such as copper and aluminum are used for electrical wiring. Point out that metals are useful because they are good conductors of electricity. Define conductors as materials that allow electric charges to flow through them easily.

DEMONSTRATION Display examples of insulators, such as rubber, cork, and plastic, in front of the classroom. Explain that insulators prevent the flow of electric charges.

DEMONSTRATION Use a conductivity tester to test various common materials for conductivity.

CLASS ACTIVITY Have students look around the classroom to find examples of conductors and insulators. List student responses on the chalkboard.

RETEACHING OPTION Pick up an electric cord in the classroom and describe the three main parts. Be sure students understand the importance of rubber or plastic as an insulator.

REINFORCEMENT Emphasize the danger of frayed or worn electric cords.

Answers to Challenges (p. 179)
Apply

8. A lightning rod is made of a conductor. The electricity from the lightning flows through the rod and into the ground.

InfoSearch

You may wish to discuss students' questions and answers as a class.

T39

Teaching Tips for Career in Physical Science
Television Repairperson

CLASS ACTIVITY To help students understand the definition of a semiconductor, have them look up the meaning of the prefix "semi-" in a dictionary. Then ask students to write in their own words how the meaning of the prefix "semi-" relates to its use in the term "semiconductor."

Question

DEFINE What is electronics? (study of electricity as it is used in helpful devices such as televisions)

LESSON 10-3
What are two kinds of electric current? (p. 180)

Teaching Strategies

PREVIEW Have students write the lesson title and objective in their notebooks. As students read the lesson, tell them to write down the sentence or sentences that provide the information needed in the objective.

DISCUSSION Refer students to Figure 1 on page 180 and explain the difference between direct current and alternating current.

REINFORCEMENT To help students remember the definition of an alternating current, have them look up the meaning of the word "alternating" in a dictionary. (alternating: changing by turn)

Answers to Challenges (p. 181)
Apply

6. alternating current

7. because the waves carrying the current would overheat

8. direct current

Ideas in Action

The positive pole of a battery is at one end. The negative pole is at the opposite end. Current flows from the negative pole to the positive pole.

Teaching Tips for Leisure Activity
Building Model Electric Trains

EXTENSION Obtain a variety of model railroad magazines to show your class.

DEMONSTRATION Ask any students who have train sets and run model trains to show their models to the rest of the class.

Question

How can you find out more about building model railroad trains? (magazines, local model railroad club)

LESSON 10-4
What is a battery? (p. 182)

Teaching Strategies

PREVIEW Before beginning this lesson, have students review the TechTerms and their definitions and read the Lesson Summary.

DEMONSTRATION Bring the parts of a copper–zinc wet cell to class. Assemble the cell and fill it with diluted sulfuric acid. Connect the terminals to a galvanometer to show that electricity is being produced. Then to show students the importance of the electrolyte, remove the acid from the cell. Wash the electrodes clean and refill the cell with distilled water. Students will observe that no current is produced.

DISCUSSION Bring a variety of dry cells to class. Show students the negative and positive terminals. Then refer students to the diagram of a dry cell on p. 182 and describe the electrodes and electrolyte of a dry cell.

DEMONSTRATION Remove the outer covering from a used radio battery and cut the battery in half. Show students the inside of the battery. Point out the zinc case, the carbon electrode inside the battery, and the moist paste, which separates the two electrodes. Students will observe that the zinc has been destroyed.

REINFORCEMENT Emphasize that in both a wet cell and a dry cell, electrons move from a negative electrode to a positive electrode.

Answers to Challenges (p. 183)
Apply

9. It stores energy.

10. The cold slows down the chemical reactions inside the cell.

Designing an Experiment

Have volunteers describe their experiments to the rest of the class.

Teaching Tips for Activity
Making a Lemon Wet Cell

Skills: *modeling, identifying*

COOPERATIVE/COLLABORATIVE LEARNING Have students work in pairs.

Questions

1. the copper strip

2. the zinc strip

3. the juice of the lemon

LESSON 10-5
What is a series circuit? (p. 184)

Teaching Strategies

PREVIEW Have students read the lesson feature before beginning the lesson. Discuss the concepts presented in the feature.

Ask students how they think this information relates to topics they have already studied or to the lesson they are about to study.

DEMONSTRATION Use a simple series circuit with a light bulb, a 6-volt dry cell, and a switch to establish the definitions of a series circuit and a closed circuit. Trace the single path that the electric current follows in a series circuit.

REINFORCEMENT Emphasize that when one part of a series circuit fails, all other parts also stop operating.

COOPERATIVE/COLLABORATIVE LEARNING Have a volunteer name and describe the four parts of an electric circuit to the rest of the class.

Answers to Challenges (p. 185)

Apply

6. to turn the current on or off

7. the water

Skill Builder: Diagramming

Check students' diagrams.

Teaching Tips for Looking Back in Science

Computers

CLASS ACTIVITY Have students work in small groups to draw time lines summarizing the information in this feature.

Questions

1. What were two disadvantages of early computers? (They were very large and had only a limited memory.)

2. What are three applications of computers? (Answers will vary. Accept all logical responses.)

LESSON 10-6
What is a parallel circuit?

(p. 186)

Teaching Strategies

PREVIEW Before beginning this lesson, write the title of each section in the lesson on the chalkboard. Have students copy these titles in their notebooks using an outline format. As students read each section, they should write the topic sentence for each paragraph in the section in their notebooks.

DISCUSSION Describe a parallel circuit and compare a parallel circuit and a series circuit. Have students trace the three paths of electric current in the diagram on p. 186 to reinforce the concept of a parallel circuit.

DEMONSTRATION Construct a parallel circuit containing a 6-volt battery, 3 small 6-volt lamps, and a switch. Trace the three paths that the electric current may follow for students. Then open the switch and replace one of the lamps with a burned-out lamp. Challenge the students to predict what will happen when you close the switch. Students will observe that the current still is able to flow through the two remaining paths. Contrast this behavior with a series circuit.

Answers to Challenges (p. 187)

Apply

5. series

6. A and B are in parallel; C, D, and E are in series.

7. parallel, possible answer: If the headlights stop working, the radio and blinkers still work.

Ideas in Action

Answers will vary. Check students' circuit diagrams.

Teaching Tips for Career in Physical Science

Computer Programmer

EXTENSION Obtain a simple computer program to show students. Point out that it is the job of a computer programmer to write programs. If a computer is available, you may wish to run the computer program.

Questions

1. **DEFINE** What is a flow chart? (a list of instructions for the computer to follow)

2. **LIST** What are three computer languages? (BASIC, FORTRAN, COBOL)

LESSON 10-7
What are volts, amps and ohms? (p. 188)

Teaching Strategies

PREVIEW Before beginning the lesson, read the list of Tech-Terms aloud so students can hear their pronunciations.

DISCUSSION Develop the analogy between water flowing in pipes and electrons moving in electric circuits. This comparison will help students understand voltage and electric currents.

DEMONSTRATION Bring a DC voltmeter with several ranges to class. Demonstrate the operation of the voltmeter and how each range is listed. The higher range of the voltmeter should be used first in making a measurement to avoid a possible overload of the instrument.

DEMONSTRATION Bring an ammeter to class and demonstrate how it works. Show students how to read the scales and how the different ranges operate.

DEMONSTRATION Show some of the factors that affect the resistance of a wire. Use a large ammeter, a number 6 dry cell, and different lengths and thicknesses of copper wire.

Answers to Challenges (p. 189)

Apply

8. an insulator, because wires made of poor conductors have more resistance than wires made of good conductors

InfoSearch

You may wish to discuss students' questions and answers as a class.

Teaching Tips for Technology and Society
Superconductors

EXTENSION Have interested students find out more about superconductors. Tell students to use library references to find out about current research with superconductors. Have students write their findings in a report.

Questions

1. IDENTIFY What is the coldest possible temperature? (-273 °C, or absolute zero)

2. RELATE What is the relationship between temperature and resistance? (as temperature decreases, resistance decreases)

Teaching Tips for Career in Physical Science
Electrician

EXTENSION To give students practice reading the classified sections of newspapers, have students study the help-wanted sections of newspapers to look for electrician jobs.

Questions

1. Why must an electrician make sure the correct type of wiring is used in a home? (to prevent electrical fires)

2. What voltage is supplied to most homes? (120 volts)

LESSON 10-8
What is Ohm's law? (p. 190)

Teaching Strategies

PREVIEW Before beginning this section, have students scan the lesson looking for words which they are unfamiliar. Have students wok in pairs or small groups to define each of the words on their lists.

REINFORCEMENT Be sure students understand that the symbol for electric current is I. Remind students that electric current, or I, is measured in amperes, or amps.

REINFORCEMENT Redraw Figure 2 on p. 190 on the chalkboard. Show students how to use the diagram to find the value for electric current, voltage, or resistance given the other two values.

CLASS ACTIVITY Give students several practice problems for finding electric current, voltage, and resistance using Ohm's law.

DEMONSTRATION To demonstrate Ohm's law, connect two dry cells together to make a 3-volt battery. Connect the battery to a 1-ohm resistor. Ask students to predict what the current will be using Ohm's law (3 amps). Then add an ammeter to the circuit. Students will observe that there is a current of 3 amps.

RETEACHING OPTION Prepare an "unknown voltage." Use three 1.4-volt dry cells in series. Place the cells in a box so that only the leads are exposed. Using a known resistance, an ammeter, and the unknown voltage, set up a series circuit. Ask students to find the voltage using Ohm's law. (Resistances should be selected that keep the current under 1 ampere. Otherwise, the internal resistance of the battery may affect the numerical results.)

Answers to Challenges (p. 191)
Apply

Check students' tables. ($I = 0.14$; $R = 16.6$; $V = 125$; $R = 22$)

Skill Builder: Analyzing

2 amps

LESSON 10-9
How can you use electricity safely? (p. 192)

Teaching Strategies

PREVIEW Have students read the lesson feature before beginning the lesson. Discuss the concepts presented in the feature. Ask students how they think this information relates to topics they have already studied or to the lesson they are about to study.

DISCUSSION Many students probably are aware that if too many electrical appliances are connected to one outlet, a fuse may blow. Use student knowledge to stimulate a discussion about short circuits, fuses, and circuit breakers.

EXTENSION Have students find out whether fuses or circuit breakers are used in their homes. Tell students whose homes have fuses to find out what types of fuses they are. Discuss students' findings as a class.

CLASS ACTIVITY Divide the class into small groups. Have each group make a poster illustrating electrical safety rules. Display the posters throughout the school so all students may observe electrical safety rules.

EXTENSION Arrange for a member of your local fire department to visit your class to discuss electrical safety.

Answers to Challenges (p. 193)
Apply

7. series circuit

8. The increased length of the connected cords increases the resistance; the cords may become overheated.

State the Problem

What is the electric current?

Teaching Tips for Science Connection
High-voltage Wires

EXTENSION If possible, arrange for your class to tour an electric power plant.

Questions

1. Why are high-voltage wires used? (to carry electric current long distances)

2. INFER Why do you think contact with the third rail of railroad tracks can be deadly? (The rail has very high voltage.)

Answers to Unit Challenges (pp. 194–196)
Understanding the Features: Reading Critically

1. They have no resistance.

2. a list of instructions written in computer language

3. mathematics, physics, and chemistry

4. 1946

5. both

6. the study of electricity as it is used in devices such as televisions

7. The electricity has to travel long distances through the wire; the high voltage is needed to give the electrons a big enough push.

Critical Thinking

1. No, because the wires may still be carrying a very high voltage.

2. Batteries provide direct current. Wall outlets provide alternating current.

3. They both have 6 amps.

4. No; a current can follow more than one path, but only travel one direction.

Interpreting a Diagram

1. series

2. parallel

3. electric current

4. voltage

5. resistances

6. 12 ohms

7. 2 amps

8. $I_1 = 5$ amps, $I_2 = 10$ amps

Previewing the Unit

Read the titles of the lessons in this unit aloud. Identify terms in the titles with which students are unfamiliar. Write these terms on the chalkboard. Using the Glossary of the text or a dictionary, define the terms as a class. Have students write the terms and their definitions in their notebooks.

Bulletin Board Suggestions

1. Assemble a bulletin board showing the earth's magnetic and geographic poles. Using the diagram on p. 204 as a guide, draw an outline of the earth on the bulletin board. Label both the magnetic and geographic poles. Use the bulletin board when explaining how the earth acts like a magnet.

2. Construct a collage on the bulletin board illustrating the many applications of magnetism in people's everyday lives. Attach pictures of common magnets, compasses, transformers, motors, generators, power plants, and so on. Use the bulletin board to emphasize the importance of magnetism and its relationship to electricity.

LESSON 11-1
What is a magnet? (p. 198)

Teaching Strategies

PREVIEW Have students write the lesson title and objective in their notebooks. As students read the lesson, tell them to write down the sentence or sentences that provide the information needed in the objective.

CLASS ACTIVITY To motivate this lesson, display a collection of materials in front of the classroom. Include samples of iron, cobalt, nickel, and other metals. Also include a paper envelope with a small strip of iron concealed inside. Challenge the students to predict which materials will be attracted by a magnet. Have students take turns testing the materials. Have the envelope tested last. List the materials that are attracted to a magnet on the chalkboard. Students will be surprised that the envelope is attracted by a magnet. The surprise of the class at the behavior of the envelope provides you with an opportunity to emphasize the value of careful observation. The three common magnetic elements, iron, cobalt, and nickel, should be identified.

DEMONSTRATION Demonstrate the poles of a magnet. Make a mark at one end of a magnet, and hang the magnet from a string. Tie the string around the middle of the magnet. Have students note in which direction the marked end of the magnet is pointing when the magnet stops swinging. Then give the magnet a little push and let it come to rest again. Students will observe that the magnet points in the same direction again. Identify the pole as the south or north pole, depending upon the direction in which it is pointing.

CLASS ACTIVITY Divide the class into small groups. Distribute several bar magnets to each group to allow students to explore the properties of a magnet. Have students bring like poles and unlike poles near each other to observe how unlike poles attract each other and like poles repel each other.

RETEACHING OPTION Obtain a set of "magnetic dogs." These are small plastic dogs with magnets in their bases. Place them on the stage of an overhead projector. Demonstrate their behavior to the class when you move one dog near the other dog. Sometimes the dogs jump together; other times, one dog may spin around or jump away. Challenge the students to explain the movement of the plastic dogs in terms of magnetic poles.

Answers to Challenges (p. 199)
Apply

8. No. A magnet always has two poles.

9. The south pole of a magnet points south.

10. The poles will repel each other.

Designing an Experiment

Have volunteers describe their experiments to the rest of the class.

Teaching Tips for Looking Back in Science
Magnetic Compasses

DEMONSTRATION Bring samples of lodestone to class, and demonstrate its magnetic properties. Explain that these natural magnets were most likely people's first contact with magnetism.

DEMONSTRATION Show students how to use and read a compass.

Questions

1. INFER Why do you think the Greeks called iron ore magnetic? (It was discovered in Magnesia, Turkey.)

2. RELATE Why did sailors call magnetite lodestone? (It always pointed towards the North Star, which was also called the lodestar.)

LESSON 11-2
What is a magnetic field?
(p. 200)

Teaching Strategies

PREVIEW Before beginning this lesson, write the title of each section in the lesson on the chalkboard. Have students copy these titles in their notebooks using an outline format. As students read each section, they should write the topic sentence for each paragraph in the section in their notebooks.

DEMONSTRATION Place a bar magnet on an overhead projector. Cover the magnet with a clear sheet of plastic. Sprinkle iron filings around the magnet to show students the shape of the magnetic field.

CLASS ACTIVITY Place two bar magnets on an overhead projector with the north pole of one magnet pointed toward the south pole of the other magnet. Cover the magnets with a sheet of plastic. Sprinkle iron filings around the magnets. Have students sketch the shape of the magnetic field in their notebooks. Then repeat the demonstration with like poles pointed toward each other. Again, have students sketch the shape of the magnetic field. Tell students to compare their drawings with the diagrams shown in Figure 2 on p. 200.

DEMONSTRATION The magnetic field around a magnet is three-dimensional. The lines of force seen in the demonstration with iron filings and magnets give a two-dimensional picture of the field. You can show the three-dimensional nature of the magnetic field by filling a large test tube with clear oil or glycerine. Add a quantity of iron filings to the tube and seal it with a stopper. Shake the tube and then place it between the poles of a pair of alnico magnets. Have students observe the pattern of the iron filings in the test tube. (Repeated shaking of the test tube may be necessary.)

Answers to Challenges (p. 201)

Check

1. area around a magnet where magnetic forces can act

2. The magnetic force will attract the iron.

3. lines that show the shape of a magnetic field

4. When iron filings are sprinkled around a magnet, they will line up along the magnetic lines of force.

5. The magnetic force is strongest at the poles.

Apply

6. They will curve in opposite directions.

7. north pole; the lines of force bend away from each other.

InfoSearch

You may wish to discuss students' questions and answers as a class.

Teaching Tips for Activity
Observing Magnetic Lines of Force

Skills: *observing, modeling*

SAFETY TIP Be sure students wear their safety goggles.

COOPERATIVE/COLLABORATIVE LEARNING Have students work in pairs.

Questions

1. at the poles

2. They curve in opposite directions.

3. They curve from the end of one pole to the other.

4. Check students' diagrams.

LESSON 11-3
How can you make a magnet? (p. 202)

Teaching Strategies

PREVIEW Have students read the lesson feature before beginning the lesson. Discuss the concepts presented in the feature. Ask students how they think this information relates to topics they have already studied or to the lesson they are about to study.

DEMONSTRATION Demonstrate magnetic induction. Hold an iron nail over some iron filings. Students will observe that the nail is not a natural magnet. Then make the nail magnetic by stroking it with a magnet in one direction. Demonstrate the

---

nail's induced magnetism by again holding it over the iron filings. Students will observe that the nail now attracts the filings.

REINFORCEMENT Be sure students understand the inverse relationship between the ease with which a material is magnetized and the length of time the material stays magnetized. After comparing the magnetic induction of iron and aluminum, and identifying the two metals as temporary and permanent magnets, ask students whether nickel is a temporary or permanent magnet. (temporary)

CLASS ACTIVITY To help students understand and remember the definition of magnetic induction, have them look up the definition of "induction" in a dictionary. (induction: bringing about; causing)

Answers to Challenges (p. 203)

Apply

7. Yes. Magnets always have two poles.

8. A temporary magnet is easily magnetized, but tends to lose its magnetism quickly. A permanent magnet is hard to magnetize, but tends to keep its magnetism.

9. No. Materials that are easy to magnetize tend to lose their magnetism quickly.

Skill Builder: Experimenting

Check students' reports.

Teaching Tips for Technology and Society
Magnetic Levitation

DISCUSSION Before students read this feature, remind students that friction is a force that opposes the motion of an object.

EXTENSION Have interested students find out more about the use of maglev trains. Tell students to research what progress has been made toward the worldwide use of maglev trains. Tell students to write their findings in a report.

Questions

1. **DEFINE** What is magnetic levitation? (the process by which an object can be kept floating above a magnet)

2. **LIST** What are two advantages of maglev trains? (They can reach very high speeds. They do not cause pollution.)

LESSON 11-4
How is the earth like a magnet? (p. 204)

Teaching Strategies

PREVIEW Before beginning this lesson, read the TechTerm aloud so students can hear its pronunciation.

DEMONSTRATION Repeat William Gilbert's experiment for the class. Obtain a bar magnet with its poles labeled. Tie a string around the middle of the magnet and suspend it from a ring stand. Students will observe that the north pole of the magnet points North, and the south pole points South.

Discussion Describe how the earth acts like a magnet. To illustrate your description, point out the earth's geographic poles and magnetic poles on a globe or world map. Be sure students understand that the earth's poles are not named for their magnetic polarity, but for their geographic location.

Extension Northern lights are also known as the aurora borealis. Southern lights are known as the aurora australis.

Answers to Challenges (p. 205)
Apply

7. Permanent. The earth has retained its magnetic field.

8. Gravity; the bar falls to the ground.

InfoSearch

You may wish to discuss students' questions and answers as a class.

Teaching Tips for Activity
Making a Magnetic Compass

Skills: *observing, modeling*

Cooperative/collaborative learning Answer the questions as a class.

Questions

1. The needle becomes magnetic.

2. North and South

LESSON 11-5
How are electricity and magnetism related? (p. 206)

Teaching Strategies

Preview Before beginning this section, have students scan the lesson looking for words with which they are unfamiliar. Have students work in pairs or small groups to define each of the words on their lists.

Demonstration Demonstrate Oersted's experiment for the class. Set up a simple series circuit with a battery and a wire. Bring a compass near the circuit. Have students observe that the compass needle moves.

Demonstration Show the effect of moving a magnet near a wire. Use an ordinary bar magnet. Connect the wire directly to a large demonstration galvanometer. A weak current will flow through the wire when the magnet is moved past it. Show that the current stops when the magnet stops moving.

Reinforcement Be sure students understand the difference between Oersted's and Faraday's discoveries. Oersted discovered that an electric current causes a magnetic field. Faraday discovered that a magnetic field causes an electric current.

Cooperative/collaborative learning Have students quiz each other on the material in this section.

Answers to Challenges (p. 207)
Apply

7. It would increase the strength because an electric current produces a greater magnetic field in a coil of wire.

8. Yes. Moving a wire in a magnetic field causes an electric current.

Skill Builder: Researching

Check students' reports. Oersted: oersted; unit of magnetic intensity; Ampere: amp; unit of electric current; Tesla: tesla; unit of magnetic field strength; Henry: henry; unit of electromagnetic induction; Faraday: farad; unit of capacitance.

Teaching Tips for Looking Back in Science
Invention of the Telegraph

Extension To help students understand how codes were and are still used to send messages, have students work in small groups to make up their own codes. Tell students to write messages using their secret codes. After each group is finished, try to decipher all the messages as a class.

Question

Using Morse's telegraph, how could an operator send a coded signal? (by pressing the key in different combinations of dots and dashes)

LESSON 11-6
What is an electromagnet?
(p. 208)

Teaching Strategies

Preview Before beginning this lesson, have students scan the lesson for the science process skill symbols. Have students identify each skill used in the lesson. Students should then review the definition given for each skill on page 4 of their text.

Demonstration Construct electromagnets with 20, 40, and 60 coils. Use large iron nails or bolts as cores. Connect these electromagnets in turn to a No. 6 dry cell. Elicit from students that increasing the number of coils increases the strength of an electromagnet. Test each electromagnet's strength by seeing how many paper clips it will pick up when the current is flowing.

Demonstration Show students that increasing the current through a wire makes an electromagnet stronger. First, test how many paper clips an electromagnet can pick up with only one dry cell in the circuit. Then, wire two cells together in a circuit. Point out that this increases the voltage and thus the current. Again, test how many paper clips the electromagnet will pick up. Students will observe that increasing the current increases the strength of the electromagnet. (**Note:** In order to get meaningful results in these demonstrations it is necessary to load the circuit with a small resistance. Otherwise, the current will be limited mainly by the internal resistance of the dry cells, and there will be no observable change in current or magnetic strength when the second cell is added to the circuit. To make resistance suitable for use with No. 6 cells, wind about 20 feet of #24 copper wire on a spool, and insert it in series with the electromagnets. The current flow will be about 3 amperes with 1.5 volts and 6 amperes with 3.0 volts, enough to produce adequate field strengths without overloading the cells or excessive heating of conductors.)

Answers to Challenges (p. 209)

Check

1. temporary magnet made by wrapping a current-carrying wire around an iron core

2. When the current is turned off, there is no longer a magnetic field.

3. add more coils; increase current

4. core

5. radios, telephones, computers

Apply

6. When a current is turned off, an electromagnet does not have a magnetic field. A permanent magnet always has a magnetic field.

7. **a.** 40 coils **b.** 3 amps

InfoSearch

You may wish to discuss students' questions and answers as a class.

Teaching Tips for Activity

Making an Electromagnet

Skills: *modeling, comparing, measuring*

COOPERATIVE/COLLABORATIVE LEARNING Have students work in pairs.

Questions

1. Answers will vary. Accept all logical responses.

2. two batteries; The strength of an electromagnet increases with increasing current.

LESSON 11-7
What is a transformer? (p. 210)

Teaching Strategies

PREVIEW Have students write the lesson title and objectives in their notebooks. As students read the lesson, tell them to write down the sentence or sentences that provide the information needed in the objectives.

REINFORCEMENT Before beginning this lesson, review the definitions of a direct current and an alternating current. Ask students how electrons flow in a direct current (in one direction) and in an alternating current (flow of electrons changes direction regularly).

DEMONSTRATION Demonstrate a dissectible transformer. Show and identify the various parts. Use AC voltmeters and ammeters to demonstrate step-up and step-down transformer circuits.

DISCUSSION Refer students to the diagrams on p. 210 showing a step-up and step-down transformer. Describe each transformer and how it is used.

CLASS ACTIVITY If possible, take the class outside and show them a transformer at the top of a utility pole.

Answers to Challenges (p. 211)

Check

1. device in which alternating current in one coil of wire induces a current in a second coil

2. primary coil

3. A current is induced.

4. step-up

5. step-down

Apply

6. A step-up transformer increases voltage; a step-down transformer decreases voltage.

7. alternating; it is induced by alternating current in the primary coil

8. It was decreased by a step-down transformer.

9. step-up: a; step-down: b, c

Ideas in Action

Answers will vary. Accept all logical responses.

Teaching Tips for People in Science
Joseph Henry

DISCUSSION You may wish to use this feature to motivate a discussion of the concurrent discoveries of scientists. Point out that over the years, many scientific discoveries have been made by different scientists at about the same time. Ask students why they think that only one scientist is often credited with the discovery. Discuss all student responses.

INTRASCIENCE CONNECTION The U.S. Weather Service is a national weather bureau that tracks weather conditions and distributes weather information.

Question

IDENTIFY What is the unit for measuring electromagnetic induction? (the henry)

LESSON 11-8
What is the motor? (p. 212)

PREVIEW Before beginning this lesson, have students scan the lesson for the science process skill symbols. Have students identify each skill used in the lesson. Students should then review the definitions given for each skill on page 4 of their text.

DEMONSTRATION To show students that changing the direction of an electric current will reverse the poles of an electromagnet, connect an electromagnet to a dry cell. Use a compass to identify the poles. Then switch the connections to the dry cell so that the direction of the current is reversed. Again, use a compass to identify the poles. Students will observe that the poles are reversed.

DISCUSSION Refer students to the picture of an electric motor on p. 212 and describe how it works. Be sure students understand that an electric motor changes electrical energy into mechanical energy.

DEMONSTRATION If possible, bring in a small electric motor to show the class.

Answers to Challenges (p. 213)

Apply

7. to reverse the poles of the electromagnet so it keeps spinning

8. it would spin more slowly

9. An electric motor uses the force of magnetism to change electric energy into useful mechanical energy.

Skill Builder: Modeling

Check students' diagrams.

Teaching Tips for Career in Physical Science

Automobile Mechanic

EXTENSION Have interested students interview an automobile mechanic at a local service station about his or her job. Tell students to present the results of their interviews in oral reports.

Questions

1. LIST What are three electrical parts that make up an automobile engine? (starting motor, battery, generator)

2. DESCRIBE What does a specialty mechanic do? (works on only one part of a car's engine)

LESSON 11-9
What is a generator? (p. 214)

Teaching Strategies

PREVIEW Before beginning this lesson, have students scan the lesson looking for words with which they are unfamiliar. Have students work in pairs or small groups to define each of the words on their lists.

DISCUSSION Refer students to the diagram of a generator on p. 214, and describe how a generator operates.

EXTENSION You may wish to tell students that electricity produced using the energy of running water is called hydroelectric power.

REINFORCEMENT Use this lesson to emphasize the importance of turning off unnecessary lights and other electrical appliances as a measure of conserving nonrenewable fossil fuels.

COOPERATIVE/COLLABORATIVE LEARNING Have students work in pairs to answer the Check and Apply questions.

Answers to Challenges (p. 215)

Check

1. A current is produced.

2. alternating current

3. device that changes mechanical energy into electrical energy

4. turbines

5. moving steam or water

6. coal, oil, nuclear energy

Apply

7. A generator changes mechanical energy to electrical energy. An electric motor changes electric energy to mechanical energy.

8. A current would not be produced.

9. Chemical energy is used to produce mechanical energy, which is used to produce electricity.

InfoSearch

You may wish to discuss students' questions and answers as a class.

Teaching Tips for Science Connection

Electric Power Plants

CLASS ACTIVITY Arrange for a field trip to a local electric power plant.

REINFORCEMENT To help students remember the definition of a hydroelectric plant, have them look up the meaning of the prefix "hydro-" in a dictionary. (hydro-: water)

Question

EXPLAIN Why can hydroelectric plants provide only a limited supply of electrical energy? (They must be built near waterfalls.)

Answers to Unit Challenges (pp. 216–218)

Understanding the Features: Reading Critically

1. Lodestone is magnetite, a magnetic material. It pointed toward the North and could be used to find direction.

2. There is no friction between train and tracks to slow it down.

3. unit of electromagnetic induction

4. hydroelectric plant

5. automobile service stations; the service departments of automobile dealers

6. Samuel Morse

Critical Thinking

1. Magnetic induction is making a material magnetic. Electromagnetic induction is the production of magnetism by an electric current.

2. In an electric motor, electrical energy is changed to mechanical energy. In a generator, mechanical energy is changed to electrical energy.

3. The north pole of the compass would point to the south pole of the magnet.

4. As the loop moves up and down, the current flows in opposite directions.

5. Answers will vary. Accept all logical responses.

Interpreting a Diagram

1. 2; 10 **2.** 10; 2

3. diagram A; diagram B

4. step-up; A step-up transformer can be used to increase voltage.

5. step-down; A step-down transformer can be used to decrease voltage.

6. step-down; Step-down transformers lower the high voltage carried by power lines.

Previewing the Unit

Read the titles of the lessons in this unit aloud. Identify terms in the titles with which students are unfamiliar. Write these terms on the chalkboard. Using the Glossary of the text or a dictionary, define the terms as a class. Have students write the terms and their definitions in their notebooks.

Bulletin Board Suggestions

1. Have students find and collect pictures showing different elements. Make a collage on the bulletin board with all the pictures. Label each picture with the name and chemical symbol of the element that is shown.

2. Assemble a bulletin board showing the periodic table of the elements. Attach a poster size periodic table to the bulletin board. Use the bulletin board to show students how to use the periodic table and what information can be obtained from the periodic table.

LESSON 12-1
What are the properties of matter? (p. 220)

Teaching Strategies

PREVIEW Before beginning this lesson, have students review the TechTerms and their definitions and read the Lesson Summary.

DISCUSSION To motivate this lesson, have each student name a different object. List student responses on the chalkboard. After each student has had a turn, ask students what all the objects listed on the chalkboard have in common. Point out that all the objects are made up of matter. Define matter as anything that has mass and takes up space. Remind students that mass is the amount of matter an object contains and the amount of space an object takes up is its volume.

DEMONSTRATION Use a balance to demonstrate that water has mass by comparing the masses of an empty glass and a glass filled with water.

DEMONSTRATION You can show students that air has mass using two balloons and a meter stick. Take the empty balloons and a meter stick, and suspend the meter stick by a string tied around its center. Balance the balloons. Then fill one of the balloons with air. Students will observe that the meter stick will no longer balance.

DISCUSSION Hold an apple up in front of the classroom. Ask students to describe the apple. Write their descriptions on the chalkboard. Then write the following head above the list of descriptions: Properties of an Apple. Explain that properties are characteristics which describe an object. Identify the four basic properties of matter: mass, volume, weight and density.

Answers to Challenges (p. 221)
Apply

7. Mass is a measure of the amount of matter an object contains. Weight is a measure of the pull of gravity on an object.

8. Answers will vary. Accept all logical responses.

Skill Builder: Calculating

0.8 g/cm³; 11.3 g/cm³; 0.9 g/cm³

Ideas in Action

Answers will vary. Accept all logical responses.

Teaching Tips for Activity
Observing that Air is Matter

Skills: *observing, relating*

COOPERATIVE/COLLABORATIVE LEARNING Answer the questions as a class.

Questions

1. No.

2. Air prevented water from entering the glass.

3. It shows that air takes up space.

LESSON 12-2
What are three phases of matter? (p. 222)

Teaching Strategies

PREVIEW Before beginning this lesson, write the title of each section in the lesson on the chalkboard. Have students copy these titles in their notebooks using an outline format. As students read each section, they should write the topic sentence for each paragraph in the section in their notebooks.

DISCUSSION Ask students to name different solids, liquids, and gases. Write correct responses on the chalkboard. Then describe the three phases of matter and the movement of particles in each phase. You may wish to tell students that the particles that make up matter are molecules.

CLASS ACTIVITY Bring a plastic bag filled with water to class. Have students gently squeeze the bag to show that a liquid does not have a definite shape.

DEMONSTRATION Demonstrate that water has a definite volume by pouring 100 mL of water from a graduated beaker into a graduated cylinder and then into a 100-mL volumetric flask.

DEMONSTRATION Show students the properties of a gas. Bring in a basketball, a football, a balloon, a bicycle tire, and so on. Have students note the different shapes of these objects. Ask students what fills each object (air). Point out that air is a gas and that a gas does not have a definite shape. Then show that a gas has no definite volume. Slowly fill a beaker with water. Elicit that the beaker may be partly filled or completely filled. Slowly blow up a balloon. Elicit that the balloon is always completely full. Point out that gases always completely fill the containers they are in.

Answers to Challenges (p. 223)
Check

1. solids, liquids, gases

2. solids

3. It takes the shape of the container.

4. its container

5. gas

Apply

6. The particles will spread out to completely fill the container.

7. solid: a, f, i, j; liquid: b, c, h; gas: d, e, g

InfoSearch

You may wish to discuss students' questions and answers as a class.

Teaching Tips for Science Connection
The Earth's Mantle

CLASS ACTIVITY Divide the class into small groups. Have each group make a model of the earth's layers using different colored modeling clay. Tell students to use the diagram on p. 223 as a guide.

INTRASCIENCE CONNECTION Most scientists think that the crust of the earth and the solid part of the mantle are broken into crustal plates. Crustal plates float on the lower part of the mantle. Continents move because they are carried along on the moving plates.

Questions

1. OBSERVING What are the three layers of the earth? (crust, mantle, core)

2. DEFINE What is a plastic material? (material that has properties of both solids and liquids)

DEMONSTRATION Show students a beaker of boiling water. Tell students that the steam represents the gas phase of water. Define evaporation as a change from a liquid to a gas.

DISCUSSION Ask students if they have ever noticed drops of water on their bathroom mirrors after taking a hot shower. Ask students where they think the drops of water come from. Discuss student responses. Then describe the process of condensation.

COOPERATIVE/COLLABORATIVE LEARNING Have students work in pairs to answer the Check and Apply questions.

Answers to Challenges (p. 225)
Apply

6. They slow down and can only vibrate in place.

7. They spread out.

8. a. condensation **b.** evaporation **c.** melting

Ideas in Action

Answers will vary. Accept all logical responses.

Teaching Tips for Career in Physical Science
Refrigeration Technician

REINFORCEMENT Be sure students understand that refrigeration technicians may repair air conditioners, as well as refrigerators and freezers.

Question

RELATE Why must a refrigeration technician understand the principles of heat transfer? (Cooling occurs when heat is drawn away from a substance.)

LESSON 12-3
How does matter change phase? (p. 224)

Teaching Strategies

PREVIEW Before beginning this section, have students scan the lesson looking for words with which they are unfamiliar. Have students work in pairs or small groups to define each of the words on their lists.

DISCUSSION Exhibit a piece of ice in a pan, and a glass of water. Draw attention to the liquid collecting in the pan. Be sure students understand that the ice is melting and changing to liquid water. Point out that ice and water are different phases of the same substance. Then ask why the ice melts in the classroom, but freezes in the freezer section of a refrigerator. Elicit the fact that it is warmer in the classroom than in the freezer. Tell students that to melt a solid, heat must be added to it. The air in the room is warmer than the ice. Heat flows from the air into the ice, and melts it. When water is put into the freezer, the water is warmer than the air in the freezer. Heat flows out of the water and liquid water changes to ice.

EXTENSION Not all solids change directly into liquids when heat is added. Some solids (for example, dry ice or frozen carbon dioxide, and iodine crystals) change into gases. This process is called sublimation.

LESSON 12-4
What are physical and chemical changes? (p. 226)

Teaching Strategies

PREVIEW Before beginning this lesson, have students review the TechTerms and their definitions and read the Lesson Summary.

DEMONSTRATION Introduce this lesson by performing some simple physical changes and chemical changes. To show a physical change, cut an apple in half or crumple up a piece of paper. To demonstrate a chemical change, burn a match, or a small piece of paper. Point out that a chemical change results in new substances.

REINFORCEMENT Be sure students understand the difference between a physical change and a chemical change. Describe a variety of both physical and chemical changes, such as rusting, burning, cutting, freezing, and so on. Have students classify each example as a physical or chemical change.

INTRASCIENCE CONNECTION The digestion of food involves both physical and chemical changes. Large pieces of food are broken down into smaller pieces (physical change). Proteins, carbohydrates and fats are broken down into smaller molecules (chemical changes).

Answers to Challenges (p. 227)

Apply

7. Chemical changes occur.

8. chemical change; A new substance is produced.

Skill Builder: Classifying

physical change: b, c, e, f; chemical change: a, d

Teaching Tips for Activity
Observing Physical Changes

Skill: *observing*

COOPERATIVE/COLLABORATIVE LEARNING Have students work in pairs.

Questions

1. a. steam **b.** The water was evaporating.

2. a. Water droplets forming. **b.** condensation

3. a. Answers will vary. Accept all logical responses. **b.** Answers will vary. Accept all logical responses.

LESSON 12-5
What are elements? (p. 228)

Teaching Strategies

PREVIEW Before beginning this lesson, have students scan the lesson for the science process skill symbols. Have students identify each skill used in the lesson. Students should then review the definition given for each skill on page 4 of their text.

DISCUSSION Refer students to the chart on p. 228 showing the elements that make up water, sugar, and salt. Define elements as simple substances that cannot be broken down into simpler substances.

DEMONSTRATION If possible, show students the electrolysis of water to illustrate that water is made up of the elements hydrogen and oxygen.

REINFORCEMENT Bring in samples of solid elements, such as iron, tin, lead, silver, and aluminium for students to observe. Show students mercury in a thermometer as an example of an element that is a liquid. Point out that other elements, such as oxygen and nitrogen, are gases.

RETEACHING OPTION Display a glass of water, a sugar cube, and some salt in front of the classroom. Point to each substance, and ask students to identify it. Then ask students if they know what substances make up water, sugar, and salt. List correct responses on the chalkboard.

INTERDISCIPLINARY CONNECTION To help students remember that elements are simple substances, point out how the term "elementary" often is used in common language. For example, you may wish to quote Sherlock Holmes' famous remark, "Elementary my dear Watson." Ask students what they think Holmes meant (the answer is simple, or obvious).

Answers to Challenge (p. 229)

Apply

7. Yes. Mercuric oxide is not an element. It could be broken down into its elements, mercury and oxygen.

8. a, c, d, f, g, j

Health and Safety Tip

You may wish to display students' posters in the classrooms.

Teaching Tips for People in Science
Marie Curie

INTERDISCIPLINARY CONNECTION You may wish to discuss some of the practical uses of radioactive elements. Radioactivity is used to sterilize food and drugs, to fight cancer, to find the causes of disease, and to study chemical and biological reactions.

EXTENSION Have students find out about the life and work of Irene Curie, the daughter of Marie and Pierre Curie. She continued her parents' work. Tell students to write their findings in a report.

Questions

1. LIST What radioactive elements did the Curies discover? (polonium and radium)

2. NAME What French scientist influenced the work of Marie and Pierre Curie? (Henri Biecquerel)

LESSON 12-6
What are chemical symbols?

(p. 230)

Teaching Strategies

PREVIEW Have students write the lesson title and objective in their notebooks. As students read the lesson, tell them to write down the sentence or sentences that provide the information needed in the objective.

DISCUSSION To introduce this lesson, ask students if they have ever seen a secretary taking dictation (on a television program, or so on). Ask students how a secretary keeps up with his or her boss since the boss talks without pausing (The secretary writes in shorthand.) Point out that shorthand writing helps a person write a lot of words quickly and in a small space. If possible, show students a stenographer's pad to illustrate your description. Then explain the chemical symbols used by scientists. Emphasize the importance of using a universal language that all chemists understand. Refer students to Table 1 on p. 230 to observe the symbols of some common elements.

REINFORCEMENT Chemical symbols are used to describe chemical reactions. Therefore, it is important for students to learn the symbols for the elements. Being able to use the symbols will help make their work in science easier. You may wish to develop flash cards listing the names and symbols of some common elements on opposite sides. Allow students to review chemical symbols using the cards. Hold up the cards with the elements' name facing the class. Ask students to write the symbols in their notebooks. You can vary this activity by showing the symbols and asking students to write the names of the elements.

COOPERATIVE/COLLABORATIVE LEARNING Have a volunteer describe the rules for writing chemical symbols.

Answers to Challenges (p. 231)
Apply
7. a. Ca **b.** I

8. iron

9. ferrum

InfoSearch

You may wish to discuss students' questions and answers as a class.

Teaching Tips for People in Science
Jons Jakob Berzeluis

EXTENSION Write the chemical formula for water, H_2O, on the chalkboard. Explain to students that the chemical formula for water shows that a molecule of water is made up of two atoms of hydrogen and one atom of oxygen.

INTERDISCIPLINARY CONNECTION The use of symbols in algebra influenced Berzeluis' development of writing chemical symbols.

Questions

1. LIST What three elements did Berzeluis discover? (selenium, cerium, and thorium)

2. What were two of Berzeluis' contributions to science? (Students should state two of the contributions given in the feature.)

LESSON 12-7
What is the periodic table?
(p. 232)

Teaching Strategies

PREVIEW Have students read the lesson feature before beginning the lesson. Discuss the concepts presented in the feature. Ask students how they think this information relates to topics they have already studied or to the lesson they are about to study.

DISCUSSION Describe the development of the periodic table of the elements. Have students compare Mendeleev's periodic table shown on p. 232 and the modern periodic table of the elements shown on pp. 234 and 235. Emphasize that the modern periodic table is arranged by atomic number.

REINFORCEMENT To help students understand the meaning of periodic, and arranging things based on a certain pattern, show students a calendar. Point out that the days of the week repeat in a certain pattern. Compare the days of the week to elements with similar properties. Mendeleev arranged elements with similar properties in rows in a chart, just as the same days in a week are listed in rows in a calendar.

EXTENSION The atomic number of an element is equal to the number of protons in an atom of that element. The number of protons in an atom is equal to its number of electrons. The number of electrons in an atom's outermost energy level determines the chemical activity or properties of an atom. Therefore, arranging the periodic table according to atomic number was an improvement. Students will learn more about atomic numbers in Unit 14.

CLASS ACTIVITY Give students as much practice as possible using the periodic table of the elements. Ask students as many questions that require observing the periodic table to answer as possible.

Answers to Challenges (p. 233)
Check
1. by atomic number

2. periods

3. They have similar physical and chemical properties.

4. by atomic mass

5. Some elements did not seem to fit the pattern.

Apply
6. fluorine, chlorine, bromine, iodine, astatine

7. lithium, beryllium, boron, carbon, nitrogen, oxygen, fluorine, neon

8. Yes. They are in the same group.

9. 6

10. 23

Skill Builder: Analyzing

Answers will vary. Accept all logical responses.

Teaching Tips for Leisure Activity
Amateur Chemistry

EXTENSION If possible, bring in a beginner chemistry set to show the class. Perform some simple tests using the chemistry set.

REINFORCEMENT Emphasize the safety tip listed in the feature.

Question

What are two hobbies that use a basic knowledge of chemistry? (gardening and raising tropical fish)

LESSON 12-8
What are metals and nonmetals? (p. 236)

Teaching Strategies

PREVIEW Before beginning this lesson, read the list of Tech-Terms aloud so students can hear their pronunciations.

DISCUSSION Refer students to the periodic table on pp. 234 and 235. Have them locate the dark zig-zag line that separates metals from nonmetals. Then describe the properties of metals. Point out that nonmetals do not have these properties.

REINFORCEMENT Go around the classroom to review metals and nonmetals. Beginning with the first student, have each student read out the name of an element using the periodic table on pp. 234 and 235. Have the next student state whether the element is a metal or a nonmetal, and then read out the name of another element. End this activity with the first student.

DEMONSTRATION Show students some of the properties of metals. Display several samples of metallic elements, such as copper, iron, tin, aluminum, zinc, lead, nickel, and mercury. Point out that all metals, except mercury, are solids at room temperature. Rub any metals that appear dull with sandpaper to show students that all metals have luster. Strike some of the samples with a hammer. Have students observe that the samples flatten out, rather than shatter. Identify this property as malleability.

DEMONSTRATION Set up a simple open circuit containing a dry cell and a lamp. Use samples of metals to close the circuit. Elicit that all metals are good conductors of electricity.

DEMONSTRATION Exhibit several samples of nonmetallic elements, such as sulfur, iodine, oxygen, hydrogen, and nitrogen. Elicit from the students that these elements do not resemble the metallic elements. Demonstrate the properties of nonmetals in the same fashion metals were demonstrated. Elicit that nonmetals do not have a shiny luster, and that nonmetals may be solids, liquids, or gases. Test a piece of fused sulfur in the open circuit to show that it is a poor conductor of electricity. Strike a piece of fused sulfur with a hammer. Elicit that the sulfur is brittle and shatters when hit.

COOPERATIVE/COLLABORATIVE LEARNING Have students work in small groups to construct charts comparing the properties of metals and nonmetals.

Answers to Challenges (p. 237)

Apply

7. Metals are ductile and good conductors of electricity.

8. malleability

9. They are poor conductors of heat.

10. metals: a, d, f, g, h, i; nonmetals: b, c, e, j

Designing an Experiment

Have volunteers describe their experiments to the rest of the class.

Teaching Tips for Technology and Society

Plastic and Graphite

EXTENSION The conservation of metals is important because metals are nonrenewable resources. Replacement is one way to conserve metals. Recycling is another way to conserve metals. Emphasize the importance of returning metal cans if there is a deposit in your state.

EXTENSION Hold up a pencil in front of the class. Point out that the "lead" in pencils actually is graphite. Graphite is made up only of carbon atoms.

Questions

1. **STATE** What are three reasons scientists are developing materials to replace metals? (Metals are heavy, not very flexible, and nonrenewable.)

2. **HYPOTHESIZE** What things to you think will someday be made of materials other than metals? (Answers will vary. Accept all logical responses.)

LESSON 12-9
What are the inert gases?
(p. 238)

Teaching Strategies

PREVIEW Before beginning this lesson, write the title of each section in the lesson on the chalkboard. Have students copy these titles in their notebooks using an outline format. As students read each section, they should write the topic sentence for each paragraph in the section in their notebooks.

CLASS ACTIVITY To introduce this lesson, refer students to the periodic table on pp. 234 and 235. Have students identify the elements that make up group 18 (helium, neon, argon, krypton, xenon, and radon). Tell students that these elements are the inert gases. Have students look up the meaning of the word "inert" in a dictionary (inert: inactive). Point out that the inert gases are not chemically active.

EXTENSION The inert gases are not chemically active because they all have complete outer energy levels. Therefore, they have no tendency to lose, gain, or share electrons.

Answers to Challenges (p. 239)

Check

1. group 18

2. inactive

3. They do not combine readily with other elements.

4. argon

5. helium

Apply

6. Helium is lighter than air.

7. Answers will vary. Accept all logical responses.

8. Radioactive elements can destroy cancer cells.

InfoSearch

You may wish to discuss students' questions and answers as a class.

Teaching Tips for Career in Physical Science

Environmental Chemist

EXTENSION Tell students to collect rainwater on the next rainy day. Using an indicator, such as litmus paper, have students test the pH of their samples. Point out that acid rain has a pH below 5.6.

T53

Questions

1. DESCRIBE What do environmental chemists do? (study problems of the environment)

2. RELATE Why do you think some knowledge of biology is important for an environmental chemist? (Answers will vary. Accept all logical responses.)

Answers to Unit Challenges (pp. 240–242)

Understanding the Features: Reading Critically

1. Certain pollutants combine with water and form drops of acid.

2. polonium and radium

3. Students should list three contributions given in the feature.

4. gardening; raising tropical fish

5. Both are based on the principles of cooling and heat transfers.

6. Metals tend to be heavy, are not flexible, and are not renewable.

7. plastic

Critical Thinking

1. Physical properties can be observed or measured without changing the makeup of a substance. Chemical properties only can be seen during chemical reactions.

2. The particles absorb heat and spread out.

3. Answers will vary. Accept all logical responses.

4. Particles must absorb enough heat energy to escape into the gas phase.

5. Density is equal to mass per unit volume.

Interpreting a Chart

1. silver

2. oxygen, selenium, tellurium

3. boron, aluminum, gallium, indium, thallium

4. Group 13, Period 3

5. 50

6. germanium, tin, lead

7. Yes. It is a metal.

Previewing the Unit

Read the titles of the lessons in this unit aloud. Identify terms in the titles with which students are unfamiliar. Write these terms on the chalkboard. Using the Glossary of the text or a dictionary, define the terms as a class. Have students write the terms and their definitions in their notebooks.

Bulletin Board Suggestions

1. Assemble a bulletin board showing the densities of some common substances. Copy the information in Table 1 on p. 244 onto the bulletin board. Illustrate the information with pictures showing each substance, and some of its uses.

2. Assemble a bulletin board entitled "Ships and Other Sailing Vessels." Attach interesting pictures to the bulletin board showing big ships, sailboats, and so on, floating in water. Use the bulletin board to motivate Lesson 13-5. Point to the bulletin board and pose the question, "How do ships float?"

LESSON 13-1
What is density? (p. 244)

Teaching Strategies

PREVIEW Before beginning this lesson, have students scan the lesson for the science process skill symbols. Have students identify each skill used in the lesson. Students should then review the definition given for each skill on page 4 of their text.

DISCUSSION Begin this lesson by asking students the famous riddle, "Which is heavier, a pound of feathers or a pound of lead?" Elicit that both weigh the same amount (one pound). Then ask students which would take up more space, a pound of feathers or a pound of lead (a pound of feathers). Point out that density accounts for the difference between feathers and lead. Define density as mass per unit volume.

REINFORCEMENT Remind students that one cubic centimeter is equal to one milliliter. Therefore, the density of any substance can be given in g/cm^3.

CLASS ACTIVITY Randomly read out the densities of the substances listed in Table 1 on p. 244. As you read the densities, have students identify the substances.

DEMONSTRATION If possible, bring in samples of some of the substances listed in Table 1 on p. 244. The samples must be in the form of rectangular solids. Measure the volumes and weights of the solids in order to determine their densities.

Answers to Challenges (p. 245)
Apply

6. 13.6 g/cm^3

7. **a.** 7.9 g/cm^3 **b.** 7.8 g/cm^3 **c.** 13.6 g/cm^3

8. 800 cm^3

9. air, water, aluminum, steel, iron, silver, gold

Designing an Experiment

Have volunteers describe their experiments to the rest of the class.

Teaching Tips for Science Connection
Neutron Stars

CLASS ACTIVITY Have students work in small groups to make models showing the formation of a neutron star.

EXTENSION The process by which hydrogen is changed into helium in a star is called fusion.

Questions

1. **SEQUENCE** What happens after a massive star blows off its outer layers? (The star becomes a supernova.)

2. **DEFINE** What is a red giant? (large, cool star)

LESSON 13-2
How is density measured?
(p. 246)

Teaching Strategies

PREVIEW Have students write the lesson title and objective in their notebooks. As students read the lesson, tell them to write down the sentence or sentences that provide the information needed in the objective.

REINFORCEMENT Give students several practice problems on calculating density. Tell students to make sure they remember to carry units when doing the problems.

DEMONSTRATION Demonstrate the procedure for finding the density of a liquid. Remind students that they should read the meniscus when finding the volume of a liquid.

DEMONSTRATION Show students how to find the volume of a solid with a regular shape.

RETEACHING OPTION Display three beakers in front of the classroom, one containing water, one containing salt water, and one containing alcohol. Ask students how they could safely determine which beaker contains pure water. Discuss student responses. Lead students to conclude that by determining the density of each liquid, the pure water can be identified. Find the density of each liquid. Have students observe which liquid has a density of 1 g/cm^3, and is therefore pure water.

Answers to Challenges (p. 247)
Check

1. mass and volume

2. g/mL, or g/cm^3

3. graduated cylinder, balance

4. mass of graduated cylinder alone; mass of graduated cylinder and liquid; volume of liquid

5. measure its mass and its volume

Apply

6. 2 g/mL, or 2 g/cm^3

7. Answers will vary. Accept all logical responses.

8. 400 cm³

9. 0.012 g/cm³

Skill Builder: Analyzing

Determine the materials' densities.

Teaching Tips for Activity
Finding the Densities of Different liquids

Skills: *modeling, observing*

COOPERATIVE/COLLABORATIVE LEARNING Have students work in pairs.

Questions

1. The liquids have different densities.

2. a. vegetable oil **b.** water

3. water, glycerine, corn syrup, vegetable oil

LESSON 13-3
What is specific gravity? (p. 248)

Teaching Strategies

PREVIEW Before beginning this lesson, read the list of Tech-Terms aloud so students can hear their pronunciations.

DISCUSSION Define specific gravity and explain how to find the specific gravity of a substance. Some students may be confused by the difference between the density of a substance and its specific gravity. Be sure students understand that density and specific gravity are equal, but that specific gravity has no units. Work through the problem on p. 248 to show students that the units cancel each other out.

DEMONSTRATION If possible, bring a hydrometer to class and demonstrate how it works.

COOPERATIVE/COLLABORATIVE LEARNING Have students quiz each other on the material in this lesson.

Answers to Challenges (p. 249)
Apply

6. gasoline; Corn syrup has a higher specific gravity than gasoline. The higher the specific gravity of a liquid, the higher a hydrometer floats.

7. rubber

8. 10.5

9. Units cancel each other out.

Skill Builder: Researching

Check students' reports. In each situation, specific gravity could be used to identify a pure substance.

Teaching Tips for Career in Physical Science
Mineralogist

INTRASCIENCE CONNECTION Bring in samples of various minerals for students to examine. If any students have mineral collections, ask them to show their collections to the rest of the class.

Questions

1. RELATE Why would specific gravity be useful in identifying minerals? (Each mineral has its own specific gravity.)

2. IDENTIFY What is one use of sulfur? (Sulfur is used to make medicines.)

LESSON 13-4
What is displacement? (p. 250)
Teaching Strategies

PREVIEW Before beginning this section, have students search the lesson looking for words with which they are unfamiliar. Have students work in pairs or small groups to define each of the words on their lists.

DISCUSSION Drop an ice cube or some other object into a glass of water. Ask students to observe what happens to the water level. (The water level rises.) Tell students that they have made the same observation Archimedes made 2000 years ago. Ask students why they think the water rises. Discuss all student responses, and then define and discuss displacement.

DEMONSTRATION Show students how to find the volume of an irregularly shaped object by displacement.

RETEACHING OPTION Bring in a can of shortening, a spoon, and a 3-cup liquid measure. Ask students how they could measure out 1 cup of shortening without getting the measuring cup dirty. Allow students to try their ideas. (You can measure one cup of shortening without getting the measuring cup dirty by displacement.)

EXTENSION Archimedes was a Greek mathematician and inventor who is credited with many discoveries and inventions. For example, Archimedes discovered the laws of the lever and pulley, as well as how to find volume by displacement. Have interested students find out more about the life and work of Archimedes. Tell students to summarize their findings in a written report.

Answers to Challenges (p. 251)
Apply

6. No, it depends only on its volume.

7. The dimensions of an irregularly shaped object cannot easily be measured.

8. 20 mL

9. 20 mL

Ideas in Action

Answer will vary. Accept all logical responses.

Teaching Tips for Activity
Measuring Displacement

Skills: *measuring, analyzing*

COOPERATIVE/COLLABORATIVE LEARNING Have students work in pairs.

Questions

1. a. Answers will vary. Accept all logical responses. **b.** Answers will vary. Accept all logical responses.

2. No. Answers will vary. Accept all logical responses.

What is buoyancy? (p. 252)

Teaching Strategies

PREVIEW Have students read the lesson feature before beginning the lesson. Discuss the concepts presented in the feature. Ask students how they think this information relates to topics they have already studied or to the lesson they are about to study.

DISCUSSION Motivate this lesson by asking students to recall experiences they may have had while swimming. Ask students how the weight of their bodies feels in water. Point out that in water, a person's body is lighter. Tell students that just one finger will support an adult immersed in water. Then describe Archimedes' principle and its relationship to buoyancy. Be sure students understand that the buoyant force on an object is equal to the object's weight in water. The object's weight in water is equal to the weight of the displaced water.

INTERDISCIPLINARY CONNECTION A buoy is a floating object anchored in a lake or river to warn ships of rocks, or to mark a channel. Most buoys are equipped with a bell or light.

DEMONSTRATION Using a rectangular fish tank and a model ship, demonstrate the relationship between Archimedes principle, buoyancy, and floating. On the side of the tank, mark the water level before and after placing the model in the tank. Calculate the volume of water displaced by the ship. The weight of the displaced water should equal the weight of the model ship. Point out that the buoyant force on the ship is equal to the weight of the ship in water, and that is why the ship floats.

Answers to Challenges (p. 253)

Apply

6. a. 150 mL **b.** 150 N

7. The buoyant force on an object is equal to the weight of the water that the object displaces.

Health and Safety Tip

If possible, arrange for the whole class to visit a local swimming pool. Ask a swimming instructor to demonstrate how people are taught to float.

Teaching Tips for Leisure Activity
Ballooning

EXTENSION In 1978, three American balloonists, Isen Abrucco, Maxie Anderson, and Larry Newman, crossed the Atlantic Ocean. They took off from Presque Isle, Maine and landed just west of Paris, France. In all, the three balloonists traveled 5173 kilometers. The trip took them 137 hours, 6 minutes.

Questions

1. NAME What are two events in balloon competitions? (long-distance races and spot-landing matches)

2. HYPOTHESIZE Why do you think most balloon competitions take place in the morning or late afternoon. (Answers will vary. Accept all logical responses.)

Answers to Unit Challenges (pp. 254–256)
Understanding the Features: Reading Critically

1. All the matter of a supernova shrinks into a dense ball.

2. long-distance races, spot-landing matches

3. to identify minerals and determine their purity

Critical Thinking

1. Iron has a greater density than wood.

2. Marble and plastic each have a different specific gravity.

3. One milliliter is equal to one cubic centimeter.

4. They are the same, expect specific gravity has no units.

5. The buoyant force on an object is equal to the object's weight in water. The object's weight in water is equal to the weight of the displaced water.

Interpreting a Table

1. 19.3

2. silver

3. gasoline, ice

4. A layer of gasoline will float on top of the water.

5. a kilogram of rubber

6. in glycerine, because it has a higher specific gravity than gasoline

Previewing the Unit

Have students carry out the task in the Study Hint. Compare students' lists and definitions as a class.

Bulletin Board Suggestion

Assemble a working bulletin board to use while teaching this unit. Cut out large circles to represent atoms. Cut out smaller circles to represent electrons, protons, and neutrons. Use different colored circles for each type of atomic particle. Write a plus sign on all the circles representing protons and a minus sign on all the circles representing electrons. Put the different circles into separate envelopes. Then, attach the envelopes to the bottom of the bulletin board. Use the circles with push pins to illustrate descriptions of atomic models, atomic number, mass number, isotopes, and so on.

LESSON 14-1
What are atoms? (p. 258)

Teaching Strategies

PREVIEW Have students write the lesson title and objectives in their notebooks. As students read the lesson, tell them to write down the sentence or sentences that provide the information needed in the objectives.

DISCUSSION Begin this lesson by asking students what they think matter is made of. Discuss student responses. Remind students that all matter is made up of elements. Point out that elements are made up of atoms. Then describe Democritus' theory about matter. Democritus concluded that if matter were divided in half again and again, a smallest piece would be reached. Democritus stated, "Matter is composed of tiny, hard round individual bits called atoms." He was the first person to use the term "atoms." Explain to students that Democritus' ideas were not based on experimental observations.

CLASS ACTIVITY Divide the class into small groups. Have each group make a model showing a molecule of water and its component atoms. Tell students to use the diagram of water on p. 258 as a guide. Modeling clay, styrofoam balls, or paper circles could be used to make models.

EXTENSION In addition to his atomic theory of matter, John Dalton also formulated the law of partial pressures for gases. This law states that with an ideal gas, the total pressure of a gas mixture on a container equals the sum of the pressures each gas would exert alone.

COOPERATIVE/COLLABORATIVE LEARNING Have a volunteer describe the four main parts of Dalton's atomic theory of matter to the rest of the class.

Answers to Challenges (p. 259)
Apply

6. Both men believed that matter was made up of atoms, and that atoms could be joined together.

7. Answers will vary. Accept all logical responses.

8. Answers will vary. Accept all logical responses.

InfoSearch

You may wish to discuss students' questions and answers as a class.

Teaching Tips for Technology and Society
Particle Accelerators

REINFORCEMENT Remind students that kinetic energy is energy of motion. Be sure students understand that particle accelerators must be so large in order for atoms to attain a huge amount of kinetic energy.

EXTENSION There are two types of particle accelerators. In a linear accelerator, particles are accelerated in a straight line. In a doughnut-shaped accelerator called a synchrotron, particles travel around and around in a circle.

Question

HYPOTHESIZE What do you think scientists can learn about matter using particle accelerators? (Answers will vary. Accept all logical responses.)

LESSON 14-2
What are the parts of an atom? (p. 260)

Teaching Strategies

PREVIEW Before beginning this lesson, read the list of TechTerms aloud so students can hear their pronunciations.

REINFORCEMENT Draw a time line on the chalkboard summarizing the development of the modern atomic model. Include dates and brief descriptions of Thompson's model, Rutherford's model, and Bohr's model. Have students copy the time line in their notebooks.

CLASS ACTIVITY Divide the class into four groups. Assign one group to make a model showing Thompson's atomic model, one group Rutherford's atomic model, one group Bohr's atomic model, and one group the present model of the atom. After students complete their models, have a representative from each group demonstrate and describe the group's model for the rest of the class.

RETEACHING OPTION Write the following head on the chalkboard: *Atom.* From this head, draw arrows to each of the following heads: *Protons, Neutrons, Electrons.* Identify these three particles as the basic parts of an atom. Describe the charge of each particle. Have students copy the chart in their notebooks.

COOPERATIVE/COLLABORATIVE LEARNING Answer the Check and Apply questions as a class.

Answers to Challenges (p. 261)
Check

1. negatively charged particles

2. in the nucleus

3. neutrons

4. Thomson

5. Rutherford

6. Bohr

Apply

7. Rutherford believed protons were found in a small central core. Thomson's model had positively charged material throughout.

8. a cloud

Skill Builder: Analyzing

a. Most of the particles went straight through the gold foil. **b.** Some particles were bounced straight back; a few particles were deflected. **c.** Positively charged particles were repelled.

Teaching Tips for Science Connection
Quarks and Leptons

INTERDISCIPLINARY CONNECTION Have students read James Joyce's novel *Finnegans Wake* to see how he used the word "quark" in his book.

Questions

1. HYPOTHESIZE Do you think quarks and leptons could be broken into smaller particles? (Answers will vary. Accept all logical responses.)

2. LIST What are names of the six types of quarks? (strange, charm, up, down, top, and bottom)

LESSON 14-3
What is atomic number? (p. 262)

Teaching Strategies

PREVIEW Before beginning this lesson, write the title of each section in the lesson on the chalkboard. Have students copy these titles in their notebooks using an outline format. As students read each section, they should write the topic sentence for each paragraph in the section in their notebooks.

DISCUSSION To introduce this lesson, remind students that the modern periodic table of the elements is arranged by atomic number. Define atomic number as the number of protons in the nucleus of an atom. Explain that no two elements have the same atomic number. Be sure students understand that the number of electrons in an atom is always equal to the number of protons.

REINFORCEMENT Refer students to Table 1 on p. 262. Randomly read out the atomic numbers of the elements listed. Ask students to state how many protons and electrons each element has.

EXTENSION Tell each student to construct a table similar to Table 1 on p. 262 with ten elements not listed in Table 1. Tell students to check the periodic table of the elements on pp. 234 and 235 for atomic numbers.

RETEACHING OPTION On the chalkboard, redraw the diagram on p. 262 showing the nuclei of helium, beryllium, and neon. Use different colored chalk to show neutrons and protons. Have students observe the diagrams and state the atomic number for each element. Then ask students how many electrons each atom of the element has. (same number of electrons as protons)

Answers to Challenges (p. 263)
Apply

7. chlorine

8. calcium

9. oxygen

10. iron

11. Iron has 26 protons and 26 electrons.

Skill Builder: Interpreting a Chart

1. 29 **2.** 24 **3.** 14 **4.** 3, 11, 19 **5.** Group 9; 109

Teaching Tips for People in Science
Chien-Shiung Wu

EXTENSION Another form of radioactive decay is alpha decay. In alpha decay, the nucleus of an atom gives of a particle made up of two protons and two neutrons.

EXTENSION Have interested students find out about the theory proposed by Tsung Dao Lee and Chen Ning Yang and how Chien-Shiung Wu's work helped confirm the theory. Tell students to present their findings to the class in an oral report.

Questions

1. What happens when beta decay takes place? (The nucleus of an atom gives off electrons.)

2. INFER If an atom of carbon underwent beta decay, would it still be a carbon atom? Explain (No. In beta decay, an atom changes into another element.)

LESSON 14-4
What is atomic mass? (p. 264)

Teaching Strategies

PREVIEW Have students read the lesson feature before beginning the lesson. Discuss the concepts presented in the feature. Ask students how they think this information relates to topics they have already studied or to the lesson they are about to study.

DISCUSSION Review the definition of atomic number. Then describe the atomic mass unit, atomic mass, and mass number. To help students understand the minuscule mass of an electron, ask students how many electrons it would take to equal the mass of one proton, or one neutron. (1836)

DEMONSTRATION Using different color marbles to represent protons and neutrons, demonstrate how to find the number of neutrons in an atom, given its mass number and atomic number.

CLASS ACTIVITY Give students several practice problems on finding the number of protons, neutrons, and electrons in an atom given atomic number, mass number, number of protons, and so on.

Answers to Challenges (p. 265)
Apply

6. protons: 30; neutrons: 35; electrons: 30

7. The elements have different numbers of neutrons.

Skill Builder: Interpreting a Table

a. hydrogen b. calcium c. iron d. helium, carbon, nitrogen, oxygen, sulfur, and calcium; Mass number equals atomic number multiplied by two. 3. Lead has the highest mass number.

Teaching Tips for People in Science
Dmitri Mendeleev

REINFORCEMENT   You may wish to refer students back to p. 232 in Unit 12 to observe Mendeleev's periodic table of the elements.

Question

CONTRAST   How does Mendeleev's periodic table of the elements differ from the modern periodic table of the elements? (Mendeleev arranged elements according to increasing atomic mass. Today, elements are arranged according to increasing atomic number.)

LESSON 14-5
What are isotopes? (p. 266)

Teaching Strategies

PREVIEW   Have students write the lesson title and objective in their notebooks. As students read the lesson, tell them to write down the sentence or sentences that provide the information needed in the objective.

DISCUSSION   To introduce this lesson, refer students to the periodic table of the elements on pp. 234 and 235. Ask students to identify the atomic masses of calcium (40.08), carbon (12.01), and other common elements. List the atomic masses on the chalkboard. Have students observe that the atomic masses are not whole numbers. Explain the reason for this. Describe isotopes of an element. Explain that the atomic mass of an element is an average of the atomic masses of all the isotopes of the element.

INTRASCIENCE CONNECTION   Isotopes of hydrogen are involved in the fusion reactions that release tremendous amounts of energy from the sun. In the sun, hydrogen-2 and hydrogen-3 nuclei fuse to form helium. A neutron and energy are released by the reaction. On the earth, scientists are still researching fusion as an alternative energy source. If fusion becomes a practical source of energy, the ocean could supply the fuel in the form of hydrogen-3. The ocean contains large amounts of this hydrogen isotope.

Answers to Challenges (p. 267)
Check

1. All atoms of the same element may not have the same number of neutrons in the nucleus.

2. atom of an element with the same number of protons but a different number of neutrons.

3. one or more

4. hydrogen, carbon, nitrogen

Apply

5. 0; 1; 2

6. It is an average of the atomic masses of all the isotopes of the element.

---

7. 6; 8

8. Atomic mass is a fraction less than 16.

InfoSearch

You may wish to discuss students' questions and answers as a class.

Teaching Tips for Career in Physical Science
Radiation Therapist

DISCUSSION   Students often are confused by radiation therapy as a form of cancer treatment since radiation also can cause cancer. Point out that large amounts of radioactivity are harmful to living things and can cause cells to mutate and become cancer cells. However, radiation also can be used to kill cancer cells. A proper balance is used in cancer treatment.

EXTENSION   If possible, arrange for a radiation therapist to visit your class. Tell students to prepare questions beforehand that they would like to ask the radiation therapist.

Questions

1. DESCRIBE   What do radiation therapists do? (prepare patients for treatment, position patient's properly under radiation equipment, and operate the equipment)

2. INFER   Why must radioactive substances be handled with great care? (Radiation can harm living things.)

LESSON 14-6
How are electrons arranged in an atom? (p. 268)

Teaching Strategies

PREVIEW   Before beginning this lesson, have students scan the lesson for the science process skill symbols. Have students identify each skill used in the lesson. Students should then review the definition given for each skill on page 4 of their text.

DISCUSSION   Redraw the diagrams on p. 268 showing electrons in energy levels onto the chalkboard. Describe the arrangement of electrons in an atom. Emphasize that an energy level is the place where an electron is most likely to be at any one time.

REINFORCEMENT   Have students draw atoms of helium, nitrogen, and carbon to show electrons in energy levels.

Answers to Challenges (p. 269)
Apply

7. 10; The first energy level holds a minimum of two electrons and the second holds a maximum of eight electrons.

8. The greater the distance of an electron from the nucleus, the more energy the electron has.

InfoSearch

You may wish to discuss students' questions and answers as a class.

Teaching Tips for People in Science
Maria Goeppert Mayer

EXTENSION Have interested students find out more about the life and work of Maria Goeppert Mayer. Tell students to write a biography about her life.

Questions

1. Why do you think there was only one school to prepare women in science when Maria Goeppert Mayer was a young woman? (Answers will vary. Accept all logical responses.)

2. Who was the first woman to win a Nobel Prize in physics? (Marie Curie)

Answers to Unit Challenges (pp. 270–272)
Understanding the Features: Reading Critically

1. It speeds them up.

2. A certain pattern of the elements must exist.

3. quarks and leptons

4. theoretical physics

5. Radiation can harm living things.

6. It was not viewed as important.

Critical Thinking

1. Answers will vary. Accept all logical responses.

2. Answers will vary. Accept all logical responses.

3. No, no two atoms have the same atomic number.

4. Atomic mass is an average of the atomic masses of all the isotopes of an element. Mass number is equal to the number of protons and neutrons an atom has.

5. 5

6. They would have the same properties, since they have the same number of electrons.

Interpreting a Diagram

1. B; It has a smaller atomic number.

2. 8; 6

3. 20; 10

4. A

5. Electrons in the second energy level.; A

COMPOUNDS AND MIXTURES (p. 273)

Previewing the Unit

Have students complete the task in the Study Hint.

Bulletin Board Suggestions

1. Assemble a bulletin board that illustrates the differences among compounds and the elements that make them up. Attach pictures showing elements, such as iron and sulfur, or sodium and chlorine, at the top of the bulletin board. Beneath these pictures attach pictures showing iron sulfide (the compound formed from iron and sulfur) and sodium chloride, or table salt (the compound formed from sodium and chlorine). Label all the pictures. Use the bulletin board to show students that compounds differ from the elements that make them up.

2. Assemble a bulletin board illustrating the organic compounds needed by living things. Divide the bulletin board into three sections. Label each section with one of the following heads: *Carbohydrates, Lipids, Proteins*. Under each section attach pictures showing foods that are high in these nutrients or that illustrate how the body uses them.

LESSON 15-1
What is a compound? (p. 274)

Teaching Strategies

PREVIEW Before beginning this section, have students scan the lesson looking for words with which they are unfamiliar. Have students work in pairs or small groups to define each of the words on their lists.

DEMONSTRATION To introduce this lesson, heat a small sample of sugar in a test tube. Tell students to observe what happens to the sugar when it is heated. Have students note that first the sugar melts into a liquid, and then only a black solid remains. Point out that the black solid is carbon. Carbon is left in the test tube when hydrogen and oxygen gas are released into the air. Elicit from students that carbon, hydrogen, and oxygen are the three elements that combine to form sugar. Point out that sugar is a compound.

DISCUSSION Ask students if they would add the black char left when sugar is burned to their cereals. Use this question to illustrate how the properties of a compound are very different from the properties of the elements that make it up. Compare the properties of table salt and sodium and chlorine as an alternative example.

REINFORCEMENT Show the class some iron filings and some sulfur. With a magnet, show students that iron is magnetic. Point out that sulfur is yellow, and can be ground to a fine powder. Then show students a piece of pyrite. Tell students that pyrite is a compound made up of the elements iron and sulfur. Try to attract the pyrite to a magnet. Students will note that pyrite is not magnetic. Point out that in pyrite, iron is combined with sulfur to form iron sulfide. The properties of the compound differ from the properties of the element that make it up.

Answers to Challenges (p. 275)
Apply

7. Hydrogen and oxygen would be released and only carbon would be left.

8. compound; At least three different elements formed when the solid was heated.

Ideas in Action

Answers will vary. Accept all logical responses.

Teaching Tips for Activity
Breaking Down Hydrogen Peroxide

Skills: *observing, analyzing*

SAFETY TIP Caution students to be careful when using a knife.

COOPERATIVE/COLLABORATIVE LEARNING Answer the questions as a class.

Questions

1. Bubbles rose from the liquid.

2. Hydrogen and oxygen gas were released.

LESSON 15-2
What are molecules? (p. 276)

Teaching Strategies

PREVIEW Before beginning this lesson, have students review the TechTerms and their definitions and read the Lesson Summary.

CLASS ACTIVITY Divide the class into small groups to make models of the molecules shown in the diagrams on p. 276. Tell students to cut out different colored circles to represent atoms and to join the atoms together to make molecules.

EXTENSION There are no particles that could be called molecules of an ionic compound, such as sodium chloride. The concept applies only to covalently bonded molecules.

REINFORCEMENT Be sure students understand the differences among atoms, elements, molecules, and compounds. Using water as an example, point out that water is a compound that is made by combining hydrogen and oxygen in the ratio of two atoms of hydrogen to one atom of oxygen. The smallest particle of water that has all the properties of water is one molecule. Then remind students that hydrogen and oxygen are elements. The smallest particle of hydrogen is one hydrogen atom. The smallest particle of oxygen is one oxygen atom.

REINFORCEMENT To help students remember that diatomic molecules contain two atoms of the same element joined together, have students look up the meaning of the prefix "di-" in a dictionary. (di-: two)

Answers to Challenges (p. 277)
Check

1. molecule

2. atoms

3. three

4. salt

5. carbon, hydrogen, oxygen

6. molecule that contains two atoms of the same element joined together

Apply

7. nine

8. carbon: 2; hydrogen: 6; oxygen: 1

InfoSearch

You may wish to discuss students' questions and answers as a class.

Teaching Tips for Science Connection

Mass Spectroscopy

EXTENSION Show students the mass spectrum of a simple molecule, such as ammonia. Mass spectra usually are published as bar graphs or in tabular form. You can find mass spectra in most college chemistry textbooks.

Questions

1. EXPLAIN How can scientists identify elements and molecules using mass spectroscopy? (by the mass spectra they form)

2. IDENTIFY What part of a mass spectrometer separates charged particles according to their mass? (magnetic field)

LESSON 15-3
What are mixtures? (p. 278)

Teaching Strategies

PREVIEW Before beginning this lesson, have students scan the lesson for the science process skill symbols. Have students identify each skill used in the lesson. Students should then review the definition given for each skill on page 4 of their text.

INTRASCIENCE CONNECTION Divide the class into small groups and distribute soil samples to each group. Have students examine the soil and describe what they see. Elicit from students that soil is a mixture of different materials. Define a mixture as two or more substances that are combined, but not chemically changed.

DEMONSTRATION Make a mixture of salt water. Heat the salt water to show students how a mixture can be separated by evaporation.

DEMONSTRATION Demonstrate filtration by filtering a mixture of sand and water.

RETEACHING OPTION Display a sample of pyrite in front of the class. Remind students that pyrite is a compound made up of iron and sulfur. Have students recall that pyrite is not magnetic. Then show students a mixture of iron filings and sulfur. Hold a magnet over the mixture. Students will observe that the magnet attracts the iron filings. Point out that in a mixture, the substances keep their own properties.

Answers to Challenges (p. 279)

Apply

7. with a magnet

8. mixture; The coffee and water are not chemically changed.

Designing an Experiment

Have volunteers describe their experiments to the rest of the class.

Teaching Tip for Leisure Activity

Cooking

EXTENSION Ask the home economics teacher in your school to give your class a cooking lesson.

EXTENSION Ask each student to write down a favorite recipe. Compile all the recipes to make a "Class Cookbook."

Question

LIST What are three ways people can learn to cook? (reading cookbooks; watching cooking programs; taking a cooking class)

LESSON 15-4
How are compounds and mixtures different? (p. 280)

Teaching Strategies

PREVIEW Have students write the lesson title and objective in their notebooks. As students read the lesson, tell them to write down the sentence or sentences that provide the information needed in the objective.

REINFORCEMENT Review the definitions of the terms "compound" and "mixture." Refer students to Table 1 on p. 280 to observe the differences between mixtures and compounds.

DEMONSTRATION Show the class a mixture of iron and sulfur, and then heat the mixture to produce the compound iron sulfide. Explain the differences between a mixture of iron and sulfur, and the compound iron sulfide.

DEMONSTRATION Show students mixtures of iron and sulfur in varying proportions. Point out that a mixture does not have a definite chemical composition. Emphasize that compounds always have a definite chemical composition.

CLASS ACTIVITY Randomly read out the descriptions in Table 1 on p. 280. Have students state whether you are describing a mixture or a compound.

COOPERATIVE/COLLABORATIVE LEARNING Have students work in pairs to answer the Check and Apply questions.

Answers to Challenges (p. 281)

Apply

7. compound; The substances lost their properties.

8. No, because the amount of salt in different salt water samples may vary.

Skill Builder: Building Vocabulary

oxide: oxygen; bromide: bromine; carbide: carbon

Teaching Tips for Science Connection
Chemistry of Cooking

DISCUSSION Write the ingredients needed to make pancakes on the chalkboard. Ask students what they would do with the ingredients to get the final product. (Mix the ingredients together and then cook the mixture.) Point out that when ingredients are mixed, a mixture is formed. When the mixture is heated, a compound is formed.

Questions

1. APPLY When you cook cake batter, do you form a mixture or a compound? (compound)

2. APPLY If you made a chicken salad by stirring together pieces of chicken, mayonnaise, celery, and vinegar, would you have formed a mixture or a compound? (mixture)

LESSON 15-5
What are ionic bonds? (p. 282)

Teaching Strategies

PREVIEW Before beginning this lesson, read the list of Tech-Terms aloud so students can hear their pronunciations.

REINFORCEMENT Review the structure of an atom before discussing ionic bonds. Ask students what atoms are made up of (protons, neutrons, electrons). Then ask what is the charge of each particle. (Electrons have a negative charge. Protons have a positive charge. Neutrons are neutral.) Remind students that in an atom, the number of protons is equal to the number of atoms, and, therefore, an atom is neutral.

EXTENSION The nature of a chemical bond can be explained by the tendency of an atom to lose or gain electrons. Atoms tend to complete their outermost energy level in order to become stable. Atoms complete their outermost energy level by gaining, losing, or sharing electrons.

DISCUSSION Describe how ions and ionic bonds form using sodium chloride as an example. Refer students to the diagram on p. 282 showing the formation of an ionic bond. Ask students which atom loses an electron and which atom gains an electron. (Sodium loses an electron, chlorine gains one.)

CLASS ACTIVITY Make salt crystals by suspending a string in a saturated salt water solution. Have students observe the string each day. After about a week, salt crystals will form on the string. Have students observe the crystals with a hand lens. Ask students to describe the shape of the crystals that have formed. (cubic)

Answers to Challenges (p. 283)
Apply

7. Answers will vary. Accept all logical responses.

8. It is the same.

Skill Builder: Modeling

Check students' diagrams.

Teaching Tips for Looking Back in Science
Crystallography

CLASS ACTIVITY Have students work in small groups to make models showing the six basic types of crystal lattices. Tell students to use the diagrams on p. 283 as a guide.

Questions

1. When were X-rays discovered? (1895)

2. INFER What are some properties used to classify minerals? (shape, color, crystal lattices)

LESSON 15-6
What are covalent bonds?
(p. 284)

Teaching Strategies

PREVIEW Before beginning this lesson, write the title of each section in the lesson on the chalkboard. Have students copy these titles in their notebooks using an outline format. As students read each section, they should write the topic sentence for each paragraph in the section in their notebooks.

DEMONSTRATION You may wish to obtain a ball-and-stick molecular model kit to show students covalent bonding.

REINFORCEMENT Draw a variety of electron dot models showing covalent bonds on the chalkboard. For example, you may wish to draw electron-dot models for HF, HCl, NH_3, CH_3, or CH_4. Be sure students understand that atoms share electrons in order to complete their outer energy levels. Show how covalent bonding completes the outer energy levels.

CLASS ACTIVITY To help students remember that in a covalent bond, atoms share electrons, have them look up the meaning of the prefix "co-" in a dictionary (co-: together with, joint). Ask them how the prefix relates to the meaning of the term "covalent." (Two atoms join together to share electrons.)

Answers to Challenges (p. 285)
Apply

7. covalent; The atoms are neutral.

8. four

Skill Builder: Predicting

Ionic compounds are good conductors of electricity. Covalent compounds are poor conductors of electricity.

Teaching Tips for Activity
Making a Molecular Model

Skills: *modeling, analyzing*

COOPERATIVE/COLLABORATIVE LEARNING Have students work in pairs.

Questions

1. hydrogen

2. carbon

3. covalent

4. No. Atoms in a covalent bond are neutral.

LESSON 15-7
What are organic compounds? (p. 286)

Teaching Strategies

PREVIEW Before beginning this section, have students scan the lesson looking for words with which they are unfamiliar. Have students work in pairs or small groups to define each of the words on their lists.

DISCUSSION Display a wide variety of organic compounds in front of the classroom, such as wool, wood, alcohol, and so on. Ask students what all the items have in common. Discuss student responses. Point out that all the items are organic compounds. Define an organic compound as a compound that contains carbon. Tell students that the study of organic compounds is called organic chemistry.

EXTENSION Carbon is unique in its ability to form very strong covalent bonds with other carbon atoms, while at the same time being able to form strong bonds with the atoms of other elements. This property, more than any other, accounts for the existence of the field called organic chemistry.

DISCUSSION Refer students to the diagrams on p. 286 showing the covalent bonding of carbon atoms. Describe how carbon is able to form single, double, and triple bonds. Be sure students understand that the double and triple bonds are between the two atoms of carbon. You may wish to tell students that all of the compounds shown are hydrocarbons. Hydrocarbons are compounds whose molecules contain only carbon and hydrogen.

EXTENSION Hydrocarbons whose molecules contain a single bond between carbon atoms are called alkanes. Hydrocarbons with a double bond between carbon atoms are called alkenes. Those with a triple bond are called alkynes. Students probably are familiar with the alkane octane, which is used in gasoline.

COOPERATIVE/COLLABORATIVE LEARNING Answer the Check and Apply questions as a class.

Answers to Challenges (p. 287)
Apply

7. No, because it could not share three pairs of electrons.

8. single bonds; Each atom shares one pair of electrons.

InfoSearch

You may wish to discuss students' questions and answers as a class.

Teaching Tips for Career in Physical Science
Pharmacist

EXTENSION If possible, arrange for a pharmacist to visit your class. Ask the pharmacist to describe the education necessary to become a pharmacist and the daily work of a pharmacist.

Questions

1. INFER Why would a pharmacist need to be able to talk to new people easily? (A pharmacist's job requires interacting with people.)

2. How do pharmacists and doctors work together? (Pharmacists prepare medications according to the directions of a doctor.)

LESSON 15-8
What compounds are needed by living things? (p. 288)

Teaching Strategies

PREVIEW Have students read the lesson feature before beginning the lesson. Discuss the concepts presented in the feature. Ask students how they think this information relates to topics they have already studied or to the lesson they are about to study.

RETEACHING OPTION Write the following head on the chalkboard: *Organic Compounds Needed by Living Things*. From this head, draw arrows to each of the following heads: *Carbohydrates, Lipids, Proteins*. From the head Carbohydrates, draw arrows to the following heads: *Sugars, Starches*. From the head Lipids, draw arrows to the following heads: *Fats, Oils*. Have students copy the charts in their notebooks.

EXTENSION Vitamins and minerals also are compounds needed by living things. Vitamins are organic nutrients that are needed for the body to function normally. Minerals are inorganic nutrients.

DISCUSSION Students probably are familiar with the term "cholesterol" as it relates to health. Have students discuss what they know about cholesterol. Describe the problems that may arise when too much cholesterol builds up in the body.

Answers to Challenges (p. 289)
Apply

7. Spaghetti is a good source of carbohydrates, which provide energy.

8. Peanut butter sandwich; a cheeseburger contains cholesterol, which may be harmful to the body.

Skill Builder: Organizing Information

Check students' tables for a logical organization of food types.

Teaching Tips for Science Connection
A Balanced Diet

CLASS ACTIVITY Have students find out the recommended daily servings from the four food groups. Tell students to organize their findings in a chart.

CLASS ACTIVITY Bring in pictures of foods from each of the four food groups. Show each picture to the class. Have students identify in which food group each food belongs and name some of the organic compounds the food contains.

Question

LIST What are the four food groups? (milk group, meat group, fruit and vegetable group, grain group)

Answers to Unit Challenges (pp. 290–292)

Understanding the Features: Reading Critically

1. mass spectrum

2. Answers will vary. Accept all logical responses.

3. reading cookbooks; watching cooking programs on television; attending a cooking school

4. milk group, meat group, fruit and vegetable group, grain group

5. community pharmacies or drug stores

6. Scientists could examine the structure of crystals with X-rays.

Critical Thinking

1. **a.** Answers will vary. Accept all logical responses. **b.** Answers will vary. Accept all logical responses. **c.** Answers will vary. Accept all logical responses. **d.** Answers will vary. Accept all logical responses.

2. Students should list the differences included in Table 1 on p. 284.

3. Sodium is a metal that burns easily. Chlorine is a poisonous gas. Sodium chloride is neither a metal nor a poisonous gas. It is a harmless, white crystal.

4. Answers will vary. Accept all logical responses.

5. It can share four electrons with many different atoms.

Interpreting a Diagram

1. Ionic: top; Covalent: bottom

2. 1; 7

3. sodium; chlorine

4. 1; 6

5. 2

Previewing the Unit

Read the titles of the lessons in this unit aloud. Identify terms in the titles with which students are unfamiliar. Write these terms on the chalkboard. Using the Glossary of the text or a dictionary, define the terms as a class. Have students write the terms and their definitions in their notebooks.

Bulletin Board Suggestion

Assemble a working bulletin board covering the concepts introduced in this unit. Using stencils, write and cut out the letters of the alphabet and numbers from 1 to 10. Place the letters and numbers in envelopes attached to the bottom of the bulletin board. Use the letters and numbers to show the chemical formulas of compounds discussed in the unit. Use the bulletin board to review subscripts, oxidation numbers, polyatomic ions, and so on.

LESSON 16-1
What is a chemical formula?

(p. 294)

Teaching Strategies

Preview Before beginning this lesson, have students scan the lesson for the science process skill symbols. Have students identify each skill used in the lesson. Students should then review the definition given for each skill on page 4 of their text.

Discussion Provide motivation for this lesson by explaining that just as chemical symbols are a shorthand way of writing the names of the elements, chemical formulas are a shorthand way of writing the names of compounds. Refer students to Table 1 on p. 294 to have them observe the chemical formulas of some common compounds. Write the name of each compound and its chemical formula on the chalkboard. Then have students name the elements that combine to form each compound.

Reinforcement Give students practice reading chemical formulas. Write the chemical formulas of a variety of compounds on the chalkboard. Ask students what elements make up each compound, and how many atoms of each element make up one molecule of the compound. Be sure students understand that the subscript 1 is never written in a chemical formula.

Reinforcement Have students look up the meaning of the prefix "sub-" in a dictionary, to help them understand the placement of a subscript in a chemical formula. (sub-: below, beneath)

Answers to Challenges (p. 295)
Apply

6. NH_3: 1 atom of nitrogen, 3 atoms of hydrogen
$BaCl_2$: 1 atom of barium; 2 atoms of chlorine

7. A molecule of water contains two atoms of hydrogen and one atom of oxygen. A molecule of hydrogen peroxide contains two atoms each of hydrogen and oxygen.

8. the elements that make it up and how many atoms of each element make up one molecule of the compound

Skill Builder: Interpreting Chemical Formulas

a. carbon, oxygen **b.** nitrogen, hydrogen **c.** calcium, carbon, oxygen **d.** potassium, oxygen, hydrogen **e.** sodium, hydrogen, carbon, oxygen

Teaching Tips for Science Connection
Destruction of the Ozone Layer

Extension CFC is the abbreviation for chloroflourocarbon.

Intrascience connection The atmosphere is made up of three main layers called the troposphere, the stratosphere, and the ionosphere. The troposphere is the layer of the atmosphere closest to the earth. The stratosphere is the middle layer of the atmosphere. The ionosphere is the top layer of the atmosphere. The ozone layer is found in the upper stratosphere.

Questions

1. apply What is the chemical formula of ozone? (O_3)

2. What happens when CFCs reach the ozone layer? (Ozone molecules are destroyed.)

LESSON 16-2
What is an oxidation number? (p. 296)

Teaching Strategies

Preview Before beginning this lesson, read the list of Tech-Terms aloud so students can hear their pronunciations.

Discussion Redraw the diagrams on p. 296 showing the valence electrons in sodium and chlorine atoms on the chalkboard. Define valence electrons as the electrons in an atom's outer energy level. Remind students that atoms tend to complete their outer energy levels by gaining, losing, or sharing electrons. Point out that chlorine and sodium complete their outer energy levels by forming an ionic bond. Refer students to the diagrams on the chalkboard and ask them to identify which atom loses an electron and which atom gains an electron. (Chlorine gains an electron; sodium loses an electron.)

Reinforcement Students often associate the term "gains" with a plus sign and the term "loses" with a minus sign. Be sure students understand that an atom that gains electrons has a negative oxidation number and an atom that loses electrons has a positive oxidation number.

Class activity Show students a periodic table of the elements that includes oxidation numbers. Write the names and oxidation numbers of different elements on the chalkboard. Have volunteers come up to the chalkboard and write whether each element gains or loses electrons when it bonds with other atoms.

Reinforcement Give students lots of practice in using valence electrons and oxidation numbers to write chemical formulas.

Answers to Challenges (p. 297)

Check

1. oxidation number

2. valence

3. negative

4. 7

5. 1−

6. They must add up to zero.

Apply

7. Aluminum; Elements with positive oxidation numbers are written first.

8. 3+; 1−; 3 × (1−) = 3; 3− + 3 = 0

Ideas in Action

Answers will vary. Accept all logical responses.

Teaching Tips for Activity

Writing Chemical Formulas

Skill: *analyzing*

COOPERATIVE/COLLABORATIVE LEARNING Answer the questions as a class.
HBr; HI, H_2O; H_2S
$NaCl$; $NaBr$; NaI, Na_2O; Na_2S
$AuCl$; $AuBr$; AuI; Au_2O; Au_2S
KCl; KBr; KI, K_2O, K_2S
$FeCl_3$; $FeBr_3$; FeI_3; Fe_2O_3; Fe_2S_3
$AlCl_3$; $AlBr_3$; AlI_3; Al_2O_3; Al_2O_3

Questions

1. metal

2. The oxidation numbers of the elements in a compound must add up to zero.

DEMONSTRATION To show students that some elements have more than one oxidation number, bring a periodic table of the elements listing oxidation numbers to class. Point out the different oxidation numbers of elements such as iron, tin, gold, and copper.

REINFORCEMENT Be sure students understand that iron (II) chloride and iron (III) chloride are different compounds. Emphasize the importance of including the roman numeral in the name of the compound.

COOPERATIVE/COLLABORATIVE LEARNING Have students work in pairs to answer the Check and Apply questions.

Answers to Challenges (p. 299)

Apply

6. No. It does not contain two elements.

7. No. Mercury (I) chloride contains only one atom of chlorine. Mercury (II) chloride has two atoms of chlorine.

InfoSearch

You may wish to discuss students' questions and answers as a class.

Teaching Tips for Career in Physical Science

Chemical Technician

DISCUSSION Use this feature to discuss the importance of technicians to scientific work. Point out that the job of technicians is invaluable to scientific research, industry, and so on.

Questions

1. **DESCRIBE** What does a chemical technician do? (helps develop, test, and manufacture chemical products)

2. Why do you think it is important for a chemical technician to keep detailed records? (Answers will vary. Accept all logical responses.)

LESSON 16-3
How are chemical compounds named? (p. 298)

Teaching Strategies

PREVIEW Have students write the lesson title and objective in their notebooks. As students read the lesson, tell them to write down the sentence or sentences that provide the information needed in the objective.

REINFORCEMENT To help students remember that a binary compound is made up of only two elements, point out that the term "binary" means "two."

CLASS ACTIVITY Write the names of a variety of binary compounds, such as sodium chloride, aluminum oxide, and calcium chloride, on the chalkboard. You may wish to use some of the compounds listed in the Activity on p. 297. Ask students to name the two elements that make up each compound and which element of each compound has a positive oxidation number based on its position in the compound's name. (The name of the element with a positive oxidation number is written first.)

LESSON 16-4
What is a polyatomic ion?
(p. 300)

Teaching Strategies

PREVIEW Before beginning this lesson, write the title of each section in the lesson on the chalkboard. Have students copy these titles in their notebooks using an outline format. As students read each section, they should write the topic sentence for each paragraph in the section in their notebooks.

CLASS ACTIVITY Have students look up the meaning of the prefix "poly-" in a dictionary. (poly-: many) Then relate the meaning of the prefix "poly-" to its usage in the term "polyatomic ion."

REINFORCEMENT Be sure students understand that when polyatomic ions take part in a chemical reaction, these groups of atoms behave as if they were a single unit. Tell students to treat polyatomic ions and their oxidation numbers in the same way as a single atom when writing chemical formulas.

Answers to Challenges (p. 301)
Apply

6. Bases: a, b, e; Nonbases: c, d

7. ionic; They have a positive or negative charge.

Skill Builder: Building Vocabulary

Check students' definitions. (polygon: geometric figure with many sides; polymer: large molecule made by combining many small molecules; polysyllabic: word containing many syllables)

Teaching Tips for Science Connection
Nitrates

DISCUSSION Bring in the food labels from packages of ham, bacon, hot dogs, and bologna. Read out the list of ingredients in each food. Have students note that each of these meats contain nitrates. Then discuss the reasons nitrates are added to some foods, and why people should limit their intakes of nitrates.

Questions

1. Why have some nutrition experts questioned the safety of adding nitrates to meat? (They may cause cancer.)

2. Why is sodium nitrate added to fresh meat? (to prevent the growth of harmful organisms in the meat)

LESSON 16-5
What are diatomic molecules? (p. 302)

Teaching Strategies

PREVIEW Have students read the lesson feature before beginning the lesson. Discuss the concepts presented in the feature. Ask students how they think this information relates to topics they have already studied or to the lesson they are about to study.

RETEACHING OPTION On an overhead transparency, draw diagrams showing how the atoms in the diatomic molecules listed in Table 1 on p. 302 share electrons. Be sure students understand that diatomic molecules are formed by covalent bonding.

REINFORCEMENT To emphasize the importance of the ozone layer in the earth's atmosphere, have students reread the feature on p. 295.

CLASS ACTIVITY Have students work in small groups to make models of the diatomic molecules listed in Table 1 on p. 302. Tell students to cut out circles to represent each atom.

Answers to Challenges (p. 303)
Apply

6. a, d

7. Fl_2; Cl_2; Br_2; I_2; At_2

Skill Builder: Building Vocabulary

Check students' definitions. (di-: two; diagonal: extending between the corners of any two nonadjacent angles; dicotyledon: flowering plant with two seed leaves; dibromide: containing two bromine atoms; dichloride: containing two chlorine atoms)

Teaching Tips for Science Connection
The Nitrogen Cycle

CLASS ACTIVITY Divide the class into small groups. Have students in each group draw pictures of plants, animals, soil, the sun, clouds, and so on. Have students cut out each picture and then assemble all the pictures in a way that illustrates the nitrogen cycle. Have students paste the pictures on a posterboard and then add arrows to show the flow of nitrogen through the environment.

Questions

1. INFER What is the most abundant gas in the earth's atmosphere? (nitrogen)

2. SEQUENCE How do animals obtain nitrates? (by eating plants)

LESSON 16-6
What is formula mass? (p. 304)

Teaching Strategies

PREVIEW Before beginning this section, have students scan the lesson looking for words with which they are unfamiliar. Have students work in pairs or small groups to define each of the words on their lists.

REINFORCEMENT Review the concept of mass number. Have students recall that the mass number of an atom is equal to the total number of protons and neutrons in its nucleus.

DISCUSSION Describe the formula mass of a molecule. Point out that a molecule contains one or more atoms. From this, develop the idea that the mass of one molecule is equivalent to the total mass of the atoms that make up the molecule.

CLASS ACTIVITY Write the chemical formulas for a variety of compounds on the chalkboard. Have students find the formula mass of one molecule of each compound by using the four steps outlined on p. 304.

RETEACHING OPTION Use styrofoam balls to represent atoms. The symbol and atomic mass of the atom represented can be marked on each ball. By joining the balls together with toothpicks, models of different molecules can be constructed. Then have students calculate the formula mass of each model.

Answers to Challenges (p. 305)
Apply

6. It contains only one atom of the element.

7. No.

8. 342

Skill Builder: Predicting

The formula mass of CO_2 is greater than the formula mass of CO. CO: 28; CO_2: 44

Teaching Tips for Activity
Calculating Formula Mass

Skill: *calculating*

Cooperative/collaborative learning Before students begin this activity, have a volunteer describe the four steps to follow when calculating formula mass.

Formula Masses: H_2O_2: 34; $CHCl_3$: 118; NO_2: 46; Na_2CO_3: 106; CaO: 56; $MgCl_2$: 94; $CaCO_3$: 100

Questions

1. Atomic mass is equal to the total mass of the protons and neutrons in an atom. Mass number is equal to the number of protons and neutrons in the nucleus of an atom. Mass number is used to calculate formula mass.

2. 22.990; 23

Answers to Unit Challenges (pp. 306–308)
Understanding the Features: Reading Critically

1. It can cause skin cancer.

2. path by which nitrogen is removed from the air, used by organisms, and returned to the air

3. to prevent the growth of harmful organisms in meat and to give meat flavor and color

4. help develop, test, and manufacture chemical products

Critical Thinking

1. They are universal and can be understood by all scientists.

2. **a.** 40 **b.** 36 **c.** 213 **d.** 102

3. The number of valence electrons an atom has determines its oxidation number.

4. Carbon has a tendency to lose or gain its four electrons.

Interpreting a Table

1. Metallic elements: Na, Al, Ag; Nonmetallic elements: Cl, S, O

2. metallic

3. 3+; 3+

4. 1−; 1−

Previewing the Unit

Have students work in pairs to carry out the task in the Study Hint.

Bulletin Board Suggestion

Assemble a bulletin board showing the four types of chemical reactions. Write representative chemical equations for a synthesis reaction, a decomposition reaction, a single-replacement reaction, and a double-replacement reaction on the bulletin board. Include an example of a redox reaction. Label the type of reaction, and the reactants and products for each equation. Be sure all the equations are balanced. Use the bulletin board to review chemical equations, balancing equations, and the different types of chemical reactions.

LESSON 17-1
What is conservation of matter? (p. 310)

Teaching Strategies

PREVIEW Have students write the lesson title and objective in their notebooks. As students read the lesson, tell them to write down the sentence or sentences that provide the information needed in the objective.

REINFORCEMENT Review the differences between a physical change and a chemical change. Tear up a piece of paper and burn a match. Ask students to identify the physical change and the chemical change. (physical change: tearing up the piece of paper; chemical change: burning the match) Then ask students which type of change results in new substances. (chemical change) Define a chemical reaction as the process by which new substances are formed.

DEMONSTRATION Demonstrate the conservation of mass with a flashbulb. Weigh the flashbulb before and after flashing it to show students that the weight of the flashbulb does not change. Point out that in a flashbulb, magnesium combines with oxygen to produce magnesium oxide. During the reaction, matter is neither created nor destroyed.

COOPERATIVE/COLLABORATIVE LEARNING Have students work in pairs to answer the Check and Apply questions.

Answers to Challenges (p. 311)
Apply

6. New substances are formed.

7. A new substance is present.

InfoSearch

You may wish to discuss students' questions and answers as a class.

Teaching Tips for Activity
Observing a Chemical Reaction

Skills: *hypothesizing, comparing*

SAFETY TIP Caution students to be careful when using matches and to discard used matches in a container of water.

COOPERATIVE/COLLABORATIVE LEARNING Have students work in pairs.

Questions

1. The ashes should have less mass than the paper.

2. When paper burns, water and carbon dioxide also are produced.

LESSON 17-2
What are chemical equations? (p. 312)

Teaching Strategies

PREVIEW Before beginning this lesson, have students scan the lesson for the science process skill symbols. Have students identify each skill used in the lesson. Students should then review the definition given for each skill on p. 4 of their text.

DISCUSSION On the chalkboard, write out in words the chemical reaction for photosynthesis: six molecules of carbon dioxide plus six molecules of water yields one molecule of glucose and six molecules of oxygen. Write the chemical equation for photosynthesis underneath its word description:

$$6CO_2 + 6H_2O \longrightarrow C_6H_{12}O_6 + 6O_2.$$

Then discuss the use of chemical equations. Point out that chemical equations are a shorthand, universal way of describing chemical reactions.

REINFORCEMENT Remind students to use oxidation numbers to write correct chemical formulas.

INTERDISCIPLINARY CONNECTION To help students understand chemical equations, you may wish to compare chemical equations with addition problems. Relate the reactants to the addends, the arrow to the equal sign, and the products to the sum.

CLASS ACTIVITY Give students a lot of practice in balancing chemical equations. Write out a variety of simple unbalanced equations on separate index cards. Place all the index cards into a bag. Go around the classroom, having each student pick one card. Tell students to balance the chemical equations on their cards. Then go over all the equations as a class.

REINFORCEMENT Some students may have difficulty balancing chemical equations. Tell students to keep playing around with coefficients until their equations are balanced.

Answers to Challenges (p. 313)
Apply

6. c and d

7. $2H_2O_2 \longrightarrow 2H_2 + O_2$
$2CO + O_2 \longrightarrow 2CO_2$
$N_2 + 3H_2 \longrightarrow 2NH_3$

8. 1

State the Problem

a. removed subscripts from the diatomic molecules chlorine and bromine **b.** removed subscript from oxygen **c.** did not multiply coefficients by subscripts

Teaching Tips for Activity
Balancing Chemical Equations

Skill: *analyzing*

COOPERATIVE/COLLABORATIVE LEARNING Have volunteers demonstrate how they balanced the equations for the rest of the class.

$C + 2Br_2 \longrightarrow CBr_4$
$2Fe + O_2 \longrightarrow 2FeO$
$2B + 3Cl_2 \longrightarrow 2BCl_3$
$2Mg + O_2 \longrightarrow 2MgO$
$BaCl_2 + H_2SO_4 \longrightarrow BaSO_4 + 2HCl$
$4P + 5O_2 \longrightarrow P_4O_{10}$
$2Zn + 2HCl \longrightarrow 2ZnCl + H_2$
$2PbO_2 \longrightarrow 2PbO + O_2$
$CH_4 + 2O_2 \longrightarrow CO_2 + 2H_2O$
$2Na + 2H_2O \longrightarrow 2NaOH + H_2$

Questions

1. An unbalanced equation does not have the correct number of atoms for each element involved in the chemical reaction.

2. conservation of matter

LESSON 17-3
What are oxidation and reduction? (p. 314)

Teaching Strategies

PREVIEW Before beginning this lesson, read the list of Tech-Terms aloud so students can hear their pronunciations.

DEMONSTRATION Demonstrate the oxidation of iron. Use a pair of tongs to hold a piece of steel wool over a flame. Have students observe the new substance formed by this chemical reaction (iron oxide). Then describe redox reactions using the formation of iron oxide as an example. Tell students that iron is oxidized, or loses electrons, and oxygen is reduced, or gains electrons.

DISCUSSION Students may be confused by the definition of the term "reduction" because the terms "reduced" and "gains" are usually thought of as opposites. Associate the term "reduction" with the negative charge of an electron to help students better understand its scientific meaning.

REINFORCEMENT Be sure students understand that oxidation and reduction can take place without oxygen taking part in the reaction. Emphasize this concept by writing the redox reaction between sodium and chlorine on the chalkboard: $2Na + Cl_2 \longrightarrow 2NaCl$.

EXTENSION Have interested students draw electron-dot models showing the redox reactions between iron and oxygen, sodium and chlorine, and magnesium and oxygen. Tell students to label the reactants and products on each of their models. Have students explain their models to the rest of the class.

Answers to Challenges (p. 315)
Apply

6. Electrons lost by one reactant must be gained by another.

7. No. Oxidation occurs whenever a reactant loses electrons.

InfoSearch

You may wish to discuss students' questions and answers as a class.

Teaching Tips for Science Connection
Respiration

INTRASCIENCE CONNECTION Write the chemical equations for respiration and photosynthesis on the chalkboard, one on top of the other:
respiration: $C_6H_{12}O_6 + 6O_2 \longrightarrow 6H_2O + 6CO_2 + energy$
photosynthesis: $6H_2O + 6CO_2 + energy \longrightarrow C_6H_{12}O_6 + 6O_2$
Ask students what they observe about the two equations. (They are opposite reactions.)

REINFORCEMENT Be sure students understand that plants carry out both respiration and photosynthesis. Students may tend to think of plants as only carrying out the process of photosynthesis.

Questions

1. LOCATE Where does respiration take place? (inside body cells)

2. INFER What are the waste products of respiration? (carbon dioxide and water)

LESSON 17-4
What is a synthesis reaction?
(p. 316)

Teaching Strategies

PREVIEW Before beginning this lesson, write the title of each section in the lesson on the chalkboard. Have students copy these titles in their notebooks using an outline format. As students read each section, they should write the topic sentence for each paragraph in the section in their notebooks.

DISCUSSION Introduce this lesson by reviewing the reaction between iron and oxygen. Point out that during the reaction, two elements combine to produce a new substance, the compound iron oxide. Tell students that many reactions involving the oxidation of metals are synthesis reactions. Develop the understanding that in a synthesis reaction, two or more simple substances combine, or synthesize, to form a more complex substance.

RETEACHING OPTION Using common materials such as childrens' building blocks or pegs, make models illustrating synthesis reactions.

Answers to Challenges (p. 317)
Apply

6. b, c, d, e, f

7. Two reactants combine to form a product.

Ideas in Action

Answers will vary.

Teaching Tips for Looking Back in Science

Development of Synthetic Fabrics

Discussion Illustrate the wide use of synthetic fabrics by asking students how many are wearing clothing made of nylon, polyester, rayon, or some other synthetic fabric. All students probably will be wearing some type of synthetic fabric.

Questions

1. **identify** What was the first synthetic fiber to be developed? (rayon)

2. **list** What are three natural fibers? (silk, cotton, wool)

Designing an Experiment

Have volunteers describe their experiments to the rest of the class.

Teaching Tips for Technology and Society

Disposal of Chemical Waste

Extension Have interested students research the efforts being made to clean up toxic waste sites. Tell students to present their findings in an oral report.

Questions

1. **explain** Why are chemical wastes dangerous? (They contain chemicals that can harm or kill living things.)

2. How can you help reduce the amount of hazardous waste that is produced? (Answers will vary. Accept all logical responses.)

LESSON 17-5
What is a decomposition reaction? (p. 318)

Teaching Strategies

Preview Have students read the lesson feature before beginning the lesson. Discuss the concepts presented in the feature. Ask students how they think this information relates to topics they have already studied or to the lesson they are about to study.

Demonstration Decompose water with a Hoffman apparatus. (A small amount of sulfuric acid should be added to the water in the apparatus to speed up the reaction.) After the apparatus has operated for some time, point out that one tube of the apparatus contains about twice as much gas as the other tube. Refer to the equation for the decomposition of water shown on p. 318. Tell students that the tube with more gas in it contains hydrogen and the other tube contains oxygen.

Reinforcement Compare decomposition reactions with synthesis reactions. Point out that the two reactions are opposite. In synthesis reactions, simple substances combine to produce more complex substances. In decomposition reactions, more complex substances break down to produce simpler substances.

Answers to Challenges (p. 319)

Check

1. synthesis reaction

2. elements, compounds, or both

3. two or more simpler substances

4. electric current

5. energy

Apply

6. H_2CO_3

7. An element cannot be broken down into simpler substances.

LESSON 17-6
What is a single-replacement reaction? (p. 320)

Teaching Strategies

Preview Before beginning this lesson, have students scan the lesson looking for words with which they are unfamiliar. Have students work in pairs or small groups to define each of the words on their lists.

Class activity Have three students "act" out the examples of single-replacement reactions described on p. 320. Have two students hold onto the ends of a string. These students represent the compound involved in each reaction. The other student represents the free element. Give each student a piece of white paper labeled with the symbol of the element the student represents. Be sure the chemical formulas are correct and the equations are balanced. Have the student representing the free element replace the correct element in the compound. Then write new labels for the new compound and element that are produced.

Reinforcement Guide students through the examples of single-replacement reactions on p. 320.

Answers to Challenges (p. 321)

Apply

6. A single-replacement reaction; iron replaces copper in the compound copper sulfate to form iron sulfate and copper.

7. Copper replaces silver in silver nitrate to form copper nitrate and silver.

InfoSearch

You may wish to discuss students' questions and answers as a class.

Teaching Tips for Activity

Replacing Metals

Skills: *observing, analyzing*

Cooperative/collaborative learning Have students work in pairs.

Questions

1. Copper formed on it.

2. Silver formed on it.

What is a double-replacement reaction? (p. 322)

Teaching Strategies

PREVIEW Before beginning this lesson, have students review the TechTerms and their definitions and read the Lesson Summary.

DISCUSSION Describe what happens in a double-replacement reaction and provide students with different examples. Remind students that polyatomic ions react with other substances as a single unit.

DEMONSTRATION Neutralization reactions provide excellent demonstrations of double-replacement reactions.

DEMONSTRATION A solution of $AgNO_3$ can be mixed with NaCl, NaBr, or NaI to show various double-replacement reactions. Each of the salts formed will precipitate out of solution. Each of the precipitates is also light-sensitive, and will darken when exposed to light. You may wish to tell students that such light-sensitive compounds are used extensively in photographic emulsions.

Answers to Challenges (p. 323)
Apply

6. KCl and H_2O

7. c and e

InfoSearch

You may wish to discuss students' questions and answers as a class.

Teaching Tips for Activity
Formation of a Precipitate

Skills: *observing, inferring*

COOPERATIVE/COLLABORATIVE LEARNING Have students work in pairs.

Questions

1. A white precipitate is formed.

2. The two compounds reacted to form a new compound.

3. $K_2CO_3 + CaCl_2 \longrightarrow CaCO_3 + 2KCl$

Answers to Unit Challenges (pp. 324–326)
Understanding the Features: Reading Critically

1. food, oxygen; carbon dioxide, water

2. They harm living things.

3. rayon

Critical Thinking

1. The same number of atoms that react are produced by a chemical reaction.

2. New substances are formed.

3. Oxidation means combining with oxygen, or losing electrons.

4. In a single-replacement reaction a free element takes the place of an element in a compound. In a double-replacement reaction, two compounds switch elements.

5. Silver chloride and sodium nitrate are formed. Two solutions react to form a precipitate.

6. **a.** synthesis **b.** double-replacement **c.** synthesis **d.** single-replacement **e.** decomposition **f.** double-replacement **g.** synthesis **h.** single-replacement **i.** decomposition **j.** double-replacement

Interpreting a Diagram

1. decomposition

2. electrolysis of water

3. water and energy

4. hydrogen and oxygen

5. $2H_2O \longrightarrow 2H_2 + O_2$

UNIT 18
METALS (p. 327)

Previewing the Unit

Read the titles of the lessons in this unit aloud. Identify terms in the titles with which students are unfamiliar. Write these terms on the chalkboard. Using the Glossary of the text or a dictionary, define the terms as a class. Have students write the terms and their definitions in their notebooks. Then, have students work in small groups to carry out the task in the Study Hint.

Bulletin Board Suggestions

1. Assemble a bulletin board illustrating metals. Make a collage on the bulletin board with pictures showing mineral ores and their metals, mining operations, alloys, rusted metals, plated metals, and so on. Use the bulletin board to emphasize the importance of metals in people's daily lives.

2. On a large sheet of white paper, redraw the diagram on p. 338 showing copper being electroplated. Use different colored markers to draw the carbon rods, the copper sulfate solution, the batteries, and the labels for ions. Attach the sheet of paper to the bulletin board. Use the bulletin board to review the process of electroplating.

LESSON 18-1
What is an ore? (p. 328)
Teaching Strategies

PREVIEW Have students read the lesson feature before beginning the lesson. Discuss the concepts presented in the feature. Ask students how they think this information relates to topics they have already studied or to the lesson they are about to study.

DISCUSSION Motivate this lesson by displaying several ores and the metals derived from each. Have students note the color and luster of each ore. Define an ore as a rock or mineral from which a useful metal can be removed. Then discuss the countless uses of metals in people's everyday lives.

REINFORCEMENT Review chemical formulas. For each ore listed in Table 1 on p. 328, ask students to identify the elements in the compound and how many atoms of each element are in one molecule of the compound.

EXTENSION Have interested students research methods that are used to take ores from the ground. Tell students to write their findings in a report. Have students include a discussion of how each mining method affects the land.

Answers to Challenges (p. 329)
Apply

5. oxygen

6. carbon and oxygen

7. Gold and silver are not oxides, sulfides, or carbonates. They are not combined with other elements.

8. a. FeS_2 b. Cu_2O c. $MgCO_3$ d. ZnS e. Fe_2O_3 f. PbO g. PbS h. Al_2O_3

Ideas in Action

Answers will vary.

Teaching Tips for Leisure Activity
Designing and Making Jewelry

CLASS ACTIVITY Students may have jewelry made of metals, beads, shells, or other materials. If so, have students identify the materials and, if possible, show them to the rest of the class.

INTERDISCIPLINARY CONNECTION Many craft and hobby shops offer courses in jewelry-making. Tell students who would like to learn more about designing and making jewelry to contact a local craft store in the area.

Questions

1. **SEQUENCE** What is the first step in making jewelry? (creating a design)

2. What was the first jewelry made from? (objects found in nature)

LESSON 18-2
How are metals removed from their ores? (p. 330)

Teaching Strategies

PREVIEW Before beginning this lesson, have students scan the lesson for the science process skill symbols. Have students identify each skill used in the lesson. Students should then review the definition given for each skill on p. 4 of their text.

REINFORCEMENT Students will benefit from a brief review of chemical reactions and balancing equations before learning about the processes used to remove metals from their ores. Remind students that reduction involves the gain of electrons. A metal gains electrons when oxygen is removed from an ore.

DEMONSTRATION Demonstrate the two steps needed to remove a metal from a sulfate or a carbonate ore. Be sure students understand that both roasting and reduction are necessary to remove the metal from a sulfide or carbonate ore, whereas only reduction is needed to remove metals from oxide ores.

COOPERATIVE/COLLABORATIVE LEARNING Have a volunteer describe two methods of removing metals from their ores for the rest of the class.

Answers to Challenges (p. 331)
Apply

7. to combine with the oxygen in the ore; 2

8. $2ZnS + 2O_2 \longrightarrow 2ZnO + 2SO_2$; $2ZnO + C \longrightarrow 2Zn + CO_2$

9. 2; 2

Skill Builder: Balancing Equations

$C + 2Cu_2O \longrightarrow 4Cu + CO_2$

$C + 2FeO \longrightarrow 2Fe + CO_2$

$C + 2ZnO \longrightarrow 2Zn + CO_2$

Teaching Tips for Science Connection
Metallurgy

REINFORCEMENT Read the term "metallurgy" aloud so students can hear its pronunciation.

INTRASCIENCE CONNECTION Alchemists were people who tried to turn less costly metals, such as copper, into gold. Of course, they were never successful. However, the work of alchemists was important to modern metallurgy because alchemists learned a great deal about metals and their uses.

Questions

1. **DEFINE** What is metallurgy? (the science of removing metals from their ores)

2. **HYPOTHESIZE** Why do you think mining may damage the environment? (Answers will vary. Accept all logical responses.)

LESSON 18-3
What are alloys? (p. 332)

Teaching Strategies

PREVIEW Have students write the lesson title and objective in their notebooks. As students read the lesson, tell them to write down the sentence or sentences that provide the information needed in the objective.

DEMONSTRATION Review the properties of metals. Set up a simple open circuit to demonstrate conductivity. Use samples of metals to close the circuit. Strike some of the samples with a hammer to demonstrate malleability. Hold up strips of copper wire to show ductility. Then point out that alloys are produced to take advantage of the best properties of each of the metals they contain.

CLASS ACTIVITY Randomly read out the names of the alloys listed in Table 1 on p. 332. Have students name the metals that make up each alloy and some of the alloy's uses.

RETEACHING OPTION Bring in a variety of objects made up of the alloys steel, brass, bronze, pewter, and sterling silver. Hold up each object in front of the classroom and identify each alloy. Ask students to name the metals that make up each alloy.

Answers to Challenges (p. 333)
Apply

11. A pure metal is a better conductor of electricity.

InfoSearch

You may wish to discuss students' questions and answers as a class.

Teaching Tips for Technology and Society
Alloys of Steel

DISCUSSION Show students a rusted iron nail and a stainless steel utensil. Elicit from students that stainless steel resists rusting. Point out that stainless steel is just one alloy of steel. Then discuss some of the other alloys of steel and their uses.

Questions

1. **COMPARE** What is the difference between low-alloy steel and high-alloy steel? (High-alloy steel contains larger amounts of nickel, chromium, and the other metals found in low-alloy steel.)

2. What is the main element used to make steel? (iron)

LESSON 18-4
Why are some metals more active than others? (p. 334)

Teaching Strategies

PREVIEW Before beginning this lesson, read the list of TechTerms aloud so students can hear their pronunciations.

CLASS ACTIVITY Refer students to the periodic table of the elements on pp. 234 and 235. Ask students to name the elements found in groups 1 and 2. List the elements in group 1 on the chalkboard under the head "Alkali Metals." List the elements in group 2 on the chalkboard under the head "Alkaline Earth Metals." Then describe the properties of group 1 and group 2 metals. Have students copy the information in their notebooks.

REINFORCEMENT Randomly read out the names of any two metals listed in Table 1 on p. 334. Have students state which metal would replace the other in a chemical reaction.

DEMONSTRATION Place an iron nail in a solution of copper sulfate. After several minutes, remove the nail and exhibit it to the class. Students will observe that iron is more active than copper, and therefore iron replaced the copper. Have students note that iron sulfate forms a green solution.

Answers to Challenges (p. 335)
Apply

8. Iron is more active than copper.

9. It is the least active metal.

Skill Builder: Sequencing

lithium, barium, calcium, sodium, magnesium, aluminum, zinc, copper, silver, gold

Teaching Tips for Career in Physical Science
Welder

REINFORCEMENT Use this feature to review some of the uses of laser light and ultrasonic sounds.

Questions

1. **INFER** What is the most common welding technique? (arc welding)

2. **DEFINE** What is welding? (a process used to join two pieces of metal)

LESSON 18-5
What is corrosion? (p. 336)

Teaching Strategies

PREVIEW Before beginning this lesson, write the title of each section in the lesson on the chalkboard. Have students copy these titles in their notebooks using an outlining format. As students read each section, they should write the topic sentence for each paragraph in the section in their notebooks.

DISCUSSION Provide motivation for this lesson by exhibiting several samples of metals showing corrosion, such as rusted iron, tarnished silver, and oxidized aluminum. Describe the chemical reactions that caused these corrosive changes. Have students note the color change and dullness of a corroded metal.

DEMONSTRATION Demonstrate the corrosion of iron. Moisten a piece of plain steel wool with acetic acid, and stuff it tightly into the bottom of a test tube. (The acid speeds up the reaction.) Insert a 15-cm piece of glass tubing halfway though a 1-hole stopper. Stopper the test tube and invert it in a beaker of water so that the glass tubing projects through the water for a length of about 5 cm. Use a ring stand and clamp to hold the inverted test tube. Have students check the apparatus periodically over the course of several days for evidence of rusting and for the rise of water in the glass tubing. The rise of water indicates oxidation is taking place. Oxygen from the air in the test tube is combining with iron in the steel wool.

DEMONSTRATION Show students an oxidized piece of aluminum. Clean the surface of the metal with a piece of sandpaper to show that metals are oxidized from the outside in and that only exposed surfaces are oxidized. This demonstration will help students understand how coating the surface of metals with paint or oil helps protect metals from corrosion.

DEMONSTRATION Coat iron nails with petroleum jelly, oil, paint, nail polish, or lacquer, and immerse the nails in a very dilute (less than 1%) acetic acid solution. Then place two unprotected nails in the same solution. Have students observe the nails for several days to determine the relative effectiveness of each substance as an anticorrosive agent.

CLASS ACTIVITY Take your class on a walk around the school to look for and observe any signs of corrosion.

Answers to Challenges (p. 337)

Check

1. wearing away of a metal

2. Fe_2O_3

3. oxidation

4. Corrosion wears away and weakens metals.

5. Painting keeps air and moisture away from the surface of a metal.

6. Rust forms.

Apply

7. No. It is not exposed to oxygen.

8. The bottom is in water.

Health and Safety Tip

Rusty objects may have harmful bacteria growing on them. The bacteria cause the deadly disease tetanus. A person who is cut with a rusted metal object should see a doctor immediately to get a tetanus shot if the person's tetanus vaccine is not up-to-date. People should receive a tetanus vaccine every 10 years.

Teaching Tips for Activity
Observing the Tarnishing of Silver

Skills: *observing, analyzing*

COOPERATIVE/COLLABORATIVE LEARNING Have students work in pairs.

Questions

1. dull; tarnished

2. Silver and sulfur combined.

3. by polishing the spoon

LESSON 18-6
How are metals plated? (p. 338)

Teaching Strategies

PREVIEW Before beginning this lesson, have students review the TechTerms and their definitions and read the Lesson Summary.

DISCUSSION Show samples of plated metals to the class. Describe coating with a corrosion-resistant metal as a method of protecting another metal from corrosion. The coating of steel automobile bumpers with corrosion-resistant chromium is an excellent example of plating.

DEMONSTRATION If possible, set up an apparatus to demonstrate electroplating.

REINFORCEMENT Refer students to the diagram on p. 338 showing the process of electroplating with copper. Some students may have difficulty understanding this process. Be sure students realize that in a solution of copper sulfate, ions of copper and sulfate form.

EXTENSION In the commercial process of copper electroplating, a piece of pure copper is used as the positive electrode, rather than the carbon rod shown in the illustration on p. 338. In the commercial process, the loss and gain of electrons can be described as follows:
at the cathode (negative electrode)
$Cu^{+2} + 2$ electrons $\longrightarrow Cu^0$ (metallic copper)
at the anode (positive electrode)
$Cu^0 - 2$ electrons $\longrightarrow Cu^{+2}$ (copper ion)

Answers to Challenges (p. 339)

Apply

6. negative electrode

7. They will not rust when exposed to air, rain, and snow.

8. negative electrode; The positive silver ions will be attracted to the negative electrode.

Designing an Experiment

Have volunteers describe their experiments to the rest of the class.

Teaching Tips for Activity
Protecting Metals from Corrosion

Skills: *observing, comparing*

COOPERATIVE/COLLABORATIVE LEARNING Have students work in pairs.

Questions

1. the nail without any coating

2. They reduce the corrosion of the iron nails.

3. Answers will vary. Accept all logical responses.

Answers to Unit Challenges (pp. 340–342)

Understanding the Features: Reading Critically

1. gold and silver

2. Answers will vary. Accept all logical responses.

3. High-alloy steel contains larger amounts of nickel, chromium, and the other metals found in low-alloy steel.

4. to protect themselves from heat, metal fragments, and so on

Critical Thinking

1. electrons; electricity is a flow of electrons

2. The technology needed to produce bronze was developed before the technology needed to produce iron.

3. Copper is being oxidized.

4. It has only one valence electron.

Interpreting a Diagram

1. electroplating

2. carbon; carbon

3. copper sulfate solution

4. Copper will be plated onto the carbon rod.

1. **STATE** What are three things analytical chemists do on the job? (Students should restate three things described in the feature.)

2. Where do most analytical chemists work? (in laboratories)

UNIT 19
SOLUTIONS (p. 343)

Previewing the Unit

Read the titles of the lessons in this unit aloud. Identify terms in the titles with which students are unfamiliar. Write these terms on the chalkboard. Using the Glossary of the text or a dictionary, define the terms as a class. Have students write the terms and their definitions in their notebooks. Then have students work in small groups to carry out the task in the Study Hint.

Bulletin Board Suggestion

Assemble a bulletin board showing the solubility curve of sodium nitrate. Use the graph on p. 364 as a guide. Use the bulletin board to explain how temperature affects the amount of solute that can be dissolved in a solvent.

LESSON 19-1
What is a solution? (p. 344)

Teaching Strategies

PREVIEW Before beginning this lesson, write the title of each section in the lesson on the chalkboard. Have students copy these titles in their notebooks using an outline format. As students read each section, they should write the topic sentence for each paragraph in the section in their notebooks.

REINFORCEMENT Remind students that a mixture is made up of two or more substances that are not chemically combined.

DEMONSTRATION Add some salt to water in a test tube. Add sand to water in another test tube. Have students observe the difference in the two mixtures. Elicit from students that the salt dissolved in the water, but the sand did not. Use this demonstration to introduce the definition of a solution.

RETEACHING OPTION Exhibit different types of solutions, such as alcohol in water and club soda. Point out that these solutions are examples of liquid solutions. Then refer students to Table 1 on p. 344 to see examples of other types of solutions.

Answers to Challenges (p. 345)
Apply
6. c, e, h

7. liquid, liquid, gas
Skill Builder: Organizing Information
Check students' tables.

Teaching Tips for Career in Physical Science
Analytical Chemist

DISCUSSION Before students read this feature, ask them what they think an analytical chemist does on the job. (As a hint, remind students that to analyze means to study.) List student responses on the chalkboard. Then have students read the feature to see which responses were correct.

LESSON 19-2
What are the parts of a solution? (p. 346)

Teaching Strategies

PREVIEW Before beginning this lesson, read the list of Tech-Terms aloud so students can hear their pronunciations.

DEMONSTRATION Point out that in a solution the substance that dissolves is called the solute and the substance in which the solute dissolves is called the solvent. Then prepare a solution of sugar and water. Ask students to identify the solute and solvent of the solution. (solute: sugar; solvent: water)

RETEACHING OPTION Exhibit the following mixtures in front of the classroom: sugar and water, iron filings and water, sulfur and water. Have students observe the mixtures. Point out that not all substances dissolve in water. Explain that substances that dissolve in water are soluble in water. Substances that do not dissolve in water are insoluble in water.

REINFORCEMENT Be sure students understand that substances are soluble or insoluble only in terms of another substance.

Answers to Challenges (p. 347)
Apply
6. solute: coffee powder; solvent: hot water

7. No. Wood floats on water. It does no mix evenly with water.

Ideas in Action
Answer will vary.

Teaching Tips for Science Connection
Chemical Weathering

EXTENSION To show students the results of the chemical weathering of limestone, bring in colorful pictures showing Karst topography. An area with Karst topography has caverns, sinkholes, and natural bridges.

DEMONSTRATION Demonstrate the effect of carbonic acid on calcite, or limestone, by placing a small piece of calcite into some club soda. Have students note the formation of small bubbles on the calcite as the acid reacts with it.

Questions

1. **DEFINE** What is weathering? (the breaking up of rocks and minerals by natural forces)

2. **RELATE** How does chemical weathering affect mechanical weathering? (Rocks that have been broken down chemically are more likely to be broken apart by mechanical weathering.)

LESSON 19-3
Why is water a good solvent?
(p. 348)

Teaching Strategies

PREVIEW Have students read the lesson feature before beginning the lesson. Discuss the concepts presented in the feature. Ask students how they think this information relates to topics they have already studied or to the lesson they are about to study.

CLASS ACTIVITY Have students work in small groups to make models of water molecules using clay, styrofoam, or paper circles. Tell students to label each atom on their models and to identify each atom's charge.

EXTENSION A polar molecule results from a covalent bond in which there is an unequal sharing of electrons. Oxygen has a stronger attraction for electrons in a water molecule; thus, the oxygen atom has a partial negative charge and the hydrogen atoms have a partial positive charge.

DEMONSTRATION You can demonstrate the polarity of water by rubbing a balloon against your hair and then holding the balloon near a thin stream of water. The stream of water will be bent as the two like charges repel each other.

INTRASCIENCE CONNECTION The structure of its molecules gives water a number of properties that are essential to life, including its ability to dissolve many different types of substances. Emphasize the health benefits of drinking at least eight glasses of water each day.

Answers to Challenges (p. 349)
Apply

6. The force of attraction between particles of the substance is greater than the force of attraction between the substance and water.

7. No. A solution is formed when one substance dissolves in another substance.

Skill Builder: Building Vocabulary

polar: end, one side; The North Pole and South Pole of the earth are at opposite ends of the earth. The poles of a magnet are at opposite ends of the magnet.

Teaching Tips for Career in Physical Science
Water Purification Chemist

EXTENSION If possible, arrange a field trip to a water treatment facility so students can observe some of the steps in purifying water.

DEMONSTRATION Bring a pool tester kit to class. Show students how it is used to check the amount of chemicals in a pool. Relate the demonstration to the work of a water purification technician.

Questions

1. HYPOTHESIZE What do you think might make water unsafe to drink? (Answers will vary, but may include germs and chemicals.)

LESSON 19-4
How can you change the rate of dissolving? (p. 350)

Teaching Strategies

PREVIEW Have students write the lesson title and objective in their notebooks. As students read the lesson, tell them to write down the sentence or sentences that provide the information needed in the objective.

DEMONSTRATION Demonstrate that stirring will speed up the rate of dissolving. Place a spoonful of sugar or a sugar cube into two separate beakers of water. Stir only one beaker of water.

DEMONSTRATION Demonstrate how temperature affects the rate at which a solid solute dissolves by placing a spoonful of sugar into each of two beakers, one containing hot water and the other containing cold water. Stir the beakers simultaneously until the solute completely dissolves in one of the beakers. Students will observe that sugar dissolves faster in hot water.

DEMONSTRATION Demonstrate that particle size affects the rate of dissolving by placing some large crystals of copper sulfate into one beaker of water and some pulverized crystals into a second beaker of water. Stir both beakers until the solute completely dissolves in one beaker.

Answers to Challenges (p. 351)
Apply

6. Smaller particles dissolve faster.

7. Substances dissolve faster in boiling water than in cold water.

8. b: particle size is smaller; e: increasing temperature increases the rate of dissolving

InfoSearch

You may wish to discuss students' questions and answers as a class.

Teaching Tips for Activity
Changing the Rate of Dissolving

Skills: *comparing, hypothesizing*

COOPERATIVE/COLLABORATIVE LEARNING Have students work in pairs.

Questions

1. Stirring increased the rate.

2. Crushing increased the rate.

3. crush the cube; heat the water; stir the water after adding the crushed cube

LESSON 19-5

What is the concentration of a solution? (p. 352)

Teaching Strategies

PREVIEW Before beginning this lesson, have students scan the lesson looking for words with which they are unfamiliar. Have students work in pairs or small groups to define each of the words on their lists.

DISCUSSION Prepare solutions of either instant coffee or tea in varying concentrations and display the solutions in front of the classroom. Ask students to state how the solutions differ. Elicit the response that the darker solutions contain more solute and the lighter solutions contain less solute. Point out that weak solutions are called dilute solutions and strong solutions are called concentrated solutions.

RETEACHING OPTION Adding several grams of solute at a time, slowly dissolve enough KNO_3 in 100 mL of water to make a saturated solution. Undissolved solute will remain on the bottom after vigorous stirring. While preparing the solution, explain that a solution in which more solute can be dissolved is called an unsaturated solution. Define a saturated solution as one that contains all the solute it can hold. Then heat the saturated solution of KNO_3. Stirring the heated solution will cause the solute at the bottom of the beaker to dissolve. Dissolve a small amount of additional solute in the heated solution. Point out that the solution is no longer saturated because heating the solution causes it to become unsaturated.

REINFORCEMENT Show students various solubility curves to reinforce the concept that warmer solutions can usually hold more solute.

INTERDISCIPLINARY CONNECTION Some students may be familiar with the terms "saturated" and "unsaturated" as they relate to fats. Saturated fats are organic compounds that are solids at room temperature. Saturated fats increase the amount of cholesterol in the body. Unsaturated and polyunsaturated fats are liquids at room temperature. They do not raise cholesterol levels. Be sure students understand that use of the terms "saturated" and "unsaturated" to describe fats is not related to the way the terms are used to describe solutions.

Answers to Challenges (p. 353)

Apply

6. concentrated solution; water must be added to dilute the solution

7. dilute solution; adding more water makes the solution weaker

Ideas in Action

Answers will vary. Accept all logical responses.

Teaching Tips for Activity
Making a Supersaturated Solution

Skills: *analyzing, observing*

COOPERATIVE/COLLABORATIVE LEARNING Answer the questions as a class.

LESSON 19-6

How do solutes affect freezing point? (p. 354)

Teaching Strategies

PREVIEW Before beginning this lesson, have students scan the lesson for the science process skill symbols. Have students identify each skill used in the lesson. Students should then review the definition given for each skill on p. 4 of their text.

DISCUSSION Provide motivation for this lesson by exhibiting a bag of rock salt in front of the classroom. Ask students when rock salt is used. (when snowy, icy conditions exist on sidewalks, driveways, and so on) Then ask students how they think rock salt affects ice. (Many students may think that the rock salt causes the ice to melt.) Point out that the salt itself does not cause the ice to melt. Instead, the salt dissolves in the thin film of water that forms when the ice begins to melt. Point out that as the salt dissolves in the film of water, the freezing point of water is lowered. Thus, the ice is less likely to refreeze. Identify rock salt as one application of freezing point depression.

COOPERATIVE/COLLABORATIVE LEARNING Have students quiz each other on the material in this section.

Answers to Challenges (p. 355)
Apply

6. beaker B; It has more solute. Adding solute to a solution lowers the freezing point.

7. beaker 1; There is more solute per volume of solvent in beaker 1.

InfoSearch

You may wish to discuss students' questions and answers as a class.

Teaching Tips for Science Connection
Fishes With Antifreeze

INTRASCIENCE CONNECTION Use this feature to discuss how organisms are adapted to their environments. Define an adaptation as any characteristic of an organism that helps the organism survive in its environment. Point out that adaptations make it possible for one kind of organism to survive in an environment where other kinds of organisms cannot survive. For example, tropical fish could not survive in the Arctic Ocean.

Question

PREDICT What would happen to a fish in the Arctic Ocean that lacked the protein found in the blood of Arctic fishes? (The blood of the fish would freeze and the fish would die.)

How do solutes affect boiling point? (p. 356)

Teaching Strategies

PREVIEW Before beginning this lesson, have students review the TechTerms and their definitions and read the Lesson Summary.

DEMONSTRATION Pour 100 mL of distilled water into each of two beakers. Add 10 g of salt to one of the beakers. Put a thermometer into each beaker. Then heat the beakers until the solutions boil. Have students note that the presence of salt raises the boiling point of water. Identify this property as boiling point elevation.

DEMONSTRATION Pour 100 mL of distilled water into each of three beakers. With a grease pencil, label the beakers 1, 2, and 3. Place 5 g of salt in the first beaker, 10 g of salt in the second beaker, and 20 g of salt in the third beaker. Then find the boiling points of the three solutions. Students will observe that the more solute there is in a solution, the higher the boiling point.

DISCUSSION Bring a bottle of antifreeze to class. Ask students why antifreeze is used in cars. (to prevent overheating and engine freezing) Ask students how they think antifreeze works. Discuss all student responses. Then describe the effects of antifreeze in terms of boiling point elevation and freezing point depression. Some students may be confused by the use of antifreeze to prevent overheating as well as freezing. Emphasize that adding solutes both lowers the freezing point of water and raises its boiling point.

Answers to Challenges (p. 357)
Apply

6. solution B; Adding solute increases the boiling point.

7. beaker 2; There is more solute per volume of solvent in beaker 2.

Designing an Experiment

Have volunteers describe their experiments to the rest of the class.

Teaching Tips for Activity
Observing Boiling Point Elevation

Skills: *observing, analyzing*

COOPERATIVE/COLLABORATIVE LEARNING Have students work in pairs.

Questions

1. 101 °C

2. 102 °C

3. 104 °C

4. Increasing the amount of solute in a solution increases the boiling point of the solution.

How can a solution be separated? (p. 358)

Teaching Strategies

PREVIEW Read the list of TechTerms aloud so students can hear their pronunciations.

DISCUSSION Provide motivation for this lesson by demonstrating that solutions cannot be separated by filtering. Then ask students how they think a solute and solvent can be separated. Allow students to suggest possible methods. List their responses on the chalkboard.

DEMONSTRATION Place 10 mL of a copper sulfate solution in an evaporating dish. Heat the solution until the water boils away. Exhibit the residue of $CuSO_4$ crystals. Point out that a solute can be recovered from a solution by evaporating the solvent.

RETEACHING OPTION Show students a distillation apparatus and, if possible, demonstrate distillation. Point out the tube-within-a-tube structure of the condenser. Explain that water in the outside tube cools the gases passing though the inside tube, thus causing the gases to condense.

Answers to Challenges (p. 359)
Apply

6. evaporation

7. condensation

Skill Builder: Researching

Check students' reports. Sometimes it is important to use distilled water instead of ordinary tap water because tap water may contain solutes, such as minerals.

Teaching Tips for Technology and Society
Fractional Distillation of Petroleum

EXTENSION In a sense, petroleum is stored sunlight. Solar energy was absorbed by organisms in the oceans 300 to 400 million years ago. The organisms used the sun's energy to make organic compounds. When they died, the organisms settled on the ocean floor and were covered by layers of sediments. Under great heat and pressure, the sediments became rock and their organic compounds were converted into petroleum.

Questions

1. LIST What are three products that are found in petroleum? (gasoline, kerosene, heating oil)

2. EXPLAIN What is the purpose of a condenser in fractional distillation? (to cool substances until they become liquids)

How are crystals formed?

(p. 360)

Teaching Strategies

PREVIEW Before beginning this lesson, have students read the Lesson Summary.

CLASS ACTIVITY Have students examine salt crystals with a magnifying lens and sketch the shape of salt crystals in their notebooks. (Salt crystals are cubic.)

DEMONSTRATION Show students three-dimensional models of crystals. Point out that crystals have an orderly arrangement of atoms and molecules.

CLASS ACTIVITY Make salt crystals by making a saturated solution of salt water. Place a string in the solution. Have students observe the string for the next few days to look for crystal formation.

Answers to Challenges (p. 361)
Apply

6. Sugar crystals will form.

7. salt; Salt crystals are cubic.

Skill Builder: Organizing Information

The six basic crystal shapes are cubic, hexagonal, monoclinic, tetragonal, triclinic, and orthorhombic.

Teaching Tips for Activity
Growing Sugar Crystals

Skills: *observing, analyzing*

SAFETY TIP Caution students to be careful not to spill the boiling water.

COOPERATIVE/COLLABORATIVE LEARNING Have students work in pairs.

Questions

1. sugar crystals

2. supersaturated; Boiling the water allowed more sugar to dissolve.

3. As the solution cooled, sugar crystals came out of solution.

Answers to Unit Challenges (pp. 362–364)
Understanding the Features: Reading Critically

1. analyze chemicals; perform experiments; test substances to see what will happen when different substances are combined

2. They have a protein in their blood which lowers the freezing point of the blood.

3. In mechanical weathering, rocks are broken down into smaller pieces. Chemical weathering causes substances in rocks to dissolve.

4. Chemical purification technicians ensure the quality of people's drinking water.

5. fractional distillation

Critical Thinking

1. Water changes from one form to another when it evaporates.

2. An increased amount of solute raises the boiling point and lowers the freezing point.

3. Substances have specific, known crystal shapes.

4. Evaporation increases the concentration of salt.

5. The carbon dioxide escapes into the air.

Interpreting a Graph

1. 110 g

2. 90 °C

3. 180 g

4. 70 g

5. The higher the temperature of the solvent, the more solute it can dissolve.

Previewing the Unit

Have students work in pairs to carry out the task in the Study Hint.

Bulletin Board Suggestion

Assemble a bulletin board illustrating the suspensions, emulsions, and colloids discussed in this unit. Have students find and collect pictures showing salad dressings, mud, milk, paint, shaving cream, whipped cream, mayonnaise, clouds, fog, smoke, butter, jellies, and so on. Tell students to label each picture as a suspension, emulsion, or colloid. Attach all the pictures to the bulletin board. Use the bulletin board to review the properties of each type of substance. You may also wish to include pictures of solutions in order for students to compare the properties of solutions and suspensions.

LESSON 20-1
What is a suspension? (p. 366)

Teaching Strategies

PREVIEW Have students write the lesson title and objective in their notebooks. As students read the lesson, tell them to write down the sentence or sentences that provide the information needed in the objective.

DEMONSTRATION Add some soil to a jar of water to make a suspension. Ask students to describe the mixture. (It will be cloudy.) Then allow the mixture to stand for awhile and have students observe what happens. Students will observe that the soil particles settle to the bottom.

DISCUSSION Develop this lesson by comparing the properties of suspensions with the properties of solutions. Rewrite Table 1 on p. 366 on the chalkboard. Have students copy the table in their notebooks.

DEMONSTRATION Darken the classroom and shine a flashlight through a mixture of milk and water to demonstrate how the particles of a suspension scatter light.

COOPERATIVE/COLLABORATIVE LEARNING Have a volunteer compare the properties of solutions and suspensions for the rest of the class.

Answers to Challenges (p. 367)
Apply

6. The particles in a suspension can be seen. They also settle on standing. The particles in a solution are too small to be seen and are evenly mixed. They do not settle on standing.

7. physical; they do not involve a change in the substances

Ideas in Action

Answers will vary. Accept all logical responses.

Teaching Tips for Activity
Reading Medicine Labels

Skill: *observing*

COOPERATIVE/COLLABORATIVE LEARNING Answer the questions as a class.

Questions

1. amount to be taken; interval at which the drug should be taken

2. when the drug should no longer be used; A drug is useless after its expiration date.

3. A prescription drug label has a doctor's name, the date the prescription is filled, and the person's name for whom the drug is prescribed.

4. Answers will vary. Warnings often are given for pregnant or nursing women, people who have certain medical problems or allergies, or people who are taking other drugs. All drug labels warn people to keep all drugs out of the reach of children.

LESSON 20-2
How can a suspension be separated? (p. 368)

Teaching Strategies

PREVIEW Before beginning this lesson, read the list of Tech-Terms aloud so students can hear their pronunciations.

DISCUSSION Begin this lesson by asking students how they think a suspension can be separated. Discuss all student responses. Then describe filtration, coagulation, and separation by centrifuge.

DEMONSTRATION Demonstrate that the size of particles determines the rate at which the particles settle. Place clay, sand, and 0.025-cm glass beads in a 100-mL graduated cylinder. Fill the graduated cylinder with water, cover the top, and shake the mixture. Allow the mixture to stand overnight. Have students observe the mixture the next day. Students will observe three layers of sediment: clay on top, sand in the middle, and the glass beads on the bottom. Ask students which substance has the smallest particles. (clay)

DEMONSTRATION Demonstrate filtration by filtering some clay out of a water suspension.

INTRASCIENCE CONNECTION Obtain a blood agglutination kit from a biological supply company. Describe agglutination to the class.

EXTENSION Spinning in a high-speed centrifuge creates increased g forces due to the centrifugal, or centripetal, forces. The increased forces speed up the settling process, thus separating the suspension.

Answers to Challenges (p. 369)
Check

1. The particles settle.

2. the removal of particles in a suspension by passing the suspension through a filter

3. particles that are smaller than the holes in the filter

4. process in which chemicals are used to make the particles in a suspension clump together

5. blood clotting

6. A centrifuge spins a mixture at a high speed. As the suspension is spun around, the particles settle.

Apply

7. filtration: b; coagulation: a, d; centrifuge: c

Skill Builder: Modeling

The clay particles in the test tube to which alum solution was added settle faster because the particles that clump together are heavier than separate particles.

Teaching Tips for Science Connection
The Mississippi Delta

DEMONSTRATION Demonstrate deposition with a stream table. Have students note the amount of sediment that forms at the bottom of the stream table.

EXTENSION Point out the Mississippi River on a map of the United States. Have students locate its source and its mouth.

Questions

1. IDENTIFY Where is the mouth of the Mississippi River located? (at the Gulf of Mexico)

2. INFER What type of soil is good for farming? (topsoil)

LESSON 20-3
What is an emulsion? (p. 370)

Teaching Strategies

PREVIEW Read the list of TechTerms aloud so student can hear their pronunciations.

DEMONSTRATION To provide motivation for this lesson, mix some cooking oil with water and then shake the mixture. Allow the mixture to stand and ask students to describe what happens. (The oil and water separate.) Point out that a mixture of oil and water is an emulsion. Define an emulsion as a mixture of two liquids.

REINFORCEMENT Be sure students understand that an emulsion is a suspension and that some emulsions are temporary and others are permanent.

DISCUSSION Bring a variety of cleaning agents to class and display them in front of the classroom. Ask students how they think cleaning products wash away dirt and grease. Discuss all student responses. Then identify cleaning products as emulsifying agents. Define an emulsifying agent as a substance that keeps an emulsion from separating.

Answers to Challenges (p. 371)
Apply

6. By breaking fats into smaller droplets, bile makes it easier for digestive enzymes to act on them.

7. Paint is an emulsion. The liquids that make up the paint have settled into layers.

Skill Builder: Modeling

The mixture to which soap was added does not separate, because soap is an emulsifying agent.

Teaching Tips for Activity
Making an Emulsion

Skills: *observing, predicting*

COOPERATIVE/COLLABORATIVE LEARNING Have students work in pairs.

1. when oil was added

2. No.

3. egg yolk

4. It would separate.

LESSON 20-4
What is a colloid? (p. 372)

Teaching Strategies

PREVIEW Before beginning this lesson, write the title of each section in the lesson on the chalkboard. Have students copy these titles in their notebooks using an outline format. As students read each section, they should write the topic sentence for each paragraph in the section in their notebooks.

DISCUSSION Define a colloid as a suspension in which the particles are permanently suspended. Identify particle size and movement as the reasons a colloid is a permanent suspension.

REINFORCEMENT Be sure students understand that some emulsions, such as homogenized milk, are also colloids since they are permanent suspensions.

CLASS ACTIVITY Exhibit a variety of colloids, such as shaving cream, whipped cream, paint, dyes, mayonnaise, jelly, and butter, in front of the classroom. Randomly hold up each colloid. Have students classify the colloid as a foam, sol, emulsion, or gel. Then have students identify the phase of each colloid. Tell students to use Table 1 on p. 372 as a guide.

DEMONSTRATION Pour some milk into a glass. Add water to dilute the milk. Then shine a bright beam of light from a flashlight or a slide projector through the glass of milk. Students will observe that the beam of light passes through the mixture of water and milk. Point out that the beam is visible because light is reflected by molecules of milk, which are mixed in with molecules of water. Have students note that this effect is similar to seeing the beam of car headlights in a fog.

Answers to Challenges (p. 373)
Apply

6. Fog is a colloid. The particles in colloids scatter light.

7. In colloids and solutions, substances are mixed. The particles in a colloid are not as small as the particles in a solution.

8. a. foam **b.** fog **c.** gel **d.** sol

InfoSearch

You may wish to discuss students' questions and answers as a class.

Teaching Tips for Science Connection
Homogenized Milk

DISCUSSION Ask any students who have seen separated fresh milk describe its appearance to the rest of the class. (Students who live or work on dairy farms probably have seen fresh milk.)

REINFORCEMENT Have students look up the definition of the term "homogeneous" in a dictionary. Tell students to write in their own words how the meaning of the term "homogeneous" relates to homogenized milk.

Questions

1. DEFINE What is homogenization? (process in which skim milk and cream are mixed together to form a colloid)

2. RELATE Why does cream float on top when skim milk and cream separate? (Cream is less dense than the watery parts of skim milk.)

LESSON 20-5
What are air and water pollution? (p. 374)

Teaching Strategies

PREVIEW Before beginning this lesson, have students scan the lesson for the science process skill symbols. Have students identify each skill used in the lesson. Students should then review the definition given for each skill on p. 4 of their text.

DISCUSSION Write the term *pollution* on the chalkboard. Discuss what students think of when they hear the word "pollution." Emphasize that pollution is upsetting the balance of the environment.

CLASS ACTIVITY Have students find and bring in articles related to air and water pollution. Read and discuss the articles as a class.

REINFORCEMENT Emphasize the relationship between conserving fossil fuels and reducing air pollution. Point out that every time a person shuts off an unnecessary light, the person is conserving the fuels used to produce electricity. The less fuel that is burned, the less the amount of poisonous gases that are released into the air.

Answers to Challenges (p. 375)
Apply

6. Air pollution damages the trees.

7. Pollution in water may form suspensions and colloids.

Health and Safety Tip

Water can be purified by boiling.

Teaching Tips for Activity
Observing Pollutants in Air

Skills: *observing, hypothesizing*

COOPERATIVE/COLLABORATIVE LEARNING Have students observe each other's slides.

Questions

1. Answers will vary.

2. Answers will vary.

3. Answers will vary.

4. Answers will vary. Accept all logical responses.

Answers to Unit Challenges (pp. 376–378)
Understanding the Features: Reading Critically

1. land formed at the mouth of a river

2. New topsoil is always being produced.

3. Fresh milk is a suspension that separates on standing. Cream is less dense than water, so it floats on top when skim milk and cream separate.

4. A homogenizer uses a high-pressure pump to force the milk-and-cream mixture through small openings in a metal plate. This breaks fat droplets into smaller pieces.

5. Sediments do not have time to settle.

Critical Thinking

1. An emulsifying agent keeps an emulsion from separating by breaking particles in an emulsion into smaller particles.

2. The particles in a colloid are permanently suspended.

3. Particles in a colloid are kept in suspension because they are bombarded by the molecules around them.

4. Allow a mixture to settle to see if it is a suspension. Then test to see if it scatters light. If it does, it is a colloid. If it does not scatter light, it is a solution.

Interpreting a Diagram

1. Answers will vary. Accept all logical responses.

2. Answers will vary. Accept all logical responses.

3. Answers will vary. Accept all logical responses.

4. Answers will vary. Accept all logical responses.

5. No. Pollutants in the water would harm the fish and plants.

6. Chemical wastes from the factory flow downstream to the rural area.

Previewing the Unit

Have students work in pairs to carry out the task in the Study Hint.

Bulletin Board Suggestion

Using the diagram on p. 386 as a guide, assemble a bulletin board showing the pH scale. Label the acidic and basic ranges on the scale. The background color of the scale should go from red at the acidic end of the scale to blue at the basic end of the scale. Have students find out the pH value of substances such as milk, blood, ammonia, vinegar, gastric juice, lemon juice, orange juice, and baking soda. Tell students to find pictures of these substances. Using thumbtacks or tape, attach the pictures to their proper place on the pH scale.

8. a, c, e, h; The symbol for hydrogen appears first in each chemical formula.

Health and Safety Tip

Students should use extreme care when working with acids or bases in the laboratory. They should never pour water into an acid or a base. An acid or a base should always be poured into water.

Teaching Tips for Science Connection
Acid Rain

EXTENSION Have students collect samples of rainwater on the next rainy day. Tell students to bring their samples to class. Test the samples with pH paper to determine the pH of the rainwater. Acid rain has a pH of less than 5.6.

Questions

1. RELATE Why do you think international cooperation is needed to reduce the harmful effects of acid rain? (Acid rain is a worldwide problem.)

2. What is the cause of acid rain? (the release of sulfur dioxide and nitrogen oxides into the atmosphere)

LESSON 21-1
What is an acid? (p. 380)

Teaching Strategies

PREVIEW Have students read the lesson feature before beginning the lesson. Discuss the concepts presented in the feature. Ask students how they think this information relates to topics they have already studied or to the lesson they are about to study.

DISCUSSION Provide motivation for this lesson by exhibiting several substances that contain acids, such as vinegar, lemons, and sour cream. Ask students to describe the taste of these different substances. (Do not allow students to taste the substances.) Elicit the response that each substance tastes sour. Point out that each substance tastes sour because it contains acids. Define an acid as a substance that reacts with a metal to release hydrogen. Then describe the other properties of acids. Relate the properties of acids to the presence of hydrogen.

REINFORCEMENT Review the acids listed in Table 1 on p. 380. Ask students to identify the elements that make up each acid and the number of atoms of each element that combine to form one molecule of the acid. Emphasize that acids can be recognized by the fact that the symbol for hydrogen appears first in the chemical formula.

DEMONSTRATION Combine some hydrochloric acid and zinc. Have students note the bubbles that form during the reaction. Point out that the bubbles indicate that hydrogen gas is being released.

Answers to Challenges (p. 381)
Apply

7. They release hydrogen ions, which are the same as protons.

LESSON 21-2
What is a base? (p. 382)

Teaching Strategies

PREVIEW Before beginning this lesson, read the list of Tech-Terms aloud so students can hear their pronunciations.

DISCUSSION Exhibit some substances containing bases in front of the classroom, such as a bar of soap, milk of magnesia, ammonia, and limewater. Tell students that each substance contains a base. Define a base as a substance that is formed when a metal reacts with water. Wet the bar of soap and hand it to a student. Ask the student to describe the feel of the soap. Explain that the slippery feel of soap is caused by the base it contains.

RETEACHING OPTION Refer students to Table 1 on p. 382. Tell students to study the chemical formulas of the bases listed in the table. Ask students what all the bases have in common. (They all contain a hydroxyl ion [OH^-].) Explain that all bases contain hydroxyl ions, which they release in water.

COOPERATIVE/COLLABORATIVE LEARNING Have students work in pairs to answer the Check and Apply questions.

Answers to Challenges (p. 383)
Apply

8. The greater the concentration of hydroxyl ions, the stronger the base.

9. a. 2 **b.** 4 **c.** 1 **d.** 3

Skill Builder: Experimenting

Answers will vary.

Teaching Tips for Science Connection
Soap

EXTENSION The end of a soap molecule that is soluble in water is hydrophilic. The end of a soap molecule that is soluble in oil is hydrophobic. Have students look up the definitions of the terms "hydrophilic" and "hydrophobic" in a dictionary. Tell students to write in their own words how the meaning of these two terms relates to the action of a soap molecule.

Questions

1. What is the equation for the soap-making process? (fat + lye $\longrightarrow$ soap + glycerine)

2. What are two uses of glycerine? (Glycerine is used to make cellophane, printer's ink, cosmetics, medicines, and soap.)

LESSON 21-3
What are indicators? (p. 384)

Teaching Strategies

PREVIEW Have students write the lesson title and objective in their notebooks. As students read the lesson, tell them to write down the sentence or sentences that provide the information needed in the objective.

CLASS ACTIVITY Divide the class into small groups. Distribute various indicators, such as litmus paper and bromthymol blue, and different samples of weak acids and bases to each group. Have students test each sample to determine if it is an acid or a base. Tell students to use Table 1 on p. 384 as a guide. After all groups have completed their tests, have a representative from each group write the group's data on the chalkboard. Review all the data as a class.

REINFORCEMENT Be sure students understand that acids and bases only exhibit their properties when ionization occurs in solutions.

COOPERATIVE/COLLABORATIVE LEARNING Have a volunteer compare the properties of acids and bases for the rest of the class.

Answers to Challenges (p. 385)
Apply

7. base; Bromthymol blue turns blue in bases.

8. a. base **b.** acid **c.** base **d.** base **e.** acid **f.** base

Designing an Experiment

Have volunteers describe their experiments to the rest of the class.

Teaching Tips for Activity
Making an Indicator

Skills: *observing, analyzing*

SAFETY TIP Caution students to be careful when boiling liquids.

COOPERATIVE/COLLABORATIVE LEARNING Have students work in pairs.

Questions

1. Answers will vary according to the samples used.

2. bright red

3. blue-purple

LESSON 21-4
What is the pH scale? (p. 386)

Teaching Strategies

PREVIEW Before beginning this lesson, write the title of each section in the lesson on the chalkboard. Have students copy these titles in their notebooks using an outline format. As students read each section, they should write the topic sentence for each paragraph in the section in their notebooks.

DISCUSSION Explain what the pH scale is used to measure. Explain that the pH scale is used to determine whether a substance is an acid or a base. A pH number below 7 describes an acidic solution. A pH number above 7 describes a basic solution. Make sure students understand that the higher the number, the more basic the substance; the lower the number, the more acidic the substance.

DEMONSTRATION Using pH paper, find the pH of various acids and bases.

INTRASCIENCE CONNECTION As students may have inferred from the description of natural indicators, some soils are acidic and some are basic. Demonstrate how soils are tested. Soil test kits can be obtained from scientific supply houses or local nurseries. Have students collect samples of soil from different areas. If possible, use some soil found under evergreens and other soil from open grassy fields. Indicators such as litmus or pH paper can be used in testing soils.

CLASS ACTIVITY Write the numbers from 0 to 14 on separate index cards. Place all the cards in a bag and mix them up. Have students take turns picking cards from the bag. Tell each student to state whether the number on the card they picked indicates an acid, a base, or a neutral substance. You also may wish to vary this activity by having each student pick two cards and state which number indicates the stronger acid or base.

Answers to Challenges (p. 387)
Apply

1. a. 6.8 **b.** 7.0 **c.** 13.0 **d.** 1.0 **e.** 2.0 **f.** 10.5 **g.** 5.0

9. a. hydrochloric acid; lemon juice; cabbage; milk **b.** milk of magnesia; lye **c.** water

Ideas in Action

Answers will vary.

Teaching Tips for Activity
Testing the Acidity of Foods

Skills: *analyzing, observing*

COOPERATIVE/COLLABORATIVE LEARNING Answer the questions as a class.

Questions

1. Answers will vary.

2. Answers will vary.

3. Answers will vary. Any foods with a pH of 7.0 are neutral.

LESSON 21-5
What is neutralization? (p. 388)

Teaching Strategies

PREVIEW Before beginning this lesson, have students scan the lesson for the science process skill symbols. Have students identify each skill used in the lesson. Students should then review the definition given for each skill on page 4 of their text.

REINFORCEMENT Before beginning this lesson, you may wish to review double-replacement reactions. This review will help students understand a neutralization reaction.

DISCUSSION Display a bottle of table salt on your desk. Point out that table salt is just one of many different salts. Explain that a salt is produced when an acid reacts with a base. Write the neutralization reaction that takes place between hydrochloric acid and sodium hydroxide on the chalkboard and guide students through the equation.

DEMONSTRATION To demonstrate that a neutralization reaction results in a neutral substance, test a solution of table salt and water with red litmus paper and blue litmus paper. Students will observe that the indicators do not change color.

RETEACHING OPTION Bring an antacid to class. Hold it up in front of the classroom and ask students what antacids are used to treat. (indigestion caused by excess stomach acids) Point out that an antacid is a base, which neutralizes stomach acids. Then describe neutralization reactions in general.

Answers to Challenges (p. 389)
Check

1. substance with a pH of 7.0

2. It can cause burns if it is spilled on the skin.

3. Sodium chloride and water are formed.

4. table salt

5. reaction between an acid and a base to produce a salt and water

6. substance formed from the negative ion of an acid and the positive ion of a base

7. They do not change color.

Apply

8. NaCl

9. $CaSO_4$

10. NH_4NO_3

11. K_3PO_4

Skill Builder: Experimenting

You may wish to perform this experiment as a class.

Teaching Tips for Looking Back in Science
History of Salt

EXTENSION Have interested students research the development of desalinization plants. Tell students to present their findings in an oral report.

Questions

1. **IDENTIFY** Where are deposits of salt found? (where prehistoric inland seas have dried up)

2. **INFER** What effect do you think brine has on the growth of germs? (Brine prevents their growth.)

LESSON 21-6
What are electrolytes? (p. 390)

Teaching Strategies

PREVIEW Before beginning this lesson, have students review the TechTerms and their definitions and read the Lesson Summary.

DEMONSTRATION Set up a conductivity tester as shown in the diagrams on p. 390. Place some distilled water in the tester and have students observe that the bulb does not light. Then add some acetic acid to the water. Students will observe that the bulb lights when acetic acid is added. Use this demonstration to introduce electrolytes and ionization.

EXTENSION Different compounds dissociate into ions at different rates. Strong electrolytes have higher rates of ionization than weak electrolytes. When strong electrolytes dissolve, there are more ions in solution. Such a solution is a better conductor of electricity than a solution containing a weak electrolyte.

Answers to Challenges (p. 391)
Apply

7. **a.** K^+, Cl^- **b.** Ca^{+2}, OH^- **c.** NH_4^+, OH^- **d.** H^+, NO_3^- **e.** Ca^{+2}, Cl^- **f.** Na^+, Br^- **g.** H^+, I^- **h.** Mg^{+2}, SO_4^{-2}

Designing an Experiment

Have volunteers describe their experiments to the rest of the class.

Teaching Tips for Activity
Classifying Acids and Bases

Skills: *analyzing, interpreting*

COOPERATIVE/COLLABORATIVE LEARNING Have students work in pairs.

Questions

1. Answers will vary. The pH is less than 7.0.

2. Answers will vary. The pH is greater than 7.0.

3. Answers will vary. The pH is 7.0.

Answers to Unit Challenges (pp. 392–394)
Understanding the Features: Reading Critically

1. sulfuric and nitric acids

2. fat + lye $\longrightarrow$ soap + glycerine

3. Salt was too valuable to be wasted.

4. Answers will vary. Accept all logical responses.

Critical Thinking

1. It could result in injury or death by burning or poisoning.

2. acid

3. Yes. An acid conducts electricity.

4. Answers will vary. Accept all logical responses.

Interpreting a Table

1. hydrogen

2. $HC_2H_3O_2$

3. nitric acid

4. $H_3C_6H_5O_7$; citric acid

5. It is a weak acid.

TEACHER'S NOTES

CONCEPTS AND CHALLENGES IN
Physical Science

Leonard Bernstein • Martin Schachter • Alan Winkler • Stanley Wolfe
STANLEY WOLFE
Project Coordinator

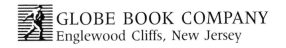 GLOBE BOOK COMPANY
Englewood Cliffs, New Jersey

AUTHORS
Alan Winkler
Leonard Bernstein
Martin Schachter
Stanley Wolfe
   Project Coordinator

Cover Photos
Background: Fireworks (Super Stock)
Inset: Brass gear (Dick Luria/Photo Researchers)

Illustrations: Ebet Dudley
Technical Art: Gary Tong
Photo Researchers: Rhoda Sidney

Content Reviewers
Jerry Faughn
Professor of Physics
Eastern Kentucky University
Richmond, Kentucky

 Globe Book Company
A division of Simon & Schuster
Englewood Cliffs, New Jersey 07632

10 9 8 7 6 5
ISBN: 1-55675-744-1

Contents

SCIENTIFIC SKILLS AND METHODS

CONTENTS

STUDY HINT As you read each lesson in Unit 1, write the lesson title and lesson objective on a sheet of paper. After you complete each lesson, write the sentence or sentences that answer each objective.

What is physical science?

Objective ▶ Identify and describe the main branches of physical science.

TechTerm

▶ **specialization** (SPESH-uh-lih-zay-shun): studying or working in one part of a subject.

Studying Physical Science Physical science is one of the major fields of science. It is the study of matter and energy. Everything around you is either matter or energy.

Physical science has two main branches. The two main branches are chemistry and physics. Chemistry is the study of all forms of matter and changes in matter. Physics is the study of energy and changes among forms of energy.

�▶*Name:* What are the two main branches of physical science? chemistry and physics

Specialization A specialist is a person who studies or works in only one part of a subject. Working in one part of a subject is called **specialization** (SPESH-uh-lih-zay-shun). Some of the specialized fields in physical science are listed in Table 1.

Table 1 Specialized Fields in Physical Science	
FIELD	**WHAT IS STUDIED**
Organic chemistry	Most substances containing carbon
Inorganic chemistry	Substances that do not contain carbon
Biochemistry	Chemical make-up of living things
Mechanics	Motion of objects
Optics	Light
Acoustics	Sound
Nuclear physics	Nucleus of the atom and its changes
Astrophysics	Outer space
Thermodynamics	Heat

▬*Analyze:* Use Table 1. What does a nuclear physicist study? nucleus and its changes

Importance of Physical Science Why should you study physical science? Physical science is an important part of your everyday life. It is difficult to think of anything that does not involve physical science and the discoveries of physical scientists. For example, each year seat belts save thousands of lives. Seat belts are based on the laws of motion.

The discoveries of physical scientists also have resulted in nuclear energy. Nuclear energy has both problems and benefits. Physical scientists are constantly working to solve the problems related to nuclear energy. Their solutions may someday solve the world's energy problems.

IIII▶*Identify:* On what laws are seat belts based? laws of motion

LESSON SUMMARY

▶ Physical science is the study of matter and energy.

▶ The two main branches of physical science are chemistry and physics.

▶ Specialization is the study of only one part of a subject.

▶ Physical science is important in your daily life.

▶ The discoveries of physical scientists affect people's everyday lives.

CHECK *Complete the following.*

1. The study of sound is called _____ . acoustics

2. The two main branches of physical science are chemistry and _____ . physics

3. A person who works in only one part of a subject is a _____ . specialist

4. Thermodynamics is the study of _____ . heat

5. Physical science is the study of matter and _____ . energy

6. The study of substances containing carbon is called _____ chemistry. organic

APPLY *Complete the following.*

7. **Analyze:** The number of specialized fields in physical science has grown during the past few decades. Why do you think this growth has occurred?

8. What are three ways in which physical science is important in your everyday life?

9. **Infer:** What two branches of science does biochemistry combine?

....

Skill Builder ....

Building Vocabulary Science words often are made up of many word parts. The word "thermodynamics," for example, is made up of the root word "dynamics" and the prefix "thermo-." Use a dictionary to find the meaning of the prefix "thermo-." Write the definition. Find five words that begin with the prefix "thermo-." Write their definitions in your own words. Circle the part of the definition that relates to the meaning of the prefix "thermo-."

♦♦♦♦♦ CAREER IN PHYSICAL SCIENCE ♦♦♦♦♦♦♦♦♦♦♦♦♦♦♦♦♦♦♦♦♦♦♦♦♦♦♦♦♦

SCIENTIFIC ILLUSTRATOR

Do you like to draw? If so, you may be interested in a career as a scientific illustrator. You can see the work of scientific illustrators on many pages of this book. Scientific illustrators make drawings, diagrams, and sketches to illustrate scientific processes, structures, cycles, equations, and so on. Their drawings help people to understand scientific concepts.

Scientific illustrators are employed by publishers of science books and journals. They also may be employed by advertisers. The drawings of scientific illustrators are used in sales brochures and advertisements. In the catalogues of scientific supply houses and pharmaceutical companies, many drawings are used to demonstrate equipment and other supplies.

If you are interested in becoming a scientific illustrator, you need a college degree in graphics or fine arts. Many scientific illustrators also find a background in the sciences helpful in their careers. For more information, you may write to Art Student's League of New York, 215 West 57th Street, New York, NY 10019.

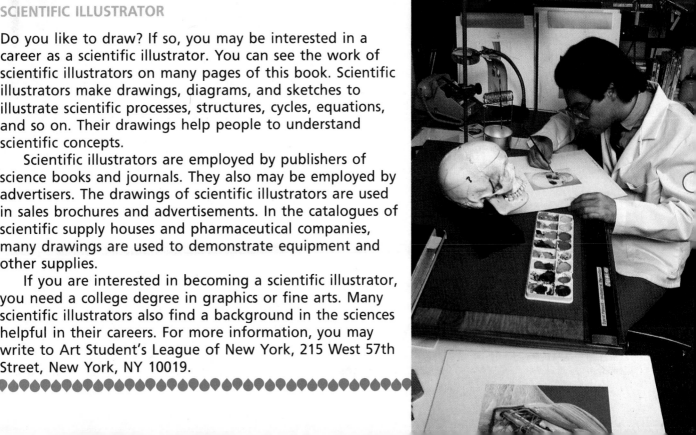

1-2 What are science skills?

Objective ▶ Identify and use science skills to solve problems and answer questions.

TechTerm

▶ **hypothesis** (hy-PAHTH-uh-sis): suggested solution to a problem

Science Skills Scientists use many skills to gather information. These skills are sometimes called science skills. You use science skills, too. You probably used some science skills today. When you use most science skills you use five senses. Your senses are seeing, hearing, touching, smelling, and tasting.

Ten science skills are used in this book. You will see skills symbols. These symbols will let you know when you are using a skill. Soon, you will be thinking like a scientist.

▐▌▌▶*Explain:* Why do scientists use science skills?
to gather information

Researching Do you know what the word "research" means? The prefix "re-" means again. "Search" means to look for or to find out. When you do research you look for something again. If you go to the library to find information, you are doing research. You can do research by reading books, magazines, newspapers, and reports. Another kind of research is experimenting. You can perform experiments to gather information.

▐▌▌▶*Identify:* What are two ways to do research?
reading and experimenting

Communicating When you talk to your best friend you are communicating, or sharing information. If you write a letter, you are communicating. Part of the job of a scientist is communicating. Scientists communicate by teaching, talking, and writing about their work. Sharing information is very important to the progress of science.

▐▌▌▶*Describe:* How do scientists communicate?
by teaching, talking, and writing about their work

Think like a Scientist

Observing When you observe, you use your senses. You must pay close attention to everything that happens.

Measuring When you measure, you compare an unknown value with a known value. Measuring makes observations more exact.

Inferring When you infer, you form a conclusion based upon facts without making observations.

Classifying When you classify, you group things based upon how they are alike.

Organizing When you organize, you work in an orderly way. You put your information in order.

Predicting When you predict, you state ahead of time what you think will happen based upon what you already know.

Hypothesizing When you hypothesize, you state or suggest a solution to a problem. A **hypothesis** (hy-PAHTH-uh-sis) is a suggested solution to a problem based upon what is already known about the problem.

Modeling When you model, you use a copy of what you are studying to help explain it. A model can be a three-dimensional copy, a drawing, or a diagram. The model can be larger or smaller than the original.

Analyzing When you analyze, you study information carefully.

LESSON SUMMARY

▶ Scientists use science skills to gather information.

▶ Ten science skills are used in this book.

▶ Researching includes reading and experimenting.

▶ Communicating means sharing information.

▶ Other science skills are observing, measuring, classifying, organizing, predicting, hypothesizing, modeling, and analyzing.

CHECK *Write true if the statement is true. If the statement is false, change the underlined term to make the statement true.*

1. When you <u>predict</u>, you compare an unknown value to a known value. measure

2. A <u>prediction</u> is a suggested solution to a problem. hypothesis

3. When you <u>classify</u>, you group things based upon how they are alike. true

4. Your senses are seeing, hearing, touching, smelling, and <u>thinking</u>. tasting

5. When you <u>analyze</u>, you study information carefully. true

6. When you <u>infer</u>, you study information carefully. analyze

APPLY *Complete the following.*

7. Choose two science skills. How do you use these science skills every day?

8. **Explain:** Why do you think measurements make observations more exact?

▶ 9. **Predict:** What team do you predict will win the World Series this year? Why do you think they will win?

Ideas in Action

IDEA: Many things are classified in your everyday life.

ACTION: Name three things that are classified in your everyday life. How are they classified?

◄►◄►◄ PEOPLE IN SCIENCE ►◄►◄►◄►◄►◄►◄►◄►◄►◄►◄►◄►◄►◄►◄►◄►◄►◄

STEPHEN W. HAWKING (1942–PRESENT)

Stephen Hawking is considered one of the greatest scientists of all time. He is a physicist who studies a branch of physics called cosmology (cahz-MAHL-uh-jee). Cosmology is the study of how the universe began and how it may develop in the future. Stephen Hawking's goal is a complete understanding of the universe and its basic laws. To find answers about the universe, Hawking has studied black holes. A black hole is a dense region of space. Black holes have such strong gravities that even light cannot escape a black hole.

Stephen Hawking also is considered one of the most remarkable scientists. Since 1958, Hawking has suffered from a serious disease of the nervous system and has been confined to a wheelchair. From this wheelchair, Hawking has used his mind to explore the universe. In 1974, Hawking was made a member of the Royal Society of London. Isaac Newton was one of its first members. Today, Hawking is a professor of mathematics at Cambridge University. Hawking holds the same position that Newton once held.

1-3 What is scientific method?

Objective ▶ Describe how to use scientific method to solve a problem.

TechTerms

- ▶ **data** (DAY-tuh): information
- ▶ **scientific method:** model or guide used to gather information and solve problems

Scientific Method Scientists solve problems in the same way that you do. Suppose that the pen you are writing with stops working. You might shake the pen. You might rub the point back and forth several times on your paper. You are trying to solve the problem by using **scientific method.** Scientific method is a model, or guide, used to solve problems.

Scientists do not have one scientific method. Every problem is different and needs to be solved in a different way. Scientists combine some or all of the science skills to solve different problems. They also use certain steps. The steps can be used in any order. You can write a laboratory report using these steps.

▐▶*Define:* What is scientific method?

model, or guide, used to solve problems

Work like a Scientist

- ▶ **Identify and State the Problem** Scientists often state a problem as a question.

- ▶ **Gather Information** Scientists read and communicate with one another. In this way, they learn about work that has already been done. A student may do library research to find out more about a problem.

- ▶ **State a Hypothesis** Scientists state clearly a hypothesis, which is a suggested solution to a problem.

- ▶ **Design an Experiment** To test a hypothesis, scientists design an experiment.

- ▶ **Make Observations and Record Data** During an experiment, scientists make careful observations. The information that they get is their **data** (DAY-tuh). Scientists also keep careful records of the data.

- ▶ **Organize and Analyze Data** Scientists organize their data. Scientists often use graphs, charts, tables, and diagrams to organize data. Then the data can be analyzed, or studied.

- ▶ **State a Conclusion** A conclusion is a summary that explains the data. It states whether or not the data support the hypothesis. It answers the question stated in the problem.

6

LESSON SUMMARY

▶ Scientific method is a model, or guide, used to solve problems.

▶ Scientists do not have one scientific method.

▶ The steps of scientific method are: Identify and state the problem; Gather information; State a hypothesis; Design an experiment; Make observations and record data; Organize and analyze data; State a conclusion.

CHECK *Complete the following.*

1. What is data?
2. Why do scientists design experiments?
3. What is scientific method?
4. How do scientists organize data?
5. How do scientists gather information?

APPLY *Complete the following.*

▶ 6. **Infer:** Why do you think it is important for scientists to organize their data?

7. Do you think it is necessary to use all parts of scientific method to solve a problem? Explain.

8. Scientists usually perform an experiment several times before stating a conclusion. Why do you think this is important?

Skill Builder ....

Writing a Laboratory Report Write a laboratory report describing the following experiment. Be sure to include the steps of scientific method in your report.

A shiny plastic balloon is taken into a warm room and filled with helium. It is then taken into a cold room. When the balloon is moved from the warm room into the cold room it becomes smaller. This happens because when helium is cooled it contracts, or shrinks.

Skill Builder ....

Organizing Data Conduct a survey to find out people's favorite television programs. Ask your family and friends to name their favorite television program. Be sure to ask people in different age groups. Organize the data you collect in the way that you think best. Be sure to give each table, graph, or diagram a title. Tables should have headings for each column.

ACTIVITY

CONTROLLING VARIABLES

You will need 2 beakers, salt, a stirrer, a tablespoon, ice, water, and a wax pencil.

A well-designed experiment tests only one variable at a time. A variable is anything that can affect the results of an experiment.

1. Using the wax pencil, label one of the beakers A and the other B.
2. Fill each beaker half full with cold water. Add three ice cubes to each beaker.
3. Stir a tablespoon of salt into Beaker A.
4. Observe how long it takes for the ice to melt.

Questions

1. **Compare:** In which beaker did the ice melt faster?
2. **Identify:** What was the variable being tested in this experiment?
3. **Analyze:** How does salt affect the rate at which ice melts?

1-4 How are length and area measured?

Objective ▸ Identify the SI and metric units used to measure length and area.

TechTerms

▸ **meter** (MEE-tur): basic SI and metric unit of length

▸ **unit** (YOU-nit): amount used to measure something

Scientific Measurements Measuring is an important part of science. The metric system is an international system of measurement. Most countries use the metric system of measurement. Their everyday measurements are given in metric **units** (YOU-nits). A unit is an amount used to measure something.

Since 1960, scientists have used a modern form of the metric system called Systeme International, or SI. Most of the units in SI are the same as metric units.

▸ *Define:* What is a unit? amount used to measure something

Changing Units Both the metric system and SI are based on units of ten. This makes these measurement systems easy to use. Each unit is ten times smaller or larger than the next unit. Prefixes are used to change the size of a unit.

Table 1	Prefixes and Their Meanings
PREFIX	MEANING
kilo-	one thousand (1000)
hecto-	one hundred (100)
deca-	ten (10)
deci-	one tenth (1/10 or 0.1)
centi-	one hundredth (1/100 or 0.01)
milli-	one thousandth (1/1000 or 0.001)

▸ *Analyze:* Use Table 1. What is the meaning of the prefix hecto-? one hundred

Units of Length In the metric system, length and distance are measured in **meters** (MEE-turs). The meter is the basic unit of length. A doorknob is about one meter from the floor. You can use prefixes to make longer or shorter lengths. For example, the prefix kilo- means one thousand. One kilometer is equal to 1000 meters. Table 2 shows how prefixes can be used to make different lengths. Table 2 also shows the symbols for each unit.

Table 2	Units of Length
1000 meters (m) = 1 kilometer (km)	
100 centimeters (cm) = 1 meter (m)	
1000 millimeters (mm) = 1 meter (m)	

▸ *Calculate:* How many millimeters are there in one centimeter? 10

Measuring Length and Area Length and distance can be measured with a meter stick. A meter stick is one meter long. A meter stick is divided into 100 equal lengths by numbered lines. The distance between each of these lines is equal to one centimeter. Each centimeter is divided into 10 equal parts. Each one of these parts is equal to one millimeter.

Suppose you wanted to wallpaper a room. How would you know how much wallpaper to buy? You would need to know the area of each wall. You can find the area of a rectangle by multiplying its length by its width. Area is measured in square units, such as square meters (m^2).

▸ *Explain:* How do you find the area of a rectangle? by multiplying its length by its width

LESSON SUMMARY

▶ The metric system is an international system of measurement.

▶ Scientists use SI.

▶ The metric system and SI are based on units of ten.

▶ The meter is the basic unit of length in the metric system and SI.

▶ A meter stick is used to measure length.

▶ The area of a rectangle is found by multiplying its length times its width.

CHECK *Complete the following.*

1. A _____ is an amount used to measure something. unit

2. In the metric system, the basic unit of length is the _____ . meter

3. Each unit in the metric system is _____ times smaller or larger than the next unit. ten

4. You can find the _____ of a rectangle by multiplying its length by its width. area

5. Scientists use a modern form of the metric system called _____ . SI

6. Most countries use the _____ system of measurement. metric

7. A meter stick is divided into _____ equal lengths by numbered lines. 100

APPLY *Complete the following.*

8. **Calculate:** What is the area of a rug that is 12 m long and 10 m wide?

9. Match each prefix with its correct meaning.
   1. deca-   a. one tenth
   2. milli-   b. one thousand
   3. centi-   c. one hundred
   4. deci-   d. one thousandth
   5. hecto-   e. ten
   6. kilo-   f. one hundredth

10. **Describe:** What are the advantages of using the metric system or SI?

Ideas in Action

IDEA: Most countries use the metric system.
ACTION: Imagine that you are traveling in a foreign country where the metric system is used. What are three reasons it would be helpful to know the metric system?

SCIENCE CONNECTION

NATIONAL INSTITUTE OF STANDARDS AND TECHNOLOGY

Standard measurements are important to avoid confusion in science, commerce, and industry. In the United States, it is the job of the National Institute of Standards and Technology (NIST) to establish standard measurements. The NIST was formed in 1901. It is part of the Department of Commerce.

The NIST has developed measurement standards for units of length, mass, and time. It also maintains standards of measurement for volume, temperature, light, color, electric energy, radioactivity, X-ray intensity, and sound.

The NIST is divided into different institutes, each with its own job. Among the institutes are the Institute for Basic Standards, the Institute for Materials Research, and the Institutes for Computer Sciences and Technology. If you would like more information about the NIST, write to National Institute of Standards and Technology, U.S. Department of Commerce, Gaithersburg, Maryland.

How are mass and weight measured?

Objectives ▶ Identify the SI unit of mass.
▶ Compare mass and weight.

TechTerms

▶ **kilogram** (KIL-uh-gram): basic unit of mass
▶ **mass:** amount of matter in an object
▶ **weight:** measure of the pull of gravity on an object

Mass The amount of matter in an object is its mass. There is more matter in a bag of potatoes than in a bag of popcorn. The bag of potatoes has more mass.

The basic unit of mass in the metric system is the **kilogram** (KIL-uh-gram)(kg). The gram (g) is a smaller unit of mass. Chemists often use grams when measuring chemicals. Remember that the prefix "kilo-" means 1000. There are 1000 g in 1 kg.

▶**Identify:** What is the basic metric unit of mass? kilogram

Mass and Weight Mass and **weight** are related, but they are not the same. Weight is a measure of the pull of gravity on an object. Gravity is a force of attraction between all objects. The force of gravity depends on the mass of the objects and how far apart they are. Objects with a large mass have a strong force of gravity. As objects move farther away from each other, the force of gravity between them becomes less. On earth, the force of gravity pulls all objects toward the center of the earth.

Weight can change because the pull of gravity is not the same everywhere. For example, the moon has less mass and less gravity than the earth. An astronaut on the moon would weigh less than on the earth. However, the astronaut's mass would not change. The mass of an object always remains the same.

▶**Describe:** What does weight measure?
pull of gravity on an object

Measuring Mass Mass is measured with an instrument called a balance. A balance works like a seesaw. It compares an unknown mass with a known mass. One kind of balance is a triple-beam balance. A triple-beam balance has a pan on which the object being measured is placed. It also has three beams. Weights, or riders, are moved along each beam until the object on the pan is balanced. Each rider gives a reading in grams. The mass of the object is equal to the total readings of all three riders.

Pointer

Pan   Riders (weights)   Beams

▶**Infer:** What is the known mass with which the unknown mass is compared on a triple-beam balance? total reading of the riders

LESSON SUMMARY

▶ Mass is the amount of matter in an object.

▶ The basic unit of mass in the metric system is the kilogram.

▶ Weight is a measure of the pull of gravity on an object.

▶ The weight of an object can change, but its mass remains the same.

▶ A balance is used to measure mass.

CHECK *Write true if the statement is true. If the statement is false, change the underlined term to make the statement true.*

1. The amount of matter in an object is its <u>weight</u>. mass

2. The basic unit of mass in the metric system is the <u>gram</u>. kilogram

3. A balance is used to measure <u>mass</u>. true

4. There are <u>100</u> grams in one kilogram. 1000

5. An astronaut would weigh <u>less</u> on the moon than on the earth. true

6. <u>Weight</u> is a force of attraction between all objects. gravity

APPLY *Complete the following.*

7. **Compare:** What is the difference between mass and weight?

8. **Calculate:** Your weight on the moon would be only about one-sixth your weight on the earth. How much would you weigh on the moon?

▶ 9. **Infer:** The mass of Mars is about one-tenth the mass of Earth. Would an astronaut weigh more on Mars or on Earth? Explain.

10. What unit would you use to measure the mass of a potato?

Designing an Experiment

Design an experiment to solve the problem.

PROBLEM: Does a metal's mass change when it is heated?

Your experiment should:

1. List the materials you need.

2. Identify safety precautions that should be followed.

3. List a step-by-step procedure.

4. Describe how you would record your data.

LOOKING BACK IN SCIENCE

WEIGHT MEASUREMENTS OF THE PAST

Throughout history, people have used many units for measuring weight. The first standard measurements of weight were made in terms of easily available materials. Weights were either stone or metal in various shapes and sizes. Some of the weights had their value carved on the top. In England, a unit of weight called a stone is still used. One stone is equal to 14 pounds.

The oldest known weights appear in records buried in ancient Egyptian graves. The records date back to 4000 B.C. The oldest records of weighing date back to about 2500 B.C. These records show stone weights being used on balances. Ancient Egyptians developed balances to weigh gold and grain.

The ancient Hebrews used a unit of weight called a talent. One talent was equal to the load that could be comfortably carried by an adult man. The Hebrews also used a unit called the drachma, which was equal to a handful.

How is volume measured?

Objective ▶ Explain how volume is measured.

TechTerms

▶ **liter** (LEE-tur): basic metric unit of volume

▶ **meniscus** (mi-NIS-kus): curved surface of a liquid in a graduated cylinder

▶ **volume:** amount of space something takes up

Volume The amount of space that an object takes up is called its **volume.** The volume of liquids is often measured in **liters** (LEE-turs). The liter (L) is the basic unit of volume in the metric system. Smaller volumes can be measured in milliliters (mL). There are 1000 mL in 1 L.

▶*Name:* What is the basic unit of volume in the metric system? liter

Measuring Volume A graduated cylinder is used to measure the volume of a liquid. A graduated cylinder is a glass tube that is marked with divisions to show the amount of liquid in it. It is like a measuring cup.

When a liquid is poured into a graduated cylinder, the surface of the liquid is curved. This curve is called the **meniscus** (mi-NIS-kus). To read the volume of a liquid, the surface of the liquid should be at your eye level. You should read the mark on the graduated cylinder closest to the bottom of the meniscus.

▶*Define:* What is the meniscus?
curved surface of a liquid

Cubic Centimeters Volume of a solid can be measured in cubic centimeters. Look at the cube. Each side is 1 cm long. The cube has a volume of one cubic centimeter (cm^3). One cubic centimeter is the same as one milliliter. A cube with sides that are 10 cm long has a volume of 1000 cm^3. This is equal to the volume of one liter. Now look at the drawing of the rectangle. You can find the volume of a cube or a rectangle by multiplying its length by its width by its height. The volume of the rectangle is 12 cm^3.

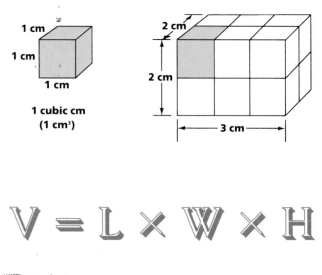

1 cm
1 cm
1 cm
1 cubic cm
(1 cm³)

2 cm
2 cm
3 cm

$$V = L \times W \times H$$

▶*Relate:* How many cubic centimeters is equal to 5 milliliters? 5

LESSON SUMMARY

▶ Volume is the amount of space an object takes up.

▶ A graduated cylinder is used to measure the volume of a liquid.

▶ The curved surface of a liquid is called the meniscus.

▶ Volume can be measured in cubic centimeters.

CHECK *Complete the following.*

1. The basic unit of volume in the metric system is the _____ . liter

2. There are _____ milliliters in one liter. 1000

3. You can find the volume of a rectangle by multiplying its length by its width by its _height_ .

4. The curved surface of a liquid is the _meniscus_ .

5. One cubic centimeter is the same as _____ milliliter. one

1 cm
2 cm
4 cm

APPLY *Complete the following.*

6. **Calculate:** What is the volume of the blue box in the diagram?

7. Use the photograph on page 12 to answer the following. What is the volume of the red liquid shown?

8. **Hypothesize:** How could you find the volume of an irregularly shaped object using a graduated cylinder?

Ideas in Action

IDEA: In the United States, liquids often are sold in liter bottles.

ACTION: Find three bottles that have liter measurements on their labels.

Skill Builder

Calculating Collect three different sized boxes. Measure the length, the width, and the height of each box. Calculate the volume of each box. Then determine how much liquid each box could hold. Organize your data in a table.

ACTIVITY

USING A GRADUATED CYLINDER

You will need a 100-mL graduated cylinder, water, a beaker, 3 different-sized test tubes, and a test tube rack.

1. Fill the beaker with water. Using the beaker, fill each of the test tubes with water.

2. Pour the water from one of the test tubes into the graduated cylinder.

3. Record the volume of the water. Be sure the graduated cylinder is at eye level when making your reading. Be sure to read the bottom of the meniscus.

4. Measure the volume of the water in the other test tubes.

5. Repeat steps 1 to 4 two times.

Questions

1. What is the volume of the water in each test tube?

2. **a.** Did you get the same readings for each of your three measurements? **b. Analyze:** Why might there be differences among measurements?

How is temperature measured?

Objective ▸ Explain how temperature is measured.

TechTerms

▸ **degree Celsius** (SEL-see-us): metric unit of temperature

▸ **temperature:** measure of how hot or cold something is

Temperature The **temperature** of anything is a measure of how hot or cold it is. It is a measure of how much heat energy something contains. For example, warm water contains more heat energy than the same mass of ice.

▌▶*Define:* What is temperature?
measure of how hot or cold something is

The Thermometer Temperature is measured with an instrument called a thermometer. Many thermometers are made of glass tubes. At the bottom of the tube is a wider part called the bulb. The bulb is filled with a liquid. Some liquids that can be used are mercury, colored alcohol, or colored water. All the air is removed from the thermometer tube and it is sealed. When the bulb is heated, the liquid in the bulb expands, or gets larger. It rises in the glass tube. When the bulb is cooled, the liquid contracts, or gets smaller. It falls in the tube. On the sides of the thermometer are a series of marks. You read the temperature by looking at the mark on the tube where the liquid stops.

▌▶*List:* What are some liquids that are used in thermometers? mercury, colored alcohol, colored water

Measuring Temperature Temperature is usually measured on one of two scales. They are the Fahrenheit (FER-un-hyt) scale and the Celsius (SEL-see-us) scale. The Fahrenheit (F) scale is used in the United States for everyday measurements. Most other countries use the Celsius (C) scale. The Celsius scale is usually used in science. Each unit on a temperature scale is called a degree. The **degree Celsius** (°C) is the metric unit of temperature.

Scientists working with very low temperatures use a different temperature scale. They use the Kelvin (K) scale. The Kelvin scale is part of SI. The Kelvin scale begins at absolute zero, or 0 K. At 0 K no more heat energy can be removed from matter. 0 K is the lowest possible temperature.

Table 1 Comparing Temperatures			
	°C	K	°F
Absolute zero	−273	0	−459
Freezing point of water	0	273	32
Room temperature	22	295	72
Human body temperature	37	310	98.6
Boiling point of water	100	373	212

◉*Observe:* What is the boiling point of water on the Celsius scale? 100 °C

LESSON SUMMARY

▶ Temperature is a measure of how hot or cold something is.

▶ Temperature is measured with a thermometer.

▶ Temperatures usually are measured on the Fahrenheit scale or the Celsius scale.

▶ Scientists working with very low temperatures use the Kelvin scale.

CHECK *Complete the following.*

1. What temperature scale is used in the United States for everyday measurements?

2. How do you read a thermometer?

3. What does temperature measure?

4. What happens when the bulb of a thermometer is heated?

5. What is the metric unit of temperature?

6. What is the lowest possible temperature?

APPLY *Use the Figures A, B, and C and the table on page 14 to answer the following questions.*

7. What is the temperature reading in Figure A? What is the temperature on the Fahrenheit scale? On the Kelvin scale?

8. What is the temperature reading in Figure B?

9. What is the temperature in Figure C?

Complete the following.

 10. **Analyze:** How do you change from Celsius to Kelvin?

Health and Safety Tip ....

Normal body temperature is 37 degrees Celsius. Your body temperature often rises when you are sick. Using library references, find out why your body temperature rises when you are sick, and why high fevers can be very dangerous. Write your findings in a report. Include a description of proper ways to bring a fever down.

ACTIVITY

CALIBRATING A THERMOMETER

You will need an unmarked thermometer, 2 beakers, a heat source, ice, a wax pencil, a ruler, and a standard Celsius thermometer.

1. Boil some water in a beaker. Place your unmarked thermometer in the beaker. Wait until the mercury rises as far as it will go. Mark this point.

2. Fill a beaker with ice water. Place the unmarked thermometer into this beaker. Wait until the mercury goes as low as it will go. Mark this point.

3. Using your ruler, divide the space between the marks into ten equal parts. Mark these divisions.

Questions

1. **a.** What is the temperature at which the mercury rose as high as it would go?   **b.** What is the temperature at which the mercury went as low as it would go?

2. **Analyze:** How many degrees are there between the marks on your thermometer?

UNIT 1 Challenges

STUDY HINT Before you begin the Unit Challenges, review the TechTerms and Lesson Summary for each lesson in this unit.

TechTerms

data (6)
degree Celsius (14)
hypothesis (4)
kilogram (10)
liter (12)

mass (10)
meniscus (12)
meter (8)
scientific method (6)
specialization (2)

temperature (14)
unit (8)
volume (12)
weight (10)

TechTerm Challenges

Matching *Write the TechTerm that matches each description.*

1. suggested solution to a problem hypothesis
2. basic unit of mass kilogram
3. amount used to measure something unit
4. studying or working in only one part of a subject specialization
5. curved surface of a liquid in a graduated cylinder meniscus
6. basic SI and metric unit of length meter

Identifying Word Relationships *Explain how the words in each pair are related. Write your answers in complete sentences.* Answers will vary. Accept all logical responses.

1. mass, weight
2. degree Celsius, temperature
3. data, scientific method
4. liter, volume

Content Challenges

Multiple Choice *Write the letter of the term or phrase that best completes each statement.*

1. The basic SI unit of length is the
   a. liter. **b.** meter. **c.** gram. **d.** centimeter. b

2. Scientists working with very low temperatures use the
   a. Kelvin scale. **b.** Celsius scale. **c.** Fahrenheit scale. **d.** Centigrade scale. a

3. The two main branches of physical science are
   a. chemistry and biology. **b.** physics and optics. **c.** chemistry and physics.
   d. mechanics and biochemistry. c

4. When you state ahead of time what you think will happen, you are
   a. inferring. **b.** concluding. **c.** measuring. **d.** predicting. d

5. Weight is a measure of the
   a. amount of matter in an object. **b.** amount of space something takes up.
   c. how hot or cold something is. **d.** pull of gravity on an object. d

6. The metric system is based on units of
   a. two. **b.** five. **c.** ten. **d.** one hundred. c

7. A suggested solution to a problem is a
   a. prediction. **b.** hypothesis. **c.** unit. **d.** conclusion. b

8. Area is measured in
   a. square units. b. liters. c. grams. d. cubic centimeters. a
9. Acoustics is the study of
   a. light. b. heat. c. sound. d. carbon. c
10. The prefix kilo- means
   a. one hundred. b. one tenth. c. one thousand. d. one thousandth. c
11. When the bulb of a thermometer is heated, the liquid in the bulb
   a. gets smaller. b. contracts. c. falls. d. gets larger. d
12. Physical science is the study of matter and
   a. energy. b. changes in matter. c. light. d. living things. a
13. Length can be measured with a
   a. balance. b. meter stick. c. thermometer. d. graduated cylinder. b
14. The metric unit of temperature is the
   a. degree Kelvin. b. degree Celsius. c. degree Fahrenheit. d. kilogram. b
15. The distance between each numbered line on a meter stick is equal to one
   a. kilometer. b. meter. c. centimeter. d. millimeter. c

Think like a Scientist

Completion *Write the term that best completes each statement.*

1. The information scientists gather during an experiment is called _____ . data
2. You can find the area of a rectangle by multiplying its length by its _____ . width
3. Everything around you is either matter or _____ . energy
4. Your senses are seeing, hearing, touching, smelling, and _____ . tasting
5. To read the volume of a liquid, the surface of the liquid should be at your _____ level. eye
6. Temperature is a measure of how much _____ energy something contains. heat
7. As objects move farther away from each other, the gravity between them becomes _____ . less
8. Two ways to do research are reading and _____ . experimenting
9. A _____ is a person who studies or works in only one part of a subject. specialist
10. In the United States, the _____ scale is used for everyday measurements. Fahrenheit

Understanding the Features

Reading Critically *Use the feature reading selections to answer the following. Page numbers for the features are shown in parentheses.*

1. **Analyze:** What problems do you think arose from using the drachma as a unit of weight? (11)
2. **List:** What are three institutes that are part of the National Institute of Standards and Technology? (9)
3. What does a scientific illustrator do? (3)
4. **Define:** What is cosmology? (5)

Concept Challenges

Interpreting a Table *Use the table to answer the following.*

Table 1 Airline Distances in the United States (km)					
	Chicago	Denver	Houston	Miami	St. Louis
Atlanta	970	1933	1102	952	774
Boston	1388	2827	2565	2013	1674
Dallas	1277	1046	347	1776	874
Kansas City	645	869	1029	1982	366
Seattle	752	1630	2998	4360	2734

1. What metric unit is used in the table?
2. **Observe:** How far is Miami from Boston?
3. What is the basic metric unit for length?
4. **Calculate:** How many meters is it from Kansas City to Chicago?
5. **Analyze:** Why are tables listing distances in kilometers usually not used in the United States?

Critical Thinking *Answer each of the following in complete sentences.*

1. Why are standards of measurement important in science?
2. **Name:** What are three areas in which people you come in contact with in your everyday life specialize?
3. **Calculate:** How many liters of liquid would fit in a carton that has a volume of 2000 cubic centimeters?
4. Why do scientists not have one scientific method?
5. How could you use temperatures to identify different substances?

Finding Out More

1. Temperature scales were named after the scientists who developed them. Using library references, find out about the life of Fahrenheit, Celsius, or Kelvin. Write a brief biography on the scientist you choose.
2. Using library references, find out about the problems and benefits of nuclear energy. Research the efforts being made to make nuclear energy a safe energy source. Present your findings in an oral report.
3. Find out about efforts in the United States to convert to the metric system. Present your findings to the class. Then have a class debate on whether or not the United States should convert to the metric system.

FORCE

CONTENTS

STUDY HINT Before studying Unit 2, scan through the lessons in the unit looking for words that you do not know. On a sheet of paper list these words. Work with a classmate to try to define each word on your list.

2-1 What is a force?

Objective ▶ Define force and give some examples of forces in nature.

TechTerm

▶ **force:** push or pull

Force A **force** is a push or a pull. When you open a door, you have to either push or pull the door. To open the door, you have to use a force. A force always acts in a certain direction. When you push a door, the force is in the direction of the push. When you pull on a rope, the force is in the direction of the pull. When you lift a book, the force is in the direction of the lift.

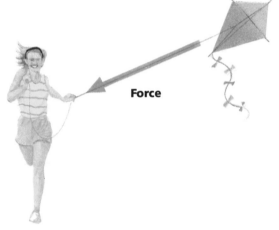

Force

▶**Define:** What is a force? a push or a pull

Forces in Nature There are many different kinds of forces that you can see in everyday life. Here are a few examples.

▶ If you walk across a carpeted floor and touch a doorknob, you will feel a small shock. This shock is caused by static electricity. Static electricity is one type of electrical force.

▶ If you hold a magnet near a metal paper clip, the paper clip will move toward the magnet. The force that pulls the paper clip toward the magnet is called a magnetic force.

▶ The water in a waterfall is pulled downward by a force. This force is called gravity.

▶**Identify:** When raindrops fall to the ground, what is the force pulling on the drops? gravity

Weight Weight is a force. The weight of an object tells you how heavy it is. In the diagram, the balance scale on the left has nothing on it. Each side of the scale is at the same position. When an object is placed on the left side of the scale, the object's weight pushes down on the left side of the scale. Suppose you put another weight on the right side of the scale. The force of the second weight will push down on the right side of the scale. If each side has the same amount of weight, the scale will be balanced by equal forces.

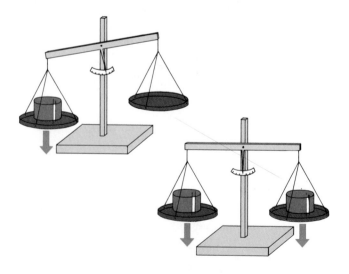

▶**Identify:** What is weight? a force

LESSON SUMMARY

▶ A force is a push or pull.

▶ A force always acts in a certain direction.

▶ There are many examples of forces in nature.

▶ Forces in nature include electrical forces, magnetic forces, and the force of gravity.

▶ Weight is a force.

CHECK *Complete the following.*

1. When you push or pull on something, you are using a _____ . force

2. A force always __acts__ in a specific direction.

3. When you walk across a carpeted floor and touch a metal doorknob, you will feel an _____ force. electrical

4. When a piece of metal moves toward a magnet, the metal is being pulled by a _____ force. magnetic

5. When an object falls to the ground, it is being pulled by the force of _____ . gravity

6. Weight is a _____ . force

7. A __heavy__ object requires more force to lift it.

APPLY *Complete the following.*

▶ 8. **Infer:** If you are flying a kite and pull on the string, in which direction is the force?

▶ 9. **Analyze:** Lift a pencil. Now lift your textbook. Which object needs more force to lift it? Why?

10. Suppose you push your textbook across your desk, and it falls over the edge of the desk to the floor. Describe the forces that make the book move.

State the Problem

Take a compass and a small magnet. Move the magnet around the compass. What happens to the compass needle as you move the magnet? What is the problem for this experiment?

Ideas in Action

Idea: Weight is a force.

Action: Look at two people on a seesaw. How does their weight affect the position of the seesaw?

ACTIVITY

INTERPRETING FORCE DIAGRAMS

You will need a metric ruler.

1. Look at Figure 1. The arrow represents the force used to push a chair across the floor. Use a metric ruler to measure the length of the arrow.

2. Look at Figure 2. It shows a second force helping to push the chair. Measure the total length of the two arrows.

3. Look at Figure 3. It shows a force pushing in the opposite direction. Measure the length of the arrow pointing in the opposite direction.

Questions

1. Force is measured in newtons (N). If 1 cm = 1 N, what force was used to push the chair in Figure 1?

2. What was the total force used to push the chair in Figure 2?

3. Look at Figure 3.   **a.** What force was used to push the chair in the opposite direction?   **b.** What total force was used to push the chair?

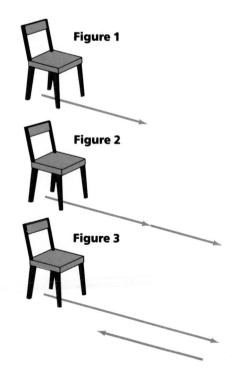

Figure 1

Figure 2

Figure 3

2-2 What is gravity?

Objective ▶ Explain Newton's law of gravity.

TechTerm

▶ **gravity:** force of attraction between all objects in the universe

Newton Isaac Newton was a famous scientist. He lived in England about 300 years ago. Newton wondered why objects fall to the ground. Newton discovered that there is a force that makes all objects move toward each other. This force is called **gravity.** All objects in the universe are attracted to one another because of the force of gravity between them. This idea is now known as Newton's law of gravity, or universal gravitation.

▨▶*Name:* What scientific law explains why objects fall? the law of gravity, or universal gravitation

Gravity Gravity is a force of attraction between all objects in the universe. On the earth, all objects fall to the ground because of the earth's gravity. When an apple falls to the ground, the earth's gravity pulls the apple in the direction of the center of the earth. No matter where on the earth you stand, an object falls in the direction of the earth's center when you drop it.

▨▶*Describe:* In which direction does an object fall on Earth? toward the earth's center

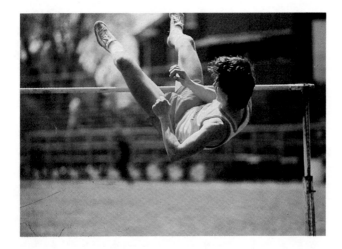

Gravity and Mass The amount of gravitational force between two objects depends on the mass of each object. When an apple falls to the ground, the earth's gravitational force pulls on the apple. The gravitational force of the apple also pulls on the earth. Because the earth has much more mass than the apple, the gravitational force of the earth makes the apple move.

▨▶*Explain:* Why is the force of the earth's gravity so strong? because of the earth's mass

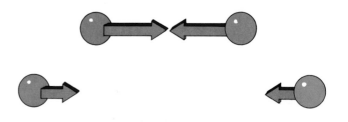

Gravity and Distance The force of gravity between two objects decreases as the distance between them increases. When you stand on the earth at sea level, the amount of gravitational force you feel is your weight. If you were far away from the earth, the gravitational force of the earth would be less. You would weigh less the farther away you were from the earth's surface. The force of gravity decreases by an amount equal to one divided by the distance (d) squared, or $1/d^2$. For example, if the moon were twice as far away from the earth, the pull of the earth's gravity on the moon would be $1/2^2$, or $1/4$.

▨▶*Calculate:* If you were twice as far from the center of the earth as you are now, how much would you weigh? 1/4 normal weight

LESSON SUMMARY

▶ Isaac Newton discovered the law of gravity.

▶ All objects in the universe move toward each other because of the force of gravity between them.

▶ The amount of gravitational force between two objects depends on their mass.

▶ The force of gravity between two objects decreases as the distance between them increases.

CHECK *Write true if the statement is true. If the statement is false, change the underlined term to make the statement true.*

1. Newton discovered the <u>law</u> of gravity. true

2. There is a force that makes all objects move <u>away</u> from each other. toward

3. All objects in the universe are attracted to each other because of the force of <u>gravity</u>. true

4. On the earth, all objects fall in the direction of the earth's <u>center</u>. true

5. The farther away you are from the earth the <u>more</u> you weigh. less

6. Gravity decreases as distance <u>increases</u>. true

APPLY *Complete the following.*

◀ 7. **Analyze:** Why does an apple fall toward the earth, instead of the earth moving toward the apple?

◀ 8. **Infer:** The sun has a much greater mass than the earth. What does this tell you about the force of gravity between the earth and the sun?

InfoSearch

Read the passage. Ask two questions about the topic that you cannot answer from the information in the passage.

Sir Isaac Newton Isaac Newton was born in England in 1652. Newton had a very long career in science. He made scientific discovers in areas such as light, color, and planetary orbits, as well as gravity. Newton also invented a new branch of mathematics called calculus.

Newton's most famous book was called the *Principia*. Because of his many contributions to science and mathematics, Newton was knighted in 1705. He died in 1727.

SEARCH: Use library references to find answers to your questions.

CAREER IN PHYSICAL SCIENCE

ASTRONAUT

An astronaut is trained to live and work in space. An astronaut does a lot of work many kilometers above the earth's surface. Since astronauts are in orbit around the earth with their spacecraft, they experience zero gravity. An astronaut has to learn how objects move in this new environment. An astronaut has to learn how to move around in zero gravity.

Most spacecraft are built for scientific purposes, but not always for comfort. Astronauts work in a very small space with several other astronauts. Part of their training is to get used to living and working in a cramped space with other people. A spacecraft is a complicated technological tool. Astronauts also have to learn how to operate the spacecraft. They must also know how to work the computers that control the craft.

If you think you might be interested in a career as an astronaut, you should write to the National Aeronautics and Space Administration (NASA), 600 Independence Avenue SW, Washington, DC 20546.

2-3 How does a spring scale work?

Objective ▶ Describe how a spring scale is used to measure weight.

TechTerm

▶ **newton:** metric unit of force

The Unit of Force In the metric system, there are different units for different types of measurements. For example, the basic unit of distance is the meter (m). The basic unit of mass is the gram (g). The **newton** (N) is the basic unit of force in the metric system. The unit is named in honor of Isaac Newton. On the earth, it takes a force equal to 9.8 N to lift a 1-kg mass.

▦▶*Identify:* What is the basic unit of force in the metric system? newton

Measuring Weight Weight is a force. An object's weight is a measure of the force of gravity on the object. When you weigh an object, you are measuring the pull of gravity on the object. Because weight is a force, an object's weight is measured in newtons. For example, an object that weighs 15 N is heavier than an object that weighs 10 N.

▦▶*Identify:* What are you measuring when you weigh an object? the force of gravity on the object

Using a Spring Scale A spring scale is used to measure weight. A spring scale measures the force of gravity on an object. The diagram shows the main parts of a spring scale.

To use the scale, attach a mass to the hook. The weight of the mass stretches the spring. The pointer moves down along the scale. The pointer stops at one of the numbers on the scale. That number is the object's weight. For example, suppose one mass moves the pointer to 2, and another mass moves the pointer to 1. The first mass weighs twice as much as the other mass.

▦▶*Identify:* What does a spring scale measure? weight

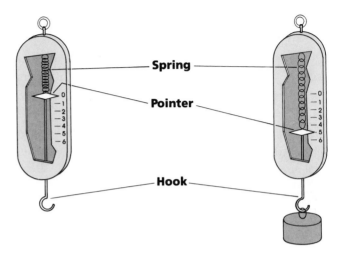

Types of Spring Scales There are many types of spring scales. A bathroom scale is a type of spring scale. When you stand on the scale, your weight pushes on a spring. The spring lines up the number on the circular scale with the pointer. Another type of spring scale is the one you see at the market. If you place several apples on the scale, the weight of the apples pulls a spring. The spring turns the pointer to show the weight of the apples.

▦▶*List:* What are two types of spring scale? bathroom scale and market scale

24

LESSON SUMMARY

▶ The newton (N) is the metric unit of force.

▶ Weight is a force that is measured in newtons.

▶ A spring scale can be used to measure weight.

▶ There are many types of spring scales.

CHECK *Complete the following.*

1. The metric unit of force is the _____ . newton

2. Weight is a _____ that is measured in newtons. force

3. The heavier an object, the _____ it weighs. more

4. A spring scale is used to measure _____ . weight or force

5. A bathroom scale is a kind of _____ . spring scale

6. When you put an object on a spring scale, the pointer shows the object's _____ . weight

APPLY *Complete the following.*

▶ 7. **Infer:** It takes 10 N to lift an object. What is the object's weight? How do you know?

8. **Identify:** What force is being measured when you use a spring scale to measure weight?

▶ 9. **Infer:** On the moon, an object weighs 1/6 of its weight on the earth. What does this tell you about the moon's gravity?

InfoSearch

Read the passage. Ask two questions about the topic that you cannot answer from the information in the passage.

Springs A spring stretches the same amount for each unit of force added to it. The amount that the spring stretches for each unit of force is called the spring constant. Different springs have different spring constants. When you add a weight to a spring, the spring stretches. When you remove the weight, the spring returns to its normal shape. If you stretch a spring too far, it may not return to its normal shape. The spring has been stretched beyond its elastic limit.

SEARCH: Use library references to find answers to your questions.

ACTIVITY

USING A SPRING SCALE

You will need a spring scale, a small mass, and several small objects.

1. Be sure that the pointer of the spring scale is at zero.

2. Carefully place the mass on the hook of the spring scale. Do not let the mass drop. This might damage the spring scale.

3. Observe where the pointer stops on the scale. If the pointer stops between two numbers, round off the number to the nearest half. Record your measurement.

4. Use the spring scale to weigh several small objects. Record and compare your measurements.

Questions

1. What is the weight of the mass in newtons?

2. What force are you measuring when you weigh an object?

3. How much does a 1-kg mass weigh in newtons?

2-4 What is air resistance?

Objective ▶ Explain how air resistance affects falling objects.

TechTerms

▶ **air resistance:** force that opposes the downward motion of a falling object
▶ **terminal velocity:** speed at which air resistance and gravity acting on a falling object are equal
▶ **vacuum:** empty space

Falling Objects The force that opposes the downward motion of falling objects is called **air resistance.** Suppose you have a ball that weighs 5 N and a sheet of paper that weighs 0.1 N. If you drop each object, which one will hit the ground first? As each object falls to the ground, air pushes up against the surface of each object. The sheet of paper has a larger surface than the ball. As a result, the air pushes the sheet of paper with more force. The ball hits the ground first.

A lighter object feels more air resistance than a heavier object. Air resistance slows the lighter object as it falls to the ground. The heavier object also feels air resistance, but not as much as the lighter object.

▶*Predict:* Which object will hit the ground first, a ball or a sheet of paper? ball

Terminal Velocity When you drop an object, it falls to the ground at a certain speed. The force of gravity pulls the object, causing its speed to change. When it is first dropped, the object has a speed of zero. As it falls, the object's speed increases at a steady rate. At some point, the upward force of air resistance is equal to the downward pull of gravity. At this point, the speed of the object stops increasing. This constant speed is the object's **terminal velocity.**

▶*Describe:* What happens to an object's speed as it falls? It increases at a steady rate.

Free Fall There is no air in a **vacuum.** A vacuum is empty space. In a vacuum, there is no air resistance. If a ball and a sheet of paper are dropped from the same height in a vacuum, they will hit the ground at the same time. There is no air resistance to slow the objects as they fall.

All objects fall at the same speed in a vacuum. When the Apollo astronauts landed on the moon, they tested this idea. There is no air on the moon. The astronauts dropped a feather and a hammer from the same height. What do you think happened? Both the hammer and the feather hit the ground at the same time!

▶*Define:* What is a vacuum? empty space

LESSON SUMMARY

► Air resistance is a force that opposes the downward motion of a falling object.

► When air resistance equals the pull of gravity, a falling object reaches its terminal velocity.

► A vacuum is empty space.

► In a vacuum, all objects fall at the same speed.

CHECK *Complete the following.*

1. When you drop an object, it falls with a certain _____ . speed

2. As an object falls, it will reach its _____ because of air resistance. terminal velocity

3. When you drop an object, _____ slows it down. air resistance

4. There is no air resistance in a _____ . vacuum

5. All objects fall at the same speed in a _____ . vacuum

APPLY *Complete the following.*

6. **Analyze:** A feather and a hammer are dropped to the earth's surface. Will they hit the earth at the same speed? Explain your answer?

7. **Hypothesize:** Certain birds spread out their wings before they land. Can you think of a reason why they would do this?

8. **Compare:** Describe what would happen if you dropped a hammer and a feather on the earth and on the moon.

9. **Predict:** What effect would opening a parachute have on a sky diver's fall?

· ·
Designing an Experiment · · · · · · · · · · · ·

Design an experiment to solve the problem.

PROBLEM: How can you show that air resistance slows down objects as they fall?

Your experiment should:

1. List the materials you would need.

2. Identify safety precautions that should be followed.

3. List a step-by-step procedure.

4. Describe how you would record your data.

◼◄◇▶◼◄ PEOPLE IN SCIENCE ▶◼◄◇▶◼◄◇▶◼◄◇▶◼◄◇▶◼◄◇▶◼◄◇▶◼◄◇▶◼◄◇▶◼◄◇▶◼◄◇▶◼◄◇▶◼◄◇▶◼

GALILEO GALILEI (1564–1642)

Galileo was born and educated in Italy. Many of his scientific theories challenged the beliefs of his day. Because of Galileo's theories about the earth, he was arrested. Galileo spent the last eight years of his life under house arrest.

Galileo made many contributions to science. He was one of the first scientists to use mathematics to describe the motion of objects. He proposed the idea that all objects fall with the same acceleration (ak-sel-uh-RAY-shun), or rate of change in speed. This idea later became an important part of Newton's theory of gravity. Galileo was the first person to use a telescope to study the stars and planets. He used it to see craters on the moon, the phases of Mercury, and the moons of Jupiter. Galileo was able to show that the Milky Way is made of stars.

Perhaps Galileo's most important contribution to science was his use of the experimental method. He introduced an organized and complete way of performing an experiment. Galileo served as a model for all later scientists.

Objective ▶ Identify examples of friction.

TechTerm

▶ **friction:** force that opposes the motion of an object

Forces and Motion To stop a moving object, a force must act in the opposite direction to the direction of motion. If you push your book across your desk, the book will move. The force of the push moves the book. As the book slides across the desk, it slows down and stops moving.

👁**Observe:** Give your textbook a slight push across your desk. What must you do to keep the book moving? keep applying a force, or keep pushing the book

Friction The force that opposes the motion of an object is called **friction.** Look at Figure 1. At first, the book is at rest. A push causes the book to slide across the desk. The force of the push (F) keeps the book moving. As the book slides across the desk, a force of friction (f) acts in the opposite direction. The friction slows down the motion of the book. Finally, the book is once again at rest.

Figure 1

▏▶**Define:** What is friction? force that opposes the motion of an object

Types of Friction There are different types of friction. A book moving across the desk is an example of sliding friction. As the book slides across the desk, the bottom of the book is touching the

desk. The source of the friction is the contact between the surface of the book and the desk. Air resistance is a type of friction. As an object falls, air resistance pushes up on the object. When you ride a bicycle, the contact between the wheel and the road is an example of rolling friction.

▏▶**State:** When does friction occur? when two objects are in contact

Uses of Friction On the earth, friction can be very helpful. When you ride a bicycle, the friction between the road and the bicycle wheels keeps the bike in motion. Without friction, you would not be able to move the bicycle. If you did move it, you would not be able to stop it. The same applies to cars, trains, and even your own two feet.

Figure 2

Sometimes friction is not helpful. For example, metal parts are always touching in a car's engine. Too much friction between the metal parts wears out the parts and overheats the engine.

▏▶**Explain:** Why is friction necessary? Without friction, motion would not be possible.

LESSON SUMMARY

▶ To stop a moving object, a force must act in the opposite direction.

▶ Friction is a force that opposes the motion of an object.

▶ There are different types of friction.

▶ Friction makes motion possible.

▶ Friction can sometimes be problem.

CHECK *Complete the following.*

1. A book sliding across a desk will come to a stop because of the force of _____ . friction

2. The book sliding across the desk is an example of _____ friction. sliding

3. The type of friction that acts on a falling object is called _____ . air resistance

4. Without friction, _____ would not be possible on the earth. motion

APPLY *Complete the following.*

▶ 5. **Infer:** When a car's tires are stuck in snow or mud, is it better to have more or less friction? Why?

6. **Hypothesize:** Sand is often placed on top of snow on roads and highways. Why do you think the sand is used?

7. **Classify:** Decide which of the following is an example of sliding friction, rolling friction, or air resistance. **a.** an airplane descending **b.** a person roller skating **c.** a person ice skating **d.** a falling leaf

8. Copy the diagram on a sheet of paper. The object is moving from left to right after being pushed with a 10-N force. There is a 3-N force of friction. Complete the diagram using arrows to show all the forces.

10N

3N

State the Problem

Study the illustration below. State the problem for this experiment.

F

LEISURE ACTIVITY

ICE SKATING

A popular winter sport in many parts of the country is ice skating. When you ice skate, sliding friction between the skates and the ice is very low and allows you to move across the ice. Too much friction would keep you from moving fast enough to keep your balance.

Would you like to try ice skating? The first thing you will need is a pair of ice skates. Unlike roller skates, ice skates have a thin piece of metal that touches the ice. This piece of metal is known as a blade. The blade is the only part of the skate that touches the ice.

If you go ice skating outdoors, there are some rules that you should follow. Always be sure that your skates are securely fastened. Start slowly and do not go beyond a speed that you are comfortable with. Never skate on a pond unless you know that the ice is thick enough to support your weight, and the weight of other skaters.

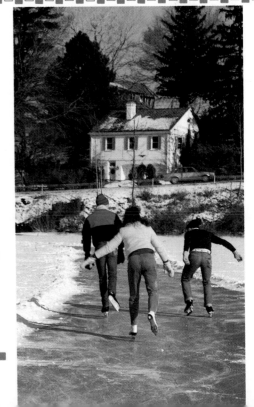

2-6 How can friction be reduced?

Objective ▶ Describe some ways to reduce friction.

TechTerm

▶ **lubricants** (LOO-bruh-kunts): substances that reduce friction

Moving Against Friction Friction makes it hard to move objects. Force is needed to overcome the force of friction. Suppose you wanted to push a heavy wooden box across the floor. As you push against the box, friction equal to a force of 5 N pushes in the opposite direction. This means that it will take a total force of 5 N to push the box.

▌▌▌▶ **Calculate:** If the force of friction is 16 N, how much force is needed to move the object?

16 N

Reducing Friction Reducing friction makes it easier to move an object. One way to reduce fric-

tion is to change sliding friction to rolling friction. When you try to push a wooden box, the bottom of the box is in contact with the floor. If you put the box on a wheeled cart, there will be much less friction. There is less friction because only part of each wheel is in contact with the floor. You can now use less force to push the box.

▌▌▌▶ **Explain:** Why must you reduce friction in order to move certain objects? so that less force will be needed to move the objects

Lubricants You can reduce friction by using **lubricants** (LOO-bruh-kunts). A lubricant is any substance that reduces friction. Lubricants reduce the amount of contact between two surfaces. For example, in a car's engine metal parts called pistons are in contact with other metal parts. When two pieces of metal touch, there is a lot of friction. Oil is used to reduce the friction between the metal parts. Oil is a lubricant. It makes the metal parts slide against each other. Without the oil, the metal parts would scrape against each other. This might cause the engine to overheat.

▌▌▌▶ **Identify:** When would it be helpful to use a lubricant? when friction needs to be reduced

LESSON SUMMARY

▶ Friction makes it harder to move objects because more force is needed to overcome the force of friction.

▶ Reducing friction makes it easier to move an object.

▶ One way to reduce friction is to change sliding friction to rolling friction.

▶ Another way to reduce friction is to use a lubricant.

CHECK *Complete the following.*

1. Because of friction, it takes _____ force to move an object. more

2. When friction is _____ , it is easier to move an object. reduced

3. Waxing a floor makes the floor slippery because wax is a _____ . lubricant

4. If you wear roller skates instead of shoes, you are _____ the friction between your feet and the floor. reducing

APPLY *Complete the following.*

5. Is it easier to push a chair across a carpeted floor or across a polished wooden floor? Why?

6. **Calculate:** Find the amount of force in each example. **a.** If it takes 20 N to push a wooden crate across the floor, what is the force of friction? **b.** By putting casters on the crate, the force of friction is reduced to 2 N. How much force is now needed to push the crate now?

7. On rainy days, the amount of friction between the road and the wheels of a car can sometimes be reduced by half. Why would reducing friction not be helpful in this case?

8. **Hypothesize:** A hydrofoil is a special type of boat. It skims over the surface of the water and can travel at very high speeds. Why do you think the hydrofoil can reach such high speeds?

Ideas in Action

Idea: Without friction, motion would not be possible.

Action: Watch what happens to pedestrians on wet or icy streets. Why do people walk more carefully? Why does it seem to take more effort to walk on a slippery sidewalk?

ACTIVITY

MEASURING FRICTION

You will need a spring scale, a small object, a long sheet of sandpaper, and tape.

1. Weigh the small object with the spring scale.
2. Place the object on a table top. Attach the spring scale to the object. Pull the object across the table. Record the amount of force shown on the spring scale.
3. Subtract the result in Step 1 from the result in Step 2. The difference is the force of friction for the table top.
4. Repeat Steps 1 to 3. This time use the sheet of sandpaper. Tape the sandpaper to the table top. Use the spring scale to pull the object across the sandpaper.

Questions

1. What was the force of friction for the table top?
2. What was the force of friction for the sandpaper?
3. Does a smooth surface have less friction or more friction than a rough surface?

2-7 What is pressure?

Objective ▶ Identify pressure as a force acting on a certain area.

TechTerms

- ▶ **fluid pressure:** pressure in gases and liquids
- ▶ **pressure:** force per unit area

Force and Area The amount of force acting on a surface is called **pressure.** Pressure is equal to force divided by area. Suppose you hold a book on the palm of your hand. If the book weighs 10 N, it presses down on your hand with 10 N of force. Suppose your hand has an area of 100 cm^2. The pressure caused by the weight of the book on your hand can be found by using the equation:

pressure = force/area
pressure = 10 N/100 cm^2
pressure = 0.1 N/cm^2

The pressure on your hand is 0.1 N/cm^2.

Weight of book = 10 N

Surface area of hand = 100 cm^2

▶**Calculate:** How much pressure is used when a 30-N force is applied over an area of 10 m^2?

3 N/m^2

Changing Pressure Pressure can be changed by changing the force. When the area stays the same, increasing the amount of force increases the pressure. Think of a pencil point. Press the point gently against the palm of your hand. You can feel the pressure on your hand. If you press a little harder,

More pressure

Less pressure

you increase the force. You can feel an increase in pressure.

Pressure can also be changed by changing the area. If you apply the same amount of force to a smaller area, you increase the pressure. Think of a pencil again. Press the pencil eraser against the palm of your hand. You can feel the pressure on your hand. Now press the pencil point against your hand. You can feel more pressure. The point has a smaller area than the eraser. The same force acting on a smaller area causes more pressure.

▶**List:** What are two ways to change the amount of pressure?

change the amount of force or the area

Fluid Pressure Pressure acts in gases and liquids, as well as solids. Pressure in gases and liquids is called **fluid pressure.** Gases and liquids are fluids. Water is a liquid. When you swim under water, you can feel the pressure of the water in your ears. Air is a mixture of gases. Air is all around you. At sea level, air pressure is about 10 N/cm^2.

▶**Define:** What is fluid pressure?

pressure in gases and liquids

LESSON SUMMARY

► Pressure is the amount of force acting on a surface.

► Pressure can be changed by changing the amount of force acting on an area.

► Pressure can be changing by changing the area on which a force acts.

► The pressure exerted by gases and liquids is called fluid pressure.

CHECK *Complete the following.*

1. Pressure is equal to force divided by _____ . area

2. Pressure can be measured in _____ /cm². N

3. You can decrease pressure by _____ the amount of force on the same area. decreasing

4. If you increase the area on which a force acts, you will _____ the pressure. decrease

5. The pressure exerted by gases and liquids is called _____ . fluid pressure

APPLY *Complete the following.*

▬ 6. **Analyze:** You can get a stronger spray of water from a garden hose if you make the opening of the hose smaller. Explain.

7. **Calculate:** Find the amount of pressure for each of the following examples. **a.** How much pressure is applied when a 50-N force acts over an area of 10 m²? **b.** What will happen to the pressure if the force is increased to 60 N? What is the new pressure? **c.** How could you decrease the pressure from 6 N/m² to 3 N/m²?

Skill Builder . . . .

Measuring When you measure, you compare an unknown quantity with a known quantity. You can use the equation pressure = force/area to find the force. To find force, rearrange the equation as follows:

$$\text{force} = \text{pressure} \times \text{area}$$

Use this equation to find the force for each of these examples:

1. A pressure of 12 N/m² is applied over an area of 10 m².

2. The contents of a spray can are at a pressure of 15 N/m². The opening of the spray nozzle has an area of 0.2 m².

3. A pressure of 25 N/m² is applied to a nail. The area of contact between the nail and the wood is 0.32 m².

SCIENCE CONNECTION ◆○◆○◆○◆○◆○◆○◆○◆○◆○◆○◆○◆○◆○◆○◆○◆○◆○◆

HYDRAULICS

If two containers hold the same amount of liquid, the narrow container has more pressure. Suppose two containers are connected by a pipe. If you apply pressure on the narrow container, the water will rise in the wide container. The pressure on both sides is the same. But the amount of force is greater in the wide container. Why?

The wider container has a larger area. Remember that force = pressure × area. Because the wider container has a larger area, it will feel more force.

Using fluid pressure to increase a force is the idea behind the study of hydraulics. A hydraulic lift works in this way. By applying a small force at one end, a larger force is obtained at the other end. For example, a barber's chair uses a hydraulic lift. The barber presses a pedal with a small force. This small force makes the chair rise even when someone is sitting on it. With a small force, even a heavy person can be lifted.

There are many uses for hydraulic lifts. What other devices can you name that use hydraulics?

2-8 ▶ How is fluid pressure measured?

Objective ▶ Describe how manometers and barometers are used to measure fluid pressure.

TechTerms

- ▶ **barometer** (buh-RAHM-uh-tur): instrument used to measure air pressure
- ▶ **manometer** (muh-NAHM-uh-tur): instrument used to measure pressure in a liquid

Air Pressure Air pressure is caused by the weight of particles in the air. You are not aware of the weight of the air around you. You do not feel the weight of the air because the pressure inside your body is equal to the air pressure around you. Air pressure changes with altitude, or height above sea level. There are fewer air particles at higher altitudes. As a result, the air pressure is lower. The higher you are above the ground, the lower the air pressure.

Low pressure

High pressure

▶*Compare:* Is the air pressure higher at sea level or on top of a mountain? at sea level

Measuring Air Pressure Air pressure is measured with a **barometer** (buh-RAHM-uh-tur). In a barometer, air pressure holds up a column of liquid. Because air pressure changes with altitude, barometers can also be used to find altitude.

▶*Identify:* What does a barometer measure? air pressure

Water Pressure Water exerts pressure because of the weight of the water particles. If you place an object in a container of water, the water applies pressure to the object. The force of the water pressure on the object is the same in all directions.

Low pressure

High pressure

Water pressure changes with depth. There is more pressure on a submarine the deeper it goes under water. At certain depths, the pressure from the water is so great that it can crush an object. Submarines are built to withstand a great deal of force from water pressure.

When you dive to the bottom of a swimming pool, you can feel the water pressure against your ears. The pressure of the water is greater than the pressure inside your body.

▶*Describe:* How does water pressure change the deeper an object is in the water?
The pressure increases.

Measuring Water Pressure The instrument used to measure water pressure, or the pressure of any liquid, is a **manometer** (muh-NAHM-uh-tur). A manometer is made up of a U-shaped tube and a piece of rubber tubing. The U-shaped tube is filled with colored water. When the end of the rubber tubing is placed under water, the water in the U-shaped tube rises. The greater the water pressure, the higher the water rises in the tube.

▶*Identify:* What does a manometer measure?
pressure in liquids

34

LESSON SUMMARY

▶ Air pressure is caused by the weight of particles in the air.

▶ Air pressure changes with altitude.

▶ A barometer measures air pressure.

▶ Water pressure changes with depth.

▶ A manometer measures the pressure in liquids.

CHECK *Complete the following.*

1. Air pressure is caused by the _____ of particles in the air. weight

2. A barometer is used to measure _____ . air pressure

3. The higher the altitude, the _____ the air pressure. lower

4. The pressure in a liquid is measured with a _____ . manometer

5. Water pressure _____ as the depth increases. increases

APPLY *Complete the following.*

6. As you climb a mountain, what happens to the air pressure?

7. What happens to a submarine as it descends?

8. **Hypothesize:** Standard air pressure at sea level is about 10 N/cm². This means that a force of 10 N is pushing on every square centimeter of your body. Why do you not feel this force?

Health and Safety Tip

At high altitudes, there is less air pressure. If you fly in an airplane, you can feel the change in the pressure as the plane takes off or descends for a landing. You can feel the pressure as a popping in your ears. The pressure inside the plane is the same as the pressure on the surface of the earth. However, your ears are very sensitive to changes in pressure. Find out what you should do to avoid the discomfort in your ears.

SCIENCE CONNECTION

THE "BENDS"

The deeper you dive under water, the more the water pressure increases. Deep sea divers often reach very great depths. They can stay under water for a long time because they have air tanks in order to breathe.

The longer a diver stays under water, the more his or her body gets used to the increased pressure. The air the diver breathes is also under increased pressure. When the diver comes up to the surface, the pressure decreases. If the diver comes to the surface too fast, there is too much of a change in pressure. Gas bubbles may get trapped in the bloodstream and in other parts of the body. This is known as decompression sickness, or the "bends." The bends can cause serious damage to the body. To avoid the bends, divers must return to the surface slowly. They must change the pressure slowly, so that their bodies can adjust to the change.

Divers are sometimes put into decompression chambers to prevent the bends. They remain in these sealed chambers until their bodies adapt to normal air pressure.

2-9 What is Bernoulli's principle?

Objective ▶ Explain how Bernoulli's principle is related to flight.

TechTerm

▶ **Bernoulli's principle:** principle that states that as the speed of a fluid increases, its pressure decreases

Bernoulli's Principle You know that fluids, or gases and liquids, exert pressure. **Bernoulli's principle** states that as the speed of a fluid increases, its pressure decreases. You can try a simple experiment to show Bernoulli's principle. Hold a sheet of paper in front of you. Blow over the top of the paper. What happens to the sheet of paper? The sheet of paper will rise.

Figure 1

▶ *Define:* What is Bernoulli's principle?
As the speed of a fluid increases, its pressure decreases.

Air Pressure and Wing Shape Airplane wings are designed to use Bernoulli's principle. Look at the airplane wing in Figure 2. Air goes over and under the wing. Because of the wing's curved shape, the air moving over the top of the wing speeds up. According to Bernoulli's principle, there is less pressure on the top of the wing than on the bottom. Normal air pressure under the wing pushes up on the wing. This unbalanced force pushes the wing up.

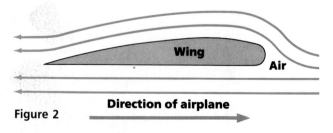

Figure 2

▶ *Identify:* What force pushes up on an airplane wing? air pressure

Flight Three forces combine to make an airplane fly. Figure 3 shows a side view of an airplane. The arrows show the directions of the three forces.

Figure 3

The upward force on the bottom of the airplane's wings is called the lift. This force pushes the wings up. In normal flight, the lift is equal to the weight of the airplane.

The forward force on the plane is called thrust. The normal flow of air over the wings cannot give an airplane enough lift to take off. An airplane must reach a certain speed and create enough air flow around the wings before it can fly. The plane's engines provide the extra force, or thrust, needed.

The flow of air over the wings is a form of air resistance. This frictional force is called drag. Drag slows an airplane down. Too much drag causes the plane to use too much fuel. Airplane designers build planes that will reduce drag. The process of reducing drag is called streamlining.

▶ *Identify:* What is the forward force a plane needs in order to fly? thrust

LESSON SUMMARY

▶ Bernoulli's principle says that as the speed of a fluid increases, its pressure decreases.

▶ Airplane wings are designed to use Bernoulli's principle.

▶ Three forces combine to help an airplane fly.

▶ The upward force on a plane's wings is called lift.

▶ A forward force, or thrust, helps the plane take off.

▶ The air resistance on a plane is called drag.

CHECK *Complete the following.*

1. Like all fluids, the water flowing in pipes has _____ . pressure

2. The faster the water flows through the pipes, the __lower__ the pressure.

3. The pressure on the bottom of an airplane wing is __greater__ than the pressure on the top.

4. A streamlined airplane will feel less __drag__ .

5. A plane's engines provide enough _____ to lift the plane off the ground. thrust

APPLY *Complete the following.*

6. A plane flying into a stiff wind feels more drag than in calm air. Explain.

7. **Hypothesize:** The jet stream is a fast-moving stream of air found at high altitudes. Airplane pilots like to fly their planes in the direction of the jet stream. How does the jet stream help the plane?

8. When a plane lands, the flaps under the wings open up to give the plane more drag. Why is this necessary?

Ideas in Action

Idea: As the speed of a fluid increases, the pressure decreases.

Action: In your home, what do you notice when several faucets are on at the same time? Is there an increase or a decrease in water pressure? What about when a washing machine, a dishwasher, and other appliances that use water are on at the same time? How do they affect the water pressure in your home?

CAREER IN PHYSICAL SCIENCE

AUTOMOBILE DESIGNER

An automobile designer is like an artist. A car's design has to be attractive enough so that people want to buy the car. Over the years, there have been many different types of car designs. In the past, a car's design was more important than performance. Today, the car's performance is just as important as its design.

An automobile designer has to know about modern technology. A car is a technological tool. It is constantly being updated. The fuel efficiency, the safety, and the life of a car have become important issues for car manufacturers. A car designer does not build the engine of the car. However, the design of the car can affect engine performance.

One of the main concerns of automobile designers is to create a streamlined car. Because air resistance can make a car use more fuel, a streamlined design can save fuel. However, the design must also be attractive. Finding a balance between what are called "form and function" is part of the challenge of being an automobile designer.

STUDY HINT Before you begin the Unit Challenges, review the TechTerms and Lesson Summary for each lesson in this unit.

TechTerms

barometer (34)
Bernoulli's principle (36)
fluid pressure (32)
force (20)

friction (28)
gravity (22)
lubricants (30)
manometer (34)

newton (24)
pressure (32)
terminal velocity (26)
vacuum (26)

TechTerm Challenges

Matching *Write the TechTerm that matches each description.*

1. metric unit of force newton
2. substances that reduce friction lubricants
3. empty space vacuum
4. pressure in gases and liquids fluid pressure
5. force divided by area pressure
6. instrument for measuring air pressure barometer
7. push or pull force

Low pressure

High pressure

Fill in *Write the TechTerm that best completes each statement.*

1. When you walk across the floor, there is _____ between your shoes and the floor. friction
2. As water flows through pipes, the pressure on the pipes from the water is decreased. This is an example of _____ . Bernoulli's principle
3. When skydivers jump from an airplane, they will reach a _____ before landing. terminal velocity
4. Objects fall to the ground because of the force of _____ . gravity
5. The pressure in a container of water is measured with a _____ . manometer
6. The earth's force of _____ pulls objects in the direction of the earth's center. gravity

Content Challenges

Multiple Choice *Write the letter of the term or phrase that best completes each statement.*

1. The wheels of a bicycle move over the surface of the road because of
   a. sliding friction. **b.** gravity. **c.** rolling friction. **d.** air resistance.

2. The electric current in a wire is pushed by
   a. a magnetic force. **b.** an electric force. **c.** a gravitational force. **d.** a force of friction.   b

3. If it takes 25 N to push a car, the friction between the road and the car's tires is
   a. 20 N. **b.** 25 N. **c.** 15 N. **d.** 5 N. b

4. A hydraulic lift uses the force from
   a. weight. **b.** fluid pressure. **c.** air pressure. **d.** friction. b

5. A submarine rising to the surface of the water from deep in the ocean is going from
   a. low pressure to high pressure.   b. high pressure to low pressure.
   c. zero pressure to high pressure.   d. high pressure to zero pressure. b

6. A ball rolls off a table and hits the floor. The force that caused the ball to hit the floor is
   a. gravity.   b. friction.   c. air pressure.   d. a magnetic force. a

7. As an airplane lands, it lowers the flaps under the wings in order to
   a. increase air resistance.   b. decrease air resistance.
   c. increase the plane's weight.   d. decrease the plane's weight. a

8. A manometer is used to measure
   a. friction.   b. air pressure.   c. gravity.   d. water pressure. d

9. A barometer is used to measure
   a. friction.   b. air pressure.   c. gravity.   d. water pressure. b

10. **An airplane can fly because the pressure on the plane's wings is**
    a. greater on the top of the wings.   b. zero.
    c. greater on the bottom of the wings.   d. equal on both sides of the wings. c

True/False *Write true if the statement is true. If the statement is false, change the underlined term to make the statement true.*

1. Weight is a <u>force</u>. true

2. Pressure is equal to force <u>times</u> area. divided by

3. On the earth, all objects fall in the direction of the earth's <u>surface</u>. center

4. Air pressure <u>decreases</u> at higher altitude. true

5. Sliding a chair across the floor is an example of <u>rolling</u> friction. sliding

6. Water pressure <u>decreases</u> the deeper you go into the water. increases

7. Riding a bicycle is an example of <u>sliding</u> friction. rolling

8. A spring scale measures <u>weight</u>. true

9. Air resistance <u>slows</u> the speed of a falling object. true

10. A lubricant reduces <u>pressure</u>. friction

11. Friction <u>slows down</u> the motion of objects. true

12. <u>Fluid</u> pressure is measured with a manometer. true

Understanding the Features

Reading Critically *Use the feature reading selections to answer the following. Page numbers for the features are shown in parentheses.*

1. Why does an astronaut work under changed conditions? (23)

2. What did Galileo use to show that the Milky Way is made of stars? (27)

3. What kind of friction is used in ice skating? (29)

4. A car's shock absorbers make a car ride more comfortable by absorbing the force from bumps on the road. Use the principle of hydraulics to explain how shock absorbers work. (33)

5. Why must you rise to the surface slowly after diving into deep water? (35)

6. Why do designers try to decrease a car's air resistance? (37)

Concept Challenges

Interpreting a Diagram *Use the following force diagram showing a crate sliding down a ramp to complete the following.*

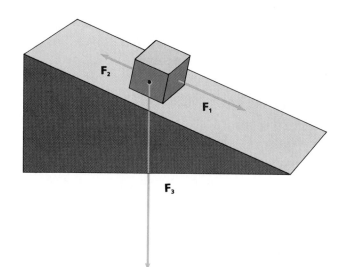

1. Which force represents the weight of the crate?
2. Which force represents the force pushing the crate down the ramp?
3. Which force represents the force of sliding friction that opposes the motion of the crate?

Let 1 cm = 10 N. Use a metric ruler to measure the length of each arrow representing a force.

4. How much does the crate weigh?
5. What is the force of friction?
6. How large is the force pushing the crate down the ramp?

Critical Thinking *Answer each of the following in complete sentences.*

1. How is the atmosphere on the surface of the moon similar to a vacuum?

2. What effect does altitude have on air pressure? Explain.

3. **Hypothesize:** Why would it not be practical to live in a world without friction?
4. **Predict:** Will you exert more pressure when you stand on your toes or when you stand on your feet? Explain.

Finding Out More

1. There are different metric units for measuring pressure. Use library references to find out how the following units are used: pascals; millimeters of mercury.
2. Visit an automobile tire dealer to find out about different types of tires. Here are some questions you might ask: How do the tire grooves affect friction? What kind of materials are used to make tires? How do snow and rain affect the tires? What are all-weather tires? Do they have more or less friction with the road than other tires?
3. Research the effects of the moon's gravity on the earth. How are the earth's oceans affected by the moon's gravity?

4. Decompression sickness, or the bends, was first discovered during the building of the Brooklyn Bridge in 1888. How could building a bridge cause the bends? Use library references to answer this question. Write a report of your findings.

ENERGY AND WORK

CONTENTS

STUDY HINT Before beginning Unit 3, scan through the lessons in the unit looking for words that you do not know. On a sheet of paper, list these words. Work with a class-mate to try to define each word on your list.

What are the two basic kinds of energy?

Objective ► Compare potential energy and kinetic energy.

TechTerms

- ► **energy:** ability to do work
- ► **kinetic** (ki-NET-ik) **energy:** energy of motion
- ► **potential** (puh-TEN-shul) **energy:** stored energy

Energy The water in a waterfall has **energy.** Energy is the ability to do work. The energy of the water rushing over a waterfall can be used to make electricity. The electricity can be used to heat and light your home.

Energy is all around you. There are two basic kinds of energy. They are potential (puh-TEN-shul) energy and kinetic (ki-NET-ik) energy.

▐▐▶ *Define:* What is energy? the ability to do work

Potential Energy A match has **potential energy.** Potential energy is stored energy. The potential energy in a match is stored in the chemicals in the match head. When you strike a match, you release this potential energy.

An object that is raised above the ground has potential energy. When the object falls, the energy is released. This kind of energy is called gravitational potential energy. Gravitational potential energy is also called energy of position.

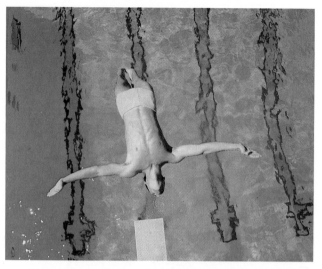

Gravitational potential energy depends on weight and height. The heavier an object is, the more gravitational potential energy it has. The higher an object is above the ground, the more gravitational potential energy it has.

▐▐▶ *Identify:* What kind of energy is released when you strike a match? potential energy

Kinetic Energy A moving object has **kinetic energy.** Kinetic energy is energy of motion. All moving objects have kinetic energy. When you walk or run, you have kinetic energy. You have more kinetic energy when you run than when you walk. The faster you run, the more kinetic energy you have. Kinetic energy also depends on mass. The more mass an object has, the more kinetic energy it has.

▐▐▶ *List:* What two factors determine how much kinetic energy a moving object has? speed and mass

42

LESSON SUMMARY

▶ Energy is the ability to do work.

▶ There are two basic kinds of energy.

▶ Potential energy is stored energy.

▶ An object that is raised above the ground has gravitational potential energy.

▶ Gravitational potential energy depends on weight and height.

▶ Kinetic energy is energy of motion.

CHECK *Complete the following.*

1. The ability to do work is _____ . energy

2. Two kinds of energy are potential energy and _____ energy. kinetic

3. Stored energy is _____ energy. potential

4. Gravitational potential energy is energy of _____ . position

5. Kinetic energy is energy of _____ . motion

6. A diver on a diving board has _____ potential energy. gravitational

APPLY *Complete the following.*

7. **Compare:** What is the difference between potential energy and kinetic energy?

8. In which drawing below does the rock have potential energy? In which does the rock have both potential and kinetic engery? Explain.

A   B

Skill Builder ....

Measuring When you measure, you compare an unknown value with a known value. Potential energy is found using the following formula:

$$PE = weight \times height$$

Remember that weight is measured in newtons (N); height is measured in meters (m). Therefore, potential energy is measured in units called newton-meters, or N-m. Use the formula to find the gravitational potential energy of each of the following objects: a 50-N brick on top of a 4-m wall; a 440-N student standing on a 2-m ladder; a 780-N diver standing on a 10-m diving board.

LEISURE ACTIVITY

ARCHERY

Millions of people in the United States take part in the sport of archery (AHR-chur-ee). Archery consists of target shooting with a bow and arrow. Archery is very popular in many schools and summer camps.

The basic equipment for an archery competition includes a curved bow. Most bows are made of wood and fiberglass. A bowstring is attached to each end of the bow. An arrow made of wood or lightweight aluminum is fitted into the bowstring. When the archer draws back the arrow, the bow has great potential energy. The greater the potential energy of the bow, the farther the arrow will travel. Targets are placed at different distances, from 30 m to 90 m. Archers must also wear safety equipment. This equipment includes an arm guard and a special glove to protect the fingers.

▶ Never point a drawn bow at any object other than the target.

▶ Never shoot an arrow when people or animals are in the area.

▶ Never shoot an arrow straight up.

What are different forms of energy?

Objective ▶ Identify and describe different forms of energy.

Forms of Energy Your body gets energy from the food you eat. An automobile uses the energy in gasoline to make it move. A clock spring stores energy to turn the hands of the clock. These are some examples of different forms of energy. There are five main forms of energy. They are mechanical energy, electromagnetic (i-lek-troh-mag-NET-ik) energy, heat energy, chemical energy, and nuclear (NOO-klee-ur) energy.

▶ **List:** What are the five main forms of energy?
mechanical, electromagnetic, heat, chemical, nuclear

▶ **Mechanical Energy** The energy in moving things is mechanical energy. Wind, moving water, and falling rocks all have mechanical energy. When you walk, run, or ride a bicycle, you are using mechanical energy. Sound is a form of mechanical energy.

▶ **Electromagnetic Energy** Moving electrons have electromagnetic energy. Electricity and light are both forms of electromagnetic energy. Radios, television sets, refrigerators, and light bulbs all use electromagnetic energy.

▶ **Heat Energy** If you rub your hands together, they become warm. Heat energy is the energy of moving particles of matter. The faster the particles move, the more heat energy they have. All things contain some heat energy.

▶ **Chemical Energy** The energy that holds particles of matter together is chemical energy. The energy stored in a match head is chemical energy. The energy in fuels such as wood or coal is chemical energy.

▶ **Nuclear Energy** Nuclear energy is the energy stored in the nucleus of the atom. When the nucleus is split, the energy is released as heat and light. Nuclear energy also is released when the nuclei of atoms combine. The heat and light from the sun are produced from nuclear energy.

LESSON SUMMARY

▶ There are five main forms of energy.

▶ Mechanical energy is the energy in moving things.

▶ Electromagnetic energy is the energy in moving electrons.

▶ Heat energy is the energy in moving particles of matter.

▶ Chemical energy is the energy that holds particles of matter together.

▶ Nuclear energy is the energy stored in the nuclei of atoms.

CHECK *Complete the following.*

1. Where does your body get energy from? food

2. How many basic forms of energy are there? five

3. What is mechanical energy? energy in moving things

4. What kind of energy is light? electromagnetic

5. What kind of energy is produced when you rub your hands together? heat

6. What kind of energy is released when wood burns? chemical

7. What is nuclear energy? energy stored in the nuclei of atoms

APPLY *Complete the following.*

8. **Classify:** Of the five basic forms of energy, which forms are potential energy? Which are kinetic energy? Explain your answers.

9. **Identify** each of the objects as a source of mechanical, electromagnetic, heat, chemical, or nuclear energy. Some of the objects may be sources of more than one form of energy. Explain your answers.
   a. gasoline   d. dynamite explosion
   b. burning wood   e. river
   c. lightning   f. sun

Designing an Experiment............

Design an experiment to solve the problem.

PROBLEM: How can you show that sound is a form of mechanical energy?

Your experiment should:

1. List the materials you would need.

2. Identify safety precautions that should be followed.

3. List a step-by-step procedure.

4. Describe how you would record your data.

TECHNOLOGY AND SOCIETY

NUCLEAR REACTORS

Nuclear energy can be released in two types of nuclear reactions. In one reaction, a large atomic nucleus is split into smaller nuclei. This is called nuclear fission. In the other reaction, two small nuclei are joined to form a larger nucleus. This is nuclear fusion. In both fission and fusion reactions, a great deal of energy is released. This energy can be used to produce electricity. Today, only nuclear fission is used as a source of energy. Scientists are still trying to find ways to control nuclear fusion reactions.

Nuclear energy is used to produce electricity in nuclear reactors. The fuel used in most nuclear reactors is uranium. Carefully controlled fission reactions in the uranium fuel release large amounts of heat energy. This heat energy is then used to make steam. The steam turns turbines and generates electricity.

Nuclear reactors have some safety problems. One major problem is that they produce dangerous waste products. For this reason, many people are against using nuclear reactors to produce electricity. What do you think?

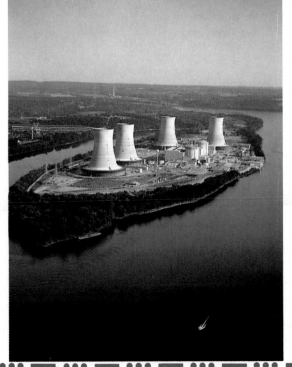

How does energy change form?

Objective ▶ Identify and give examples of energy changing form.

TechTerm

▶ **thermal** (THUR-mul) **pollution:** damage that occurs when waste heat enters the environment

Changing Potential and Kinetic Energy

Energy can change from one form to another. Potential energy and kinetic energy are always changing form. Think of a pendulum (PEN-joo-lum). Figure 1 shows a swinging pendulum. As the pendulum swings, potential energy is changed into kinetic energy and back into potential energy. The pendulum has the greatest amount of potential energy at the top of its swing. It has the greatest amount of kinetic energy at the bottom of its swing.

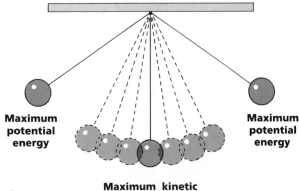

Maximum potential energy

Maximum potential energy

Maximum kinetic energy

Figure 1

◀ *Analyze:* When does a swinging pendulum have the least amount of kinetic energy? at the top of its swing

Changing Forms of Energy

You can observe many examples of changing forms of energy all around you. When you turn on an electric light, electrical energy is changed into light energy and heat energy. When you start an automobile, the engine changes the chemical energy in gasoline into mechanical energy. Nuclear reactors change nuclear energy into electrical energy. Your muscles change the chemical energy in food into mechanical energy.

Figure 2

▶ *Identify:* What energy change takes place when you turn on an electric light? Electrical energy is changed into light and heat.

Waste Heat

When energy changes form, some of the energy is always changed into heat. Most of this heat energy is wasted. When waste heat energy escapes into the environment, it causes **thermal** (THUR-mul) **pollution.** For example, the water in lakes and rivers is used to remove waste heat from power plants. The waste heat makes the water warmer. The water may become too warm for living things. If the water gets too warm, fishes in the lakes and rivers may die.

▶ *Define:* What is thermal pollution? damage caused when waste heat escapes into the environment

LESSON SUMMARY

▶ Energy can change from one form to another.

▶ Energy changes can be observed all around you.

▶ When energy changes form, some of the energy is always changed into heat energy.

CHECK *Complete the following.*

1. A pendulum has the greatest amount of <u>potential</u> energy at the top of its swing.

2. When a pendulum is at the _____ of its swing, it has the greatest amount of kinetic energy. bottom

3. When you turn on a light, electrical energy is changed into light and _____ . heat

4. An automobile engine changes _____ energy into mechanical energy. chemical

5. The _____ energy in food is changed into mechanical energy by your muscles. chemical

6. A nuclear reactor changes nuclear energy into _____ energy. electrical

7. When energy changes form, _____ energy is always produced. heat

APPLY *Complete the following.*

8. Describe how heat energy is wasted in each of the following: **a.** automobiles; **b.** television sets; **c.** electric lights.

9. **Interpret:** Look at Figure 2 on page 46. Identify and describe the different examples of energy changing form shown in the picture.

Ideas in Action ....

IDEA: Many appliances in your house probably waste heat energy.
ACTION: Look around your house. Make a list of things that waste heat energy. Does the waste heat cause any problems in your home environment? Is there any way you can reduce the amount of heat that is wasted?

Ideas in Action ....

IDEA: Energy is changing from one form to another all around you.
ACTION: Identify as many examples of energy changing form as you can.

SCIENCE CONNECTION

THERMOGRAPHY

All objects give off some heat energy. Hot objects give off more heat energy than cool objects. A technique called thermography (thur-MAHG-ruh-fee) can detect differences in heat energy. A device called a thermograph turns the invisible heat energy into a visible picture. This "heat picture" is called a thermogram. Thermography is used in medicine, industry, and many other fields.

In medicine, doctors use thermography to diagnose certain illnesses, such as breast cancer, arthritis, and circulatory system problems. Thermography can be used to find leaks in home insulation that allow heat to escape from a house. Pollution-control technicians can use thermography to find sources of thermal pollution in bodies of water.

A thermograph looks like a small television camera. Inside the thermograph, heat energy is changed into electrical energy. The electrical signals form pictures on a television screen. Different temperatures appear as different colors on the thermogram.

3-4 What is conservation of energy?

Objective ▶ State the law of conservation of energy.

TechTerms

- **law of conservation of energy:** energy cannot be made or destroyed, but only changed in form
- **scientific theory:** idea supported by evidence over a period of time

Conservation of Energy You know that energy can change from one form to another. Energy also can move from place to place. However, energy can never be lost. Energy can never be made or destroyed. Energy can only be changed in form. This is the **law of conservation of energy.**

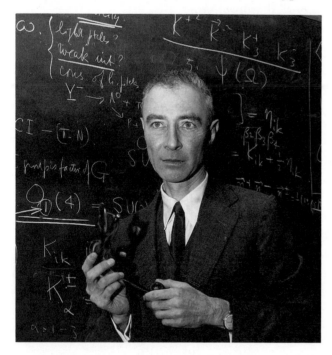

A **scientific theory,** such as the law of conservation of energy, is an idea supported by evidence over a period of time. Scientists have studied energy for many years. They have done many experiments and made many observations about energy. As a result of these experiments and observations, scientists were able to state the law of conservation of energy.

▶ **State:** What is the law of conservation of energy? Energy cannot be made or destroyed, but only changed in form.

Energy from the Sun Before 1905, the law of conservation of energy did not seem to apply to nuclear energy. In the sun, nuclear energy is changed into heat energy and light energy. The sun seemed to be producing too much energy. In 1905, Albert Einstein showed that matter and energy are two forms of the same thing. Matter can be changed into energy, and energy can be changed into matter.

The total amount of matter and energy in the universe does not change. Einstein stated this idea in the equation

$$E = mc^2$$

In this equation, E is energy, m is matter, or mass, and c is the speed of light. Einstein's equation showed that a small amount of matter could be changed into a huge amount of energy. This is what happens in the sun.

▶ **Identify:** What is Einstein's equation? $E = mc^2$

48

LESSON SUMMARY

▶ The law of conservation of energy states that energy can never be made or destroyed, but only changed in form.

▶ The law of conservation of energy is an example of a scientific theory.

▶ Matter can be changed into energy, and energy can be changed into matter.

▶ The total amount of matter and energy in the universe does not change.

CHECK *Write true if the statement is true. If the statement is false, change the underlined term to make the statement true.*

1. Energy <u>cannot</u> change from one form to another. can

2. Energy <u>can</u> move from place to place. true

3. The law of conservation of energy is an example of a scientific <u>hypothesis</u>. theory

4. A scientific <u>theory</u> is an idea supported by evidence over a period of time. true

5. The law of conservation of energy did not seem to apply to <u>chemical</u> energy. nuclear

6. In Einstein's equation, *E* stands for <u>energy</u>. true

APPLY *Complete the following.*

7. **Analyze:** Describe the changes in kinetic and potential energy that take place as a basketball is thrown through a hoop and caught again. How do these changes in energy support the law of conservation of energy?

8. **Interpret:** How does Einstein's equation explain how the sun produces energy in the form of heat and light?

State the Problem

Study the illustration below. State the problem for this experiment.

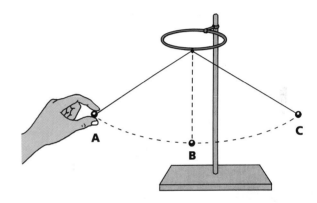

PEOPLE IN SCIENCE

ALBERT EINSTEIN (1879–1955)

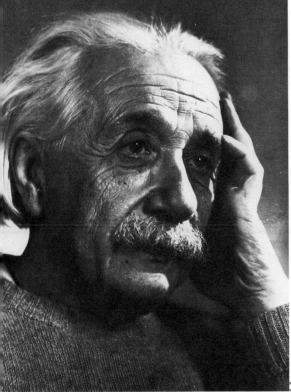

Albert Einstein was one of the greatest scientists who ever lived. His ideas changed the way people think about the universe. Einstein was born in Ulm, Germany. He became interested in science as a young boy. When Einstein was five years old, his father showed him a pocket compass. Einstein was fascinated to see that the compass needle always pointed in the same direction. This experience sparked his interest in science.

Einstein is probably best known for his theory of relativity and the famous equation, $E = mc^2$. However, he also wrote important papers about the nature of light and the movement of particles in a liquid or gas. His theories about matter and energy were the basis for controlling the release of nuclear energy from the atom. In 1921, Einstein received the Nobel Prize in physics.

Einstein worked and taught in Switzerland and Germany. Just before World War II, Einstein left Germany and came to the United States. He became an American citizen in 1940. Einstein spent the rest of his life at the Institute for Advanced Study in Princeton, New Jersey.

3-5 What is work?

Objective ▶ Relate work, force, and distance.

TechTerm

▶ **work:** force times distance

Work When are you doing **work**? Work is done when a force moves an object a certain distance. This relationship can be shown in the equation

$$\text{work} = \text{force} \times \text{distance}$$

Suppose two boys pushed a car stuck in the mud. They were not able to move the car. They were very tired afterwards. Did the boys do any work? The answer is no. For work to be done, something must be moved. The boys used a great deal of force, but the car did not move. Work was not done.

▶**Describe:** What is the relationship between work, force, and distance? work = force × distance

Work and Energy Remember that energy is the ability to do work. However, energy can be used without doing work. When a force moves an object, work is done. Anything that can make some-

thing else move has energy. A moving baseball bat has energy. It can do work. When the bat hits a ball, the ball moves. The energy stored in gasoline can do work. It can make a car move. The boys pushing the car used energy. However, no work was done because the car did not move. If you hold a heavy bag of groceries, your muscles are using energy. However, you are not doing work because the bag does not move.

▶ **Explain:** Why does a moving baseball bat have energy? because it can do work when it hits a ball

Direction of Motion For work to be done, a force must make an object move in the same direction as the force. Suppose you pushed a chair across the floor. Have you done any work? The answer is yes. The direction of motion was the same as the direction of the applied force. However, suppose you carried the chair across the floor. The direction of the applied force was upward. The direction of motion was forward. In this case, no work was done. Did you do any work when you picked up the chair? Yes, because the direction of the force and the direction of motion were both upward.

▶**Explain:** Why is work done when you push a chair across the floor? because the direction of the applied force and the direction of motion are the same

LESSON SUMMARY

▶ Work is done when a force moves an object a certain distance.

▶ Energy can be used without doing work.

▶ For work to be done, the direction of the applied force must be the same as the direction of motion.

CHECK *Complete the following.*

1. Work = force × _____ . distance

2. Work is not done unless something is _moved_ .

3. The ability to do work is _____ . energy

4. For work to be done, the direction of the _____ must be the same as the direction of motion. force

5. The stored _____ in gasoline can make a car move. energy

APPLY *Complete the following.*

6. Is work being done in each of the following examples? Explain your answers.
   a. A boy holds a heavy package for one hour.
   b. A girl coasts downhill on a bicycle.
   c. A football player kicks a field goal.
   d. A boy carries his baby sister across the room.
   e. A girl hits a tennis ball over the net.

Health and Safety Tip

Always be careful when picking up a heavy box or other object from the floor. You should bend your knees and use your leg muscles, not your back muscles, to lift the object. Use library references to find out other ways to prevent back injuries when doing work.

CAREER IN PHYSICAL SCIENCE

CONSTRUCTION WORKER

Every year, millions of homes, office buildings, and bridges are built in the United States. The construction industry provides thousands of jobs in many interesting and varied projects. Jobs available on building sites include crane operator, contractor, structural engineer, and carpenter.

Some jobs in the construction industry, such as structural engineer, require four years of college. Others, such as crane operator, involve on-the-job training or training at a technical school. Construction workers may work for large construction companies or small, private contractors.

Contractors plan buildings to be safe and practical for their intended use. Structural engineers choose the best materials for a job. Crane operators move heavy building equipment from the ground to the top of a building under construction. All construction workers must be aware of safety at all times. If you think that you might be interested in a career in the construction industry, you should write to Associated General Contractors of America, Inc., Construction Education Services, 1957 E Street NW, Washington, DC 20036.

3-6 How can work be measured?

Objective ▶ Use the proper unit to measure work.

TechTerm

▶ **joule** (JOOL): metric unit of work; equal to 1 N-m

Measuring Work To measure work, you must know two things. First, you must know the force used to move an object. Force can be measured with a spring scale. The unit of force is the newton (N). Second, you must know the distance that the object was moved. Distance is measured in meters (m).

Force = 10 newtons

Mass weighing 10 newtons

Distance = 2 meters

Work can be measured in newton-meters (N-m). Work is equal to force times distance.

$$W = F \times d$$

In this equation, W is work. F is force and d is distance. Suppose you lift an object weighing 10 N. Remember that weight is a force. You lift the object a distance of 2 m. To measure the amount of work done, multiply the force times the distance:

$$W = F \times d$$
$$W = 10 \text{ N} \times 2 \text{ m}$$
$$W = 20 \text{ N-m}$$

▶ **List:** What two things must you know in order to measure work? force used and distance moved

Unit of Work Scientists use a unit called a **joule** (JOOL) to measure work. One joule (1 J) of work is done when a force of 1 N moves an object a distance of 1 m. One joule is equal to 1 N-m of work.

▶ **Identify:** What unit is used to measure work? joule

Direction of Force To measure work, you must measure the force applied in the direction of motion. Suppose you used a spring scale to pull a 10-N box a distance of 2 m along a table top. How much work have you done? In this case, the weight of 10 N does not count. You must multiply the force shown on the spring scale times the distance. Suppose the applied force shown on the spring scale is 4 N. Then the work done equals 4 N × 2 m = 8 N-m, or 8 J.

Distance = 2 meters

Mass weighing 10 newtons

Force = 4 newtons

▶ **Calculate:** How much work is done if you use 5 N of force to push a 20-N chair 3 m across the floor? 15 J

LESSON SUMMARY

▶ To measure work, you must know the force in newtons and the distance in meters.

▶ Work can be measured in newton-meters (N-m).

▶ The unit of work is the joule (J); 1 J = 1 N-m.

▶ When measuring work, you must measure the force applied in the direction of motion.

CHECK *Complete the following.*

1. The unit of force is the _____ . newton

2. To measure work, you must know both force and _____ . distance

3. Work can be measured in newton-<u>meters</u> .

4. The unit of work is the _____ . joule

5. One joule is equal to 1 _____ . N-m

6. To measure work, you must know the amount of force applied in the direction of _____ . motion

APPLY *Complete the following.*

7. **Calculate:** How much work is done in each of the following examples? Show all of your calculations.

a. A child uses 4 N of force to pull a wagon a distance of 2 m along a sidewalk.

b. A construction worker uses 30 N of force to drag a piece of equipment a distance of 3 m.

8. **Compare:** In which case is more work done? Explain your answers.

a. You lift a 40-N object 2 m straight up.

b. You use 10 N of force to pull the same 40-N object 2 m across the floor.

InfoSearch ....

Read the passage. Ask two questions about the topic that you cannot answer from the information in the passage.

James Prescott Joule The metric unit of work, the joule, is named after James Prescott Joule. Joule was a physicist. He was born in England in 1818. Joule was one of the four scientists who helped state the law of conservation of energy. Joule's law is also named after him. This law states that heat is produced when electricity flows through a wire.

SEARCH: Use library references to find answers to your questions.

ACTIVITY

MEASURING WORK

You will need a book, string, a spring scale, and a meterstick.

1. Tie a piece of string around a book.

2. Attach a spring scale to the book using the string.

3. Using the spring scale, lift the book a distance of 1 m. Record the amount of force shown on the spring scale. Calculate the amount of work done in joules.

4. Using the spring scale, pull the book a distance of 1 m across your desk or table top. Record the amount of force shown on the spring scale. Calculate and record the amount of work done in joules.

Questions

1. How much work did you do when you lifted the book?

2. How much work did you do when you pulled the book across your desk?

3. Did you do the same amount of work in steps 3 and 4?

3-7 What is power?

Objective ▶ Explain how to measure power.

TechTerms

▶ **power:** amount of work done per unit of time
▶ **watt:** metric unit of power; equal to 1 J/sec

Power The amount of work done per unit of time is called **power**. In science, the term "power" describes the rate at which you do work. Suppose you took 30 minutes to shovel snow from a sidewalk. Your neighbor used a snowblower and cleared the sidewalk in 10 minutes. If you both did the same amount of work, which one of you used more power? Your neighbor, who did the work in less time, used more power.

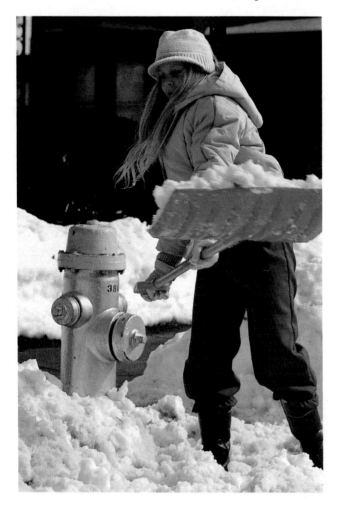

Measuring Power To measure power, you must know two things. First, you must know the amount of work done. Second, you must know the time needed to do the work. The formula used to measure power is

$$\textbf{power} = \textbf{work/time}$$

Remember that work is equal to force times distance. The formula for power can also be written as follows:

$$\textbf{power} = \textbf{(force} \times \textbf{distance)/time}$$

▐▐▐▶ **Identify:** What is the formula used to measure power? power = work/time, or power = (force × distance)/time

Unit of Power The metric unit of power is the **watt** (W). Power is equal to work divided by time. The unit of work is the newton-meter, or joule. The unit of time is the second. Therefore, one watt (1 W) is equal to 1 N-m/sec, or 1 J/sec. The watt is named after James Watt. Watt was a Scottish engineer who built the first useful steam engine.

Large amounts of power are measured in kilowatts (kW). One kilowatt (1 kW) is equal to 1000 W. You are probably familiar with watts and kilowatts as units of electric power. For example, light bulbs are rated as 60 W, 100 W, or 250 W.

▐▐▐▶ **Define:** What is power? work done per unit of time

▐▐▐▶ **Identify:** What is the unit of power? watt

LESSON SUMMARY

▶ Power is the amount of work done per unit of time.

▶ Power = work/time, or (force × distance)/time.

▶ The metric unit of power is the watt (W).

▶ Large amounts of power are measured in kilowatts (kW).

CHECK *Complete the following.*

1. The rate at which work is done is _power_ .

2. Power is the amount of work done per unit of _____ . time

3. To measure power, you must know the amount of _____ and the time needed. work

4. Power = (force × _____)/time. distance

5. The metric unit of power is the _____ . watt

6. One _____ is equal to 1000 W. kilowatt

7. One watt is equal to 1 N-m/sec, or 1 _____ /sec. J

8. The unit of power is named after James _____ . Watt

APPLY *Complete the following.*

9. **Calculate:** Find the amount of power used in each of the following examples. Show your calculations.
   a. You use a force of 10 N to move a box 100 m in 10 sec.
   b. An athlete lifting weights does 900 J of work in 1 sec.
   c. A truck does 30,000 J of work in 15 sec.
   d. A furniture mover uses a force of 150 N to push a heavy trunk 5 m across the floor in 5 sec.

· ·
Designing an Experiment············

Design an experiment to solve the problem.

PROBLEM: How much power, in watts, do you use when you climb a flight of stairs?

Your experiment should:

1. List the materials you would need.

2. Identify safety precautions that should be followed.

3. List a step-by-step procedure.

4. Describe how you would record your data.

▼ ▼ ▼ LOOKING BACK IN SCIENCE ▼

HORSEPOWER

You are probably familiar with the term "horsepower." Engines and motors are commonly rated in horsepower. An automobile engine, for example, may have about 100 horsepower. A train engine may produce 10,000 horsepower. Where does this unit of power come from?

James Watt was the first person to use the term "horsepower." Watt was a Scottish engineer and inventor. In the 1760s, he built the first practical steam engine. Watt wanted to use a unit of power for his engine that would be familiar to most people. He decided to use the power of a horse as the standard unit of power for the steam engine. Watt found that a strong horse could lift a 750-N load a distance of 1 m in 1 sec. In other words, a horse produced 750 J/sec of power. Watt defined this amount of power as one horsepower (hp).

Today, the unit of power is the watt (W). It is named in honor of James Watt. One watt is equal to 1 J/sec. Therefore, 1 hp (750 J/sec) is equal to 750 W. Real horses are no longer used as a standard of power. However, ratings for engines, motors, and power plants are still commonly given in terms of horsepower.

UNIT 3 Challenges

STUDY HINT Before you begin the Unit Challenges, review the TechTerms and Lesson Summary for each lesson in this unit.

TechTerms

energy (42)
joule (52)
kinetic energy (42)
law of conservation of energy (48)

potential energy (42)
power (54)
scientific theory (48)

thermal pollution (46)
watt (54)
work (50)

TechTerm Challenges

Matching *Write the TechTerm that matches each description.*

1. stored energy potential energy
2. ability to do work energy
3. force times distance work
4. energy of motion kinetic energy

5. metric unit of work joule
6. metric unit of power watt
7. work done per unit time power

Fill In *Write the TechTerm that best completes each statement.*

1. The moving water in a waterfall has _____ . kinetic energy
2. The _____ in a match is stored in the chemicals in the match head. energy
3. The faster you run, the more _____ you have. kinetic energy
4. Waste heat that escapes into the environment can cause _____ . thermal pollution
5. The _____ states that energy cannot be made or destroyed, but only changed in form. law of conservation of energy

6. An idea supported by evidence over a period of time is a _____ . scientific theory
7. When you use force to move an object, you are doing _____ . work
8. The _____ is the unit used to measure work. joule
9. The rate at which you do work is called _____ . power
10. The unit of power is the _____ . watt

Content Challenges

Multiple Choice *Write the letter of the term or phrase that best completes each statement.*

1. An object that is raised above the ground has
   a. heat energy. **b.** kinetic energy. **c.** potential energy. **d.** nuclear energy. c

2. All moving objects have
   a. heat energy. **b.** kinetic energy. **c.** potential energy. **d.** nuclear energy. b

3. Sound is a form of
   a. nuclear energy. **b.** electromagnetic energy. **c.** chemical energy. **d.** mechanical energy. d

4. Electromagnetic energy includes electricity and
   a. light. **b.** sound. **c.** heat. **d.** atoms. a

5. An automobile engine changes chemical energy into
   a. electricity. **b.** nuclear energy. **c.** mechanical energy. **d.** light. c

6. A scientific theory is the result of experiments and
   a. guesses.   **b.** observations.   **c.** predictions.   **d.** hypotheses. b

7. In the sun, nuclear energy is changed into light energy and
   a. sound energy.   **b.** chemical energy.   **c.** electrical energy.   **d.** heat energy. d

8. Work = force times
   a. distance.   **b.** mass.   **c.** power.   **d.** energy. a

9. Work is measured in units called
   a. watts.   **b.** meters.   **c.** joules.   **d.** newtons. c

10. One watt is equal to
   a. 1 m/sec.   **b.** 1 J/sec.   **c.** 1 N/sec.   **d.** 1 kW/sec. b

True/False *Write true if the statement is true. If the statement is false, change the underlined term to make the statement true.*

   1. Energy is the ability to do <u>work</u>. true
   2. Stored energy is <u>kinetic</u> energy. potential
   3. <u>Potential</u> energy is energy of motion. kinetic
   4. There are <u>six</u> main forms of energy. five
   5. The energy stored in atoms is <u>nuclear</u> energy. true
   6. When energy changes form, some energy is always lost as <u>sound</u>. heat
   7. Energy <u>cannot</u> be made or destroyed. true
   8. Matter and energy <u>are not</u> two forms of the same thing. are
   9. Energy <u>cannot</u> be used without doing work. can
   10. To measure work, you must know force and <u>time</u>. distance
   11. To measure power, you must know the amount of work and the <u>time</u>. true
   12. Large amounts of <u>power</u> are measured in kilowatts. true

Understanding the Features ....

Reading Critically *Use the feature reading selections to answer the following. Page numbers for the features are in parentheses.*

   1. What are two items of safety equipment worn by archers? (43)
   2. What happens in a nuclear fusion reaction? (45)
   3. What is a thermograph? (47)
   4. In what year did Albert Einstein receive the Nobel Prize for physics? (49)

   5. **Infer:** Why must construction workers be aware of safety at all times? (51)
   6. **Hypothesize:** Why do you think James Watt chose the power of a horse as a standard unit of power? (55)

Concept Challenges ....

Critical Thinking *Answer each of the following in complete sentences.*

   1. Explain the difference between potential energy and kinetic energy.
   2. Describe the changes in potential and kinetic energy that take place in a swinging pendulum.
   3. How does Einstein's equation, $E = mc^2$, support the law of conservation of energy?

   4. How is it possible for you to use energy without doing any work?
   5. What is the relationship between work and power?
   6. Why is gravitational potential energy called energy of position?

Interpreting a Diagram *Use the diagrams showing a person lifting an object and pulling an object to complete the following.*

1. What is the weight of the object in the left-hand diagram?
2. What distance is this object being lifted?
3. How much force is needed to lift this object?
4. How much work is being done to lift this object?
5. What is the weight of the object in the right-hand diagram?
6. What distance is this object being pulled?
7. How much force is needed to pull this object?
8. How much work is being done to pull this object?
9. Is the amount of work being done in the two diagrams the same or different? Explain.

Finding Out More . . . .

1. **Classify:** Collect pictures from newspapers and magazines showing different forms of energy. Label each picture with the form of energy shown. Use your pictures to make a poster illustrating all of the different forms of energy.
2. Different units can be used to measure energy. Use library references to find out how the following units are used: kilowatt-hour, calorie, BTU. Write a report of your findings.
3. The power obtained from falling water is called hydroelectric power. Find out how the water behind a dam is used to generate electricity in a hydroelectric power plant. Draw a labeled diagram of a hydroelectric plant. Where in the United States are most hydroelectric power plants located? What are some advantages and disadvantages of hydroelectric power?
4. Create a bulletin board display for your classroom showing examples of energy changing form.
5. Write or visit your local electric company. Find out how the company generates electricity for your community.

MOTION

CONTENTS

STUDY HINT As you read each lesson in Unit 4, write the topic sentence of each paragraph in the lesson on a sheet of paper. After you complete each lesson, compare your list of topic sentences with the Lesson Summary.

4-1 What are speed and velocity?

Objective ▶ Differentiate between speed and velocity.

TechTerms

▶ **motion:** change in position
▶ **speed:** distance traveled per unit of time
▶ **velocity** (vuh-LAHS-uh-tee): speed and direction

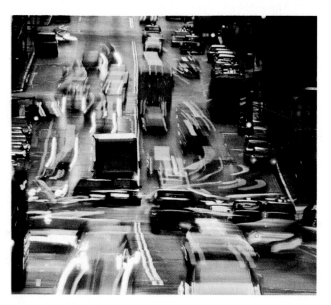

Motion and Speed Many of the things you see around you are in **motion**. Motion is a change in position. When you move from place to place, you cover a certain distance. Distance is measured in meters or kilometers. The time it takes you to cover a certain distance is called **speed**. If you travel 50 km in one hour, your speed is 50 km/hr. The equation for finding speed is

$$speed = distance/time$$

▶*Calculate:* At a speed of 80 km/hr, how far will you go in 3 hr? 240 km

Average Speed When you travel, you do not move at the same speed at all times. Suppose you are taking a car trip. During the trip, you may speed up, slow down, or stop many times. Even though you moved at different speeds during your trip, you can find your average speed for the whole trip. The average speed is equal to the total distance traveled divided by the total time for the trip.

$$average\ speed = total\ distance/total\ time$$

The speedometer of a car tells you what your speed is at any instant. This speed is called the instantaneous (in-stun-TAY-nee-us) speed. The instantaneous speed can be faster, slower, or the same as the average speed. Average speed gives information about the speed for the whole trip. Instantaneous speed gives information about the speed at one instant during the trip.

▶*Calculate:* If you traveled 360 km in 6 hr, what was your average speed? 60 km/hr

Velocity When you move from place to place, you move at a certain speed. You also move in a certain direction. **Velocity** (vuh-LAHS-uh-tee) tells you both the speed and direction of a moving object. People sometimes use the words ''speed'' and ''velocity'' as if they were the same. In science, however, velocity always includes speed and direction.

▶*Contrast:* What is the difference between speed and velocity?
Velocity includes both speed and direction of motion.

LESSON SUMMARY

▶ Motion is a change in position.

▶ Average speed is equal to the total distance traveled divided by the total time for the trip.

▶ Speed at any instant is called instantaneous speed.

▶ Velocity describes the speed and direction of a moving object.

CHECK *Complete the following.*

1. What is motion?

2. What is the formula for finding speed?

3. What is average speed?

4. What is instantaneous speed?

5. What does velocity tell you about a moving object?

APPLY *Complete the following.*

6. **Compare:** A truck is traveling east on a highway at 80 km/hr. What is its speed? What is its velocity?

7. **Analyze:** A highway speed limit is posted as 90 km/hr. Is this average speed or instantaneous speed? Explain your answer.

ACTIVITY

MEASURING AVERAGE SPEED

You will need a book, a piece of cardboard, a metric ruler, a watch or clock with a second hand, and a marble.

1. You will need about 1.5 m of floor space. Fold the piece of cardboard in half. Place one end of the cardboard on a book so that the end of the cardboard is raised 1.5 cm off the floor.

2. Hold the marble at the raised end of the cardboard. Release the marble and let it roll down the center of the cardboard and across the floor.

3. Measure the distance in centimeters the marble rolls from the end of the cardboard in 2 sec. Record your measurement in a table.

4. Repeat steps 2 and 3 three more times. Record your measurements.

Questions

1. What was the average distance the marble rolled in 2 sec?

2. What was the average speed of the marble?

What is acceleration?

Objective ▶ Explain how to calculate acceleration.

TechTerm

▶ **acceleration** (uk-sel-uh-RAY-shun): change in speed or direction

Changing Speed Speeding up and slowing down are changes in speed. Have you ever talked about the "pickup" of a car? This term describes how fast a car can reach a certain speed from a complete stop. When you step on the gas, the speed increases. When you step on the brake, the speed decreases. To find the change in speed, subtract the initial (i-NISH-ul), or starting, speed from the final speed. Suppose a car is moving at a speed of 4 m/sec. Ten seconds later its speed is 10 m/sec. The change in speed is 6 m/sec.

change in speed = final speed − initial speed
change in speed = 10 m/sec − 4 m/sec
change in speed = 6 m/sec

▶**Predict:** A car goes from 5 m/sec to 10 m/sec. What is its change in speed? 5 m/sec

Acceleration When a car changes speed, it is accelerating. **Acceleration** (uk-sel-uh-RAY-shun) is a change in speed or direction. When a car speeds up, it is accelerating. When a car slows down, it is accelerating. Slowing down is sometimes called deceleration (dee-SEL-uh-ray-shun). A car is also accelerating when it goes around a curve or turns a corner.

▶**Infer:** Why is a car accelerating when it turns a corner? Its direction is changing.

Measuring Acceleration Acceleration describes how fast the speed of a moving object is changing. To find acceleration, you must know the change in speed and the time for the change to occur. The equation for calculating acceleration is

acceleration = change in speed/time
       or
acceleration = (final speed − initial speed)/time

Suppose a car is stopped at a red light. When the light turns green, the car accelerates to a speed of 150 m/sec. The car takes 10 sec to reach this speed. What is its acceleration?

acceleration = (150 m/sec − 0 m/sec)/10 sec
acceleration = 150 m/sec/10 sec
acceleration = 15 m/sec/sec

The acceleration is 15 meters per second per second, or 15 m/sec/sec. This means that the car's speed increases 15 m/sec every second.

▶**Infer:** Why is the initial speed of the car 0 m/sec? It is not moving.

LESSON SUMMARY

▶ Speeding up and slowing down are changes in speed.

▶ A car is accelerating when it speeds up, slows down, or changes direction.

▶ To find acceleration, you must know the change in speed and the time for the change to occur.

CHECK *Complete the following.*

1. Speeding up and slowing down are changes in _____ . speed

2. The starting speed of a car is called its _____ speed. initial

3. To find change in speed, subtract initial speed from _____ speed. final

4. A change in speed or direction is called _____ . acceleration

5. Slowing down is called _____ . deceleration

6. To find acceleration, divide the change in speed by the _____ . time

APPLY *Complete the following.*

▶ 7. **Infer:** Another name for deceleration is negative acceleration. Why do you think this term is used?

8. **Classify:** Which of the following are examples of acceleration?
   a. a train sitting in the station
   b. a train pulling out of the station
   c. a train traveling at 100 km/hr on a straight stretch of track
   d. a train going around a curve in the track
   e. a train pulling into the station

Skill Builder

Interpreting Tables When an object falls to the ground, it accelerates as it falls. When it is released, its speed is zero. As it falls, its speed increases. The table shows how the speed of a falling object changes.

Time (seconds)	Speed (m/sec)
0	0
1	9.8
2	19.6
3	29.4
4	39.2
5	0

What is the object's acceleration from 0 to 1 sec? from 0 to 3 sec? from 2 sec to 4 sec? Based on your calculations, what can you say about the acceleration of a falling object?

CAREER IN PHYSICAL SCIENCE

AIRPLANE PILOT

An airplane pilot is trained to fly an airplane and to supervise the flight crew. The pilot is responsible for the safety of the passengers and crew. About half of all pilots are commercial airline pilots.

Many pilots fly their own airplanes for pleasure or for business. Test pilots fly experimental airplanes. They may test the effects of high speed or acceleration on the structure of the planes. Instructor pilots train other pilots. Special training is necessary to become an instructor pilot.

All pilots must have a high school education. Most employers also require a college degree. All pilots need a license issued by the Federal Aviation Administration (FAA). To find out more about a career as an airplane pilot, write to the Federal Aviation Administration, 800 Independence Avenue SW, Washington, DC 20591.

What are balanced and unbalanced forces?

Objectives ▶ Identify balanced and unbalanced forces. ▶ Describe how unbalanced forces affect the motion of an object.

TechTerms

▶ **balanced forces:** forces that are equal in size but opposite in direction

▶ **force:** push or pull

▶ **unbalanced forces:** forces that cause a change in the motion of an object

Balanced Forces Any push or pull is a **force.** To describe a force, you must know two things. You must know the size of the force and the direction of the force. Suppose two teams are playing tug of war. Each team is pulling with equal force, but in opposite directions. Neither team can make the other team move.

Figure 1 Balanced forces

Forces that are equal in size but opposite in direction are called **balanced forces.** Balanced forces do not cause a change in motion. When balanced forces act on an object at rest, the object will not move. If you push against a wall, the wall pushes back with an equal but opposite force. Neither you nor the wall will move.

▶**Predict:** What effect will balanced forces have on a book? The book will not move.

Unbalanced Forces Forces that cause a change in the motion of an object are **unbalanced forces.** Unbalanced forces are not equal and opposite. Suppose that one of the teams in the tug of war pulls harder than the other team. The forces would no longer be equal. One team would be able to pull the other team in the direction of the larger force.

Figure 2 Unbalanced forces

▶**Identify:** What kinds of forces cause a change in motion? unbalanced forces

Force and Motion Unbalanced forces can change the motion of an object in two ways.

▶ When unbalanced forces act on an object at rest, the object will move.

▶ When unbalanced forces act on a moving object, the velocity of the object will change. Remember that a change in velocity means a change in speed, direction, or both speed and direction.

▶**Predict:** What will happen when unbalanced forces act on a moving car?
The velocity (speed or direction) of the car will change.

LESSON SUMMARY

▶ Forces that are equal in size but opposite in direction are called balanced forces.

▶ Balanced forces do not cause a change in the motion of objects.

▶ Forces that cause a change in the motion of objects are called unbalanced forces.

▶ Unbalanced forces can change the motion of an object in two ways.

CHECK *Complete the following.*

1. A _____ is a push or a pull. force

2. To describe a force, you must know its size and _____ . direction

3. Balanced forces are _____ in size but opposite in direction. equal

4. Unbalanced forces cause moving objects to change their _____ . velocity (speed or direction)

5. Balanced forces cannot change an object's _____ . motion

APPLY *Complete the following.*

6. **Analyze:** A skater is moving from **east to west** across a frozen pond. Her speed is 4 m/sec. Someone gives her a push. As a result of the push, her speed increases to 6 m/sec in the same direction.

   a. Is the push a balanced force or an unbalanced force? How do you know?

   b. In what direction was the push applied? How do you know?

Ideas in Action

IDEA: Balanced forces do not cause a change in motion. Unbalanced forces always cause a change in motion.

ACTION: Identify and describe different examples of balanced and unbalanced forces in your everyday life. Explain how you know if the forces are balanced or unbalanced.

LEISURE ACTIVITY

RUNNING

Running is one of the oldest of all sports. One of the great appeals of running is that it is open to people of all ages and all abilities. Many people run every day for their health and for enjoyment. Others take part in organized races.

One popular type of race for long-distance runners is the marathon (MAR-uh-thahn). A marathon covers a distance of 42.2 kilometers. New York City and Boston have marathons each year.

To prepare for a race, runners must train each day. Runners may begin their training as long as eight weeks before a race. All runners wear special shoes and must be in excellent health. Marathon runners must save their strength to complete the race. Beginners should be very careful. The main goal of a beginner should be just to finish the race at his or her own speed.

The benefits of running are a lowered pulse rate and an improved circulatory system. Runners also say that running improves their general feeling of well-being.

4-4 What is Newton's first law of motion?

Objective ▶ Describe Newton's first law of motion.

TechTerm

▶ **inertia** (in-UR-shuh): tendency of an object to stay at rest or in motion

Inertia Place a book on your desk. Does the book move? Unless you move the book, it will remain where you put it without moving. Imagine a spacecraft moving through space. When the engines are turned off, the spacecraft will coast through space at the same speed and in the same direction. The book and the spacecraft have **inertia** (in-UR-shuh). Because of inertia, an object at rest tends to stay at rest. An object in motion tends to keep moving at a constant speed in a straight line.

▶*Identify:* What causes a book on a table to remain at rest? inertia

Newton's First Law Newton's first law of motion explains how inertia affects moving and non-moving objects. Newton's first law states that an object will remain at rest or move at a constant speed in a straight line unless it is acted on by an unbalanced force.

According to Newton's first law, an unbalanced force is needed to move the book on your desk. You could supply the force by pushing the book. An unbalanced force is needed to change the speed or direction of the spacecraft. This force could be supplied by the spacecraft's engines.

▶*Predict:* According to Newton's first law of motion, what will happen to an object at rest if no unbalanced force acts on it? It will remain at rest.

Effects of Inertia You can feel the effects of inertia every day. Suppose you are riding in a car. What happens if the car comes to a sudden stop? Your body has inertia. When the car stops, you keep moving forward. What happens when the car starts moving? Because of inertia, your body tends to stay at rest when the car moves forward. In baseball, inertia tends to keep a player running in a straight line. So base runners have to "round" the bases instead of making sharp turns.

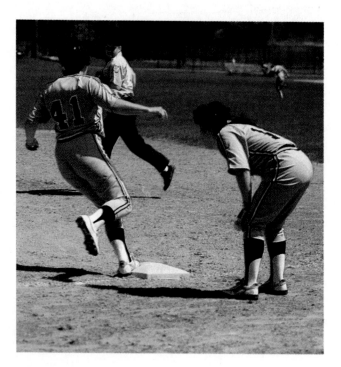

▶*Explain:* Why do you keep moving forward when your car stops suddenly?

because your body has inertia

LESSON SUMMARY

▶ Inertia is the tendency of an object to remain at rest or in motion.

▶ Newton's first law of motion states that an object will remain at rest or move at a constant speed in a straight line unless it is acted on by an unbalanced force.

▶ The effects of inertia can be felt every day.

CHECK *Complete the following.*

1. With its engines turned off, a spacecraft will move at the same speed in the same ___direction___ .

2. A book will not move by itself because it has _____ . inertia

3. A book will remain at rest unless it is acted on by an _____ force. unbalanced

4. When a car stops suddenly, your body tends to keep moving _____ . forward

5. Base runners tend to run in a straight line because of _____ . inertia

6. Newton's first law explains how inertia affects moving and _____ objects. nonmoving

APPLY *Complete the following.*

7. **Analyze:** Look at the diagram. In terms of inertia, explain what happens when the card is flicked away.

8. **Predict:** Push a rollerskate across a smooth surface. Will the skate keep moving when you stop pushing? Explain.

Skill Builder

Building Vocabulary Look up the words "inert" and "inertia" in a dictionary. The Latin root for both these words is *iners*. What does this root word mean? How is the meaning of the root word related to the definitions of inert and inertia? What is an inert gas?

SCIENCE CONNECTION

SEAT BELTS

Seat belts could be called "anti-inertia" belts. A moving car has inertia. It tends to keep moving in a straight line, even if the driver's foot is not on the gas pedal. Everyone inside the car also has inertia. They are moving at the same speed as the car.

Suppose you are riding in a car. The driver is forced to step on the brakes suddenly. Inertia keeps you moving forward at the same speed as the car was moving. You will keep moving until something stops you. This might be the car's steering wheel, dashboard, or windshield. You might be hurt if you hit these parts of the car, unless you are wearing a seat belt. A seat belt keeps you from moving forward when the car stops suddenly. Seat belts can prevent serious injuries. You should always remember to "buckle up" when you get into a car.

What is Newton's second law of motion?

Objective ▶ Describe Newton's second law of motion.

TechTerm

▶ **newton:** unit of force equal to one kilogram-meter per second per second

Effects of Unbalanced Forces Unbalanced forces cause objects to accelerate. When an unbalanced force acts on an object, the motion of the object is changed. If the object is at rest, the force makes it move. If the object is in motion, the force changes its velocity. Any change in velocity is an acceleration.

▶ **Describe:** What effect does an unbalanced force have on a moving object?

The force changes its velocity.

Force, Mass, and Acceleration The amount by which an object accelerates depends on two things. They are the size and direction of the force, and the mass of the object. If two forces act on the same object, the larger force will produce more acceleration than the smaller force. Suppose you apply the same amount of force to two objects with different masses. The object with the smaller mass will accelerate more than the object with the larger mass.

▶ **Identify:** What two things affect the acceleration of an object?

size and direction of the force, and mass of the object

Newton's Second Law The relationship among force, mass, and acceleration is explained by Newton's second law of motion. Newton's second law states that the unbalanced force acting on an object is equal to the mass of the object times its acceleration.

$$F = m \times a$$

In this equation, F is the force. The mass is m and the acceleration is a. Suppose the mass is measured in kilograms. The acceleration is measured in meters per second per second. Then the force is measured in **newtons** (N). A force of 1 N will accelerate a mass of 1 kg at 1 m/sec/sec. One newton of force is equal to one kilogram-meter per second per second (1 kg-m/sec/sec).

▶ **Define:** What is 1 N of force equal to?

1 kg-m/sec/sec

Figure 1 Equal masses, unequal forces   **Figure 2** Unequal masses, equal forces

LESSON SUMMARY

▶ Unbalanced forces cause objects to accelerate.

▶ The acceleration of an object depends on the mass of the object and the size and direction of the force acting on it.

▶ Newton's second law of motion describes the relationship among force, mass, and acceleration ($F = m \times a$).

CHECK Complete the following.

1. When it is acted on by an unbalanced force, an object will _____ . accelerate

2. When an unbalanced force acts on an object at rest, the object will _____ . move

3. A change in velocity is called _____ . acceleration

4. A large force will cause _____ acceleration than a small force. more

5. Newton's second law of motion states that force is equal to _____ times acceleration. mass

6. The _____ is a unit of force equal to 1 kg-m/sec/sec. newton

7. An object's acceleration depends on the size and direction of the force, and on the _____ of the object. mass

APPLY Complete the following.

▶ **Calculate:** Use the equation $F = m \times a$ to answer the following. Show your calculations.

8. What force is needed to accelerate a 2-kg mass at 1 m/sec/sec?

9. How hard would you have to push a 50-kg skater to increase her speed by 2 m/sec/sec?

10. What is the mass of an object if a force of 10 N causes it to accelerate at 5 m/sec/sec?

Skill Builder ....

Interpreting a Diagram Look at the two diagrams. Will the acceleration of the piano be greater in A or in B? Use Newton's second law of motion to explain your answer.

◼◄◇►◼ PEOPLE IN SCIENCE ►◼◄◇►◼◄◇►◼◄◇►◼◄◇►◼◄◇►◼◄◇►◼◄◇►◼◄◇►◼◄◇►◼◄◇►◼

SIR ISAAC NEWTON (1642–1727)

Isaac Newton was born in England on December 25, 1642. He was a physicist, an astronomer, and a mathematician. At the age of 45, Newton published his theories of motion and gravity. Newton's great book is usually called the *Principia*. It is considered one of the most important works in the history of science.

In the *Principia*, Newton explained his three laws of motion and his theory of gravitation. Newton also invented a branch of mathematics called "calculus" to help calculate motion using his three laws. Newton made many important discoveries about light and color.

Newton was a professor of mathematics at Cambridge University and a member of the Royal Society. He was knighted by Queen Anne in 1705. Newton once said about himself, "If I have seen further than others it is because I have stood on the shoulders of giants." What do you think Newton meant by this statement?

◼◄ ►◼◄ ►◼◄ ►◼◄ ►◼◄ ►◼◄ ►◼◄ ►◼◄ ►◼◄ ►◼◄ ►◼◄ ►◼◄ ►◼

4-6 What is Newton's third law of motion?

Objective ▶ Describe Newton's third law of motion.

TechTerms

- ▶ **action force:** force acting in one direction
- ▶ **reaction force:** force acting in the opposite direction

Action and Reaction Forces always act in pairs. The two forces act in opposite directions. When you push on an object, the object pushes back with an equal force. Think of a pile of books on a table. The weight of the books exerts a downward force on the table. This is the **action force.** The table exerts an equal upward force on the books. This is the **reaction force.** Notice that the two forces act on different objects. The action force acts on the table. The reaction force acts on the books.

Reaction force

Action force

▶ **Contrast:** How are action and reaction forces different? They act in opposite directions.

Newton's Third Law Newton's third law of motion explains action and reaction forces. The third law states that for every action force, there is an equal and opposite reaction force. Imagine hit-

ting a baseball. The bat exerts a force on the ball. This is the action force. The ball exerts an equal and opposite force on the bat. This is the reaction force.

▶ **State:** What does Newton's third law of motion state? For every action force, there is an equal and opposite reaction force.

Rocket Engines Newton's third law explains how rocket engines work. Hot gases are forced out of the back of the rocket. This is the action force. The gases exert an equal and opposite force on the rocket. This is the reaction force. The reaction force pushes the rocket upward.

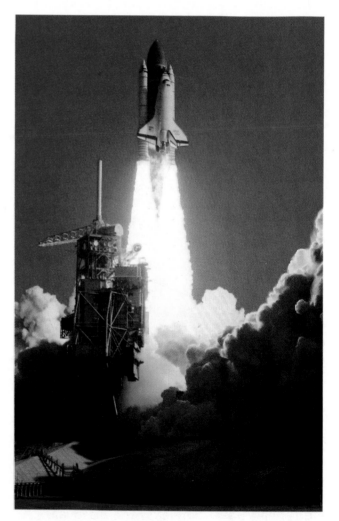

▶ **Infer:** Does the action force in a rocket engine act on the hot gases or on the rocket? on the hot gases

LESSON SUMMARY

▶ Forces always act in pairs.

▶ Newton's third law of motion states that for every action force, there is an equal and opposite reaction force.

Newton's third law explains how rocket engines work.

CHECK *Write true if the statement is true. If the statement is false, correct the underlined term to make the statement true.*

1. Forces always act <u>alone</u>. in pairs

2. Books on a table exert <u>an upward</u> force on the table. a downward

3. For every action force, there is an equal and opposite <u>reaction</u> force. true

4. When you hit a baseball, the <u>bat</u> exerts a force on the ball. true

5. In a rocket engine, the <u>action</u> force pushes the rocket upward. reaction

6. Action forces and reaction forces always act on <u>different</u> objects. true

APPLY *Complete the following.*

7. An object resting on a table weighs 100 N. With what force is the object pushing on the table? With what force is the table pushing on the object?

8. **Classify:** When you walk, your feet push against the ground. At the same time, the ground pushes against your feet. Which is the action force? Which is the reaction force?

9. **Hypothesize:** When you walk, you move forward. Does the earth move in the opposite direction? Explain your answer.

Health and Safety Tip

Newton's third law explains how many sports injuries are caused. The more force you use to hit a tennis ball, the more reaction force your arm receives from the racket. Every time your feet hit the ground when you are running, the ground hits your feet with an equal and opposite force. Use library references to find out how to protect yourself from sports injuries by wearing the proper equipment.

TECHNOLOGY AND SOCIETY

NATIONAL AEROSPACE PLANE

In 1986, the United States began research to develop a new kind of airplane. It was called the aerospace (ER-oh-spays) plane. The word "aerospace" means the earth's atmosphere and the space outside it. An aerospace plane would leave the atmosphere for part of its trip.

Unlike the space shuttle, which is launched like a rocket, an aerospace plane would take off from an ordinary airport. The plane would then increase speed to over Mach 5, or more than five times the speed of sound. The plane would go into earth orbit, and return to land at an airport. The aerospace plane might be able to fly across the United States in a little more than one hour.

Many problems must be solved in designing an aerospace plane. The high speeds reached would cause "sonic booms" near the airports used by the plane. The plane's engines would be very noisy, and would use large amounts of fuel. People must decide if the advantages of high-speed travel outweigh the disadvantages of an aerospace plane. What do you think?

STUDY HINT Before you begin the Unit Challenges, review the TechTerms and Lesson Summary for each lesson in this unit.

TechTerms

acceleration (62)
action force (70)
balanced forces (64)
force (64)

inertia (66)
motion (60)
newton (68)
reaction force (70)

speed (60)
unbalanced forces (64)
velocity (60)

TechTerm Challenges

Matching *Write the TechTerm that matches each description.*

1. speed and direction velocity
2. change in speed or direction acceleration
3. push or pull force
4. unit of force newton
5. change in position motion
6. distance traveled per unit of time speed

Applying Definitions *Explain the difference between the words in each pair. Write your answers in complete sentences.*
Answers will vary. Accept all logical responses.

1. action force, reaction force
2. speed, velocity
3. balanced forces, unbalanced forces
4. force, acceleration
5. motion, inertia

Content Challenges

Multiple Choice *Write the letter of the term or phrase that best completes each statement.*

1. When you move from place to place, you are changing your
   a. mass. **b.** inertia. **c.** position. **d.** speed. c

2. An unbalanced force causes a moving object to change
   a. speed. **b.** direction. **c.** neither speed nor direction. **d.** either speed or direction. d

3. A car's speedometer tells you
   a. average speed. **b.** instantaneous speed. **c.** acceleration. **d.** velocity. b

4. Balanced forces are always opposite in
   a. direction. **b.** size. **c.** size and direction. **d.** size or direction. a

5. Velocity includes speed and
   a. acceleration. **b.** inertia. **c.** direction. **d.** force. c

6. Average speed is equal to total distance divided by
   a. average distance. **b.** average time. **c.** instantaneous speed. **d.** total time. d

7. Action forces and reaction forces are described by Newton's
   a. first law of motion. **b.** second law of motion. **c.** third law of motion.
   d. law of gravitation. c

8. According to Newton's second law of motion, force is equal to mass times
   a. acceleration. **b.** speed. **c.** velocity. **d.** inertia. a

9. The newton is a unit of
   a. speed. **b.** force. **c.** velocity. **d.** acceleration. b

10. Inertia is described by Newton's
    a. first law of motion.   b. second law of motion.   c. third law of motion.
    d. law of gravitation. a
11. To find a car's change in speed, subtract its final speed from its
    a. total speed.   b. initial speed.   c. average speed.   d. instantaneous speed. b
12. Acceleration is equal to change in speed divided by
    a. time.   b. distance.   c. final speed.   d. initial speed. a

Completion *Write the term or phrase that best completes each sentence.*

1. Speed is the distance traveled in a certain _____ . time
2. If an action force acts in one direction, the reaction force acts in the _opposite_ direction.
3. In the equation $F = m \times a$, F equals the _____ . force
4. One newton of force is equal to _____ . 1 kg-m/sec/sec
5. Speed = _____ /time. distance
6. Acceleration is a change in speed or _____ . direction
7. Newton's third law of motion explains how _____ engines work. rocket
8. Motion is a change in _____ . position
9. Slowing down is sometimes called _____ , or negative acceleration. deceleration
10. Balanced forces are always _____ in size. equal
11. Forces always act in _____ . pairs
12. For every action force, there is an _____ reaction force. equal and opposite
13. Acceleration is equal to _____ divided by time. change in speed (final speed − initial speed)

Understanding the Features . . . .

Reading Critically *Use the feature reading selections to answer the following. Page numbers for the features are shown in parentheses.*

1. What is another possible name for seat belts? (67)
2. **Infer:** What is the distance of the New York City Marathon? (65)
3. How would an aerospace plane be different from the space shuttle? (71)
4. What is the name of the branch of mathematics invented by Isaac Newton? (69)
5. **Hypothesize:** Why do you think instructor pilots need special training? (63)

Concept Challenges . . . .

Critical Thinking *Answer the following questions in complete sentences.*

1. **Contrast:** What is the difference between average speed and instantaneous speed?
2. **Hypothesize:** When does an object have zero acceleration? Explain.
3. Two cars are stopped at a red light. When the light turns green, both cars accelerate to a speed of 150 m/sec. The first car takes 10 sec to reach this speed. The second car takes 20 sec. Which car has the greater acceleration? Explain.
4. An unbalanced force acts on a moving object. The object slows down. In what direction is the unbalanced force acting? How do you know?

Understanding a Diagram *Use the diagram to answer the questions.*

1. Does this diagram illustrate speed or velocity? Explain.
2. In what direction is the blue car traveling?
3. In what direction is the red car traveling?
4. What speed limit is shown in the diagram? Is this an average speed or an instantaneous speed?
5. If each car continues moving at the same average speed for two hours, how far will each car travel?

Finding Out More . . . .

1. The next time you ride in a car, close your eyes and try to sense changes in the speed and direction of the car. How can you tell if the car is accelerating? How can you tell if it is decelerating? How can you tell if it is changing direction? Write a report describing your observations.
2. Do library research to find out about different speed records. For example, what is the fastest land animal? What is the fastest airplane? Who is the fastest runner? Display your findings in the form of a chart comparing the speed records.
3. Scientists describe motion in terms of frames of reference. Find out what a frame of reference is. Why is it necessary to have a frame of reference when talking about motion? Explain your findings to the class.

MACHINES

CONTENTS

STUDY HINT After you read each lesson in Unit 5, write a brief summary on a sheet of paper explaining how the information in each lesson applies to your everyday life.

What is a simple machine?

Objective ▶ Describe how machines make work easier.

TechTerms

- ▶ **effort force:** force applied to a machine
- ▶ **mechanical advantage:** number of times a machine multiplies the effort force
- ▶ **resistance force:** force that opposes the effort force

Machines People use machines to make work easier. Did you ever try to take the lid off a jar using only your hands? If the lid is on very tight, your fingers cannot supply enough force to turn the lid. However, you can use a jar opener to help you remove the lid. A jar opener is one of the many machines people use to make work easier. Machines make work easier by changing the size of a force, the direction of a force, or the speed of a force.

▐▐▌▶*Explain:* How do machines make work easier? by changing the size, direction, or speed of a force

Effort Force and Resistance Force The force you apply to a machine is called the **effort force.** For example, the force you apply to a jar opener is the effort force. The force that opposes the effort force is called the **resistance force.** The jar opener multiplies your effort force to overcome the resistance force. A machine, such as a jar opener, lets you use a small force to overcome a large force.

▶*Infer:* Does a jar opener change the size, direction, or speed of a force? size

Mechanical Advantage Most machines help you do work by multiplying your effort force. The number of times a machine multiplies the effort force is called its **mechanical advantage,** or MA. A machine with a mechanical advantage of 5 multiplies the effort force 5 times.

You can find the MA of a machine by dividing the resistance force by the effort force. The resistance force is often equal to the weight of an object being moved. Suppose a machine uses 100 N of force to move an object weighing 500 N. The mechanical advantage of the machine is equal to 500 N divided by 100 N, or 5.

MA = resistance force/effort force
MA = 500 N/100 N
MA = 5

▐▐▌▶*Define:* What is mechanical advantage? amount by which a machine multiplies the effort force

Simple Machines A jar opener, a typewriter, and an eggbeater are all machines. Most machines are made up of two or more of the six simple machines. The six simple machines are the lever, the pulley, the inclined plane, the wedge, the screw, and the wheel and axle.

▐▐▌▶*Name:* What are the six simple machines? lever, pulley, inclined plane, wedge, screw, wheel and axle

LESSON SUMMARY

▶ Machines make work easier by changing the size, direction, or speed of a force.

▶ The force you apply to a machine is the effort force, and the force that opposes the effort force is the resistance force.

▶ Mechanical advantage is the number of times a machine multiplies the effort force.

▶ Mechanical advantage is equal to the resistance force divided by the effort force.

▶ There are six kinds of simple machines.

CHECK *Complete the following.*

1. How can you find the mechanical advantage of a machine?

2. What is the resistance force?

3. How do machines make work easier?

4. What is the force you apply to a machine called?

5. What is mechanical advantage?

APPLY *Complete the following.*

6. **Calculate:** Effort forces and resistance forces for five simple machines are listed below. Find the MA of each machine.

Effort force	Resistance force
a. 300 N	3000 N
b. 200 N	1000 N
c. 160 N	1600 N
d. 100 N	800 N
e. 80 N	400 N

7. How does an eggbeater help make work easier?

Ideas in Action

IDEA: You use machines every day to make work easier.

ACTION: Keep a record of all the machines you use in one day. Describe how each machine makes work easier.

TECHNOLOGY AND SOCIETY

ROBOTS

The word "robot" comes from a Czech word meaning "worker." The word was first used in 1923 in a science fiction play called "R.U.R." Robots are mechanical workers. Science fiction robots usually look and think like people. The science fiction author Isaac Asimov has written many books and stories about humanoid (HYOO-mun-oyd) robots. He invented "Asimov's three laws of robotics."

Real robots today are very different from science fiction robots. Most robots are built to do special jobs. For this reason, they do not have to look like humans. Robots have some advantages over human workers. They can work 24 hours a day without resting. They can do the same job over and over without getting tired or bored. They can work under conditions that would not be safe for people.

If you would like to learn more about robots, use your local library. Find out how robots are used in manufacturing, in medicine, and in the space program. How do you think robots might change your life in the future?

What is efficiency?

Objective ▶ Explain how to find the efficiency of a machine.

TechTerms

- **efficiency** (uh-FISH-un-see): comparison of work output to work input
- **work input:** work done on a machine
- **work output:** work done by a machine

Work Input and Work Output The work done by a machine is called **work output.** Work output is equal to the resistance force times the distance through which the force acts. This distance is called the resistance distance.

work output = resistance force
× resistance distance

The work done on a machine is called **work input.** The work input is equal to the effort force times the distance through which the force acts. This distance is the effort distance.

work input = effort force × effort distance

Machines cannot increase the amount of work you do. For this reason, the work output can never be greater than the work input.

Efficiency Not all the work put into a machine is changed into useful work. Some of the work put into a machine is used to overcome friction. This work is lost as heat energy. The **efficiency** (uh-FISH-un-see) of a machine is a comparison of work output to work input. You can find the efficiency of a machine by dividing the work output by the work input.

efficiency = work output/work input × 100

The efficiency of a machine is usually expressed as a percentage. Suppose a machine has an efficiency of 75%. In this machine, 75% of the work input is changed into useful work. The other 25% is used to overcome friction. The efficiency of a machine is always less than 100%.

▶**Name:** What is the work done by a machine called? work output

▶**Infer:** Why is the efficiency of a machine always less than 100%?
because work output is always less than work input

LESSON SUMMARY

▶ The work done by a machine is work output.

▶ The work done on a machine is work input.

▶ The efficiency of a machine is a comparison of work output and work input.

▶ The efficiency of a machine is always less than 100%.

CHECK *Complete the following.*

1. The work put out by a machine is ___less___ than the work put into a machine.

2. The efficiency of a machine is usually expressed as a _____ . percentage

3. Some of the work you put into a machine is used to overcome _____ . friction

4. The work put into a machine is equal to the effort force multiplied by the effort ___distance___ .

5. You can find the efficiency of a machine by ___dividing___ the work output by the work input.

6. The efficiency of a machine always is less than _____ . 100%

APPLY *Complete the following.*

7. Why can a machine not produce more work than is put into it?

8. **Calculate:** You put 10 J of work into a machine. The work output is 5 J. What is the efficiency of the machine?

9. **Hypothesize:** Why do you think many complex machines have very low efficiencies?

10. Which machine does more work to overcome friction, a machine with an efficiency of 75% or a machine with an efficiency of 50%? Explain your answer.

Skill Builder

Researching People have used machines for thousands of years. Beginning in about 1760, people became more and more dependent on machines. The social changes that resulted from the use of machines is often called the Industrial Revolution. Using library references, find out about the history of the Industrial Revolution. Use your findings to construct a time line showing important inventions, uses, and so on.

LEISURE ACTIVITY

BICYCLING

The first successful bicycle was built by Baron Karl Von Drais in Karlsruhe, Germany in 1816. This early bicycle had no pedals. The rider moved the bicycle forward by pushing backward against the ground with the feet. By 1839, a Scottish blacksmith, Kirkpatick Macmillian, had added pedals. The modern bicycle began to be developed.

In Europe and Asia, bicycles are a major means of transportation. In the United States, bicycles are used mainly for recreation and exercise. The sport of bicycle riding is called cycling. Special bicycles have been developed for cycling. These racing bikes are very strong and light. Standard bicycles are heavier than racing bikes and are good for everyday use. Streets in most cities have special bicycle lanes. You should be aware of bicycle safety rules before riding your bike in city traffic.

5-3 How does a lever work?

Objectives ▶ Explain how a lever makes work easier. ▶ Describe the three classes of levers.

TechTerms

- ▶ **fulcrum** (FUL-krum): point at which a lever is supported
- ▶ **lever** (LEV-ur): bar that is free to turn about a fixed point

Levers Have you ever used a shovel or a crowbar? If so, then you have used a **lever** (LEV-ur). A lever is a bar that is free to turn about a fixed point. The fixed point at which a lever turns is called the **fulcrum** (FUL-krum). The fulcrum is the point where the lever is supported. A lever can make work easier by increasing force. Levers also can change the direction of a force and the distance over which a force acts.

Every lever has two parts called an effort arm and a resistance arm. The effort arm is the distance from the effort force to the fulcrum. The resistance arm is the distance from the resistance force to the fulcrum.

▶**Name:** What are the two parts of a lever?
effort arm and resistance arm

MA of a Lever You can find the mechanical advantage of a lever by dividing the length of the effort arm by the length of the resistance arm.
  MA = effort arm length/resistance arm length

▶**Explain:** How can you find the mechanical advantage of a lever?
Divide the length of the effort arm by the length of the resistance arm.

Classes of Levers There are three classes, or kinds, of levers. The classes of levers are based on the position of the resistance force, the effort force, and the fulcrum. In a first-class lever, the fulcrum is between the effort and the resistance. A second-class lever has the resistance between the effort and the fulcrum. In a third-class lever, the effort is between the fulcrum and the resistance.

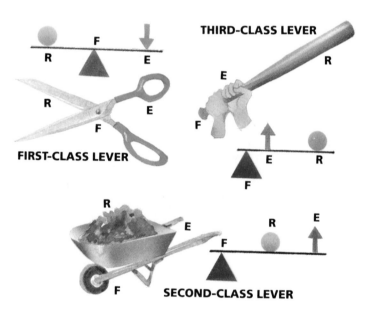

FIRST-CLASS LEVER

THIRD-CLASS LEVER

SECOND-CLASS LEVER

▶**Identify:** In what class of lever is the effort between the fulcrum and the resistance?
third-class lever

Resistance

Fulcrum

Effort

Effort arm

Resistance arm

LESSON SUMMARY

► A lever is a bar that is free to turn about a fixed point.

► A lever has two parts called an effort arm and a resistance arm.

► The MA of a lever is equal to the length of the effort arm divided by the length of the resistance arm.

► Levers are divided into three classes according to the position of the effort force, the resistance force, and the fulcrum.

CHECK Complete the following.

1. The fixed point at which a lever turns is called the _____ . fulcrum

2. There are _____ classes of levers. three

3. The mechanical advantage of a lever is equal to the length of the effort arm divided by the length of the _____ arm. resistance

4. In a _____ -class lever, the effort is between the fulcrum and the resistance. third

5. The resistance arm is the distance from the resistance to the _____ . fulcrum

APPLY Complete the following.

6. **Calculate:** What is the MA of a lever with an effort arm of 2 m and a resistance arm of 0.5 m?

7. **Hypothesize:** How could you increase the MA of a lever?

8. **Classify:** Classify each of the following as a first-, second-, or third-class lever: scissors, nutcracker, bottle opener, hammer, hockey stick.

InfoSearch ....

Read the passage. Ask two questions about the topic that you cannot answer from the information in the passage.

Levers in Your Body Different parts of the human body act as levers. Your forearm is a third-class lever. Suppose you hold a book in your hand. The book is the resistance. Your elbow is the fulcrum. The muscles in your forearm provide the effort force.

SEARCH: Use library references to find answers to your questions.

ACTIVITY

USING A LEVER

You will need a meter stick, a wood block, a spring scale, and a 500-g weight.

1. Balance the meter stick on the wood block at the 50-cm mark.

2. Place the 500-g weight on one end of the meter stick. Attach the spring scale to the other end.

3. Pull down on the spring scale. Record the effort force.

4. Move the wood block to the 45-cm mark. Repeat Step 3.

5. Move the wood block to the 55-cm mark. Repeat Step 3.

Questions

1. What was the effort force needed to lift the weight in Step 3? In Step 4? In Step 5?

2. Is it easier to lift the weight when the fulcrum is closer to the weight or farther away? Explain.

5-4 How do pulleys work?

Objectives ► Explain how pulleys make work easier. ► Compare fixed and moveable pulleys.

TechTerm

▶ **pulley:** rope wrapped around a wheel

Pulleys Look at Figure 1. A **pulley** is being used to raise a load of bricks. A pulley is a rope wrapped around a wheel. Pulleys can change either the direction or the size of a force. The pulley in Figure 1 changes the direction of a force. When the rope is pulled down, the load of bricks is pulled up.

▷ **Define:** What is a pulley?

rope wrapped around a wheel

Fixed Pulleys A pulley with a wheel that does not move is a fixed pulley. The pulley shown in Figure 1 is a fixed pulley. Fixed pulleys change the direction of the effort force. They do not increase the effort force. In a fixed pulley, the effort force is equal to the resistance force. As a result, the MA of a fixed pulley is equal to 1.

▷ **Explain:** Why is the MA of a fixed pulley equal to 1? because it does not increase the effort force

Movable Pulleys A movable pulley does not change the direction of an effort force. Instead, a movable pulley increases the size of the effort force. Figure 2 shows a movable pulley. When the rope is pulled up, the load of bricks and the pulley both move up. You can find the MA of this pulley by counting the number of ropes that lift the resistance. The MA of the movable pulley in Figure 2 is equal to 2.

▬ **Analyze:** What is the MA of a movable pulley with three ropes supporting the resistance?

3

Pulley Systems A block and tackle is a pulley system. A pulley system is made up of both fixed and movable pulleys. The pulleys are used together to increase the MA of the system. The MA of a pulley system is equal to the number of supporting ropes.

▷ **State:** Why are pulley systems used?

to increase the MA

Figure 1 Fixed pulley

Figure 2 Movable pulley

LESSON SUMMARY

▶ A pulley is a rope wrapped around a wheel.

▶ Fixed pulleys change the direction of the effort force.

▶ Movable pulleys increase the size of the effort force.

▶ A pulley system is made up of fixed and movable pulleys.

CHECK *Write true if the statement is true. If the statement is false, change the underlined term to make the statement true.*

1. A <u>fixed</u> pulley can increase the effort force.   movable

2. The mechanical advantage of a fixed pulley is <u>two</u>. one

3. A <u>block and tackle</u> is an example of a pulley system. true

4. The MA of a pulley system with four supporting ropes is <u>8</u>. 4

5. A <u>movable</u> pulley can change only the direction of a force. fixed

APPLY *Complete the following.*

6. **Classify:** What kind of pulley is used on a clothesline? Explain.

7. **Analyze:** What is the MA of each of the pulley systems shown in the diagram?

Skill Builder ....

▲***Modeling*** Design a machine that uses at least three of the six simple machines. Your machine should have a practical use. Draw a diagram or build a working model of your machine. Label each simple machine. Explain how your machine works.

ACTIVITY

USING A MOVABLE PULLEY

You will need a spring scale, string, a movable pulley, a book, and a small nail or thumbtack.

1. Tie the string around the book. Attach the book to the spring scale.

2. Use the spring scale to lift the book. Record the effort force needed to lift the book.

3. Attach the movable pulley to the book and spring scale as shown.

4. Use the pulley to lift the book the same distance. Record the effort force needed to lift the book.

Questions

1. How much force was needed to lift the book without the pulley?

2. How much force was needed to lift the book with the pulley?

3. What is the MA of the movable pulley?

How does an inclined plane work?

Objective ▶ Describe how an inclined plane makes work easier.

TechTerm

▶ **inclined plane:** slanted surface, or ramp

Inclined Planes A ramp is often used to help load barrels onto a truck. The barrels are rolled up the ramp onto the truck. The ramp is an **inclined plane.** The word "inclined" means slanted. A plane is a flat surface. Therefore, an inclined plane is a slanted surface, or ramp. Inclined planes are simple machines that help make work easier.

▶**Define:** What is an inclined plane?
slanted surface, or ramp

MA of an Inclined Plane An inclined plane makes work easier by increasing the size of the effort force. Look at the picture of the man rolling the barrel up the inclined plane. He is moving a 300-N barrel using only 100 N of effort force. The ramp has multiplied his effort force by 3. The MA of this ramp is 3. You can find the MA of an inclined plane by dividing its length by its height.

$$MA = length/height$$
$$MA = 3 \text{ m}/1 \text{ m}$$
$$MA = 3$$

▶**Explain:** How can you find the MA of an inclined plane? Divide its length by its height.

Wedges and Screws A wedge is a kind of inclined plane. A wedge is made of two inclined planes back to back. A knife, a nail, and an axe are examples of wedges.

Screw

Wedge

A screw is an inclined plane wrapped around a cylinder. A screw is like the steps wrapped around the center of a spiral staircase. Nuts and bolts are examples of screws.

▶**List:** What are three wedges?
knife blade, nail, axe

100 N

300 N

3 M

1 M

LESSON SUMMARY

► An inclined plane is a slanted surface, or ramp.
► The MA of an inclined plane is equal to its length divided by its height.
► A wedge is two inclined planes back to back.
► A screw is an inclined plane wrapped around a cylinder.

CHECK *Write true if the statement is true. If the statement is false, change the underlined term to make the statement true.*

1. A <u>wedge</u> is an inclined plane wrapped around a cylinder. screw
2. The MA of an inclined plane is equal to its length divided by its <u>height</u>. true
3. A plane is a <u>slanted</u> surface. flat
4. Nuts and bolts are examples of <u>wedges</u>. screws
5. An inclined plane <u>decreases</u> the size of the effort force. increases
6. A wedge is made up of <u>two</u> inclined planes. true

APPLY *Complete the following.*

7. How could you increase the MA of an inclined plane?
8. **Calculate:** How much effort force would be needed to push the car up the hill?

10,000 N   2,000 m   40 m

Skill Builder ....

Interpreting a Diagram The diagram shows three ramps: A, B, and C. Which ramp has the smallest MA? Explain.

A 5 m 1 m
B 4 m 8 m
C 3 m 9 m

ACTIVITY

FINDING THE MA OF AN INCLINED PLANE

You will need a wooden board, a spring scale, string, a metric ruler, and three books.

1. Stack two books one on top of the other. Place one end of the wooden board on top of the books to make an inclined plane.
2. Use the metric ruler to measure the length and height of the inclined plane. Record your measurements.
3. Tie one end of the string around a book. Tie the other end to the spring scale. Measure and record the weight of the book in newtons. The weight of the book is the resistance force.
4. Use the spring scale to pull the book up the inclined plane. Record the effort force shown on the scale.

Questions

1. What is the ideal MA of the inclined plane using the formula MA = length/height?
2. What is the actual MA of the inclined plane using the formula MA = resistance force/effort force?
3. **Hypothesize:** Why is the actual MA less than the ideal MA?

Length   Height

What is a compound machine?

Objective ▶ Name some compound machines.

TechTerm

▶ **compound machine:** machine that combines two or more simple machines

Compound Machines Most machines are made up of a combination of simple machines. Machines that combine two or more simple machines are called **compound machines.** Compound machines can do more complicated jobs than simple machines alone. They also can have large mechanical advantages. The mechanical advantage of a compound machine depends on the mechanical advantages of all of the simple machines that make it up.

▷ *Define:* What is a compound machine?
machine that combines two or more simple machines

Examples of Compound Machines Most of the machines you use every day are compound machines. For example, a pair of scissors is a compound machine. A pair of scissors is made up of two levers joined by a screw. The screw is the fulcrum of the levers. Each blade of a pair of scissors is a wedge.

A bicycle is another compound machine. What simple machines make up a bicycle? The

wheels and pedals are wheels and axles. The pedals are attached to levers. The brakes, handlebars, and gearshift controls also are levers. The chains are pulleys. Screws are used in many places to hold parts of the bicycle together.

▷ *Identify:* What are some simple machines in a bicycle? wheels and axles, levers, screws, pulleys

People and Machines Humans have been using machines for thousands of years. Early humans made simple tools from stone. Centuries later, agricultural societies used machines to water their crops, crush grain into flour, and cut wood to build homes. During the Industrial Revolution, steam power began to replace horsepower as a source of energy for machines. Today, fossil fuels and nuclear energy are used to operate modern machines. In the future, new technology may make possible even more complex machines.

▷ *Name:* What are two sources of energy for modern machines? fossil fuels and nuclear energy

LESSON SUMMARY

▶ A compound machine is composed of two or more simple machines.

▶ Most machines are compound machines.

▶ A pair of scissors and a bicycle are compound machines.

▶ Humans have been using machines for thousands of years.

CHECK *Complete the following.*

1. What are three simple machines that make up a pair of scissors?

2. What is a compound machine?

3. What type of simple machine are the wheels and handlebars of a bicycle?

4. On what does the mechanical advantage of a compound machine depend?

APPLY *Complete the following.*

5. **Compare:** How does the mechanical advantage of a compound machine compare with the mechanical advantage of each of its simple machines?

6. **Analyze:** How does a bicycle make work easier?

7. Draw a diagram of a pair of scissors. Label the simple machines that make up a pair of scissors on your diagram.

Health and Safety Tip ....

Machines are very helpful to people. However, if machines are not used properly, they can cause serious injury. It is important to use proper safety precautions when you use any kind of machine. Make a poster that illustrates "Machine Safety."

CAREER IN PHYSICAL SCIENCE ●◆●◆●◆●◆●◆●◆●◆●◆●◆●◆●◆

MACHINIST

Many new and specialized machines have been made by combining different simple machines. The parts for these machines are often standardized so that they can be used in machines that do different jobs. Special machines called machine tools are used to make these parts.

Machinists use machine tools to make metal parts for automobiles, radios, refrigerators, and televisions. The jobs done by machinists include drilling, boring, grinding, and shaping metal into different shapes and sizes.

Today, many jobs for machinists are available in the aerospace industry. Machinists may help make parts for spacecraft and airplanes. A machinist should be familiar with computers, because computers are now used to operate many machine tools. For more information, write to the International Association of Machinists and Aerospace Workers, 1300 Connecticut Avenue NW, Washington, DC 20036.

UNIT 5 Challenges

STUDY HINT Before you begin the Unit Challenges, review the TechTerms and Lesson Summary for each lesson in this unit.

TechTerms . . . .

compound machine (86)
efficiency (78)
effort force (76)
fulcrum (80)

inclined plane (84)
lever (80)
mechanical advantage (76)
pulley (82)

resistance force (76)
work input (78)
work output (78)

TechTerm Challenges . . . .

Matching *Write the TechTerm that matches each description.*

1. force applied to a machine effort force
2. rope wrapped around a wheel pulley
3. slanted surface inclined plane
4. work done by a machine work output
5. force that opposes the effort force resistance force
6. work done on a machine work input

Applying Definitions *Explain the difference between the words in each pair. Write your answers in complete sentences.*

Answers will vary. Accept all logical responses.

1. lever, fulcrum
2. simple machine, compound machine
3. mechanical advantage, efficiency
4. work input, work output
5. effort force, resistance force

Content Challenges . . . .

Multiple Choice *Write the letter of the term or phrase that best completes each statement.*

1. Machines make work easier by changing a force's
   a. size.   **b.** direction.   **c.** speed.   **d.** size, direction, or speed. d

2. A machine with a mechanical advantage of 10 multiples the effort force
   a. 5 times.   **b.** 10 times.   **c.** 10%.   **d.** 5%. b

3. The efficiency of a machine is always less than
   a. 100%.   **b.** the mechanical advantage.   **c.** work output.   **d.** work input. a

4. If the efficiency of a machine is 60%, the percentage of work input used to overcome friction is
   a. 100%.   **b.** 60%.   **c.** 40%.   **d.** 0%. c

5. In a first-class lever, the fulcrum is between the effort force and
   a. the effort arm.   **b.** the resistance arm.   **c.** the resistance force.   **d.** none of these. c

6. A wheelbarrow is an example of a
   a. first-class lever.   **b.** second-class lever.   **c.** third-class lever.   **d.** wheel and axle. b

7. A block and tackle is an example of a
   a. lever.   **b.** pulley.   **c.** pulley system.   **d.** wheel and axle. c

8. The MA of a pulley system with six ropes supporting the load is
   a. 6.   **b.** 3   **c.** 60%   **d.** 1. a

9. An inclined plane makes work easier by increasing a force's
   a. direction.   **b.** speed.   **c.** MA.   **d.** size. d

10. If you use 50 N of force to push a box weighing 200 N up an inclined plane, the MA of the inclined plane is
    a. 50. **b.** 150. **c.** 4. **d.** 0.25. c
11. A pair of scissors is an example of a
    a. lever. **b.** compound machine. **c.** simple machine. **d.** wedge. b
12. The brakes on a bicycle are
    a. pulleys. **b.** wedges. **c.** levers. **d.** screws. c

Completion *Write the term or phrase that best completes each statement.*

1. People use _____ to make work easier. machines
2. Machines can change the size, direction, or speed of a _____ . force
3. The efficiency of a machine is equal to the _____ divided by the work input. work output
4. Some of the work put into a machine is used to overcome _____ . friction
5. The _____ of a lever is the distance from the resistance force to the fulcrum. resistance arm
6. The MA of a lever with an effort arm of 60 cm and a resistance arm of 15 cm is _____ . 4
7. A pulley can change either the _____ or the size of a force. direction
8. A pulley system is made up of fixed and _____ pulleys. movable
9. Another name for an inclined plane is _____ . ramp
10. The MA of an inclined plane is equal to its _____ divided by its height. length
11. A compound machine is made up of _____ or more simple machines. two
12. A compound machine has a _____ mechanical advantage than a simple machine. larger

Understanding the Features ....

Reading Critically *Use the feature reading selections to answer the following. Page numbers for the features are shown in parentheses.*

1. **Define:** What does the word "humanoid" mean? (77)
2. Why should a machinist be familiar with computers? (87)
3. **Infer:** Why are racing bicycles stronger and lighter than standard bicycles? (79)

Concept Challenges ....

Critical Thinking *Answer each of the following in complete sentences.*

1. **Relate:** Why is work input equal to the effort force times the effort distance?
2. **Hypothesize:** Why do you think compound machines are sometimes called complex machines?
3. A perpetual motion machine would have more than 100% efficiency. Could such a machine ever be built? Explain your answer.
4. Two piano movers want to raise a piano to the fifth floor of an apartment building. Should they use a fixed pulley, a movable pulley, or a block and tackle to make the work easier? Explain.

Interpreting a Diagram *Use the diagram to answer the following questions.*

Effort

600 N
Resistance

Fulcrum

Effort arm 2 m

Resistance arm 0.5 m

1. **Observe:** What is the length of the effort arm of this lever?
2. What is the length of the resistance arm?
3. **Calculate:** What is the mechanical advantage of this lever?
4. What is the resistance force in this diagram?
5. **Predict:** How much effort force would be needed to overcome the resistance force?
6. **Classify:** Is this lever a first-, second-, or third-class lever? Explain.

Finding Out More . . . .

1. **Research:** Rube Goldberg was a cartoonist. He was famous for his drawings of humorous machines. Go to the library and find a picture of a Rube Goldberg cartoon. Make a copy of the cartoon and explain how the machine works to the class.

2. Find an example of a first-class lever, a second-class lever, and a third-class lever around your home. Show the levers to the class and demonstrate how they are used to make work easier. Point out the fulcrum on each lever.

3. **Model:** Draw or build a model of a perpetual motion machine. Describe how the machine is supposed to work. Then explain why it is not possible to build a working perpetual motion machine.

4. Visit a local garage or automobile repair shop that uses a block and tackle to lift engine blocks from cars. Ask the mechanic to explain how the block and tackle works.

HEAT

CONTENTS

STUDY HINT Before beginning each lesson in Unit 6, review the TechTerms for each lesson and read the Lesson Summary.

6-1 What is heat?

Objective ▶ Recognize heat as a form of energy.

TechTerm

▶ **heat:** form of energy in moving particles of matter

Caloric At one time, people thought that heat was a physical substance. They called this substance "caloric" (kuh-LOWR-ik). Caloric was thought to flow like a liquid. In 1798, the American scientist Benjamin Thompson (Count Rumford) performed an experiment. Rumford observed the drilling of cannons. He found that heat was produced as the cannons were drilled. Rumford concluded that heat did not flow, as stated by the caloric theory. Instead, Rumford said that the mechanical energy used to drill the cannons was being changed into heat. Therefore, **heat** is a form of energy.

▶*Identify:* Before Rumford's experiment, what did people think heat was?

a physical substance called caloric

Moving Particles Matter is made up of tiny particles. These particles are always in motion. Heat energy makes the particles of matter move faster.

As the particles move faster, they also move farther apart. Think of boiling water. When you boil water, you add heat energy to the water. The added heat energy makes the particles of water move faster. The water particles move so fast that they escape from the container. If you keep adding heat, the water will boil away.

▶*Describe:* What effect does adding heat energy have on the particles of matter?

They move faster and farther apart.

Heat and Work How do scientists know that heat is a form of energy? Energy can do work. Remember that something must be moved for work to be done. When you boil water, you can see the water bubbling and moving inside the container. Heat is doing work. A simple experiment shows that heat is a form of energy that can do work. If you hold a pinwheel over a hot light bulb, you will see the pinwheel move. Heat from the light bulb makes the air around the pinwheel move. The moving air turns the pinwheel.

▶*Infer:* How do scientists know that heat is a form of energy? Heat can do work.

LESSON SUMMARY

▶ Heat is a form of energy.

▶ Heat energy makes the particles of matter move faster and farther apart.

▶ Heat is a form of energy because it can do work.

CHECK *Complete the following.*

1. At one time, people thought that heat was a substance called _____ . caloric

2. In Rumford's experiment, _____ energy was changed into heat energy. mechanical

3. Matter is made up of tiny _____ . particles

4. Added heat energy makes particles of matter move _____ . faster

5. For work to be done, something must be _____ . moved

6. Moving _____ turns a pinwheel held over a hot light bulb. air

APPLY *Complete the following.*

7. **Classify:** Is heat a form of kinetic energy or potential energy? Explain your answer.

8. **Describe** what happens as you add heat energy to a container of water.

9. **Analyze:** Suppose you hit a piece of metal several times with a hammer. When you touch the piece of metal, it feels hot. Explain why the metal gets hot after being hit with the hammer.

PEOPLE IN SCIENCE

BENJAMIN THOMPSON, COUNT RUMFORD (1753–1814)

Benjamin Thompson was born in Woburn, Massachusetts. During the Revolutionary War, Thompson remained loyal to England. For a time, he acted as a spy for the British. He later led a British regiment. After the war, Thompson left the colonies to live in England. He was knighted by King George III. Thompson was made a count of the Holy Roman Empire in 1791.

Rumford is probably best known for showing that heat is a form of energy, not a liquid form of matter. In 1798, Rumford was in charge of a factory that made cannons. The machines used to drill the cannons were turned by horses. Rumford noticed that the cannons became very hot as they were drilled. He decided to try an experiment. He surrounded a cannon with a box filled with water. As the cannon was drilled, the water began to boil. The water continued to boil as long as the drilling went on. Rumford concluded that the work done by the horses supplied the heat to boil the water. Heat must be a form of energy.

Rumford was also an inventor. He invented a kitchen range, a double boiler, and a drip coffee pot. He was one of the first people in Europe to promote the use of James Watt's steam engine.

How is heat measured?

Objective ▸ Identify the calorie as a unit used to measure heat.

TechTerms

▸ **calorie** (KAL-uh-ree): unit of heat; amount of heat needed to raise the temperature of 1 g of water 1 °C

▸ **Calorie:** 1000 calories, or 1 kilocalorie

Adding and Removing Heat Heat affects the temperature of a substance. When heat is added, the temperature of the substance rises. When heat is removed from the substance, its temperature falls. You can tell whether heat is being added or removed by observing the change in temperature. The change in temperature depends on how much heat is added or removed.

▰▰▰▸*Identify:* How can you tell when heat is being added to a substance? Its temperature rises.

Measuring Heat Heat can be measured by observing the temperature change it causes. Suppose you heat 1 g of water until its temperature rises 1 °C. The amount of heat you have added to the water is 1 **calorie** (KAL-uh-ree). A calorie is a unit of heat. One calorie is the amount of heat that will raise the temperature of 1 g of water 1 °C. One gram is a very small amount of water. A calorie is a small amount of heat. Two calories could raise the temperature of 2 g of water 1 °C. Two calories could also raise the temperature of 1 g of water 2 °C.

▰▰▰▸*Define:* What is a calorie? the amount of heat needed to raise the temperature of 1 g of water 1 °C

Food Energy You have probably heard people talk about the number of calories in food. Calories are used to measure the amount of energy you get from food. However, a food calorie is 1000 times larger than the calorie used as a unit of heat. To show the difference, food calories are written with a capital C. One Calorie is equal to 1000 calories, or 1 kilocalorie. Remember that the prefix kilo-means 1000. One Calorie can raise the temperature of 1000 g of water 1 °C. The table shows the number of Calories in some foods.

Table 1	Calories in Food	
FOOD	AMOUNT	CALORIES
Apple	1 medium	70
Banana	1 large	200
Bread, white	1 slice	60
Butter	1 pat	50
Celery	1 stalk	3
Chili, with beans	8 oz	290
Corn flakes	1 cup	95
Egg, fried	1	110
Green beans	1/2 cup	15
Hamburger, with roll	1/4 lb	320
Milk, whole	8 oz	160
Pizza	1 slice	240
Potato chips	1	13
Potatoes, french fried	10	150

1 gram of water   **1 gram of water**

▰▰▰▸*Calculate:* How many Calories are there in a lunch consisting of a hamburger, an 8-oz glass of milk, and a banana? 680 Calories

LESSON SUMMARY

- ▶ Adding or removing heat changes the temperature of a substance.
- ▶ One calorie is the amount of heat needed to raise the temperature of 1 g of water 1 °C.
- ▶ One Calorie is equal to 1000 calories, or 1 kilocalorie.

CHECK *Complete the following.*

1. When heat is added to water, its temperature _____ . rises
2. When heat is removed from water, its temperature _____ . falls
3. Heat can be measured by observing the change in the _____ of a substance. temperature
4. A _____ is a unit of heat. calorie
5. Three calories of heat will raise the temperature of _____ g of water 1 °C. 3
6. The amount of energy in food is measured in _____ . Calories
7. One Calorie is equal to _____ calories. 1000

APPLY *Complete the following.*

8. **Compare:** What is the difference between a calorie and a Calorie?
9. **Infer:** A calorie is a unit of heat. Why can the amount of energy in food be measured in food calories, or Calories?

Skill Builder

Researching and Organizing When you do research, you gather information. When you organize information, you put the information in some kind of order. Use books on nutrition, or other library references, to find a list of the number of Calories in different foods. Use this information to prepare a menu for one day's meals. Your menu should include foods for breakfast, lunch, and dinner. The total number of Calories for each meal must be between 600 and 1000 Calories. Add up your total number of Calories for the day. Compare this number with the recommended number of Calories for someone your age.

TECHNOLOGY AND SOCIETY

CRYOGENICS

When heat is removed from a substance, the temperature of the substance goes down. When all of the available heat has been removed from a substance, its temperature cannot go down any more. At this temperature, all the particles of the substance almost stop moving. This is the lowest possible temperature that can be reached. It is called absolute zero. Scientists have not yet been able to reach absolute zero. However, they have been able to cool substances to within 0.01 °C of absolute zero. The study of such very low temperatures is called cryogenics (kry-uh-JEN-iks).

Cryogenics is important in the refrigeration of food, in space technology, and in medicine. In cryogenic surgery, the surgeon uses a supercold probe instead of a scalpel. Cryogenic surgery reduces bleeding. It also reduces much of the pain after ordinary surgery. Many cryogenic operations can be done in a doctor's office.

Hydrogen and oxygen are gases at normal temperatures. However, at very low temperatures, they become liquids. Liquid hydrogen and liquid oxygen are used as fuels in some rockets. In the future, liquid hydrogen may replace gasoline as a fuel for automobiles.

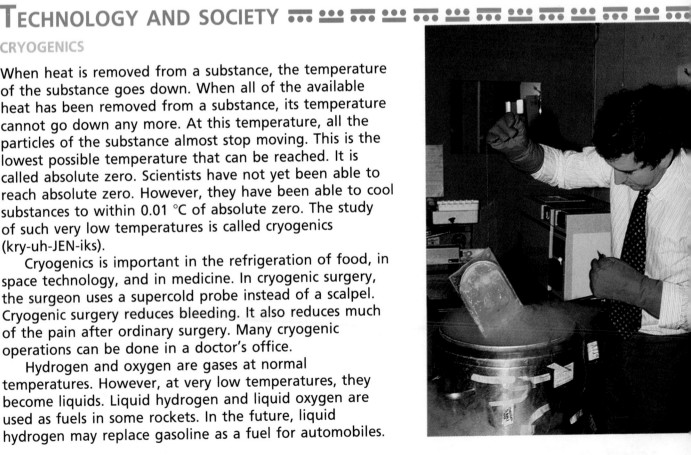

6-3 What is temperature?

Objective ▶ Differentiate between heat and temperature.

TechTerms

▶ **absolute zero:** lowest possible temperature; temperature at which particles of matter almost stop moving

▶ **temperature:** measure of the average kinetic energy of the particles of a substance

Temperature and Heat Heat is related to temperature, but they are not the same. Heat is the energy of moving particles of matter. Energy of motion is called kinetic energy. Because of their motion, moving particles of matter have kinetic energy.

The average kinetic energy of the particles of a substance is called **temperature.** When you add heat to a substance, you raise its temperature. The higher the temperature, the faster the particles of the substance are moving. When you remove heat from a substance, you lower its temperature. The lower the temperature, the slower the particles of the substance are moving.

▶**Define:** What is temperature? measure of the average kinetic energy of the particles of a substance

Absolute Zero As you remove heat from a substance, its temperature falls. The particles of the

substance move slower and slower. What happens if you keep removing heat from the substance? You will reach a temperature at which all particle motion almost stops. This temperature is called **absolute zero.** Absolute zero is the lowest possible temperature. It is equal to −273 °C.

▶**Describe:** What happens to the particles of matter at absolute zero? They stop moving.

Movement of Heat Heat moves from a place with a high temperature to a place with a lower temperature. Another way to say this is that heat moves from a hot object to a cold object. Hold a glass of ice water in your hand. After a few minutes, the ice in the water begins to melt. At the same time, your hand gets cold. The temperature of the ice water was lower than the temperature of your hand. As a result, heat moved from your hand to the ice water. The temperature of your hand got lower. The temperature of the water got higher. Heat continues to move until your hand and the water are the same temperature.

Heat

▶**Describe:** How does heat move from place to place? Heat moves from a place with a high temperature to a place with a lower temperature.

LESSON SUMMARY

▶ Heat and temperature are related, but not the same.

▶ Temperature is a measure of the average kinetic energy of the particles of a substance.

▶ **The lowest possible temperature is absolute zero.**

▶ Heat always moves from a place with a high temperature to a place with a lower temperature.

CHECK *Write true if the statement is true. If the statement is false, change the underlined term to make the statement true.*

1. Heat and temperature <u>are</u> the same thing. are not

2. Heat is the energy of <u>moving</u> particles of matter. true

3. Moving particles of matter have <u>potential</u> energy. kinetic

4. Temperature is the average <u>kinetic</u> energy of particles of matter. true

5. The higher the temperature, the <u>slower</u> the particles of matter are moving. faster

6. Temperature <u>rises</u> as you remove heat from a substance. falls

APPLY *Complete the following.*

7. **Analyze:** Use what you have learned about heat and temperature to explain what is happening in the diagram.

Skill Builder ....

Researching When you do research, you gather information. Use library references to answer the following questions.

1. What was the highest temperature ever recorded on the earth?

2. What was the lowest temperature ever recorded on the earth?

3. What was the lowest temperature ever reached in a laboratory? What was the highest?

ACTIVITY

OBSERVING TEMPERATURE DIFFERENCES

You will need three shallow pans, paper towels, hot water (not boiling), cold water, and warm water.

1. Fill one pan with very cold water. Fill the second pan with hot water. **CAUTION: Do not use boiling water.** Fill the third pan with warm water.

2. Put both of your hands into the warm water. Observe how the water temperature feels to your hands.

3. Dry your hands with a paper towel.

4. Put one hand into the cold water, and the other hand into the hot water.

5. After 1 min, put both of your hands directly into the warm water.

Questions

1. Did the temperature of the warm water feel the same to both of your hands?

2. **Hypothesize:** Why do you think your sense of temperature was fooled?

Cold

Warm

Hot

6-4 What is freezing point?

Objective ▶ Identify the freezing point and melting point of a substance.

TechTerms

▶ **freezing point:** temperature at which a liquid changes to a solid

▶ **melting point:** temperature at which a solid changes to a liquid

Freezing Water When water freezes, it changes to ice. Suppose you put a beaker of water into a freezer. Table 1 shows what might happen if you record the temperature of the water every 5 min.

Table 1	Freezing of Water
TIME	TEMPERATURE
0 min	22 °C
5 min	15 °C
10 min	10 °C
15 min	6 °C
20 min	3 °C
25 min	0 °C
30 min	0 °C
35 min	0 °C

As the temperature drops, the water loses heat. The temperature of the water drops until it reaches 0 °C. Water begins to freeze at 0 °C.

The temperature of the water stops going down as the water begins changing to ice. The water is still losing heat. The water changes to ice as more heat is lost. After 35 min, the water has changed completely to ice. The temperature at which water changes to ice is called its **freezing point.** The freezing point of water is 0 °C. Once the water has changed to ice, the temperature of the ice can drop below 0 °C.

▐▶*Define:* What is the freezing point of water? temperature at which water changes to ice

Freezing Points of Other Liquids Every liquid has its own freezing point. When heat is removed from a liquid, its temperature goes down. When the liquid reaches its freezing point, the temperature stops going down. Removing heat from a liquid at its freezing point changes the liquid into a solid. Table 2 shows the freezing points of some liquids.

Table 2	Freezing Points of Some Liquids
LIQUID	FREEZING POINT
Water	0 °C
Ethyl alcohol	−117 °C
Mercury	−39 °C
Sea water	−1 °C
Glycerine	18 °C

◣*Analyze:* What is the freezing point of ethyl alcohol? -117 °C

Melting Point The freezing point and **melting point** of a substance are the same. When heat is removed from a liquid, it changes to a solid. When heat is added to a solid, it changes back to a liquid. The temperature at which a solid changes to a liquid is called its melting point. The melting point of ice is 0 °C.

▐▶*Describe:* What happens when a solid is heated to its melting point? It changes to a liquid.

98

LESSON SUMMARY

▶ When water freezes, it changes to ice.

▶ The temperature at which water changes to ice is its freezing point.

▶ Every liquid has its own freezing point.

▶ The freezing point and melting point of a substance are the same.

CHECK *Complete the following.*

1. What happens when water freezes?

2. At what temperature does water begin to freeze?

3. What is the temperature at which water changes to ice called?

4. What is the freezing point of water?

5. Do all liquids freeze at the same temperature?

6. What is the difference between the freezing point and the melting point of a substance?

7. What happens to a solid at its melting point?

8. What is the melting point of ice?

APPLY *Complete the following.*

9. **Interpret:** Look at the graph. It shows that the temperature of a substance rises as heat is added. Explain what is happening at the part of the graph shown in red.

Skill Builder

Calculating When a solid melts, it becomes a liquid. Heat must be added to change a solid to a liquid. The amount of heat needed to change 1 g of a solid to a liquid is called heat of fusion. For water, the heat of fusion is 80 calories per gram. To melt 1 g of ice, 80 calories of heat must be added. Calculate the amount of heat needed to melt the following: 10 g of ice, 250 g of ice, 2 kg of ice.

SCIENCE CONNECTION

SUBLIMATION

Do you put your wool sweaters away for the summer? Do you pack them with moth balls for protection? If so, then you may have noticed that the moth balls were gone when you unpacked in the fall. What happened to the moth balls? Did they disappear into thin air? The answer is yes. The solid moth balls changed to a gas. The process of changing directly from a solid to a gas is called sublimation (sub-luh-MAY-shun). Substances that sublime do not melt, or change to a liquid. As the temperature of the solid rises, it reaches a point at which the solid becomes a gas. This is called the sublimation point.

Not all substances change to a gas by sublimation. Carbon dioxide is another substance that sublimes. Solid carbon dioxide is called dry ice. You have probably seen dry ice used to pack ice cream cakes. Dry ice is very useful for packing and shipping frozen foods. Because dry ice changes directly to a gas, it does not melt. This means that it can keep ice cream and other foods frozen without making a mess by melting.

6-5 What is boiling point?

Objectives ▶ Identify the boiling point of a liquid. ▶ Differentiate between boiling and evaporation.

TechTerms

- ▶ **boiling point:** temperature at which a liquid changes to a gas
- ▶ **evaporation** (i-VAP-uh-ray-shun): change from a liquid to a gas at the surface of the liquid

Boiling Water When water boils, it changes to steam. Steam is water in the form of a gas. The temperature at which water changes to steam is called its **boiling point.** As you heat water, the temperature of the water rises. When the temperature reaches 100 °C, small bubbles appear in the water. These bubbles show that a gas is being formed. As you continue to heat the water, more and more bubbles are formed. However, the temperature stays at 100 °C. The temperature of the water remains at 100 °C until all of the water has changed to steam. The boiling point of water is 100 °C.

▶*Identify:* What is steam?

water in the form of a gas

Boiling Points of Other Liquids Every liquid has its own boiling point. When heat is added to a liquid, its temperature rises. The temperature of the liquid rises until it reaches its boiling point. When the liquid reaches its boiling point, it begins to form a gas. The boiling point of the liquid is the temperature at which the liquid changes to a gas. Table 1 shows the boiling points of some liquids.

Table 1 Boiling Points of Some Liquids	
LIQUID	BOILING POINT
Water	100 °C
Mercury	357 °C
Glycerine	290 °C
Acetic acid	118 °C
Benzene	80 °C
Ethyl alcohol	78 °C
Acetone	39 °C

◢**Analyze:** Which liquid listed in the table has the highest boiling point? mercury

Evaporation When a liquid is allowed to stand uncovered at room temperature, it slowly changes to a gas. This change is called **evaporation** (i-VAP-uh-ray-shun). Evaporation happens only at the surface of the liquid. When a liquid boils, it changes to a gas all through the liquid. When a liquid evaporates, some particles at the surface of the liquid escape into the air.

▶*Define:* What is evaporation?

the change of a liquid to a gas at the surface of the liquid

LESSON SUMMARY

▶ When water boils, it changes to steam.

▶ Every liquid has its own boiling point.

▶ When a liquid stands uncovered at room temperature, it slowly changes to a gas.

CHECK *Complete the following.*

1. When water _____ , it changes to steam. boils

2. The temperature at which water changes to steam is its _____ . boiling point

3. The boiling point of water is _____ °C. 100

4. When a liquid reaches its boiling point, it begins to change to a _____ . gas

5. An uncovered liquid changes to a gas at room temperature by the process of _____ . evaporation

6. Evaporation occurs only at the _____ of a liquid. surface

7. When a liquid evaporates, _____ of the liquid escape into the air. particles

APPLY *Use Table 1 on page 100 to complete the following.*

8. What is the boiling point of mercury?

9. What is the boiling point of acetic acid?

10. Which of the liquids in the table boils at the lowest temperature?

Complete the following.

11. **Compare:** In your own words, explain the difference between boiling and evaporation.

InfoSearch

Read the passage. Ask two questions that you cannot answer from the information in the passage.

Antifreeze The engine of a car must be kept from getting too hot. Water in the radiator cools the engine. In the winter, plain water would freeze and crack the engine. To prevent this, an antifreeze is mixed with the water in the radiator. The solution of water and antifreeze has a much lower freezing point than plain water. It also has a much higher boiling point. The antifreeze solution does not evaporate as fast as plain water would in hot weather.

SEARCH: Use library references to find answers to your questions.

ACTIVITY

RELATING BOILING POINT AND ELEVATION

You will need a sheet of graph paper and a pencil.

1. Look at Table 1. It shows how the boiling point of water changes with elevation.

2. On a sheet of graph paper, plot the data in Table 1. Record elevation in meters along the horizontal axis. Record boiling point in degrees Celsius along the vertical axis.

Questions

1. What happens to the boiling point of water as elevation increases?

2. **Analyze:** Use your graph and Table 2 to answer the following.
   a. What is the boiling point of water in Atlanta?
   b. in Denver? c. in Boise? d. in Salt Lake City?
   e. in Chicago? f. in Phoenix?

Table 1 Boiling Point and Elevation	
BOILING POINT (°C)	ELEVATION (m)
100	0 (sea level)
96.6	1000
93.4	2000
90.1	3000

Table 2 Cities and Elevations	
CITY	ELEVATION (m)
Chicago, IL	183
Atlanta, GA	320
Phoenix, AR	340
Boise, ID	825
Salt Lake City, UT	1300
Denver, CO	1600

6-6 What is conduction?

Objective ▶ Describe how heat is transferred through solids.

TechTerms

- ▶ **conduction** (kun-DUK-shun): heat transfer in solids
- ▶ **conductors:** substances that conduct heat easily
- ▶ **insulators:** substances that do not conduct heat easily

Heat Transfer in Solids Heat can move from place to place. Heat moves through solids by **conduction** (kun-DUK-shun). Heat moves from a place with a high temperature to a place with a lower temperature. Think of heating a pan of water on a stove. What happens to the handle of the pan? It gets hot as heat moves from the bottom of the pan to the handle.

Heat travels by conduction when moving particles of matter bump into one another. When you place a pan of water on a flame, the bottom of the pan becomes hot first. The particles in this part of the pan begin to move faster. They bump into particles around them. Particles begin bumping into one another all through the pan. This is how heat moves from the bottom of the pan to the handle.

▶**Define:** What is conduction? heat transfer in solids

Conductors of Heat All metals are good **conductors** of heat. A conductor is a substance that allows heat to move through it easily. Copper, silver, iron, and steel are all good conductors of heat. Copper and silver are two of the best conductors of heat. Items made of copper or silver will get hot faster than items made of iron or steel.

▶**Identify:** What substances are good conductors of heat? metals, especially copper and silver

Poor Conductors of Heat Many substances are poor conductors of heat. Substances that do not conduct heat easily are called **insulators.** Wood, paper, wax, and air are poor conductors of heat. They are insulators.
Insulators prevent heat from moving from place to place. Houses are insulated to keep them warm in winter and cool in summer. Spaces are left between the inside and outside walls of the house. The spaces are filled with an insulating material. This insulation helps keep heat from escaping to the outside. During the summer, insulation helps keep heat from getting into the house.

▶**Explain:** Why are insulators used in houses? to help keep them warm in winter and cool in summer

LESSON SUMMARY

- ▶ Heat moves through solids by conduction.
- ▶ Heat travels by conduction when moving particles of matter bump into one another.
- ▶ All metals are good conductors of heat.
- ▶ Some substances are poor conductors of heat.
- ▶ Insulators prevent heat from moving from place to place.

CHECK *Complete the following.*

1. What is the process by which heat moves through solids called?

2. How does heat travel by conduction?

3. What is a conductor?

4. Name two substances that are good conductors of heat.

5. What is an insulator?

6. Name two substances that are good insulators.

7. Why are houses insulated?

APPLY *Complete the following.*

▶ 8. **Infer:** Why do you think kitchen pots and pans are sometimes made of copper?

▶ 9. **Infer:** Why do you think the handles of some pots and pans are made of plastic or wood?

Ideas in Action

IDEA: Insulators are often used to keep hot foods hot and cold foods cold.
ACTION: Look around your house and make a list of different kinds of containers that use insulation to keep foods hot or cold. If possible, describe the type of insulator used in each container.

Skill Builder

Researching When you do research, you gather information. The following is a list of different types of insulating materials: fiberglass; goose down; ceramic tiles. Use library references to find out where and how each of these insulators is used.

SCIENCE CONNECTION ◆○◆

HOME INSULATION

In some parts of the United States, winters are cold and summers are hot. In these parts of the country, houses must be insulated. Home insulation keeps houses warm in winter and cool in summer. Home insulation can help reduce the amount of fuel needed to heat or cool a home. Good insulation can cut fuel use by as much as 30%.

Insulation is needed in those parts of a house where the most heat is usually lost. In most homes, heat loss occurs through the attic floor, the ceiling of an unheated basement, and the side walls. Different types of insulation can be used in these places. For example, blankets of fiberglass can be inserted between beams in floors and ceilings. Liquid plastic foam can be sprayed into the spaces between inside and outside walls.

A number called an R-value is used to grade insulating materials. An insulating material with a high R-value is best at preventing heat loss. Choosing insulation with the best R-value can greatly reduce fuel costs.

6-7 What is convection?

Objective ▶ Describe how heat travels through gases and liquids.

TechTerms

▶ **convection** (kuhn-VEK-shun): heat transfer in gases and liquids

▶ **convection currents:** up and down movements of gases or liquids caused by heat transfer

Heat Transfer in Gases and Liquids

Heat travels through gases by **convection** (kuhn-VEK-shun). Air is a gas. When air is heated, the air particles move farther apart. As a result, the warm air becomes less dense. Warm air is less dense than cold air. The cold, dense air sinks. The warm air rises. As it rises, the warm air carries heat with it. This is how heat travels through a gas by convection.

Cool Air / Warm Air

Convection takes place in liquids as well as in gases. Cold water is denser than warm water. Cold water sinks. Warm water rises, carrying heat along with it.

▶*Identify:* How does heat travel through gases and liquids? by convection

Convection Currents

Heat is carried through the air by means of **convection currents.** Convection currents are up and down movements of the air. Sinking cool air and rising warm air move heat through the air. This is what causes convection currents. Convection currents are found in water as well as in air.

Convection current

▶*Define:* What are convection currents?
up and down movements of air or water

Uses of Convection

Heat transfer by convection is used in some home heating systems. In a hot water heating system, water is heated in a hot water heater. The hot water is then pumped through pipes to each room in the house. The hot water flows through heaters near the floor of the room. The hot water warms the air near the floor. The warm air rises, carrying heat through the room by means of convection currents. The water then returns to the heater to repeat the process.

Convection currents / Expansion tank / Pump / Hot water heater / Burner

▶*Identify:* What method of heat transfer is used in hot water heating systems? convection

LESSON SUMMARY

- ▶ Heat travels through gases by convection.
- ▶ Convection takes place in liquids as well as in gases.
- ▶ Heat is carried through the air by means of convection currents.
- ▶ Heat transfer by convection is used in some home heating systems.

CHECK *Write true if the statement is true. If the statement is false, correct the underlined term to make the statement true.*

1. Heat travels through gases by <u>conduction</u>.
   convection

2. When air is heated, the particles move <u>closer together</u>. *farther apart*

3. Warm air is <u>less</u> dense than cold air. *true*

4. <u>Sinking</u> warm air carries heat with it. *Rising*

5. Convection takes place in gases and <u>solids</u>. *liquids*

APPLY *Complete the following.*

6. **Hypothesize:** Why do you think smoke goes up a chimney?

7. **Predict:** Look at the picture. Which one of the girls would feel warmer? Explain your answer.

Ideas in Action

IDEA: Convection currents help circulate fresh air in a room.

ACTION: To get fresh air into a room at home, should you open a window from the bottom only, from the top only, or from both the bottom and the top? Try it and find out.

ACTIVITY

OBSERVING CONVECTION CURRENTS IN WATER

You will need water, dark food coloring, an eyedropper, a beaker, a ring stand, a piece of wire gauze, and a heat source.

1. Set up a ring stand, a heat source, and a beaker as shown in the diagram.
2. Fill the beaker about two thirds full of water.
3. Use the eyedropper to add a drop of dark food coloring to the water in the beaker. Be careful not to disturb the water.
4. Place the heat source to one side of the beaker. Gently heat the water in the beaker.
5. Observe what happens to the food coloring as you heat the water.

Questions

1. What happened to the food coloring as you heated the water in the beaker?
2. What is this called?
3. What caused the movement you observed?

What is radiation?

Objective ▸ Describe how heat travels through empty space.

TechTerms

▸ **radiation** (ray-dee-AY-shun): transfer of heat through space

▸ **vacuum**: empty space

Heat from the Sun The earth receives heat from the sun. The sun is 150 million kilometers from the earth. How does heat from the sun reach the earth? Heat from the sun travels through 150 million kilometers of **vacuum,** or empty space, before it reaches the earth. There are no particles of matter in a vacuum. Heat cannot travel from the sun by conduction or convection.

▶*Infer:* Why can heat from the sun not travel by conduction or convection?
Conduction and convection require particles of matter.

Heat Transfer Through Space Heat from the sun reaches the earth by **radiation** (ray-dee-AY-shun). Radiation is the transfer of heat through empty space. Particles of matter are not needed for the movement of heat by radiation. The heat from an electric heater and the heat from a fire are also forms of radiation.

▶*Define:* What is radiation?
transfer of heat through empty space

Uses of Radiation Radiation is used in some home heating systems. One kind of heating sys-

tem is called a radiant hot water system. In this system, water is heated in a hot water heater. The hot water is then sent through a coil of pipe in the floor of each room of the house. Heat radiates from the hot pipe and warms the room evenly. Solar heating systems use the heat from the sun to warm a house. In a passive solar heating system, heat from the sun heats the house directly. In an active solar heating system, heat from the sun is collected by solar panels. The heat collected in the solar panels is used to heat water. The hot water is then circulated through the house, as in a regular hot water heating system.

▶*Name:* What are two kinds of heating systems that use radiation? radiant hot water
system and solar heating system

LESSON SUMMARY

▶ The earth receives heat from the sun.

▶ Heat from the sun reaches the earth by radiation.

▶ Some home heating systems use radiation.

▶ Solar heating systems use heat from the sun to heat homes.

CHECK *Complete the following.*

1. The earth receives heat from the _____ . sun

2. Heat from the sun must travel through millions of kilometers of __vacuum__ , or empty space.

3. Heat cannot travel from the sun by _____ or convection. conduction

4. The transfer of heat through empty space is called _____ . radiation

5. The heat around an electric heater is a form of _____ . radiation

6. One home heating system that uses radiation is called a _____ hot water system. radiant

7. Two kinds of solar heating systems are active and _____ solar heating. passive

APPLY *Complete the following.*

▶ 8. **Infer:** What form of heat transfer is used to toast a marshmallow held over a campfire?

▶ 9. **Classify:** Identify each of the following examples of heat transfer as conduction, convection, or radiation.
   a. heat from a wood-burning stove warming a room   **b.** bacon cooking in a frying pan   **c.** water boiling in a tea kettle

InfoSearch

Read the passage. Ask two questions that you cannot answer from the information in the passage.

Radiant Energy   Heat from the sun travels through space as radiant energy. A hot iron and a light bulb also give off radiant energy. Like all living things, you also give off some radiant energy. You cannot see radiant energy, but you can feel its effects. When radiant energy is absorbed, it is changed to heat. This is why your skin feels warm when you stand in bright sunlight.

SEARCH: Use library references to find answers to your questions.

ACTIVITY

MEASURING THE EFFECT OF PASSIVE SOLAR HEATING

You will need two cardboard boxes, two thermometers, plastic wrap, scissors, and tape.

1. Cut a large hole, or "window," in one side of each cardboard box. **Caution: Be careful when using scissors.**

2. Cover each hole with plastic wrap. Tape the plastic wrap tightly to the box.

3. Put a thermometer inside each box.

4. Place each box in direct sunlight. Place the boxes so that one box has its "window" facing the sun and the other box has its "window" in the shade.

5. After about 20 minutes, open the boxes and read the temperature on each thermometer.

Questions

1. Which box got warmer?

2. How should the windows in a house be positioned if you want to heat the house using passive solar heating?

6-9 What is thermal expansion?

Objective ▶ Describe what happens to solids, liquids, and gases when they are heated.

TechTerm

▶ **thermal expansion:** expansion of a substance caused by heating

Expansion in Solids The expansion of solids and other substances when they are heated is called **thermal expansion.** Most solids expand, or get larger, when they are heated. Have you ever wondered why sidewalks have cracks between the squares? The cracks are there because the sidewalk expands on a hot day. Without the cracks, the sidewalk would buckle and break.

How can you explain thermal expansion? Remember what happens to the particles of a substance when it is heated. The particles of the substance move farther apart. As the particles move apart, the volume increases and the substance expands. What happens to the substance as it cools? The particles move closer together. The substance contracts, or gets smaller. Most solids contract when they are cooled.

▐▐▐▶*Describe:* What happens to most solids when they are heated? They expand.

Expansion in Gases Gases expand when they are heated and contract when they are cooled. Think of a hot-air balloon. As the air in the balloon is heated, the air particles move faster and farther apart. They hit the sides of the balloon. The volume of air inside the balloon increases and the balloon expands. As the volume of the air increases, its density decreases. This is why the balloon rises.

Expansion in Liquids Most liquids expand when they are heated. The particles of the liquid move farther apart as the liquid is heated. As the liquid is cooled, the particles move closer together. The liquid contracts. However, there is one exception to this rule. When water is cooled from 4 °C to 0 °C, it expands. Remember that water freezes at 0 °C. As water freezes, its volume increases. As the volume increases, the density decreases. Ice is less dense than liquid water. This is why ice floats.

▐▐▐▶*State:* What happens to water as it is cooled from 4 °C to 0 °C? It expands.

▐▐▐▶*Explain:* What makes a hot-air balloon expand? Heating the air increases its volume and expands the balloon.

LESSON SUMMARY

▶ Most solids expand when they are heated.

▶ Thermal expansion is the expansion of a substance caused by heating.

▶ Most liquids expand when they are heated.

▶ Gases expand when they are heated and contract when they are cooled.

CHECK *Complete the following.*

1. What would happen to a sidewalk on a hot day if there were no cracks between squares in the sidewalk?

2. What is the expansion of a solid as a result of heating called?

3. What happens to most solids when they are cooled?

4. What happens to most liquids when they are heated?

5. What happens when water is cooled from 4 °C to 0 °C?

6. Why does ice float?

7. What happens to gases when they are heated?

8. What happens to gases when they are cooled?

APPLY *Complete the following.*

9. **Relate:** What happens to the particles of a solid when the solid is heated? How is this related to thermal expansion?

10. **Hypothesize:** What would be the effect on living things if ice were more dense than liquid water?

State the Problem

Study the diagram.

State the problem for this experiment.

SCIENCE CONNECTION ◆○◆○◆○◆○◆○◆○◆○◆○

BIMETALLIC THERMOSTATS

Thermal expansion can be used to help control how a home is heated and cooled. A device called a thermostat regulates the temperature in most homes. Inside the thermostat is a strip made up of two metals. The device is called a bimetallic thermostat. The prefix "bi-" means two.

The two metals in the thermostat expand when they are heated. However, they expand at different rates. One of the metals will expand faster than the other. It will also contract faster when cooled. As a result, the metal strip bends when it is heated. It straightens out when it is cooled. This bending and unbending acts as a switch. The switch controls an electric circuit, which turns the home heating system on and off. For example, when the temperature in a room gets too high, the thermostat switches off the electric current. The heating system goes off, and the temperature drops. When the temperature gets too low, the thermostat switches the current back on. The heating system goes on, and the temperature rises.

Desired temperature

Actual temperature

Bimetallic strip

Wires to heating system

Switch

UNIT 6 Challenges

STUDY HINT Before you begin the Unit Challenges, review the TechTerms and Lesson Summary for each lesson in this unit.

TechTerms

absolute zero (96)
boiling point (100)
calorie (94)
Calorie (94)
conduction (102)
conductors (102)

convection (104)
convection currents (104)
evaporation (100)
freezing point (98)
heat (92)
insulators (102)

melting point (98)
radiation (106)
temperature (96)
thermal expansion (108)
vacuum (106)

TechTerm Challenges

Matching *Write the TechTerm that best matches each description.*

1. change from a liquid to a gas at the surface of the liquid evaporation   thermal expansion
2. expansion of a substance caused by heating
3. empty space vacuum
4. transfer of heat through space radiation
5. 1 kilocalorie Calorie
6. lowest possible temperature absolute zero

Identifying Word Relationships *Explain the difference between the words in each pair. Write your answers in complete sentences.*
Answers will vary. Accept all logical responses.

1. calorie, Calorie
2. heat, temperature
3. melting point, freezing point
4. conduction, convection
5. conductors, insulators
6. absolute zero, boiling point

Content Challenges

Multiple Choice *Write the letter of the term or phrase that best completes each statement.*

1. Before Count Rumford's experiment, people thought that heat was a
   a. force. **b.** liquid. **c.** form of energy. **d.** gas. b

2. A unit used to measure heat is the
   a. calorie. **b.** gram. **c.** degree Celsius. **d.** liter. a

3. The average kinetic energy of the particles of matter is called
   a. heat. **b.** temperature. **c.** caloric. **d.** absolute zero. b

4. Scientists now know that heat is a
   a. force. **b.** liquid. **c.** form of energy. **d.** gas. c

5. When you add or remove heat from a substance, you change its
   a. mass. **b.** weight. **c.** potential energy. **d.** temperature. d

6. You can smell perfume across a room because of
   a. sublimation. **b.** boiling. **c.** melting. **d.** evaporation. d

7. Water changes to ice at
   a. 0 °C. **b.** 22 °C. **c.** 100 °C. **d.** 212 °F. a

8. When water boils, it changes to
   a. ice. b. dry ice. c. steam. d. a solid. c
9. The freezing point of a substance is the same as its
   a. boiling point. b. sublimation point. c. evaporation point. d. melting point. d
10. Heat moves through a vacuum by
    a. conduction. b. convection. c. radiation. d. sublimation. c
11. Heat moves through a solid by
    a. conduction. b. convection. c. radiation. d. evaporation. a
12. Heat moves through gases and liquids by
    a. conduction. b. convection. c. radiation. d. sublimation. b
13. When most solids are heated, they
    a. melt. b. expand. c. burn. d. contract. b
14. Gases expand when they are
    a. heated. b. cooled. c. frozen. d. compressed. a

Completion *Write the term that best completes each sentence.*
1. Count Rumford discovered that mechanical energy could be changed into _____ . heat
2. One kilocalorie is equal to _____ Calorie(s). 1
3. Particles of matter are always in _____ . motion
4. When water boils, it changes to _____ . steam
5. When heat is added to a substance, its temperature _____ . rises
6. Temperature is a measure of the average _____ of matter. kinetic energy
7. Adding heat energy makes particles of matter move _____ . faster
8. One calorie will raise the temperature of 1 g of water from 10 °C to _____ . 11 °C
9. Absolute zero is equal to _____ °C. −273
10. Heat moves through _____ by conduction. solids
11. When water _____ , it changes to ice. freezes
12. As water changes to steam, its temperature stays at _____ °C. 100
13. Gases expand when they are heated and _____ when they are cooled. contract
14. All _____ are good conductors of heat. metals
15. Convection currents are _____ movements of air. up and down
16. The transfer of heat through _____ is called radiation. empty space
17. Heat travels through gases and _____ by convection. liquids
18. Water _____ as it freezes.
19. Heat from the sun can warm a house using a _____ heating system. solar
20. When most solids are cooled, they _____ . contract

Understanding the Features

Reading Critically *Use the feature reading selections to answer the following. Page numbers for the features are shown in parentheses.*

1. **Infer:** Why might cryogenic surgery be preferable to ordinary surgery? (95)
2. Why are devices that regulate temperature called "bimetallic" thermostats? (109)
3. **Compare:** If you were buying home insulation, would you choose a material with a high R-value or a low R-value? Why? (103)
4. In Rumford's experiment, what happened to the water around the cannon as the cannon was being drilled? (93)
5. Why is dry ice used to pack and ship frozen foods? (99)

Concept Challenges

Critical Thinking *Answer the following in complete sentences.*

1. **Compare:** What is the difference between temperature and heat?
2. Explain how thermal expansion is applied in a thermometer.
3. **Hypothesize:** Do you think it is possible to reach absolute zero in a laboratory? Why or why not?
4. Compare the transfer of heat by conduction, convection, and radiation.
5. How did Rumford's experiment disprove the caloric theory?

Interpreting a Table *Use the table to help you answer the following.*

Table 1 Freezing of Water	
TIME	TEMPERATURE
0 min	22 °C
5 min	15 °C
10 min	10 °C
15 min	6 °C
20 min	3 °C
25 min	0 °C
30 min	0 °C
35 min	0 °C

1. What does the table show?
2. What temperature scale is used in the table?
3. How else could the data in the table have been organized?
4. How long did it take for the water to begin to freeze?
5. What two science skills are you using when you find temperature?
6. **Infer:** Why did the temperature stay at 0 °C for 10 min?

Finding Out More

1. Use library references to find out how one of the following cooling systems works: air conditioner, refrigerator, dehumidifier. Present your findings in a written report.
2. Prepare a bulletin board display comparing different types of home heating systems, for example, hot water heating, radiant heating, steam heating, electric heating, solar heating.
3. Find out how a thermos bottle makes use of conduction, convection, and radiation to keep liquids hot or cold. Prepare a poster illustrating the parts of a thermos bottle.

WAVES

CONTENTS

STUDY HINT Before beginning Unit 7, scan through the lessons looking for words that you do not know. On a sheet of paper, list these words. Work with a classmate to try to define each word on your list.

7-1 What is a wave?

Objective ▶ Identify a wave as energy traveling through a medium.

TechTerms

▶ **medium:** substance through which waves can travel

▶ **waves:** disturbances that transfer energy from place to place

Waves and Energy Have you ever seen ocean **waves** crashing on a coastline? Waves are disturbances that transfer energy from place to place. Ocean waves carry energy along the surface of the water. Where does the energy come from? The energy of ocean waves comes from wind moving over the water.

Throw a stone into a still pond. What do you see? Small circular waves move outward along the surface of the pond. When the stone hits the pond, it has kinetic (ki-NET-ik) energy. Kinetic energy is energy of motion. Some of the stone's kinetic energy is transmitted to the water particles. The energy causes the particles to move. This movement produces a wave. The wave carries the energy across the surface of the pond.

▶*Define:* What are waves?
disturbances that transfer energy from place to place

Energy and Matter Water is a **medium** for waves. Any substance through which waves can travel is a medium. Air is a medium for sound waves. Some waves do not need a medium. Light waves can travel through the vacuum of space. When a wave travels through a medium, only energy moves from place to place. The particles of the medium do not move forward with the wave. Think of a cork floating on water. What happens as a wave moves past the cork? The cork moves up and down. It does not move in the same direction as the wave. The wave moves through the water.

▶*State:* Do all waves need a medium?
No, light waves can travel through a vacuum.

Particles in a Medium When a wave moves through a medium, the particles of the medium move in small circles. The diagram shows a wave moving through water. As the wave goes past, each water particle moves in a small circle. This is why a floating cork bobs up and down as a wave passes. The energy of the wave moves forward. The water does not move forward.

Direction of wave movement

▶*Describe:* What happens to water particles as a wave moves through the water?
The particles move in small circles.

LESSON SUMMARY

▶ Waves are disturbances that transfer energy from place to place.

▶ Any substance through which waves can travel is called a medium.

▶ The particles of a medium do not move in the same direction as a wave.

▶ The particles of a medium move in small circles.

CHECK *Complete the following.*

1. Water waves carry _____ . energy

2. Water is a _____ for waves. medium

3. The _____ energy of moving particles produces a wave. kinetic

4. Some waves, such as _____ waves, do not require a medium. light

5. Light waves can travel through a _____ vacuum .

6. When a wave moves through a medium, only _____ energy moves from place to place.

APPLY *Complete the following.*

7. Imagine you are sitting in a rowboat in the middle of a lake. A motorboat passes by, making waves that hit your boat. Describe what happens to your boat.

8. **Infer:** Waves can often be seen moving across fields of wheat. What is the medium for these waves?

9. **Hypothesize:** Earthquakes cause waves to travel through the earth. What is the source of energy for these waves?

Health and Safety Tip ....

During a hurricane or other large storm, a great deal of energy is carried by ocean waves. Storm waves can cause serious damage when they hit the shore. You should never be anywhere near the shore during a hurricane. Check with your local Red Cross, or use library references, to find out what safety precautions you should take during a hurricane.

LEISURE ACTIVITY

SURFING

Can you "hang ten"? Have you ever "cracked a wave"? These terms are used in surfing. When you hang ten, you hook your toes over the end of a surfboard. To crack a wave means to ride a big wave successfully.

Many people who live near the ocean enjoy the exciting sport of surfing. In surfboard riding, surfers try to catch a big wave and get their boards onto the crest of the wave. The surfers then stand up and try to ride the wave in to shore. Standing on a surfboard on top of a wave requires good balance and quick reflexes. Surfers also must be good swimmers.

Most surfboards used today are made of fiberglass. A surfboard is about 3 m long, 80 cm wide, and 8 cm thick. Surfboards can weigh from about 4 kg to 7 kg.

Surfing began in Hawaii hundreds of years ago. It is now a popular sport all over the world. In the United States, the best waves are found in Hawaii and southern California.

What are two kinds of waves?

Objective ▶ Classify waves as transverse or longitudinal.

TechTerms

▶ **compression** (kahm-PRE-shun): part of a medium where the particles are close together

▶ **crest:** high point of a wave

▶ **longitudinal** (lahn-juh-TOOD-un-ul) **wave:** wave in which the particles of the medium move back and forth in the direction of the wave motion

▶ **rarefaction** (rer-FAK-shun): part of a medium where the particles are far apart

▶ **transverse** (trans-VURS) **wave:** wave in which the particles of the medium move up and down at right angles to the direction of the wave motion

▶ **trough** (TROWF): low point of a wave

Kinds of Waves There are two kinds of waves. They are transverse (trans-VURS) waves and longitudinal (lahn-juh-TOOD-un-ul) waves. The difference between the two kinds of waves is in the way the particles of the medium move.

▌▶ *Name:* What are the two kinds of waves called? transverse and longitudinal

Direction of wave movement

Crest

Trough

Transverse Waves If you pull up and down on a rope tied to a doorknob, you can see **transverse waves.** In a transverse wave, the particles of the medium move up and down at right angles to the direction of motion of the wave.

There are two parts to a transverse wave. The **crest** is the high point of a transverse wave. The **trough** (TROWF) is the low point of the wave.

▌▶ *List:* What are the two parts of a transverse wave? crest and trough

Longitudinal Waves A clap of thunder is an example of a **longitudinal wave.** A longitudinal wave is a wave in which the particles of the medium move back and forth in the direction of the wave motion. The air is the medium that carries the energy of the thunder clap.

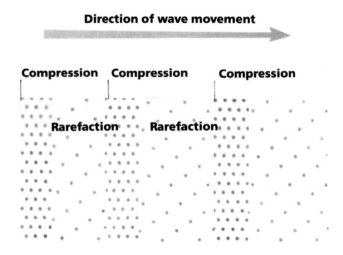

Direction of wave movement

Compression   **Compression**   **Compression**

Rarefaction   **Rarefaction**

A longitudinal wave has two parts. A clap of thunder pushes the particles of air close together. This part of the wave is called a **compression** (kahm-PRE-shun). The compressed particles move forward in the direction of the wave motion. As the particles move forward, they leave behind a part of the wave where the particles are far apart. This part of the wave is called a **rarefaction** (rer-FAK-shun). The rarefaction also moves forward.

▌▶ *Define:* What are rarefactions? parts of a longitudinal wave where the particles of the medium are far apart

LESSON SUMMARY

► The two kinds of waves are transverse and longitudinal.

► In a transverse wave, the particles of the medium move up and down at right angles to the direction of the wave motion.

► The parts of a transverse wave are the crest and the trough.

► In a longitudinal wave, the particles of the medium move back and forth in the same direction as the wave motion.

► The parts of a longitudinal wave are the compression and the rarefaction.

CHECK *Write true if the statement is true. If the statement is false, correct the underlined term to make the statement true.*

1. All waves carry <u>energy</u>. true

2. There are <u>three</u> different kinds of waves. two

3. The difference between waves depends on how the <u>particles</u> of the medium move. true

4. The particles of the medium move up and down in a <u>longitudinal</u> wave. transverse

5. The particles of the medium move back and forth in a <u>transverse</u> wave. longitudinal

6. The parts of a transverse wave are the crest and the <u>rarefaction</u>. trough

APPLY *Complete the following.*

➤ 7. **Infer:** Are ocean waves transverse waves or longitudinal waves? How do you know?

📁 8. **Classify:** Clap your hands together. What kind of wave did you make?

9. Have you ever seen fans do "the wave" at a baseball or football game? What kind of wave did they make?

Skill Builder

Building Vocabulary Knowing the meaning of certain words will help you to remember how those words are used in science. Look up the words "transverse," "longitudinal," "compression," and "rarefaction" in a dictionary. Write the definitions on a separate sheet of paper. How do the definitions of these words relate to the way they are used to describe waves?

SCIENCE CONNECTION ◆○◆○◆○◆○◆○◆○◆

EARTHQUAKE WAVES

In October 1989, a large earthquake struck San Francisco and Oakland, California. The earthquake registered 7.1 on the Richter scale. The Richter scale is a measure of how much energy an earthquake releases. The California earthquake of 1989 was very powerful.

The energy of an earthquake produces waves that travel through the earth. All earthquakes produce three main types of waves. These waves are called primary waves (P-waves), secondary waves (S-waves), and long waves (L-waves). L-waves are also called surface waves. They cause the surface of the earth to rise and fall. L-waves cause the greatest damage during an earthquake.

Scientists can record earthquake waves on an instrument called a seismograph (SYZ-muh-graf). P-waves, S-waves, and L-waves travel at different speeds. Scientists calculate the difference in arrival times of the three waves. They use this information to plot the epicenter of an earthquake.

What are the features of a wave?

Objectives ▶ Describe the features of a wave. ▶ Relate wave speed, frequency, and wavelength.

TechTerms

- ▶ **amplitude** (AM-pluh-tood): height of a wave
- ▶ **frequency** (FREE-kwun-see): number of complete waves passing a point in a given time
- ▶ **hertz** (HURTS): unit used to measure the frequency of a wave
- ▶ **wavelength:** distance between two neighboring crests or troughs

Features of Waves All waves have three basic features. These features are amplitude (AM-pluh-tood), wavelength, and frequency (FREE-kwun-see).

- ▶ When a wave moves through a medium, the particles of the medium are moved from their rest position. The distance the particles are moved is called the **amplitude,** or height, of the wave.

- ▶ All waves have a certain length. The distance from the crest or trough of one wave to the crest or trough of the next wave is the **wave-length.** Wavelength can be measured in meters or centimeters.

- ▶ A certain number of waves pass a point in a given amount of time. The number of complete waves per unit time is called the **frequency.** Frequency is measured in waves per second.

The diagram shows the relationship among amplitude, wavelength, and frequency.

▶ *List:* What are the three features of a wave?
amplitude, wavelength, and frequency

Speed of a Wave All waves move at a certain speed. The speed of a wave is related to the frequency and wavelength of the wave. Wave speed is equal to the frequency times the wavelength.

$$\text{speed} = \text{frequency} \times \text{wavelength}$$

Scientists use a unit called a **hertz** (HURTS) to measure frequency. One hertz (Hz) is equal to one wave per second. When frequency is measured in hertz and wavelength is measured in meters, speed is measured in meters per second (m/sec).

▶ *Identify:* What is the equation used to find the speed of a wave? speed = frequency × wavelength

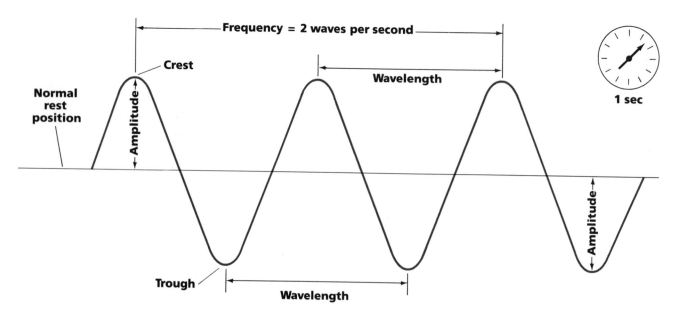

Frequency = 2 waves per second

Crest

Normal rest position

Amplitude

Wavelength

1 sec

Trough

Wavelength

Amplitude

LESSON SUMMARY

► All waves have three basic features.

► Amplitude is the height of a wave.

► Wavelength is the distance from crest to crest or from trough to trough.

► Frequency is the number of complete waves passing a point each second.

► The speed of a wave is equal to the frequency times the wavelength.

CHECK *Complete the following.*

1. All waves have amplitude, _____ , and frequency. wavelength

2. Amplitude is the _____ of a wave. height

3. Wavelength can be measured in _____ or centimeters. meters

4. The number of _____ passing a point in one second is called frequency. waves

5. The speed of a wave is equal to _____ times wavelength. frequency

6. The _____ is the unit used to measure frequency. hertz

7. One hertz is equal to one _____ per second. wave

APPLY *Use the equation speed = frequency × wavelength to complete the following.*

8. A wave has a frequency of 50 Hz and a wavelength of 10 m. What is the speed of the wave?

9. The speed of a wave is 5 m/sec. Its wavelength is 2 m. What is the frequency of the wave?

10. The frequency of a wave is 20 Hz. Its speed is 100 m/sec. What is the wavelength of the wave?

InfoSearch

Read the passage. Ask two questions about the topic that you cannot answer from the information in the passage.

Heinrich Hertz The unit used to measure the frequency of waves is named after Heinrich Hertz. Hertz was a German physicist. He discovered electromagnetic waves in the 1880s. He also showed that light waves are the same as electromagnetic waves.

SEARCH: Use library references to find answers to your questions.

ACTIVITY

OBSERVING WAVES IN A ROPE

You will need a piece of rope about 3 m long, a ribbon, and a doorknob.

1. Tie a brightly colored ribbon to the middle of a 3-m length of rope.

2. Tie one end of the rope to a doorknob.

3. Hold the other end of the rope and stand opposite the door. Quickly move your end of the rope up and down. Observe the motion of the ribbon.

4. Increase the speed at which you move the end of the rope up and down. Observe the resulting waves in the rope.

Questions

1. What happened to the ribbon when you moved your end of the rope?

2. What kind of waves did you make?

3. a. What happened to the frequency of the waves when you increased the speed of your movements? b. What happened to the wavelength?

7-4 How are waves reflected?

Objectives ▶ Describe what happens when a wave strikes a barrier. ▶ State the law of reflection.

TechTerms

- ▶ **incident wave:** wave that strikes a barrier
- ▶ **normal:** line at right angles to a barrier
- ▶ **reflected wave:** wave that bounces back from a barrier
- ▶ **reflection:** bouncing back of a wave after striking a barrier

Waves and Barriers What happens when a wave hits a barrier? Remember that all waves carry energy. Some of the wave's energy may be absorbed by the barrier. If the barrier does not absorb the wave's energy, the wave bounces back from the barrier. This bouncing back of a wave is called **reflection.**

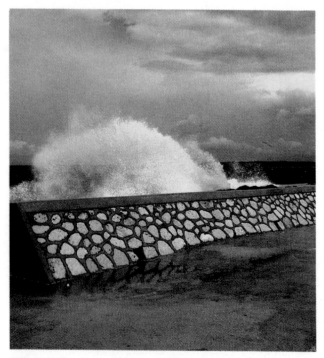

Figure 1

�'''''▶*Describe:* What happens when a wave strikes a barrier that does not absorb all of its energy? The wave bounces back from the barrier.

120

Reflection Figure 2 shows what happens when a wave strikes a barrier. The arrows show the direction of the wave. The wave that strikes the barrier is called the **incident wave.** The wave that bounces off the barrier is called the **reflected wave.**

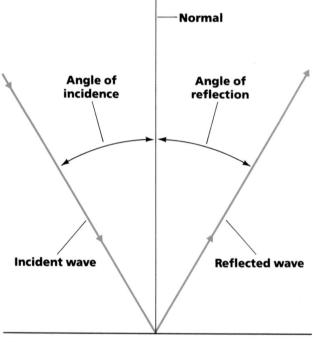

Figure 2

▐'''''▶*Define:* What is a reflected wave?
wave that bounces back from a barrier

Law of Reflection The law of reflection describes what happens when a wave is reflected from a barrier. The angle at which an incident wave strikes a barrier is called the angle of incidence, or i. The angle at which the wave is reflected is called the angle of reflection, or r. These angles are measured from a line called the **normal.** The normal is a line at a right angle to the barrier. A right angle is equal to 90°. The law of reflection states that the angle of incidence is equal to the angle of reflection. Suppose a wave strikes a barrier at a 45° angle. The reflected wave will bounce back from the barrier at a 45° angle.

▐'''''▶*State:* What is the law of reflection?
The angle of incidence is equal to the angle of reflection.

LESSON SUMMARY

▶ When a barrier does not absorb a wave's energy, the wave is reflected.

▶ A wave that strikes a barrier is called the incident wave.

▶ A wave that bounces back from a barrier is called the reflected wave.

▶ The law of reflection states that the angle of incidence is equal to the angle of reflection.

CHECK *Complete the following.*

1. What happens when a wave strikes a barrier?
2. What is reflection?
3. What is a wave that strikes a barrier called?
4. What is a wave that bounces back from a barrier called?
5. What is the normal?
6. What is the angle formed by the normal and the barrier?

APPLY *Complete the following.*

7. **Contrast:** What is the difference between the angle of incidence and the angle of reflection?

Use the diagram to complete the following.

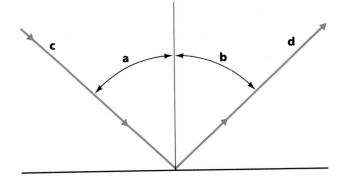

8. Which arrow represents the incident wave?
9. Which arrow represents the reflected wave?
10. Which angle is the angle of incidence?
11. Which angle is the angle of reflection?

Skill Builder ....

Researching Use library references to find out what standing waves are and how they are formed. Write a report of your findings. Include a diagram of standing waves in a rope.

ACTIVITY

MEASURING THE ANGLE OF INCIDENCE AND ANGLE OF REFLECTION

You will need a flat mirror, a flashlight, a sheet of paper, a protractor, a marking pen, and a ruler.

1. Draw a straight line across a sheet of paper.
2. Place a flat mirror on the paper so that the edge of the mirror is on the line.
3. Shine a flashlight at an angle onto the mirror.
4. Use a marking pen to trace the path of the beam.
5. Remove the flashlight and the mirror. Draw a line at a right angle to the line representing the mirror.
6. Label the angle of incidence and the angle of reflection. Use a protractor to measure the angle of incidence and the angle of reflection.

Mirror

Questions

1. What is the size of the angle of incidence?
2. What is the size of the angle of reflection?

7-5 How are waves refracted?

Objective ▶ Describe what happens to a wave when it moves from one medium to another.

TechTerm

▶ **refraction:** bending of a wave as it moves from one medium to another

Changing the Medium Waves travel in straight lines through a medium. What happens to a wave when it moves from one medium to another? Suppose a wave moves from air into water. If the wave enters the water at an angle, the wave bends. This bending of a wave as it moves from one medium to another is called **refraction.**

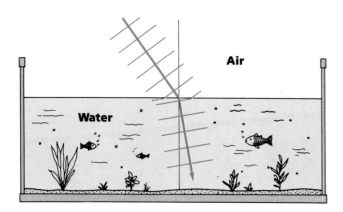

▶**Define:** What is refraction? bending of a wave as it moves at an angle from one medium to another

Refraction and Wave Speed Waves bend when they go from one medium to another because they change speed. Water is denser than air.

When a light wave moves from air into water, it slows down. When a light wave moves from water into air, it speeds up. This change in speed causes the wave to be refracted, or bent.

▶ When a wave moves at an angle from a less dense medium to a more dense medium, it is bent toward the normal.

▶ When a wave moves at an angle from a more dense medium to a less dense medium, it is bent away from the normal.

▶ When a wave moves from one medium to another along the normal, it is not bent.

You can see the results of refraction by performing a simple experiment. Place a pencil into a glass of water at an angle. The pencil appears to be broken where it enters the water. As light waves move from air into water, they slow down. This change in speed causes the light waves to bend. As a result, the pencil appears broken.

▶**Explain:** What causes refraction as waves move from one medium to another?
change in speed

Laws of Refraction The three laws of refraction describe how waves are refracted when they move from one medium to another.

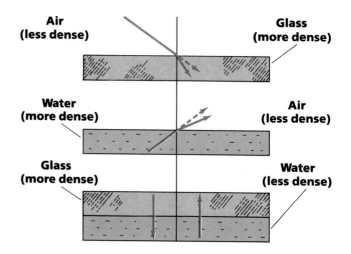

▶**Describe:** In what direction is a wave bent when it moves at an angle from a more dense medium to a less dense medium?
away from the normal

LESSON SUMMARY

▶ Waves are bent when they move at an angle from one medium to another.

▶ The bending, or refraction, of waves is caused by a change in wave speed.

▶ The laws of refraction describe how waves are refracted when they move from one medium to another.

CHECK *Complete the following.*

1. Waves travel through a medium in-_____ lines. straight

2. When a wave moves from one medium to another at an angle, it _____ . bends, or refracts

3. The bending of a wave is called _____ . refraction

4. Refraction is caused by a change in _____ . speed

5. The speed of a light wave _____ when it moves from water into air. increases

6. When a wave moves at an angle from a less dense medium to a more dense medium, it is bent _____ the normal. toward

APPLY *Complete the following.*

7. **Analyze:** Copy the diagrams onto a separate sheet of paper. Draw arrows to show how the light waves will be refracted in each example.

8. **Hypothesize:** Have you ever tried to pick up a seashell under water? Why do you think the seashell appeared closer to the surface of the water than it really was?

InfoSearch

Read the passage. Ask two questions about the topic that you cannot answer from the information in the passage.

Ocean Waves Ocean waves change direction as they come close to a shoreline. Waves almost always approach the shore at an angle. However, they usually hit the shore straight on. The speed of waves is slower in shallow water. As waves approach the shallow water near shore, they slow down. This change in speed causes the waves to change direction.

SEARCH: Use library references to find answers to your questions.

ACTIVITY

OBSERVING THE EFFECTS OF REFRACTION

You will need a small bowl, water, and a penny.

1. Place a penny into a small bowl.

2. Move away from the bowl until you can no longer see the penny.

3. Without changing your position, have a partner add water to the bowl until you can see the penny again.

Questions

1. Why could you not see the penny when you moved away from the bowl?

2. Why could you see the penny again after water was added to the bowl?

3. Draw a diagram showing how light waves reflected from the penny were refracted as they moved from the water into the air.

7-6 What is the Doppler effect?

Objective ▶ Explain what is meant by the Doppler effect.

TechTerm

▶ **Doppler effect:** apparent change in the frequency of waves

Changing Frequency The frequency of a wave can sometimes appear to change. Remember that frequency is the number of complete waves that pass a point each second. Frequency seems to change when a wave source moves toward you or away from you. Imagine you are sitting on a dock. You can count the number of waves hitting the dock. As a motorboat comes toward you, many waves hit the dock. The frequency of the waves is high. As the boat heads away from the dock, fewer waves hit the dock. The frequency of the waves is low.

▶ *Compare:* Will you be able to count more waves when a boat is approaching or heading away from a dock? approaching

Doppler Effect An apparent change in the frequency of waves is called the **Doppler effect.** The Doppler effect occurs when there is relative motion between the source of the waves and an observer. The frequency of waves appears to change when the observer is moving toward or away from the source of the waves. The frequency also seems to change when the source of the waves is moving and the observer is standing still. For the Doppler effect to take place, either the source or the observer must be moving.

▶ *Define:* What is the Doppler effect?
apparent change in the frequency of waves

Doppler Effect and Sound You are probably most familiar with the Doppler effect in sound waves. The frequency of sound waves changes as the source of the waves moves toward or away from you. Suppose you are waiting for a train to pass a crossing. You can hear the train whistle as the train approaches the crossing. The waves are pushed close together by the moving train. Many waves per second reach your ears. The sound waves appear to have a high frequency. As the train passes you, the sound waves spread out. Fewer waves reach your ears each second. The frequency of the waves appears to be lower.

Apparent frequency

▶ *Compare:* Is the frequency of sound waves from a train whistle high or low as the train comes toward you? high

LESSON SUMMARY

► The frequency of waves can sometimes appear to change.

► The Doppler effect is caused by motion of the observer or the source of the waves.

► The frequency of sound waves changes as the source of the waves moves toward or away from you.

CHECK *Complete the following.*

1. The apparent change in the frequency of waves is called the _____ . Doppler effect

2. Waves hit a dock more often when a boat is moving _____ the dock. toward

3. The Doppler effect is caused by _____ . motion

4. The frequency of waves appears to change when either the source of the waves or the _____ is moving. observer

5. The frequency of sound waves is _____ when the source of the waves is moving away from you. low

6. As a boat comes toward you, the frequency of the water waves appears _____ than if the boat was moving away from you. higher

APPLY *Complete the following.*

7. Describe how the sound of a car horn changes as the car approaches and then passes you.

8. **Predict:** Suppose you are driving by a fire station. An alarm is ringing at the station. How will the alarm sound as you approach the station? How will the sound change as you drive away?

Ideas in Action

IDEA: Many police forces use Doppler radar to measure the speed of passing cars.

ACTION: Visit your local police station. Find out how Doppler radar works. How do radar detectors in cars work? How is Doppler radar used to track the path of fast-moving storms? What are other uses of Doppler radar? Describe your findings to the class.

SCIENCE CONNECTION

RED SHIFT

Astronomers study the wavelength of light from the stars to find out about the universe. The wavelength of light determines the color of the light. Red light has a long wavelength. Blue light has a short wavelength.

The Doppler effect causes an apparent change in the frequency of a wave. It also causes an apparent change in wavelength. If a light source is moving away from an observer, the wavelength of the light appears to change. The light appears redder than it would if the source were not moving. This change in wavelength is called the red shift.

Astronomers studied the red shifts of many different galaxies. They found that all of the galaxies are moving away from the earth. Each galaxy is also moving away from every other galaxy. This means that the universe is expanding, or getting bigger. Think of a raisin in a cake. As the cake bakes, it expands. Each raisin moves away from every other raisin.

UNIT 7 Challenges

STUDY HINT Before you begin the Unit Challenges, review the TechTerms and Lesson Summary for each lesson in this unit.

TechTerms

amplitude (118)
compression (116)
crest (116)
Doppler effect (124)
frequency (118)
hertz (118)

incident wave (120)
longitudinal wave (116)
medium (114)
normal (120)
rarefaction (116)
reflected wave (120)

reflection (120)
refraction (122)
transverse wave (116)
trough (116)
wavelength (118)
waves (114)

TechTerm Challenges

Matching *Write the TechTerm that matches each description.*

1. substance through which waves can travel medium
2. high point of a wave crest
3. height of a wave amplitude
4. unit used to measure frequency hertz
5. wave that strikes a barrier incident wave
6. apparent change in frequency Doppler effect
7. line at 90° to a barrier normal
8. disturbances in a medium waves

Applying Definitions *Explain the difference between the words in each pair. Write your answers in complete sentences.* Answers will vary. Accept all logical responses.

1. compression, rarefaction
2. longitudinal wave, transverse wave
3. crest, trough
4. frequency, wavelength
5. reflected wave, incident wave
6. reflection, refraction

Content Challenges

Multiple Choice *Write the letter of the term or phrase that best completes each statement.*

1. Waves are caused by
   a. potential energy.   **b.** kinetic energy.   **c.** heat energy.   **d.** nuclear energy. b

2. Sound waves cannot travel through
   a. air.   **b.** water.   **c.** metal.   **d.** space. d

3. When a transverse wave moves through a medium, the particles of the medium move
   a. in circles.   **b.** up and down.   **c.** forward.   **d.** backward. d

4. The two kinds of waves are transverse and
   a. circular.   **b.** normal.   **c.** longitudinal.   **d.** compression. c

5. The crest of a wave is the wave's
   a. low point.   **b.** length.   **c.** speed.   **d.** high point. d

6. In a transverse wave, the particles of the medium move
   a. up and down.   **b.** back and forth.   **c.** forward.   **d.** backward. a

126

7. In a rarefaction, the particles are
   a. squeezed together.   **b.** lined up.   **c.** spread apart.   **d.** not moving. c

8. Wavelength can be measured in
   a. meters.   **b.** hertz.   **c.** angles.   **d.** number of waves. a

9. The speed of a wave is equal to the frequency of the wave times the
   a. amplitude.   **b.** wavelength.   **c.** height.   **d.** medium. b

10. The angle between the normal and a barrier is equal to
    a. 45°.   **b.** 90°.   **c.** 180°.   **d.** 360°. b

11. The angle of incidence is equal to the angle of
    a. refraction.   **b.** compression.   **c.** rarefaction.   **d.** reflection. d

12. When a wave is refracted it is
    a. bent.   **b.** bounced back.   **c.** compressed.   **d.** spread apart. a

13. A wave is refracted when its
    a. amplitude changes.   **b.** frequency changes.   **c.** speed changes.   **d.** wavelength changes. c

14. The Doppler effect is an apparent change in a wave's
    a. speed.   **b.** medium.   **c.** amplitude.   **d.** frequency. d

Completion *Write the term that best completes each statement.*

1. Waves transfer _____ from place to place. energy
2. Most waves travel through a substance called a _____ . medium
3. There are _____ kinds of waves. two
4. In a transverse wave, the particles of the medium move _____ . up and down
5. In a longitudinal wave, the particles of the medium move _____ . back and forth
6. The two parts of a longitudinal wave are the _____ and the rarefaction. compression
7. A thunder clap is an example of a _____ wave. longitudinal
8. The basic features of all waves are _____ , wavelength, and frequency. amplitude
9. Frequency is measured in waves per second, or _____ . hertz
10. A wave is refracted when a barrier does not _____ all of its energy. absorb
11. The angle at which a wave strikes a barrier is the angle of _____ . incidence
12. Waves are refracted when they move from one _____ into another. medium
13. There are _____ laws of refraction. three
14. Frequency is the number of complete _____ that pass a point each second. waves

Understanding the Features . . . .

Reading Critically *Use the feature reading selections to answer the following. Page numbers for the features are shown in parentheses.*

► 1. **Infer:** Why must surfers be good swimmers in addition to having good balance and quick reflexes? (113)

2. What did astronomers learn by studying the red shifts of different galaxies? (123)

3. How many different kinds of earthquake waves are there? What are they called? (115)

Interpreting a Diagram *Use the diagram to answer the following.*

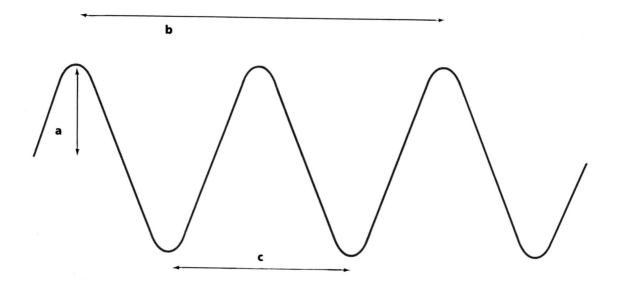

1. Which letter represents the wavelength of the wave?
2. Which letter represents the amplitude?
3. Which letter represents the frequency?
4. What is the relationship between speed, wavelength, and frequency?
5. If the frequency of a wave increases, and the speed stays the same, what happens to the wavelength?
6. If the speed of a wave does not change, but the wavelength increases, what happens to the frequency?

Critical Thinking *Answer each of the following in complete sentences.*

1. **Hypothesize:** You can make transverse waves in a rope tied to a doorknob. What happens to the waves when they reach the door?

2. **Compare:** How are the crests and troughs of a transverse wave like the compressions and rarefactions of a longitudinal wave?

3. **Hypothesize:** Suppose that light waves needed a medium to travel through. How do you think the world would be different?

1. Try this at home. Fill a fish tank with water. Add some food coloring to the water. Cut a thin slit in a piece of cardboard. Tape the cardboard over the front end of a flashlight. Shine a beam of light at an angle into the water. Then shine the beam of light straight down into the water. Describe what happens to the beam of light in each case. Explain your observations based on what you learned in this unit.

2. Use library references to find out about the three types of earthquake waves. What kind of waves are they? How does studying earthquake waves help scientists learn more about earthquakes? Write a report of your findings.

SOUND

CONTENTS

STUDY HINT As you read each lesson in Unit 8, write the lesson title and lesson objectives on a sheet of paper. After you complete each lesson, write the sentence or sentences that answer each objective.

129

8-1 What is sound?

Objective ▶ Identify sound as a form of energy that travels as waves.

TechTerms

▶ **sound:** form of energy that travels as waves
▶ **vibration:** rapid back-and-forth movement

Sound and Energy There are sounds all around you. Some sounds are loud and others are very faint. Walking in a park, you may hear the sounds of birds singing and dogs barking. In a quiet room, you may hear the sound of a wristwatch ticking. **Sound** is a form of energy. Sound energy travels in the form of waves.

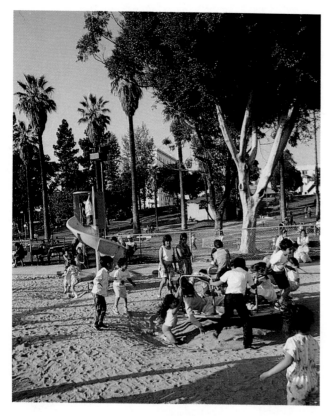

‖‖‖▶*Define:* What is sound? form of energy that travels as waves

Vibrations All sounds are caused by vibrations. A **vibration** is a rapid back-and-forth movement. Suppose you are listening to the sound from a stereo speaker. If you place your hand on the speaker, you will feel a vibration.

All vibrating objects produce sound. Place your fingers on your windpipe. When you speak, you can feel the vibration from the sound of your voice. When you feel the vibration from a sound, the sound has traveled through an object. When you hear a sound, the vibration has traveled through the air.

‖‖‖▶*Define:* What is a vibration? rapid back-and-forth movement

Source of Sound For every sound that you hear, some object is vibrating. Strum a guitar. The guitar strings vibrate to make a sound. When a string vibrates, it pushes on the air in front of it. Air particles are squeezed together. The squeezed-together particles form a compression (kum-PRESH-un). When the string moves back, the air particles spread apart. The spread-out particles form a rarefaction (rer-FAK-shun). As the string keeps vibrating, compressions and rarefactions move away from the string. They form a sound wave.

‖‖‖▶*Describe:* What forms a sound wave?
compressions and rarefactions

LESSON SUMMARY

▶ Sound is a form of energy that travels as waves.

▶ All sounds are caused by vibrations.

▶ All vibrating objects produce sound.

▶ Sound waves are caused by compressions and rarefactions.

CHECK *Complete the following.*

1. A sound is caused by a _____ . vibration

2. Sound is a form of _____ that travels as waves. energy

3. A _____ is a rapid back-and-forth movement. vibration

4. All vibrating objects produce _____ . sound

5. Sound waves are formed by compressions and _____ . rarefactions

6. If you place your fingers on your windpipe while you are speaking, you will feel a _____ . vibration

7. When you strum a guitar, the __strings__ vibrate.

APPLY *Complete the following.*

▶ **8. Predict:** What will happen if you place a glass of water on a stereo speaker?

9. Hypothesize: Is it possible to make a sound without causing a vibration? Why or why not?

10. Observe: Sit by an open window. What sounds do you hear? What is causing each sound?

Ideas in Action ....

IDEA: All vibrating objects produce sounds.
ACTION: Study the sources of different sounds. Identify the vibration that is causing the sound. Is there always a vibrating object that is the source of the sound?

Ideas in Action ....

IDEA: Sound can travel through solid objects.
ACTION: Place your ear against a wall in your home. Can you hear sounds in the next room?

ACTIVITY

OBSERVING VIBRATIONS IN A TUNING FORK

You will need a tuning fork and a glass of water.

1. Place the glass of water on a flat surface. Let the glass sit for a few minutes until the water is still.

2. Gently strike the tuning fork against the edge of a table. Observe what happens.

3. Strike the edge of the table again with the tuning fork. Put the ends of the tuning fork into the glass of water. Observe what happens to the water in the glass.

Questions

1. What happened to the tuning fork when you struck it against the table? Did you hear a sound?

2. What happened when you put the tuning fork into the glass of water?

How do sound waves travel?

Objective ► Describe how sound waves travel through a medium.

TechTerms

- **longitudinal** (lahn-juh-TOOD-un-ul) **wave:** wave in which the particles of the medium move back and forth in the direction of the wave motion
- **medium:** substance through which waves can travel

Sound Waves Sound waves are **longitudinal** (lahn-juh-TOOD-un-ul) **waves.** In a longitudinal wave, the particles of air move back and forth in the direction of the wave motion. Figure 1 shows a sound wave as it moves through the air. Compressions and rarefactions of the air particles move in the same direction as the sound wave.

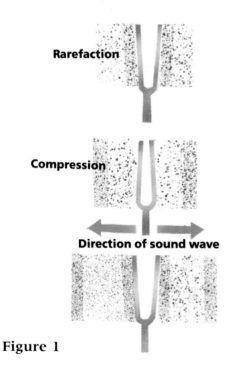

Figure 1

▐▐▐►**Describe:** Why are sound waves longitudinal waves? The particles in the air move back and forth in the direction of the wave motion.

Medium for Sound Air is a **medium** for sound waves. A medium is a substance through which

waves can travel. All sound waves need a medium through which they can travel. A sound wave can travel through gases, liquids, and solids. In each medium, the sound wave is made up of compressions and rarefactions of the particles that make up the medium.

Figure 2

Dolphins and whales use sound to communicate with each other. They can do this because sounds travel easily through water. Sounds also travel easily through solids. If you put your ear against a wall, you can hear sounds in the next room. The sounds travel through the solid wall.

▐▐▐►*Explain:* Why can dolphins use sounds to communicate? Sounds travel easily through water.

Sound in a Vacuum Astronauts on the moon could not use sound to communicate. There is no air on the moon. In a vacuum, there are no particles of a medium. The compressions and rarefactions of sound waves cannot travel in a vacuum.

►*Infer:* Would astronauts be able to hear an explosion on the moon? No, because there is no air on the moon.

LESSON SUMMARY

▶ Sound waves are longitudinal waves.

▶ All sound waves need a medium through which to travel.

▶ Sound waves cannot travel through a vacuum.

CHECK *Complete the following.*

1. In a _____ wave, the particles of the medium move in the same direction as the wave motion. longitudinal

2. A sound wave needs a _____ in which to travel. medium

3. Astronauts on the moon cannot use _____ waves to communicate. sound

4. Sound waves are made up of moving compressions and _____ . rarefactions

5. Whales can use sounds to communicate because sound waves can travel through a _____ . liquid

6. Sound cannot be heard in a vacuum because there is no _medium_ for the sound waves.

7. Sound can travel through liquid water, gases such as air, and _____ walls. solid

APPLY *Complete the following.*

▶ 8. **Infer:** If you clapped your hands on the moon, would you hear a sound? Why or why not?

9. **Classify:** Which of the following is a longitudinal wave?
   a. a vibrating guitar string **b.** an ocean wave **c.** a wave made by moving a string up and down

State the Problem

Study the illustration below.

Vacuum

State the problem for this experiment.

ACTIVITY

MAKING A STRING TELEPHONE

You will need two paper or styrofoam cups and about 10 meters of string. Work with a partner.

1. Make a hole in the bottom of each cup.

2. Insert one end of the string through the hole in each cup. Tie a knot at each end of the string so that it does not slip out of the holes.

3. You should take one of the cups, and your partner should take the other. Walk as far apart as the length of string will allow.

4. Have your partner speak into the cup while you hold your cup to your ear. Take turns speaking into the cup.

Questions

1. Could you hear your partner's voice? How clearly is the sound transmitted?

2. How does the sound travel from one cup to the other?

3. Will the string telephone still work with longer pieces of string? Try it and find out.

What is the speed of sound?

Objective ▶ Describe how the speed of sound changes in different materials.

TechTerm

▶ **supersonic:** faster than the speed of sound

Speed of Sound The speed of sound in air is about 340 m/sec. However, the speed of sound is much slower than the speed of light. In a thunderstorm, a flash of lightning is seen before the sound of thunder is heard. The light from the lightning flash travels much faster than the sound of the thunder.

| Table 1 Speed of Sound | |
MATERIAL	SPEED (m/sec)
Air	346
Water	1498
Steel	5200
Rubber	60
Glass	4540
Wood	1850
Cork	500

◾*Analyze:* In which of the materials in Table 1 does sound travel fastest? steel

Faster than Sound Jet airplanes can travel faster than sound. A **supersonic** speed is faster than the speed of sound. Some airplanes can travel two or three times the speed of sound. These planes are moving so fast that you can see them in the sky before you can hear them. Figure 2 shows the first airplane to fly faster than the speed of sound.

Figure 1

You can use the speed of sound to calculate how far away a lightning flash is. Start counting when you see a flash of lightning. Stop counting when you hear the thunder. Sound travels about 1 km in 3 sec. If it takes 3 sec for the sound of the thunder to reach you, the lightning is about 1 km away.

▐▶*State:* What is the speed of sound in air?

about 340 km/sec

Speed in Different Materials Sound travels at different speeds in different materials. In water, sound travels at about 1500 m/sec. In steel, sound travels at about 5000 m/sec. Table 1 lists the speed of sound in some different materials.

Figure 2

▐▶*Define:* What is a supersonic speed?

faster than the speed of sound

LESSON SUMMARY

▶ The speed of sound is much slower than the speed of light.

▶ Sound travels at different speeds in different materials.

▶ Jet airplanes can reach supersonic speeds.

CHECK *Complete the following.*

1. The speed of sound in air is about _____ . 340 m/sec

2. You can see a <u>supersonic</u> airplane before you hear it because it is moving faster than sound.

3. Light travels _____ than sound. faster

4. A flash of lightning can be seen _____ the sound of thunder can be heard. before

APPLY *Use the information in Table 1 on page 134 to complete the following.*

5. **Compare:** Would the sound of an approaching train travel faster through the metal railroad tracks or through the air? Explain.

6. **Hypothesize: a.** Why do you think cork is sometimes used for soundproofing a room?

b. What other solid listed in Table 1 would be good for soundproofing? Explain.

7. **Calculate:** How long would sound take to travel through 2 m of the following materials?

a. air   e. glass
b. water   f. wood
c. steel   g. cork
d. rubber

Designing an Experiment

Design an experiment to solve the problem.

PROBLEM: How can you show that the speed of sound is faster in steel than in air?

Your experiment should:

1. List the materials you would need.

2. Identify safety precautions that should be followed.

3. List a step-by-step procedure.

4. Describe how you would record your data.

TECHNOLOGY AND SOCIETY

SUPERSONIC AIRPLANES

The speed of sound is called Mach 1. Before October 14, 1947, no airplane had ever reached Mach 1. On that day, Chuck Yeager became the first person to fly faster than the speed of sound. He had broken the sound barrier.

Today, supersonic planes fly at speeds of Mach 2 or even Mach 3. Supersonic planes make it more convenient to get from one place to another. For example, the supersonic *Concorde* can cross the Atlantic Ocean in much less time than other passenger airplanes.

One disadvantage of planes that travel at supersonic speeds is that they produce loud noises called sonic booms. A sonic boom is caused when a plane breaks the sound barrier. Sonic booms can be so loud that they can break windows and knock pictures off walls. To avoid the problem of sonic booms, supersonic planes are not allowed to fly over heavily populated areas.

What is an echo?

Objective ▶ Identify an echo as reflected sound waves.

TechTerm

▶ **echo:** reflected sound waves

Reflected Waves Reflection is the bouncing back of a wave from a barrier that does not absorb its energy. If you clap your hands in a carpeted room, the sound will be absorbed by the carpeting. Suppose you clap your hands in an empty room with a wooden floor. You will hear the sound of the clap reflected from the floor and the walls. When a sound wave is reflected, the reflected sound is called an **echo.**

▶**Identify:** What is a reflected sound wave called? an echo

Echolocation Some animals use echoes to help them survive. Bats use echoes as a way to "see" objects around them. This is called echolocation (ek-oh-loh-KAY-shun). Bats can fly in the dark without bumping into anything. The bats give off high-pitched sounds. These sounds reflect off surrounding objects. With their very sensitive ears,

the bats can also locate insects from the echoes they give off. Bats also use echoes to communicate with one another.

Dolphins, like bats, give off high-pitched sounds. These sounds are reflected from objects in the ocean. Because sound travels faster in water than in air, dolphins can use echoes to communicate over long distances.

▶**Describe:** How do bats and dolphins use echoes? They give off high-pitched sounds that are reflected from objects.

Transmitter Receiver

Echo

SONAR Echoes can be used to find distances under water. This method of using echoes is called SONAR. The word "SONAR" stands for **SO**und **N**avigation **A**nd **R**anging. Sound waves are sent from a ship to the bottom of the ocean. The time the sound takes to reach the bottom and bounce back as an echo is measured. This time is then divided in half to find how long the sound took to go one way. Suppose sound waves were sent out and returned in 4 sec. The time for the sound waves to go one way is 2 sec. The speed of sound in water is about 1500 m/sec. Therefore, the water is about 3000 m deep. SONAR can be used to make maps of the ocean floor. It is also used to find schools of fish.

▶**Infer:** How can SONAR be used to find schools of fish? Sound waves bounce off the fish just as they bounce off the ocean floor.

LESSON SUMMARY

▶ A reflected sound wave is called an echo.

▶ Some animals, such as bats and dolphins, use echoes to help them survive.

▶ SONAR is a method of using echoes to find distances under water.

CHECK *Complete the following.*

1. Sound waves travel _____ in water than in air. faster

2. Bats have very sensitive _____ for hearing echoes. ears

3. Maps of the ocean floor can be drawn by using _____ . SONAR

4. Sound waves that are reflected from a surface are called _____ . echoes

5. When a sound wave is not reflected, the energy of the sound has been _____ by the surroundings. absorbed

APPLY *Complete the following.*

▶ **6. Infer:** The sounds of people in a gymnasium echo off the walls and floors of the gymnasium. What does this tell you about the material used to build the gym?

▶ **7. Predict:** Libraries are built to reduce echoes and other sounds. What kinds of materials would be used in building a library?

8. Calculate: A SONAR signal is sent from a research ship to the bottom of the ocean. The echo returns to the ship 6 sec later. How deep is the ocean at that spot?

InfoSearch

Read the passage. Ask two questions about the topic that you cannot answer from the information in the passage.

Ultrasound A technique similar to SONAR can be used to see inside solid objects. The technique is called ultrasound. Ultrasound uses very high-pitched sounds. Most of the sound waves pass through the object, but some are reflected. These echoes produce an image on a screen. Ultrasound images are similar to X-ray images. Ultrasound has many applications in medicine, engineering, and construction.

SEARCH: Use library references to find answers to your questions.

CAREER IN PHYSICAL SCIENCE

ACOUSTICS ENGINEER

Acoustics (uh-KOOS-tiks) is the science of sound. Many acoustics engineers work on the design of concert halls. In a concert hall, the sound of music has to be clear enough for everyone in the hall to be able to hear the music clearly. Many concert halls have "dead spots" where it is hard to hear the music clearly.

An acoustics engineer tries to decrease the number of echoes from the walls of a concert hall. Too many echoes would distort the music. The engineer would also make sure that not too much sound was absorbed, so that people in the back rows could hear the music. He or she would find the correct materials to use for the walls, floor, and ceiling. The design of a concert hall must allow the sound quality to be the same throughout the hall.

An acoustics engineer needs to study physics and engineering. Several years of mathematics and experience with computers are also helpful. For more information, write to the National Society of Professional Engineers, 2029 K Street NW, Washington, DC 20006.

8-5 What is intensity?

Objective ▶ Explain the relationship between the intensity and loudness of a sound.

TechTerms

- ▶ **decibel:** unit used to measure the intensity or loudness of a sound
- ▶ **intensity:** amount of energy in a sound wave

Sound Level Some sounds are loud. Others are soft. Sound is a form of energy. Loud sounds have more energy than soft sounds. Figure 1 shows the wave pattern for two sounds, one loud and one soft.

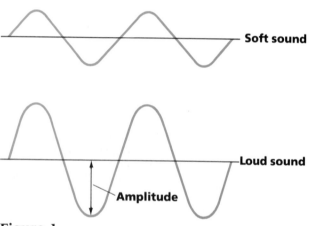

Figure 1

The pattern is the same for both waves. However, the loud sound has a larger amplitude, or height, than the soft sound. The more energy a sound has, the larger the amplitude of the sound wave.

▶ **Contrast:** How does a sound wave for a loud sound differ from that for a soft sound?
The amplitude of the wave is larger for a loud sound.

Intensity The **intensity** of a sound wave is the amount of energy it has. Intensity determines the loudness of a sound. A loud sound has a larger amplitude and more energy than a soft sound. A clap of thunder sounds louder than a drum beat. The thunder clap has more intensity than the drum beat.

Figure 2

▶ **Define:** What is intensity?
amount of energy in a sound wave

Measuring Intensity Intensity is measured in units called **decibels** (dB). The sound of people talking has an intensity of about 65 decibels. The softest sound you can hear is 0 decibels. The sound of a train going by may have an intensity of 95 decibels. Sounds louder than 120 decibels are dangerous. They can damage your ears. Table 1 shows the intensity of different sounds, measured in decibels.

▶ **Analyze:** What is the intensity of the sound at a rock concert? 115–120 decibels

Table 1 Intensity of Sound	
SOUND	INTENSITY (dB)
Whisper	10–20
Soft music	30
Conversation	60–70
Heavy traffic	70–80
Loud music	100
Thunder	110
Rock concert	115–120
Jet engine	170
Rocket engine	200

LESSON SUMMARY

▶ Loud sounds have more energy than soft sounds.

▶ A loud sound has a larger amplitude than a soft sound.

▶ The intensity of a sound wave determines the loudness of the sound.

▶ Intensity is measured in decibels (dB).

CHECK *Complete the following.*

1. When you raise the volume on a radio, you are increasing the amplitude of the ___sound waves___.

2. The _____ in a sound wave determines the amplitude of the wave. energy

3. The _____ of a sound wave is the amount of energy it has. intensity

4. The unit used to measure intensity is the _____ . decibel

5. The more intensity a sound wave has the _____ the sound. louder

6. The intensity of a normal conversation is usually _____ decibels. 60–70

APPLY *Complete the following.*

7. **Calculate:** How much louder is the sound of a jet engine than the sound of thunder?

8. **Hypothesize:** Two people hear a loud sound. One person is 10 m away from the source of the sound. The other person is 50 m away. The person standing 50 m away does not hear as loud a sound as the person standing 10 m away. What does this tell you about the intensity of a sound wave?

9. Which is more damaging to your ears, heavy traffic noise or a rock concert? Explain.

Health and Safety Tip

Your ears are very sensitive to sound. You should treat them with care. Too many loud sounds can damage your hearing. An intensity of 120 decibels is called the threshold of pain. This means that a sound louder than 120 dB is so loud that it will hurt your ears. Use library references to find out how to protect yourself from dangerously loud sounds.

SCIENCE CONNECTION

NOISE POLLUTION

The world is full of sounds. There are many sources of sound. Some of the loudest sounds come from machines. In some parts of the world, loud sounds may cause noise pollution. Any unwanted sound is called noise. In cities, the sounds of street traffic, of machines at construction sites, and of airplanes all combine to cause noise pollution.

Noise pollution is a nuisance, but it can also be a hazard. Too much noise can damage the ears. Noise can also cause stress. People who live or work around too much noise may have high blood pressure or nervous tension.

There are ways to decrease noise pollution. People are encouraged to use mass transit to decrease the amount of traffic on city streets. Laws have been passed to keep people from playing radios in public places. What can you do to help prevent noise pollution?

What are frequency and pitch?

Objective ▶ Explain how frequency and pitch are related.

TechTerm

▶ **pitch:** how high or low a sound is

▶ **ultrasonic** (ul-truh-SAHN-ik): above 20,000 Hz frequency

Frequency of a Sound Different sounds have different frequencies. The frequency of a wave is the number of complete waves per second. Each pair of compressions and rarefactions is a complete sound wave. The frequency of a sound wave is the number of compressions and rarefactions per second. Frequency is measured in hertz (Hz). One hertz is equal to one wave per second.

▶*Identify:* What is frequency?
number of complete waves per second

Frequency and Pitch Frequency and pitch are related. A sound's **pitch** tells how high or low a sound is. A high-frequency sound has a high pitch. A low-frequency sound has a low pitch. A whistle has a frequency of about 1000 Hz. It has a high pitch. Thunder has a frequency of about 50 Hz. Thunder has a low pitch.

Figure 1 High pitch

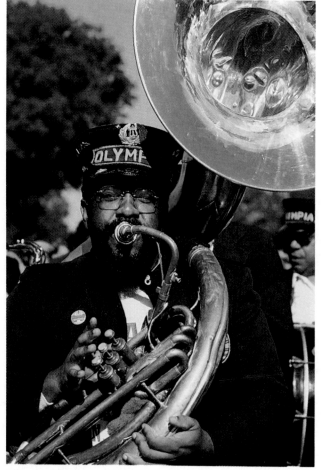

Figure 2 Low pitch

▶*Describe:* How are the frequency and pitch of a sound related? A high-frequency sound has a high pitch; a low-frequency sound has a low pitch.

Range of Hearing Humans can hear sounds with frequencies between 20 Hz and 20,000 Hz. This is called the range of human hearing. Sounds that have frequencies higher than 20,000 Hz are called **ultrasonic** (ul-truh-SAHN-iks). Ultrasonic sounds have too high a pitch to be heard by humans. The range of human hearing seems to decrease as a person gets older. Children can hear higher-frequency sounds than adults can.

Some animals can hear ultrasonic sounds. Dogs can hear sounds with frequencies up to about 50,000 Hz. Dolphins and bats can hear sounds with even higher frequencies. They use high-frequency sounds to communicate.

▶*Define:* What is an ultrasonic sound?
sound with a frequency higher than 20,000 Hz

LESSON SUMMARY

▶ Different sounds have different frequencies.

▶ A high-frequency sound has a high pitch; a low-frequency sound has a low pitch.

▶ Humans can hear sounds between 20 Hz and 20,000 Hz.

▶ Some animals can hear ultrasonic sounds.

CHECK *Complete the following.*

1. Dolphins can hear _____ sounds. ultrasonic

2. The unit used to measure frequency is the _____ . hertz (Hz)

3. Sounds that humans can hear are between 20 and _____ Hz. 20,000

4. A _____ sound has a low pitch. low-frequency

5. Even if two sounds have the same intensity they can have different _____. frequencies or pitches

6. A high-frequency sound has a _____ pitch. high

7. One hertz is the number of complete _____ per second. waves

APPLY *Complete the following.*

8. **Compare:** Which sound has a lower pitch, a sound with a frequency of 20 Hz or a sound with a frequency of 20,000 Hz? How do you know?

9. **Hypothesize:** The sound of a siren coming toward you has a higher pitch than the sound of a siren that is standing still. What does this tell you about the frequency of a sound that is moving toward you?

10. **Contrast:** What is the difference between the words "supersonic" and "ultrasonic"?

11. **Hypothesize:** Why do you think the range of human hearing decreases as a person gets older?

InfoSearch ....

Read the passage. Ask two questions about the topic that you cannot answer from the information in the passage.

Changing Pitch A guitar can make sounds of many different pitches. If you look at the strings of a guitar, you will see that some are thick and some are thin. The thick strings make low-pitched sounds. The thin strings make high-pitched sounds. A guitar also has keys that tighten or loosen the strings. Changing the tightness of the strings also changes the pitch. The frets across the neck of the guitar are another way to change the pitch.

SEARCH: Use library references to find answers to your questions.

ACTIVITY

OBSERVING CHANGES IN PITCH

You will need five test tubes, a test tube rack, a metric ruler, and a container of water.

1. Place the test tubes in the test-tube rack.

2. Pour different amounts of water into each test tube. Use the metric ruler to measure the height of the column of water in each test tube. The heights should be as follows: 2 cm, 4 cm, 6 cm, 8 cm, and 10 cm.

3. Blow air across the top of each test tube. Observe the pitch of the sound you hear from each test tube.

Questions

1. How does the height of the water affect the pitch of the sound?

2. Which test tube had the highest pitch?

3. Which test tube had the lowest pitch?

8-7 What is sound quality?

Objective ▶ Explain how overtones affect sound quality.

TechTerms

- ▶ **fundamental** (fun-duh-MEN-tul) **tone:** lowest-pitched sound produced when a whole string vibrates
- ▶ **overtones:** high-pitched sounds produced when parts of a string vibrate
- ▶ **timbre** (TAM-bur): sound quality

Sound Quality Suppose a trumpet player and a clarinet player play the same note. The sounds they make have the same frequency and intensity. Can you hear a difference between the two sounds? The sounds are different even though the frequency and the intensity are the same. The sounds differ because of their **timbre** (TAM-bur), or sound quality. The sounds of a trumpet and a clarinet have different timbres.

▶**Define:** What is timbre? sound quality

Fundamental Tone When an object vibrates, it produces a sound with a certain frequency. Figure 1 shows a vibrating string. Notice that the whole string is vibrating. When the whole string vibrates, it produces the lowest possible frequency and pitch. The low-pitched sound produced is called the **fundamental** (fun-duh-MEN-tul) **tone.**

▶**Define:** What is the fundamental tone?
the low-pitched sound produced when a whole string vibrates

Overtones Parts of the string shown in Figure 1 can vibrate faster than the whole string. When parts of the string are vibrating, the string produces sounds with a higher pitch than the fundamental tone. The high-pitched sounds produced when different parts of a string vibrate are called **overtones.**

The sound of a trumpet or a clarinet always has overtones. When you speak or sing, your voice has overtones. The timbre, or quality, of a sound is caused by these overtones. Without overtones, voices and musical instruments would all sound the same.

▶**Define:** What are overtones?
high-pitched sounds produced when parts of a string vibrate

Figure 1 Fundamental tone

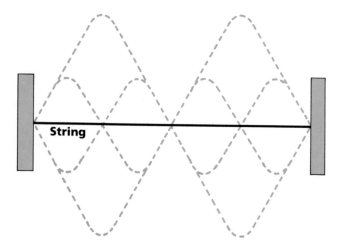

Figure 2 Overtones

LESSON SUMMARY

▶ The sounds of a trumpet and a clarinet differ because of their sound quality, or timbre.

▶ When a string vibrates as a whole, it produces a fundamental tone.

▶ When different parts of a string vibrate, they produce overtones.

▶ The quality of a sound is caused by overtones.

CHECK *Complete the following.*

1. The quality of a sound is its _____ . timbre

2. When a whole string vibrates, it produces a _____ . fundamental tone

3. All sounds are a combination of fundamental tones and _____ . overtones

4. Two sounds can have the same _____ and intensity, but still sound different. frequency

5. A fundamental tone is a _____ sound. low-pitched

6. High-pitched sounds produced when parts of a string vibrate are called _____ . overtones

APPLY *Complete the following.*

7. What would human voices sound like without overtones?

▶ 8. **Infer:** A violin and a piano are playing the same musical note with the same frequency and the same intensity. Why do they sound different?

InfoSearch

Read the passage. Ask two questions about the topic that you cannot answer from the information in the passage.

Vocal Cords Your larynx (LAR-inks) is the bulge at the top of your windpipe. Inside your larynx are your vocal cords. Usually, your vocal cords are relaxed. When you speak, the vocal cords tighten. Air passes over the vocal cords and makes them vibrate.

SEARCH: Use library references to find answers to your questions.

PEOPLE IN SCIENCE

THOMAS ALVA EDISON (1847–1931)

Thomas Edison was a great American inventor. He invented the light bulb, helped to develop the motion picture camera, and worked on the first phonograph. Edison was the first person to record a human voice.

The invention of the phonograph in 1879 made Edison famous all over the world. The earliest phonograph was called a gramophone. The gramophone was similar to modern phonographs. It consisted of a cyclinder and a stylus. A grooved record was placed on the cylinder. Sound came out of the phonograph's loudspeaker.

The earliest phonographs could reproduce sounds, but the quality of the sound was poor. Over the years, the sound quality of phonographs has improved. Today, digital phonographs can reproduce sounds with a great deal of accuracy. Modern phonograph makers owe a lot to Thomas Edison, who invented the first sound-recording device.

Objective ▶ Contrast music and noise.

TechTerms

▶ **music:** sounds combining a pleasing quality, melody, harmony, and rhythm

▶ **noise:** unpleasant sounds, with irregular patterns of vibration

Music Different musical instruments have different sound qualities. How do the different instruments in an orchestra combine to make **music?** Music is a pleasing combination of sounds that have rhythm, melody, and harmony.

▶ **Rhythm** The basic beat in music is known as the rhythm (RITH-um). When you listen to music, the rhythm you hear might come from drums or a bass guitar.

▶ **Melody** The combination of musical notes is called the tune, or the melody. Musical notes make up the melody of any piece of music. If you hum or sing a song, you are repeating the melody. If you clap your hands to the beat of the music, you are adding rhythm.

▶ **Harmony** In an orchestra, different instruments play together. Many different notes are heard at the same time. Combining the different notes is called harmony. Good harmony makes music sound better.

▶ *Identify:* What are the three parts of music? rhythm, melody, and harmony

Musical Instruments Musical instruments can be divided into three main groups.

▶ **Percussion instruments** These instruments include all types of drums and any instrument that is played by being tapped or hit.

▶ **String instruments** These instruments make music from the sounds of vibrating strings. Examples include guitars, violins, and cellos.

▶ **Wind instruments** These instruments make music when air is blown through them. Examples include trumpets, clarinets, and flutes.

▶ *Name:* What are the three main groups of musical instruments? percussion, string, and wind instruments

Noise The opposite of music is **noise.** Noise is a combination of unpleasant sounds with irregular patterns of vibration. Noise can be thought of as any unwanted sound. Unlike music, noise has no melody, rhythm, or harmony.

▶ *Define:* What is noise? a combination of unpleasant sounds with irregular patterns of vibration

▶ Music is a combination of rhythm, melody, and harmony.

▶ The three main groups of musical instruments are percussion, string, and wind instruments.

▶ Noise is an unpleasant combination of sounds with no regular pattern of vibrations.

CHECK *Complete the following.*

1. A guitar is a _____ instrument. string

2. A drum is an example of a <u>percussion</u> instrument.

3. Unwanted sounds are _____ . noise

4. The _____ is the beat of the music. rhythm

5. When you hum a tune, you are humming the _____ . melody

6. Rhythm, melody, and harmony combine to make _____ . music

APPLY *Complete the following.*

7. **Contrast:** What is the difference between music and noise?

8. **Classify:** Identify each of the following instruments as percussion instruments, string instruments, or wind instruments.
   a. tuba
   b. kettle drum
   c. harp
   d. trumpet

InfoSearch

Read the passage. Ask two questions about the topic that you cannot answer from the information in the passage.

The Musical Scale The musical scale is made up of eight notes. Each musical note has its own pitch. Each note is represented by a letter of the alphabet. A set of eight notes is called an octave (AHK-tiv). The last note in an octave has twice the frequency of the first note. The human ear can hear a total of about 10 octaves.

SEARCH: Use library references to find answers to your questions.

LEISURE ACTIVITY

SCHOOL BAND

Joining a school band is something that many students would like to do. Being a member of a band is very different from playing a musical instrument alone. There are some things you should think about if you want to be a member of your school band. The most important thing is an interest in music. What musical instrument would you like to play? Once you have decided what instrument you want to play, you will have to learn how to read music. You must then spend several hours a day practicing.

When you are part of a band, you must also practice with the entire group. Every band has a conductor. The conductor tries to keep all the instruments in harmony. The success of a band depends on how well all the different musical sounds come together as one sound. Being a band member takes a lot of work, but it is a rewarding experience.

How do you hear?

Objective ► Trace the path of sound waves through the ear to the brain.

TechTerms

► **ear:** sense organ that detects sound
► **hearing:** one of the five human senses

Hearing Sounds One of the five human senses is **hearing.** What happens when you hear a sound? For a sound to be heard, three things are needed. They are a source of the sound, a medium to transmit the sound, and a sense organ to detect the sound.

▐▐▐▶*Identify:* What three things are needed to hear a sound? a source of the sound, a medium to transmit the sound, and a sense organ to detect the sound

The Ear In humans, the sense organ that detects sounds is the **ear.** Look at the diagram of the human ear. There are three main parts to the ear: the outer ear, the middle ear, and the inner ear.

▐▐▐▶*Name:* What are the three parts of the human ear? the outer ear, the middle ear, and the inner ear

Parts of the Ear A sound wave first enters the outer ear. The outer ear funnels the sound wave into the ear. The sound wave moves through the ear canal to the eardrum. The vibrating air particles in the sound wave make the eardrum vibrate.

The vibrations from the eardrum are transferred to the middle ear. There are three small bones in the middle ear. They are the hammer, the anvil (AN-vul), and the stirrup (STUR-up). The vibrations are transferred from the hammer, to the anvil, and then to the stirrup.

The vibrations are then transmitted to the inner ear. In the inner ear, the vibrations are transferred to the cochlea (KAHK-lee-uh). The cochlea is filled with liquid and is attached to nerve fibers. The nerve fibers join to form one nerve that goes to the brain. The vibrations from the cochlea become electrical impulses in the nerve. The nerve transmits the impulses to the brain. In the brain, the electrical impulses are interpreted as sound.

▐▐▐▶*Name:* Which parts of the ear vibrate when a sound is heard? the eardrum, hammer, anvil, stirrup, and cochlea

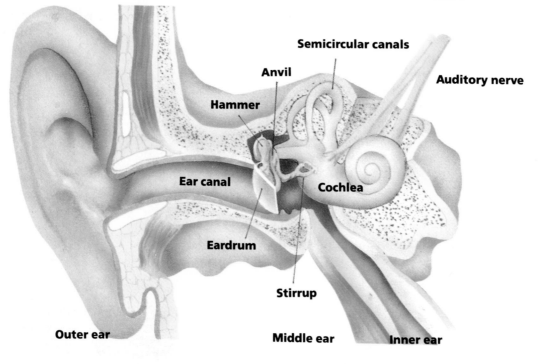

Semicircular canals

Anvil

Auditory nerve

Hammer

Ear canal

Cochlea

Eardrum

Stirrup

Outer ear

Middle ear

Inner ear

LESSON SUMMARY

▶ For a sound to be heard, there must be a source of the sound, a medium to transmit the sound, and a sense organ to detect the sound.

▶ The ear is the human sense organ that detects sound.

▶ In the outer ear, a sound wave travels through the ear canal and strikes the eardrum.

▶ In the middle ear, sound waves are transferred to the hammer, anvil, and stirrup.

▶ In the inner ear, sound waves reach the cochlea and are converted to electrical impulses that travel to the brain.

CHECK *Complete the following.*

1. The _____ is the organ that detects sound. ear

2. Of the five senses, the one that is involved with sound is _____ . hearing

3. Sound waves travel from the outer ear, to the middle ear , and then to the inner ear.

APPLY *Complete the following.*

4. **Sequence:** Arrange the following steps in the correct order.

   a. The vibrations are transferred to the hammer, anvil, and stirrup.   **b.** The vibrations of air particles in the sound wave cause the eardrum to vibrate.   **c.** The vibrations are transferred to the cochlea.   **d.** The brain interprets the electrical impulses as sounds.   **e.** A sound wave is transmitted by the source.   **f.** The vibrations are converted to electrical impulses along a nerve that goes to the brain.   **g.** The sound waves travel through the medium.   **h.** The sound waves enter the outer ear and go into the ear canal.

InfoSearch ....

Read the passage. Ask two questions about the topic that you cannot answer from the information in the passage.

Dizziness The inner ear has another important role that has nothing to do with hearing. The inner ear maintains balance. After you ride a roller coaster, you may feel dizzy. Dizziness is a temporary loss of balance. The liquid in the inner ear reacts to motion. This causes dizziness. After a while the dizziness goes away. The dizziness stops because the liquid in the inner ear stops moving.

SEARCH: Use library references to find answers to your questions.

▼▼▼ LOOKING BACK IN SCIENCE ▼▼▼▼▼▼▼▼▼▼▼▼▼▼▼▼▼▼▼▼▼▼▼▼

THE INVENTION OF THE TELEPHONE

The telephone was invented by Alexander Graham Bell. Bell was always interested in speech and communication. Before inventing the telephone, Bell invented a system of sign language for the deaf.

For Bell, the telephone was a way to improve communication. Bell worked with a partner named Thomas Watson. In 1876, they built a working telephone. An American inventor, Elisha Gray, built a telephone similar to Bell's at about the same time. However, Bell received a patent for his telephone first. Bell's telephone was not too different from the type of phone used today.

Telephone service in the United States began in 1878. The sound quality of telephones has improved over the years. Satellites and other modern technology have made it possible for people to communicate by telephone over long distances.

UNIT 8 Challenges

STUDY HINT Before you begin the Unit Challenges, review the TechTerms and Lesson Summary for each lesson in this unit.

TechTerms

decibel (138)
ear (146)
echo (136)
fundamental tone (142)
hearing (146)
intensity (138)

longitudinal wave (132)
medium (132)
music (144)
noise (144)
overtones (144)
pitch (140)

sound (130)
supersonic (134)
timbre (142)
ultrasonic (140)
vibration (130)

TechTerm Challenges

Matching *Write the TechTerm that matches each description.*

1. how high or low a sound is pitch
2. speed faster than sound supersonic
3. rapid back-and-forth movement vibration
4. frequencies above the fundamental tone overtones
5. sound quality timbre
6. sense organ that detects sound ear
7. includes rhythm, melody, and harmony music
8. one of the human senses hearing
9. form of energy that travels as waves sound
10. unit of intensity decibel

Fill in *Write the TechTerm that best completes each statement.*

1. A sound wave needs a _____ in which to travel. medium
2. A reflected sound wave is an _____ . echo
3. In a _____ , the particles of the medium move back and forth in the direction of the wave motion. longitudinal wave
4. The amount of energy in a sound wave is the _____ of the wave. intensity
5. The opposite of music is _____ . noise
6. A frequency greater than 20,000 Hz is _____ . ultrasonic
7. The _____ is the low-pitched sound produced when a whole string vibrates.
   fundamental tone

Content Challenges

Multiple Choice *Write the letter of the term or phrase that best completes each statement.*

1. To find their way in the dark, bats use
   a. overtones. **b.** ultrasound. **c.** echolocation. **d.** timbre. c

2. A high-pitched sound has a
   a. low frequency. **b.** fundamental tone. **c.** high frequency. **d.** low intensity. c

3. Two sounds with the same frequency and intensity will sound different because of
   a. timbre. **b.** pitch. **c.** vibrations. **d.** wavelength. a

4. The speed of sound in air is
   a. 125 m/sec. **b.** 340 m/sec. **c.** 430 m/sec. **d.** 540 m/sec. b

5. The unit of frequency is the
   a. hertz. **b.** decibel. **c.** m/sec. **d.** pitch. a

6. Sound travels fastest in
   a. liquids. **b.** solids. **c.** gases. **d.** a vacuum. b

7. Dolphins communicate by making
   a. low-frequency sounds.   **b.** high-intensity sounds.   **c.** ultrasonic sounds.
   d. low-intensity sounds. c

8. Music consists of rhythm, melody, and
   a. intensity.   **b.** harmony.   **c.** frequency.   **d.** percussion. b

9. Before your eardrum vibrates, the sound wave has to travel through the
   a. anvil.   **b.** ear canal.   **c.** inner ear.   **d.** cochlea. b

10. A sound wave consists of compressions and
   a. overtones.   **b.** fundamental tones.   **c.** rarefactions.   **d.** frequencies. c

True/False *Write true if the statement is true. If the statement is false, change the underlined term to make the statement true.*

1. Sound <u>cannot</u> be heard in a vacuum. true
2. A high-intensity sound wave has a lot of <u>energy</u>. true
3. Sound waves cause the <u>ear canal</u> to vibrate. eardrum
4. An echo is a sound wave that has been <u>made louder</u>. reflected
5. One hertz is equal to one <u>compression</u> per second. wave
6. A drum is an example of a <u>percussion</u> instrument. true
7. The beat of the music is the <u>melody</u>. rhythm
8. There are <u>three</u> main types of musical instruments. true
9. Vibrations from the eardrum enter the <u>outer</u> ear. inner
10. Humans can hear sounds between <u>20 Hz</u> and 20,000 Hz. true

Understanding the Features ....

Reading Critically *Use the feature reading selections to answer the following. Page numbers for the features are in parentheses.*

1. What are some ways to decrease the amount of noise pollution? (139)
2. What did Alexander Graham Bell invent before the telephone? (147)
3. **Infer:** How fast is Mach 2? (135)
4. What invention made Thomas Edison famous all over the world? (143)
5. On what does the success of a school band depend? (145)
6. Why would an acoustics engineer try to reduce the echoes in a concert hall? (137)

Concept Challenges ....

Critical Thinking *Answer each of the following in complete sentences.*

1. **Hypothesize:** Why does sound travel faster in a solid than in a gas?
2. What happens after a sound wave makes the eardrum vibrate?
3. What does a vibrating guitar string do to the air around it?
4. Which sound wave has a larger amplitude, a 20-dB sound or a 50-dB sound? How do you know?
5. **Predict:** What might happen to your sense of hearing if you were exposed to loud noises over a long period of time?

Interpreting a Diagram *Use the diagram of a human ear to complete the following.*

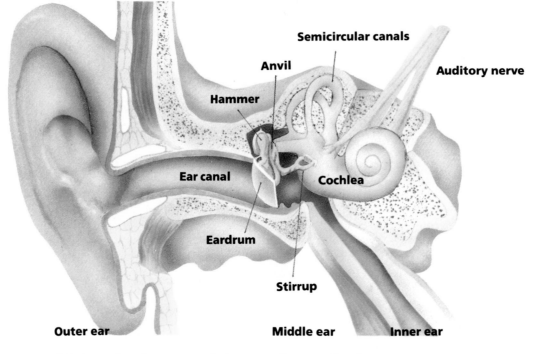

1. When a sound is heard, what part of the ear vibrates first?
2. Which parts make up the inner ear?
3. Where is the energy of a sound wave converted to electrical impulses?
4. **Hypothesize:** Why do you think the bones in the middle ear are called the hammer, the anvil, and the stirrup?
5. **Hypothesize:** Cupping your hand to your ear helps you to hear sounds more clearly. Why?

Finding Out More . . . .

1. Use library references to find out more about how sound is recorded. How does magnetic tape record sound? How is this different from the way sound is recorded on a phonograph record? How is sound recorded on a compact disc? Write a report of your findings.

2. Visit a local radio station. Ask a broadcast engineer the following questions: How accurate is the transmitted sound compared with the original sound? What is static? What is a signal-to-noise ratio? How does weather affect radio transmission?

3. Ask a doctor how a hearing aid works. Find out which parts of the ear are helped by a hearing aid. What is the quality of the sound from a hearing aid?

4. Use a tape recorder to record your voice. Play it back several times. Does your voice sound different to you? Does it sound different to other people? Use library references to find out about resonance (REZ-uh-nuns). How does resonance explain why your voice sounds different to you when you listen to a recording of it?

LIGHT

CONTENTS

STUDY HINT Before beginning Unit 9, read the title of each lesson. On a sheet of paper, write the title of each lesson. Below each title, write a short paragraph explaining what you think each lesson is about.

9-1 What is light?

Objective ▶ Recognize that light is a form of electromagnetic energy.

TechTerms

- **light:** form of electromagnetic (i-lek-troh-mag-NET-ik) energy made up of streams of photons
- **photon** (FOH-tahn): tiny bundle of energy
- **ray:** straight line that shows the direction of light

Light Energy When you sit in bright sunlight, you can feel your skin get warm. Objects warm up in sunlight. **Light** is a form of electromagnetic (i-lek-troh-mag-NET-ik) energy. Light energy can be changed into heat, electricity, and other forms of energy.

📁 *Classify:* What is light?
a form of electromagnetic energy

Particles of Light Light is made up of small bundles of energy called **photons** (FOH-tahnz). Photons are like small particles. These particles are so small that a single photon cannot be seen. A beam of light is made up of a stream of many photons. Each photon carries a certain amount of energy. Some photons have more energy than others.

▥▶ *Describe:* What is a light beam made up of?
a stream of photons

Rays of Light When you turn on a flashlight, you see a beam of light. The beam of light from the flashlight looks like a straight line. Light travels in straight lines. A **ray** of light is a straight line that shows the direction of a light beam. A ray of light will continue to travel in a straight line unless its direction is changed.

Light ray

▥▶ *Describe:* What does a ray of light look like?
a straight line

LESSON SUMMARY

▶ Light is a form of electromagnetic energy.

▶ Light is made up of a stream of photons, or small bundles of energy.

▶ Light travels in straight lines.

CHECK *Complete the following.*

1. You cannot see individual _____ of light because they are very small. photons

2. Light travels in _____ lines. straight

3. A stream of photons makes up a _____ beam. light

4. Light is a form of _____ energy. electromagnetic

5. A _____ of light shows the direction of a light beam. ray

APPLY *Complete the following.*

▶ 6. **Infer:** Turn on a lamp and hold your hand near the light bulb. Do you feel heat? What is the source of the heat?

7. Give two common examples which show that light travels in straight lines.

InfoSearch ....

Read the passage. Ask two questions about the topic that you cannot answer from the information in the passage.

Light Years The speed of light is 300,000 km/sec. Light takes less than two seconds to reach the earth from the moon. Light takes about eight minutes to travel the 150 million kilometers from the sun to the earth. The distance that light travels in one year is called a light year. One light year is equal to about 10 trillion kilometers. The nearest star is 4.2 light years from the earth. Light from this star takes 4.2 years to reach the earth.

SEARCH: Use library references to find answers to your questions.

ACTIVITY

OBSERVING THAT LIGHT TRAVELS IN STRAIGHT LINES

You will need three index cards and a flashlight.

1. Stack three index cards one on top of the other. Make a hole in the center of the three cards.

2. Fold one edge of each card vertically. The folded edge should be about 4 cm wide.

3. Separate the cards and stand them in a row as shown. Line up the cards so that you can see through the three holes.

4. Place the flashlight behind the standing cards. Turn on the flashlight and shine the beam through the holes in the cards.

5. Move the center card a few centimeters to the left or right. Observe what happens.

Questions

1. **Observe:** Can you see the flashlight beam through the three holes when the cards are lined up?

2. **Observe:** Can you still see the light when you move the center card? Why or why not?

3. What do these results tell you about the way light travels?

Objective ▶ Describe how light travels as transverse waves.

TechTerm

▶ **transverse** (trans-VURS) **wave:** wave in which the particles of the medium move up and down at right angles to the direction of wave motion

Light Waves Light is made up of a stream of particles called photons. However, light also behaves like a wave. Light is a type of electromagnetic wave. Electromagnetic waves are different from sound waves. Sound is a longitudinal (lahn-juh-TOOD-un-ul) wave. A sound wave needs a medium in which to travel. Sound cannot be heard in a vacuum, or empty space.

Light waves are different from sound waves in two ways.

▶ Light travels in **transverse** (trans-VURS) **waves.** In transverse waves, the particles of the medium move up and down at right angles to the direction of wave motion.

▶ Light waves do not need a medium in which to travel. Light can travel through a vacuum.

▷ *Contrast:* How do light waves differ from sound waves? Light waves are transverse waves; light waves do not need a medium in which to travel.

Photons and Light Waves Some experiments with light show that it is made up of photons. Other experiments show that light acts as a wave. Scientists have learned that some waves act as if they are made up of particles.

▷ *Describe:* How do light waves act?
up and down in the direction of the wave motion

Properties of Light Waves Like all waves, light waves have four properties. They are speed, wavelength, frequency, and amplitude.

▶ The speed of light is about 300,000 km/sec. All electromagnetic waves travel at the same speed. The speed of light is the fastest possible speed.

▶ The wavelength of light is the distance from the crest or trough of one wave to the crest or trough of the next wave.

▶ The number of light waves that pass by a point each second is the frequency.

▶ The amplitude is the height of a wave. A bright light has a greater amplitude than a dim light.

▷ *List:* What are the four properties of a light wave? speed, wavelength, amplitude, and frequency

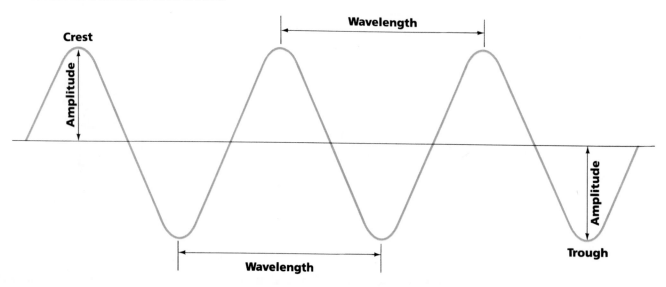

LESSON SUMMARY

▶ Light is a type of electromagnetic wave.

▶ A light wave is a transverse wave.

▶ A light wave does not need a medium in which to travel.

▶ Light waves act as if they are made of particles. All light waves have four properties: speed, wavelength, frequency, and amplitude.

CHECK *Complete the following.*

1. In a _____ wave, the particles of the medium move at right angles to the direction of the wave. transverse

2. A light wave is a moving stream of _photons_ .

3. The properties of light waves include speed, _____ , frequency, and amplitude. wavelength

4. Light waves travel best in a _____ . vacuum

5. Light waves do not need a _____ in which to travel. medium

APPLY *Complete the following.*

6. **Classify:** Identify each of the following as a sound wave or a light wave.
   a. longitudinal wave
   b. does not travel through a vacuum
   c. waves seem to be made up of particles
   d. transverse wave
   e. can travel in a vacuum

7. **Relate:** Light waves with very high frequencies are called ultraviolet light. Ultraviolet light cannot be seen by the human eye. How is ultraviolet light similar to ultrasonic sound?

Skill Builder

▶ *Predicting* When you predict, you state in advance how and why something will occur. Suppose that light waves needed a medium through which to travel. How would your life be different? Why?

LEISURE ACTIVITY

PHOTOGRAPHY

Photography is a popular activity that can teach you a lot about light. The word "photography" means "writing with light." To take a photograph, you need a camera, film, and a good source of light.

A camera has a lens and an opening for light to enter. This opening is called an aperture (AP-ur-cher). When you take a picture, the aperture is open for only a short period of time. The light goes through the lens and onto the film. Film is very sensitive to light. Too much light will cause the photograph to be too bright, or overexposed. If there is not enough light, the photograph will be too dark, or underexposed.

Many schools have amateur photography clubs. Joining a photography club is a good way to learn more about taking photographs.

9-3 What are sources of light?

Objectives ▶ Differentiate between luminous and illuminated objects. ▶ Describe what happens when light strikes different materials.

TechTerms

- ▶ **illuminated** (i-LOO-muh-nayt-ed) **objects:** objects that reflect light
- ▶ **luminous** (LOO-muh-nus) **objects:** objects that give off their own light
- ▶ **opaque** (oh-PAYK): material that blocks light
- ▶ **translucent** (trans-LOO-sunt): material that transmits some light
- ▶ **transparent** (trans-PER-unt): material that transmits light easily

Luminous Objects A flashlight, a candle, and a light bulb are sources of light. Objects that give off their own light are called **luminous** (LOO-muh-nus) **objects.** The sun is a luminous object. It is the source of light for the earth and other planets in the solar system.

▶**Define:** What is a luminous object? an object that gives off its own light

Illuminated Objects What happens when the light from a luminous object strikes another object? The light may be reflected. An object that reflects light is called an **illuminated** (i-LOO-muh-nayt-ed) **object.**

The moon is an illuminated object. The moon is not the source of its own light. Sunlight strikes the moon and is reflected from its surface. You see the moon by reflected sunlight.

▶**Identify:** What is an object that reflects light called? an illuminated object

Opaque, Transparent, and Translucent Objects When light shines on an object, three things can happen.

- ▶ An object may block the light and form a shadow. An object that blocks light is called an **opaque** (oh-PAYK) object. When you place your hand in front of a light source, you will see a shadow. Your hand is opaque.

- ▶ An object may allow all the light to pass through easily. This kind of object is called **transparent** (trans-PER-unt). A sheet of clear glass allows light to pass through. The glass is transparent.

- ▶ An object may allow some of the light to pass through. This kind of object is called **translucent** (trans-LOO-sunt). A stained glass window will allow some light to pass through, but not as much as a sheet of clear glass. The stained glass is translucent.

▶**Name:** What are three types of objects that affect light differently? opaque, transparent, or translucent

156

LESSON SUMMARY

▶ A luminous object gives off its own light.

▶ An illuminated object reflects the light from another source.

▶ An opaque object blocks all light and casts a shadow.

▶ A transparent object allows all the light to pass through easily.

▶ A translucent object allows some light to pass through.

CHECK *Complete the following.*

1. When you shine a flashlight on objects in a dark room, the objects that you see are _____ objects. illuminated

2. An object that casts a shadow is an _____ object. opaque

3. The moon is an _____ object. illuminated

4. A sheet of clear glass is _____ because it allows light to pass through easily. transparent

5. A sheet of stained glass is translucent because it allows only some light to pass through.

6. A luminous object is its own source of light.

7. An illuminated object _____ the light from another source. reflects

APPLY *Complete the following.*

▶ 8. **Infer:** You cannot see the sun on a cloudy day, even though sunlight can still illuminate objects. Are clouds transparent or translucent?

📂 9. **Classify:** Which of the following objects are luminous and which are illuminated?
 a. a candle   d. a campfire
 b. the earth   e. a sheet of paper
 c. a star

Designing an Experiment.............

Design an experiment to solve the problem.

PROBLEM: Do transparent objects allow more light to pass through than translucent objects?

Your experiment should:

1. List the materials you would need.

2. Identify safety precautions that should be followed.

3. List a step-by-step procedure.

4. Describe how you would record your data.

SCIENCE CONNECTION

SHADOWS

Opaque objects cast shadows by blocking the light shining on them. Your body is opaque. It blocks the sun's light. You have probably seen your shadow on a sunny afternoon.

During a lunar eclipse, the earth blocks the light from the sun. The moon passes through the earth's shadow. Like most shadows, the earth's shadow consists of an umbra (UM-bruh) and a penumbra (pi-NUM-bruh). The umbra is a sharp, black shadow. The penumbra is the gray, outer part of the shadow.

The only time that a shadow has only an umbra is when a point source of light shines on the object. A spotlight is an example of a point source of light. Most light comes from extended light sources. A fluorescent light is an example of an extended source of light.

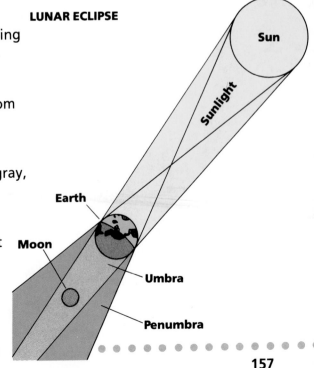

LUNAR ECLIPSE

Sun

Sunlight

Earth

Moon

Umbra

Penumbra

9-4 How do you see?

Objectives ▶ Identify the parts of the eye. ▶ Describe how the eye senses light.

TechTerms

- **eye:** sense organ that detects light
- **image:** picture formed by the eye
- **sight:** one of the five human senses

Observing Light You see an object when light from the object enters your **eye.** The eye is the human sense organ that detects light. **Sight** is one of the five human senses.

▐▶*Relate:* How is the eye related to sight?
The eye is the sense organ that is used for sight.

Parts of the Eye Look at the side view of the human eye. Identify the parts of the eye.

- ▶ The front of the eye is covered with a clear layer called the cornea (KOWR-nee-uh). The cornea is a protective layer that keeps dirt and bacteria from damaging the inner eye.

- ▶ The cornea covers the pupil. The pupil is the opening through which light enters the eye.

- ▶ The amount of light that enters the eye is controlled by the iris (Y-ris). The iris is a muscle that contracts or expands depending on the amount of light available. When there is too much light, the iris contracts. When there is not enough light, the iris expands.

- ▶ After light enters the eye, it is focused by the lens.

- ▶ Light is focused by the lens onto the retina (RET-un-uh). The lens forms an upside-down **image** on the retina. An image is a picture that is formed by the eye.

- ▶ The retina converts the image into electrical impulses. The impulses travel along the optic nerve to the brain. The brain interprets the electrical impulses as a right-side-up image.

▐▶*Name:* What are the parts of the eye involved in sight? cornea, pupil, iris, lens, retina, and optic nerve

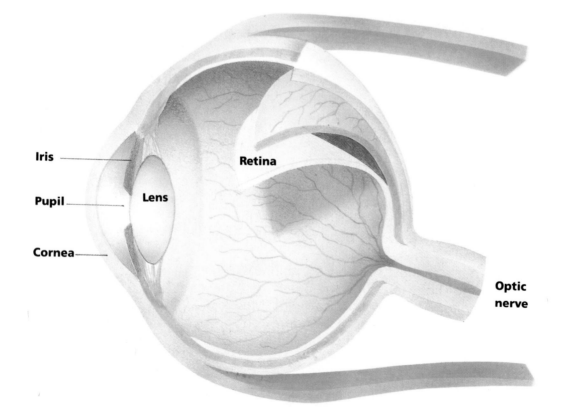

Iris

Pupil

Cornea

Lens

Retina

Optic nerve

LESSON SUMMARY

- ▶ The eye is the human sense organ used for sight.
- ▶ The parts of the eye are the cornea, pupil, iris, lens, retina, and optic nerve.
- ▶ A picture of an object formed by the eye is called an image.
- ▶ An image on the retina is upside down.
- ▶ The brain adjusts the image on the retina to make it look right-side up.

CHECK *Write true if the statement is true. If the statement is false, change the underlined term to make the statement true.*

1. Electrical impulses travel along the <u>optic nerve</u> to the brain. true

2. The <u>pupil</u> is the opening that lets light into the eye. true

3. The <u>cornea</u> controls the amount of light that can enter the eye. iris

4. The <u>iris</u> is a clear protective layer over the eye.
   cornea

5. The eye is a <u>sense</u> organ. true

6. An image on the <u>cornea</u> is upside down.
   retina

APPLY *Complete the following.*

7. What energy changes take place when you see an image?

8. **Hypothesize:** Suppose you walk into a dark room after being out in bright sunlight. Several minutes pass before you can see objects in the room clearly. What happened to your eyes during that time?

··
Health and Safety Tip ················

It is important to take proper care of your eyes. The most easily damaged part of the eye is the retina. To keep from damaging your retina, do not stare at very bright lights. You should never look directly at the sun or at a laser beam. Use library references to find out how looking into the sun or a laser beam can damage your retina.

SCIENCE CONNECTION ◆○◆○◆○◆○◆○◆○◆○◆○◆○◆○◆○◆○◆○◆○◆○

CORRECTIVE LENSES

Most people are born with good eyesight. The lenses in their eyes can always keep objects in focus. However, some people cannot see objects clearly. The objects appear blurred. These people can sometimes see objects clearly by squinting. People with blurred vision need corrective lenses to help the lenses in their eyes. Corrective lenses can be either eyeglasses or contact lenses.

People with poor eyesight are either farsighted or nearsighted. Nearsighted people can see objects clearly only if the objects are nearby. Objects that are far away look blurry. People who are farsighted can see objects clearly only if the objects are far away. Objects that are nearby look blurry. Corrective lenses allow people who are nearsighted or farsighted to see objects clearly at any distance.

9-5 What is color?

Objectives ▸ Describe how a prism forms a visible spectrum. ▸ Explain why different objects have different colors.

TechTerms

- ▸ **prism** (PRIZ-um): triangular piece of glass that breaks up white light into a band of colors
- ▸ **visible spectrum:** seven colors that make up white light

Wavelength and Color Light is an electromagnetic wave. Because it is a wave, light has a wavelength. The wavelength of light determines its color. There are seven basic colors of light. They are red, orange, yellow, green, blue, indigo, and violet. Red light has the longest wavelength. Violet light has the shortest wavelength.

▸**Describe:** How does wavelength affect the color of light?
The wavelength of light determines its color.

Visible Spectrum Most sources of light give off white light. White light is made up of all seven colors of light. When the seven colors combine, they form a beam of white light. The seven colors

that make up white light are called the **visible spectrum.**

A **prism** (PRIZ-um) can separate a beam of white light into the colors of the spectrum. A prism is a triangular piece of glass. When a beam of white light passes through a prism, a band of colors is formed.

▸**Define:** What is the visible spectrum?
the seven colors that make up white light

Color of Objects When white light strikes an object, the object appears a certain color. The object absorbs all the colors of the spectrum except the color it reflects. When white light shines on a red object, all colors except red are absorbed. The red light is reflected, and the object appears red.

White light

White light

Some objects absorb all of the light and do not reflect any color. These objects appear black. Other objects reflect all of the light. They do not absorb any of the colors. These objects appear white.

▸**Predict:** What color will an object appear if it reflects only blue light? blue

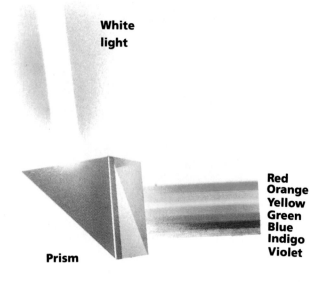

White light

Red
Orange
Yellow
Green
Blue
Indigo
Violet

Prism

LESSON SUMMARY

▶ The wavelength of light determines its color.

▶ White light includes the seven colors of the visible spectrum.

▶ A prism can be used to separate white light into the colors of the spectrum.

▶ The color of an object depends on what part of the spectrum it reflects or absorbs.

CHECK *Complete the following.*

1. A blue object absorbs all the colors of the spectrum except _____ . blue

2. A _____ separates white light into the colors of the spectrum. prism

3. Red has the _____ wavelength of all the colors of the visible spectrum. longest

4. An object appears _____ because it absorbs all the colors of the spectrum. black

APPLY *Complete the following.*

▶ **5. Infer:** The frequency of a wave increases as its wavelength decreases. Which color of light has the highest frequency? Which has the lowest?

6. Compare: How is a rainbow like a spectrum?

InfoSearch

Read the passage. Ask two questions about the topic that you cannot answer from the information in the passage.

Complementary Colors Stare at the drawing of the flag for a few minutes. Then stare at a sheet of blank white paper. What do you see?

Most people will see a red, white, and blue flag. The flag is printed in colors that are complementary (kahm-pluh-MEN-tur-ee) to the normal colors of the flag. Complementary colors are contrasting, or opposite, colors. Any two complementary colors combine to form white light.

SEARCH: Use library references to find answers to your questions.

SCIENCE CONNECTION

COLORBLINDNESS

Do you know anyone who cannot tell the color red from the color green? If so, this person is colorblind. Colorblindness is an inability to see the colors red, green, or blue. These colors appear gray. A person colorblind for red would not be able to see the number twelve below.

There are three kinds of colorblindness. A red-blind person cannot tell the difference between red and green. A blue-blind person cannot see a difference between blue and yellow. A green-blind person cannot see the green part of the visible spectrum.

Colorblindness is a genetic (juh-NET-ik) disorder. It is inherited from a person's parents. Men are colorblind more often than women. This is because the gene for colorblindness is carried on the female sex chromosome. A woman must inherit two genes, one from each parent, in order to be colorblind. A man has to inherit only one gene to be colorblind.

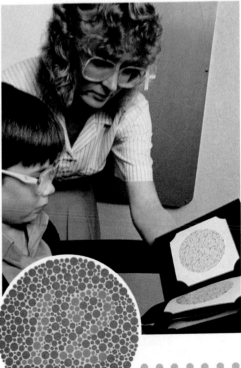

9-6 What is photosynthesis?

9-6

Objective ▶ Explain how plants use the energy of the sun to make food.

TechTerms

- ▶ **chlorophyll** (KLOWR-uh-fil): green substance in plants
- ▶ **photosynthesis** (foht-uh-SIN-thuh-sis): process by which plants use energy from the sun to make food

Light and Energy Any form of energy can be converted to another form of energy. Light energy can be changed into heat energy, electrical energy, chemical energy, or other forms of energy. Plants convert the electromagnetic energy of sunlight into chemical energy. The plants use the chemical energy to make their own food.

▶*Explain:* How do plants use sunlight to make their own food.

They convert electromagnetic energy into chemical energy.

Photosynthesis The process by which a plant uses sunlight to make food is called **photosyn-**

thesis (foht-uh-SIN-thuh-sis). The leaves of a plant absorb the most sunlight. The flat part of a leaf always faces the sun. As the sun moves across the sky, the leaves respond by moving to continue facing the sun.

▶*Identify:* What is the process by which plants use sunlight to make food? photosynthesis

Chlorophyll The substance in a leaf that absorbs the sun's light is called **chlorophyll** (KLOWR-uh-fil). Chlorophyll is a green substance that gives leaves their color. Without chlorophyll, photosynthesis could not take place.

Sunlight is made up of white light. Chlorophyll absorbs all the colors of the visible spectrum except green. The chlorophyll reflects green light. As a result, leaves look green.

The energy of the absorbed light is used by the plant to make its own food. The food is in the form of sugars and starch. Food is a form of chemical energy. The energy is stored in the plant.

▶*Predict:* What color would leaves appear if chlorophyll reflected blue light? blue

LESSON SUMMARY

▶ Because light is a form of energy, it can be converted into other forms of energy.

▶ The process by which plants use sunlight to make their own food is called photosynthesis.

▶ Chlorophyll is a green substance that gives plants their color.

▶ Chlorophyll uses the light it absorbs to make food that is stored in the plant until it is needed.

CHECK *Complete the following.*

1. In photosynthesis, electromagnetic energy is turned into _____ energy. chemical

2. The foods that a plant makes are sugars and _____ . starch

3. Food is a form of _____ energy. chemical

4. The substance that gives plants their green color is _____ . chlorophyll

5. A plant's _____ always face the sun in order to get the most amount of sunlight. leaves

6. Chlorophyll absorbs all the sun's light except _____ light. green

APPLY *Complete the following.*

7. **Infer:** Plants always make more food than they need. The extra food is stored in the plant. What happens when an animal eats a plant?

8. **Hypothesize:** What happens to the energy that plants do not store as food?

InfoSearch

Read the passage. Ask two questions about the topic that you cannot answer from the information in the passage.

Photoelectric Cells A process that is like photosynthesis is used to produce electricity from solar energy. Sunlight is absorbed by photoelectric (foht-oh-i-LEK-trik) cells. A photoelectric cell converts sunlight into electrical energy. The electrical energy is stored in batteries. This process is similar to what a plant does, except that the plant stores the energy as food. Many photoelectric cells are put together to make solar panels. Solar panels are used on homes, office buildings, and satellites.

SEARCH: Use library references to find answers to your questions.

ACTIVITY

REMOVING CHLOROPHYLL FROM GREEN LEAVES

You will need a leaf, a knife, filter paper, a beaker, and rubbing alcohol.

1. Use the knife to carefully peel off the top layer of a leaf. **Caution:** Be careful when using a knife.

2. Gently press the inner layer of the leaf to a piece of filter paper for several minutes.

3. Pour some rubbing alcohol into the beaker.

4. Dip the filter paper into the alcohol and leave it there for several minutes.

5. Remove the filter paper and let it air dry.

Questions

1. What did you see on the filter paper when you removed the leaf?

2. What colors did you see on the filter paper after soaking it in rubbing alcohol?

3. **Predict:** When would you expect to see leaves with these colors?

How do mirrors reflect light?

Objective ▶ Describe how mirrors form clear images.

TechTerms

- **diffuse reflection:** reflection that forms a fuzzy image
- **mirror:** smooth surface that reflects light and forms images
- **regular reflection:** reflection that forms a clear image

Reflected Light When light is not completely absorbed by an object, the light will be reflected. The type of surface affects the type of reflection. There are two types of reflection. They are regular reflection and diffuse reflection.

Figure 1 Regular reflection

Mirrors Regular reflections are formed by a **mirror.** A mirror is a smooth surface that reflects light and forms images. A beam of light striking a mirror at a certain angle will be reflected from the mirror at the same angle. Figure 3 shows a beam of light hitting a mirror at a 70° angle. This is the angle of incidence. The reflected beam bounces off the mirror at a 70° angle. This is the angle of reflection. According to the law of reflection, the angle of incidence is equal to the angle of reflection.

▶*Infer:* Why does a mirror form a clear image?
because it has a smooth surface

- **Regular reflection** When a beam of light strikes a smooth surface, the light is reflected at the same angle. A regular reflection forms a clear image. For example, the light reflected from the surface of a still pond will form clear images of the objects around the pond. A regular reflection is formed only when a surface is completely smooth and even.

- **Diffuse reflection** When light strikes a rough surface, the light will be reflected at different angles. The reflected light from a diffuse reflection forms a fuzzy image. On a windy day, the surface of a pond is not smooth. The light reflected from the surface of the pond will form fuzzy images. Diffuse reflections are formed when a surface is rough.

▶*Identify:* What are the two types of reflection called? diffuse reflection and regular reflection

Figure 2 Diffuse reflection

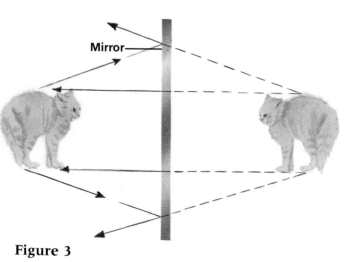

Figure 3

LESSON SUMMARY

▶ Light that is not completely absorbed by a surface will be reflected.

▶ In regular reflection, the reflected image is clear.

▶ In diffuse reflection, the reflected image is fuzzy.

▶ A mirror is a smooth surface that reflects light and forms images.

CHECK *Complete the following.*

1. On a windy day, the surface of a pond will form a _____ reflection. diffuse

2. A mirror forms _____ images. clear

3. The angle of incidence is equal to the angle of _____ . reflection

4. If the incident ray striking a mirror is at a 45° angle, the _____ ray will also be at a 45° angle. reflected

5. A _____ is a smooth surface that reflects light and forms images. mirror

6. When light strikes an object, the light will either be _____ or reflected. absorbed

APPLY *Complete the following.*

7. What happens when light rays strike a black object?

▶ 8. **Predict:** A bathroom mirror is covered with steam. Will the mirror produce a diffuse reflection or a regular reflection? Explain.

Skill Builder

Measuring Look at Figure 3 on page 164. Measure the distance from the object to the mirror. Compare this distance with the distance from the image to the mirror. Are they the same or different? Now measure the height of the object and the height of the image. Are they the same or different? What does this tell you about the image formed by a flat mirror?

ACTIVITY

OBSERVING MIRROR IMAGES

You will need a mirror, a ruler, and a book.

1. Place the ruler in front of the mirror so that the 1-cm mark is closest to the mirror.
2. Touch the 3-cm mark on the ruler. Observe the image in the mirror.
3. Move your hand to the 6-cm mark and observe the image.
4. Place the book in front of the mirror so that the title is facing the mirror. Observe the image.

Questions

1. Where did your reflection touch the ruler when you touched the 3-cm mark? the 6-cm mark?
2. Could you read the title of the book in the mirror? Why or why not?
3. What do your observations tell you about images in a mirror?

9-8 How do lenses refract light?

Objective ▶ Contrast how light rays are bent by a concave lens and a convex lens.

TechTerms

- **concave lens:** lens that curves inward
- **convex lens:** lens that curves outward
- **lens:** transparent material that bends light
- **real image:** image that can be projected onto a screen

Refraction of Light Light travels in straight lines. When light moves from one medium to another, the direction of the light changes. Suppose a beam of light passes from air into another material, such as glass. The path of the light will be bent, or refracted.

▶ *Predict:* What will happen to a beam of light when it passes from air into water?
It will be refracted.

Lenses A **lens** is a transparent material that bends, or refracts, light. The transparent material in most lenses is glass. All lenses have either one or two curved surfaces. There are two main types of lenses. They are convex lenses and concave lenses.

▶ *Hypothesize:* Why is a lens made of a transparent material?
because light must pass through the lens

Convex Lenses A **convex lens** curves outward. Figure 1 shows a convex lens. Light that passes through a convex lens is bent inward. When a convex lens refracts light, the light rays are brought together at a point called the focal point. The distance between the lens and the focal point is the focal length.

Light is refracted as it enters the lens. It also is refracted as it leaves the lens. The amount of refraction depends on how curved the lens is.

▶ *Identify:* What is the focal length of a convex lens?
the distance between the lens and the focal point

Concave Lenses A **concave lens** curves inward. Light that passes through a concave lens is bent away from the lens. The light rays are spread apart. Figure 2 shows a concave lens.

▶ *Describe:* What happens to light rays as they pass through a concave lens?
They are spread apart.

Real Images A lens can be used to form a **real image.** A real image is an image that can be projected onto a screen. The lens in your eye is a convex lens. It projects a real image onto your retina. The image projected onto the retina is upside down.

▶ *Define:* What is a real image?
an image that can be projected onto a screen

Figure 1

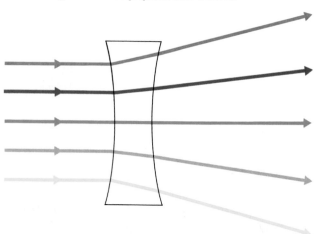

Figure 2

LESSON SUMMARY

▶ When light moves from one medium into another, it is bent, or refracted.

▶ A lens is a transparent material that bends, or refracts, light.

▶ A convex lens is a lens that curves outward.

▶ A concave lens is a lens that curves inward.

▶ A lens can be used to form real images.

CHECK *Complete the following.*

1. The _____ formed on the retina is upside down. image

2. A _____ lens curves inward. concave

3. A real image can be projected on a screen.

4. A _____ lens curves outward. convex

5. The _____ is where rays of light are brought together by a convex lens. focal point

6. The _____ is the distance from the focal point to the lens. focal length

APPLY *Use the diagram to complete the following.*

7. What type of corrective lens is used to correct nearsightedness?

8. What type of corrective lens is used to correct farsightedness?

Nearsightedness

Farsightedness

....
Ideas in Action ....

IDEA: Lenses refract light and form images.
ACTION: Make a list of optical instruments that use lenses. Describe the function of each of the instruments.

ACTIVITY

FORMING A REAL IMAGE WITH A CONVEX LENS

You will need a convex lens, a metric ruler, a large sheet of paper, and a piece of cardboard. You will need to work with a partner.

1. Have your partner stand in front of a window holding the sheet of paper.

2. Place the cardboard several meters away from your partner.

3. Hold the convex lens between the sheet of paper and the piece of cardboard.

4. Move the lens back and forth until you see an image on the cardboard. Measure the distance between the lens and the cardboard.

Questions

1. **Observe:** What do you see an image of?

2. What was the distance between the lens and the cardboard at which you could see a clear image? What is this distance equal to?

What is the electromagnetic spectrum?

Objective ▶ Identify the parts of the electromagnetic spectrum.

TechTerm

▶ **electromagnetic spectrum:** range of electromagnetic waves

Electromagnetic Waves Light is a type of electromagnetic wave. Light is made up of the colors of the visible spectrum. The color with the longest wavelength is red. The color with the shortest wavelength is violet.

Some electromagnetic waves have shorter and longer wavelengths than the visible spectrum. Light with a wavelength that is slightly longer than red light is called infrared light. Light with a wavelength that is slightly shorter than violet light is called ultraviolet light. Infrared and ultraviolet light are not part of the visible spectrum. They cannot be seen by the human eye.

▶ *Infer:* Why are infrared and ultraviolet light not part of the visible spectrum?
because they cannot be seen by the human eye

Electromagnetic Spectrum Visible light is only a small part of a larger number of electromagnetic waves. The different types of electromagnetic waves make up the **electromagnetic spectrum.** The electromagnetic spectrum includes radio waves, infrared rays, visible light, ultraviolet rays, X rays, and gamma rays.

▶ **Radio waves** Radio waves have the longest wavelength and the shortest frequency. They are used for radio, television, and radar.

▶ **Infrared rays** All objects give off infrared rays. You cannot see infrared rays, but you can feel them as heat.

▶ **Ultraviolet rays** Ultraviolet rays are present in sunlight. They can cause sunburn. Ultraviolet light is used to kill bacteria.

▶ **X rays** X rays have a very short wavelength and high frequency. X rays have a lot of energy. They can pass through most solid objects. They are used in medicine to form images of bones and internal organs.

▶ **Gamma rays** Gamma rays have a shorter wavelength and higher frequency than X rays. They are given off during nuclear reactions.

▶ *List:* What are the electromagnetic waves that make up the electromagnetic spectrum?
radio waves, infrared rays, visible light, ultraviolet rays, X rays, and gamma rays

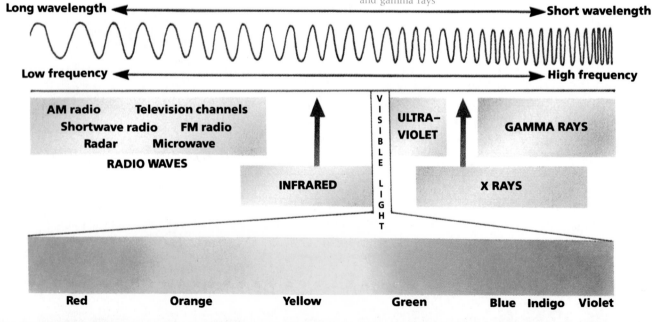

Long wavelength ◀ ▶ Short wavelength

Low frequency ◀ ▶ High frequency

AM radio	Television channels
Shortwave radio	FM radio
Radar	Microwave

RADIO WAVES

INFRARED

VISIBLE LIGHT

ULTRA-VIOLET

GAMMA RAYS

X RAYS

Red   Orange   Yellow   Green   Blue   Indigo   Violet

LESSON SUMMARY

▶ Light is made up of the colors of the visible spectrum.

▶ Visible light is a small part of a larger electromagnetic spectrum.

▶ The electromagnetic spectrum includes radio waves, infrared rays, visible light, ultraviolet rays, X rays, and gamma rays.

CHECK *Write true if the statement is true. If the statement is false, change the underlined term to make the statement true.*

1. X rays have so much energy they can go through most solid objects. true

2. Radio waves are usually given off during nuclear reactions. Gamma rays

3. Only those waves in the visible spectrum can be seen by the human eye. true

4. Ultraviolet rays have wavelengths slightly shorter than visible light. Infrared rays

5. Infrared rays are usually felt as heat. true

6. Gamma rays are used to make images of bones and internal organs. X rays

7. Television signals are examples of radio waves. true

APPLY *Complete the following.*

8. **Hypothesize:** Microwaves are high-frequency radio waves. Microwave ovens can be used to cook food quickly. The microwaves are absorbed by the food. Why do you think that food in a microwave oven gets hot, but a glass dish does not?

Long wavelength ◀

Low frequency ◀

AM radio	Television channels
Shortwave radio	FM radio
Radar	Microwave

RADIO WAVES

Health and Safety Tip

High-frequency electromagnetic waves can be dangerous. For example, too much exposure to X rays can be harmful. Use library references to find out why X-ray technicians wear lead aprons or stand behind a lead screen when taking X-ray pictures. Why are pregnant women advised not to have X rays taken?

CAREER IN PHYSICAL SCIENCE

X-RAY TECHNICIAN

Have you ever had an X-ray photograph taken in a doctor's or dentist's office? The person who operated the X-ray machine was an X-ray technician.

X rays have many uses in medicine and industry. Doctors use X-ray pictures to see broken bones and internal organs. Dentists use them to find cavities in teeth. Airport security systems use X-ray machines to examine passengers' baggage. Trained X-ray technicians are needed in all these areas.

X rays are both useful and dangerous. The high energy of X-ray photons makes them harmful to human body tissues. For this reason, X-ray technicians must take special care not to overexpose themselves or their patients to X rays. Special training for X-ray technicians is available at many technical and vocational schools.

What are lasers?

Objectives ▶ Explain how laser light is different from white light. ▶ Describe some uses of lasers.

TechTerm

▶ **laser:** device that produces a powerful beam of light

Lasers A powerful beam of light cuts through a steel plate. This light is different from ordinary white light. The light comes from a **laser**. A laser is a device that produces a very powerful beam of light.

▐▶**Define:** What is a laser?
device that produces a powerful beam of light

Laser Light Laser light is different from white light. White light is made up of many different wavelengths. It includes all the colors of the visible spectrum. Laser light is made up of only one wavelength. As a result, laser light is only one color. Unlike the light waves in white light, the waves in laser light are all in step. A beam of laser light can travel long distances in a straight line. It does not spread out as a beam of white light does.

👁**Observe:** How is laser light different from white light?
Laser light has only one color and does not spread out.

Uses of Lasers Lasers have many uses. In medicine, lasers can be used to repair a detached retina and prevent blindness. A detached retina results when the retina comes loose from the back of the eye. The heat of a laser beam can stick the retina back into place.

In industry, lasers can be used for cutting, welding, and drilling. For example, a laser beam can make a clean cut through thick layers of cloth. You are probably familiar with the use of lasers in videodiscs and compact disks, or CDs. Lasers are used in supermarkets to read the bar codes on many different products. Laser light also can be used to make three-dimensional images called holograms (HAHL-uh-grams).

▐▶*Explain:* How can a laser repair a detached retina?
The heat of the laser beam sticks the retina back into place.

White light

Laser light

LESSON SUMMARY

▶ A laser is a device that produces a powerful beam of light.

▶ Laser light is different from ordinary white light.

▶ Lasers are used in medicine and industry, for videodiscs and CDs, and to read bar codes.

CHECK *Complete the following.*

1. Laser light can be used by eye surgeons to repair a detached _____ . retina

2. A beam of light that spreads out as it travels is made up of _____ light. white

3. A beam of light that is made up of only one wavelength of light is made up of _____ light. laser

4. Waves of _____ light are all in step. laser

5. Because laser light has one wavelength, it also has only one _____ . color

6. Lasers are used in supermarkets to read _____ . bar codes

7. The three-dimensional image made by laser light is called a _____ . hologram

APPLY *Complete the following.*

8. **Contrast:** How are white light and laser light different?

9. **Hypothesize:** Scientists measured the distance from the earth to the moon by bouncing a laser beam off mirrors on the moon. They measured the time the laser beam took to reach the moon and to be reflected back to the earth. Would the scientists have been able to use ordinary white light to do the same thing? Why or why not?

10. The first laser ever made was a ruby laser. What color light do you think it produced? Why?

Skill Builder

Building Vocabulary Laser light is monochromatic (mahn-uh-kroh-MAT-ik). Use a dictionary to look up the definition of the word "monochromatic." What does the prefix "mono-" mean? What does the root word "chromatic" mean? Why is this a good description of laser light?

SCIENCE CONNECTION

HOLOGRAPHY

There is a special kind of photography that uses laser light. This process is known as holography (huh-LAHG-ruh-fee). Laser light produces a three-dimensional image. The image is known as a hologram (HAHL-uh-gram).

To make a hologram, two pictures of an object are taken. Each picture is taken from a slightly different angle. Then the two pictures are combined into one hologram.

When you look at a hologram, something amazing happens. You can see the original object from different angles. It is as if the object were in front of you, even though you are seeing only an image of the object.

There are several practical uses for holograms. Architects can use holograms to show three-dimensional views of buildings. Structural engineers can use holograms to test the sturdiness of a solid object. To learn more about holography, write to the New York Holographic Laboratory, 34 West 13th Street, New York, NY 10011.

UNIT 9 Challenges

STUDY HINT Before you begin the Unit Challenges, review the TechTerms and Lesson Summary for each lesson in this unit.

TechTerms

chlorophyll (160)
concave lens (164)
convex lens (164)
diffuse reflection (162)
electromagnetic spectrum (166)
eye (156)
illuminated objects (154)
image (156)
laser (168)

lens (164)
light (150)
luminous objects (154)
mirror (162)
opaque (154)
photon (150)
photosynthesis (160)
prism (158)

ray (150)
real image (164)
regular reflection (162)
sight (156)
translucent (154)
transparent (154)
transverse wave (152)
visible spectrum (158)

TechTerm Challenges

Matching *Write the TechTerm that matches each description.*

1. human sense organ used for sight eye
2. triangular piece of glass that separates white light into separate colors prism
3. lens that curves out convex lens
4. image that can be projected real image
5. reflection that gives fuzzy images diffuse reflection
6. seven colors in white light visible spectrum
7. electromagnetic waves including X rays and visible light electromagnetic spectrum
8. material that allows light to pass through easily transparent
9. device that produces light of only one color laser
10. small bundle of energy photon
11. made up of photons light
12. human sense involving the eyes sight
13. lens that curves inward concave lens
14. straight line that shows the direction of light ray

Fill In *Write the TechTerm that best completes each statement.*

1. By the process of _____ , a plant makes its own food from the sun's light. photosynthesis
2. Lamps are _____ because they give off their own light. luminous objects
3. In a _____ , the particles of the medium move up and down at right angles to the direction of wave motion. transverse wave
4. The green substance used in photosynthesis is _____ . chlorophyll
5. Your hand casts a shadow when placed in front of a light because it is _____ . opaque
6. A _____ is a smooth surface that reflects light and forms images. mirror
7. An upside-down _____ forms on the retina. image
8. A colored sheet of glass is _____ because it transmits only some light. translucent
9. When you shine a light in a dark room, the objects that you see are illuminated objects _____ .
10. The image formed by a smooth surface is an example of _____ . regular reflection
11. A _____ can be used to focus light and form images. lens

Content Challenges

Multiple Choice *Write the letter of the term or phrase that best completes each statement.*

1. Chlorophyll absorbs all the colors of the visible spectrum except
   a. red. **b.** green. **c.** blue. **d.** violet. b

2. The image from a convex lens is
   a. right side up. **b.** reversed. **c.** upside down. **d.** none of the above. c

3. The color that is <u>not</u> one of the seven colors of the visible spectrum is
   a. red. **b.** blue. **c.** indigo. **d.** brown. d

4. The type of electromagnetic wave that can be felt as heat is
   a. gamma rays. **b.** infrared rays. **c.** X rays. **d.** ultraviolet rays. b

5. The part of the eye that controls the amount of light coming into the eye is the
   a. pupil. **b.** cornea. **c.** iris. **d.** retina. c

6. In photosynthesis, electromagnetic energy is converted to
   a. electrical energy. **b.** nuclear energy. **c.** chemical energy. **d.** mechanical energy. c

7. Unlike a beam of white light, a laser beam has
   a. no color. **b.** one wavelength. **c.** two colors. **d.** no wavelength. b

8. The point where the light that passes through a lens is brought together is called the
   a. focal length. **b.** refraction point. **c.** focal point. **d.** retina. c

9. The type of electromagnetic wave that is often used by dentists is called
   a. an ultraviolet ray. **b.** a gamma ray. **c.** an X ray. **d.** a radio wave. c

10. An object that absorbs light and does not reflect any appears
    a. green. **b.** white. **c.** black. **d.** blue. c

True/False *Write true if the statement is true. If the statement is false, change the underlined term to make the statement true.*

1. The types of food that a plant makes during photosynthesis are <u>sugar</u> and starch. true
2. There are <u>six</u> colors in the visible spectrum. seven
3. The <u>retina</u> is a protective layer around the eye. cornea
4. The lens in the eye is a <u>concave</u> lens. convex
5. All electromagnetic waves are streams of <u>photons</u>. true
6. Television signals are transmitted by <u>radio waves</u>. true
7. The leaves of plants always face <u>away from</u> the sun. toward
8. When light travels from one <u>medium</u> to another it will be bent. true
9. The part of the visible spectrum with the shortest wavelength is <u>red</u>. violet
10. The electromagnetic waves with the shortest frequency are <u>infrared rays</u>. true
11. According to the law of <u>refraction</u>, the angle of incidence is equal to the angle of reflection. reflection

Understanding the Features

Reading Critically *Use the feature reading selections to answer the following. Page numbers for the features are in parentheses.*

1. What part of the camera allows light to enter the camera? (153)
2. How do corrective lenses help nearsighted people? (157)
3. Which is the darker part of a shadow, the umbra or the penumbra? (155)
4. **Infer:** Why is colorblindness called a genetic disorder? (159)
5. How does an architect use holograms? (169)
6. Why are X rays harmful to human tissue? (167)

Concept Challenges ....

Interpreting a Diagram *Use the diagram showing a lens to complete the following.*

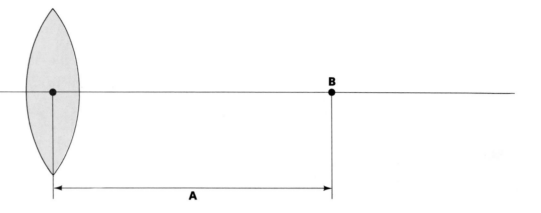

1. What type of lens is shown?
2. Which letter on the diagram represents the focal length of the lens?
3. Which letter represents the focal point?
4. Copy the diagram on a sheet of paper. Show the path of light rays passing through the lens.
5. Are the light rays brought together or spread apart by the lens?

Critical Thinking *Answer each of the following in complete sentences.*

1. **Compare:** What are two ways that a photon is similar to a particle?
2. How does a prism separate white light into a spectrum?
3. Why do electromagnetic waves travel best in a vacuum?
4. Why does a laser beam not spread out over long distances?
5. **Hypothesize:** Some animals can hear ultrasonic sounds, even though humans cannot. Do you think that some animals can see ultraviolet light, even though humans cannot? Why or why not?
6. **Model:** Think of your eye as a camera. Which part of the eye is similar to the camera's aperture? Which part is similar to the film in the camera? Which part is similar to the camera lens?

Finding Out More ....

1. Scientists have evidence that the universe began 15 billion years ago after a powerful explosion called the Big Bang. Traces of electromagnetic waves from this explosion can still be detected. Use library references to find out about this so-called "background radiation." Write a report of your findings.
2. **Classify:** Collect pictures from catalogs and magazines showing different types of lenses. Try to find as many different types of lenses as possible. Label each type of lens. Show your collection to the class.
3. Visit your local planetarium. Ask how telescopes are made. Find out how lenses and mirrors are used in modern telescopes. How do modern telescopes differ from older telescopes?
4. Use library references to find out about the evolution of the eye. What was the earliest life form with eyes? How do human eyes differ from the eyes of other animals? What animal's eyes are most like human eyes?

ELECTRICITY

CONTENTS

STUDY HINT Before beginning Unit 10, scan through the lessons in the unit looking for words that you do not know. On a sheet of paper, list these words. Work with a classmate to try to define each word on your list.

10-1 What is electricity?

Objective ▶ Distinguish between positive and negative electric charges.

TechTerms

- **electricity** (i-lek-TRIS-uh-tee): form of energy caused by moving electrons
- **electron:** atomic particle with a negative electric charge
- **neutron:** atomic particle with neither a negative nor a positive electric charge
- **proton:** atomic particle with a positive electric charge

Atoms One of the smallest particles in nature is called an atom. Matter is made up of atoms. Atoms have three basic parts. They are **neutrons, electrons,** and **protons.** The neutrons and protons are found in the center of the atom. The electrons circle around the protons and neutrons in a cloud.

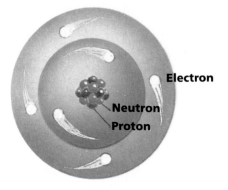

▎▶*Name:* What are the three main parts of an atom? neutrons, protons, and electrons

Electric Charge Protons and electrons have an electric charge. The electron has a negative charge. A negative charge is shown with a minus sign (−). The proton has a positive charge. A positive charge is shown with a plus sign (+). Objects that have an electric charge follow these rules:

- Objects with the same charge repel, or move away from, each other.

▶ Objects with different charges attract each other.

Because they have opposite charges, electrons and protons attract each other. Two electrons, or two protons, repel each other because they have the same charge. An object that has neither a positive nor a negative charge is neutral. Neutrons are neutral.

▎▶*State:* What kind of charge do electrons have? Electrons have a negative charge.

Neutral Objects Most objects, such as a book or a pencil, are electrically neutral. When you pick up a book, you do not feel an electric charge. However, neutral objects can become electrically charged. When a neutral object gains or loses electrons, it becomes electrically charged. When you walk across a carpet, your body picks up electrons from the carpet. If you then touch a metal doorknob, electrons jump from your hand to the doorknob. You feel an electric charge.

▎▶*Describe:* What happens to a neutral object when it gains electrons? It becomes electrically charged.

Electricity Electrons can move from place to place. Electrons always move from a negatively charged area to a positively charged area. The form of energy caused by moving electrons is called **electricity** (i-lek-TRIS-uh-tee).

▎▶*Define:* What is electricity?
form of energy caused by moving electrons

LESSON SUMMARY

- Atoms have three basic parts: neutrons, protons, and electrons.
- Protons have a positive charge, and electrons have a negative charge.
- Like charges repel; unlike charges attract.
- Most objects are electrically neutral.
- Electrons always move from a negatively charged area to a positively charged area.

CHECK *Write true if the statement is true. If the statement is false, change the underlined term to make the statement true.*

1. In an atom, the <u>protons</u> circle around the center of the atom. electrons

2. An electron has a <u>negative</u> charge. true

3. A <u>neutral</u> object has neither a positive nor a negative charge. true

4. Electricity is the energy caused by <u>stationary</u> electrons. moving

5. Two electrons will <u>repel</u> each other. true

APPLY *Complete the following.*

6. **Infer:** A bolt of lightning results when electrons jump from a cloud to the ground. What does this tell you about the cloud?

7. An atom of oxygen has eight protons and eight electrons. Because there are the same number of protons and electrons, the atom is neutral. What happens if the oxygen atom loses an electron?

8. **Hypothesize:** When you turn on a lamp switch, a flow of electrons causes the lamp to light. Why does the light go on so quickly?

ACTIVITY

OBSERVING ELECTRIC CHARGES

You will need a plastic or rubber comb and a sheet of paper. Note: this experiment will work best on a cool, dry day.

1. Tear the sheet of paper into small pieces. Spread the pieces of paper on your desk or a table.
2. Run the comb through your hair several times.
3. Hold the comb above the pieces of paper. Be sure not to touch the paper with the comb. Observe what happens to the pieces of paper.

Questions

1. **Observe:** What happened to the pieces of paper when you held the comb over them?

2. How can you explain what happened?

What are insulators and conductors?

Objective ▶ List some examples of conductors and insulators.

TechTerms

- ▶ **conductors:** materials that allow electric charges to flow through them easily
- ▶ **insulators:** materials that prevent electric charges from flowing through them easily

Conductors A material that allows electric charges to flow through it is called a **conductor.** If a metal wire is placed between two oppositely charged objects, electrons will flow through the wire toward the positively charged object. Many metals are good conductors. Suppose a metal wire is attached to the positive and negative poles of a battery. Electrons will flow toward the positive pole. If a piece of rubber is used instead of a metal wire, electrons will not flow.

Copper wire

All conductors allow electrons to flow, but some conductors are better than others. A wire made of copper or iron is one of the best conductors of electricity. A wire made of brass or magnesium is not as good a conductor of electricity.

▌▌▌▶ *Define:* What is a conductor?
a material that allows electrons to flow

Insulators A material that prevents electric charges from flowing through it is called an **insulator.** Rubber is an insulator. When a piece of rubber is placed between two charged objects, electrons will not flow. The piece of rubber keeps the electrons from flowing. Other insulators are cork, wood, and plastic.

▣ *Classify:* Is a piece of rubber an insulator or a conductor? insulator

Electric Cords An electric cord uses both insulators and conductors. There are three main parts to an electric cord. Two sets of metal wires in an electric cord carry electricity. They are kept apart by a covering of rubber. The two wires are covered with rubber so that you do not feel an electric shock when you touch the cord. Very old electric cords may have the rubber insulation worn out. These electric cords are dangerous because they can cause a short circuit. A short circuit results when two wires touch, allowing an electric charge to jump between them.

▌▌▌▶ *Explain:* Why are electric cords covered with rubber? to prevent an electric shock

LESSON SUMMARY

▶ A conductor allows electric charges to flow easily.

▶ Some conductors are better than others.

▶ An insulator prevents electric charges from flowing easily.

▶ An electric cord is made up of a conductor and an insulator.

CHECK *Complete the following.*

1. A <u>conductor</u> allows current to flow easily.

2. Rubber is an _____ because it prevents current from flowing easily. insulator

3. In an electric cord, the _____ are the conductors. metal wires

4. An electric cord is safe to touch because it is covered with _____ . rubber insulation

5. If the two wires in an electric cord touch, a _____ will result. short circuit

6. Cork is a good _____ . insulator

7. A copper wire is one of the best _____ . conductors

APPLY *Complete the following.*

▶ 8. **Infer:** A lightning rod is supposed to keep a bolt of lightning from damaging a house. The rod is placed at the top of the house and is connected to the ground. Is the lightning rod made of a material that is a conductor or an insulator? Explain.

InfoSearch

Read the passage. Ask two questions about the topic that you cannot answer from the information in the passage.

Michael Faraday Faraday was a British scientist. He contributed a great deal to the understanding of electricity. Faraday was the first to discover the principle behind the electric motor. He also studied the relationship between electricity and magnetism. The faraday is a unit of electricity that was named in honor of Michael Faraday.

SEARCH: Use library references to find answers to your questions.

CAREER IN PHYSICAL SCIENCE

TELEVISION REPAIRPERSON

More than 90% of the homes in the United States have at least one television. A television is an electronic device. A television repairperson must be familiar with electronics. Electronics is the study of electricity as it is used in helpful devices such as televisions.

Televisions use semiconductors. One type of semiconductor is similar to an ordinary conductor. It allows electric charges to flow. The difference is that this semiconductor allows electric charges to flow in only one direction. Complicated electronic devices, including televisions, can be built using semiconductors.

Modern televisions use some of the same equipment found in computers. Some televisions even have small computers built into them.

Do you think you might want to be a TV repairperson? Visit a technical school to find out more about this career.

What are two kinds of electric current?

Objective ▶ Differentiate between direct current and alternating current.

TechTerms

▶ **alternating current:** current in which electrons change direction at a regular rate

▶ **direct current:** current in which electrons always flow in the same direction

▶ **electric current:** flow of electrons through a conductor

Electric Current When a conductor is connected to two oppositely charged objects, electrons will flow through the conductor. The flow of electrons through a conductor is called **electric current.** The number of electrons flowing determines the amount of electric current.

▐▐▐▶ *Define:* What is electric current?
    flow of electrons through a conductor

Direct Current Current that always flows in the same direction is called **direct current** (DC). Figure 1 shows the positive and negative poles of a battery connected to a wire. The wire is connected to a small lamp. The electric current from the battery keeps the lamp lit. Notice that the current flows from the negative pole to the positive pole of the battery. The current flows in one direction. All batteries provide direct current.

Figure 1

▐▐▐▶ *Define:* What is direct current?
    Current that always flows in the same direction.

Alternating Current Current that changes direction at a regular rate is called **alternating current** (AC). The type of electricity used in homes is alternating current. Most of the electricity that is used in everyday life comes from alternating current, not from direct current. There is a practical reason why alternating current is used.

Figure 2

Wires carrying direct current become hot. An electric power plant sends electricity over long cables to reach your home. The heat from large amounts of direct current would damage the cables. As a result, the power plant cannot use direct current to transport the electricity. Alternating current produces less heat. For this reason, alternating current is used.

Most household appliances use alternating current. However, some appliances need direct current. A built-in converter in these appliances changes alternating current into direct current.

▐▐▐▶ *Explain:* Why is alternating current used by power plants?
    because it does not create heat the way direct current does

LESSON SUMMARY

▶ The flow of electrons through a conductor is called electric current.

▶ Direct current always flows in the same direction.

▶ Alternating current changes direction at a regular rate.

▶ Alternating current is the most commonly used type of electricity because it does not create as much heat as direct current.

CHECK *Complete the following.*

1. The flow of electrons through a conductor is called _____ . electric current

2. Current that changes direction at a regular rate is called _____ current. alternating

3. A battery is a source of _____ current. direct

4. The type of electricity supplied by electric power plants is _____ . alternating current

5. In a battery, current flows from the _____ pole. negative

APPLY *Complete the following.*

6. What type of current does a television use?

7. **Explain:** Why do electric power plants not produce direct current?

▶ 8. **Infer:** What type of current does a car battery provide?

Ideas in Action ....

IDEA: Batteries are often used as a source of direct current.

ACTION: Look at different types of batteries, from size AA to a transistor radio battery. Where is the positive pole of the battery? Where is the negative pole? In what direction does the current flow?

LEISURE ACTIVITY

BUILDING MODEL ELECTRIC TRAINS

Many people enjoy building and running model trains. Some trains use direct current. Others use alternating current from a wall outlet. Most electric trains run on AC current. A transformer changes the AC current from an outlet into the small amount of AC current used by the train.

The basic equipment for a model railroad includes train tracks, several railroad cars, and a locomotive. Electric wires connected from a control board to the train set allow one or more trains to run at the same time.

Building model trains first became a popular hobby in the 1930s. Model railroads were shown at the Chicago World's Fair in 1934. The National Model Railroad Association was organized in 1935. If you are interested in building model trains, write to your local model railroad club or a model railroad magazine.

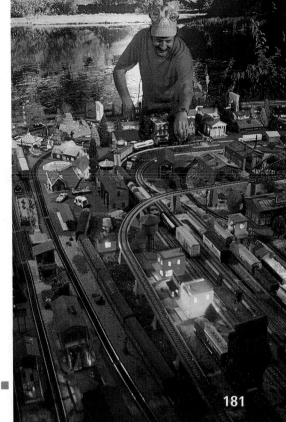

10-4 What is a battery?

Objectives ▶ Identify a battery as a series of electrochemical cells that are connected together. ▶ Contrast a wet cell and a dry cell.

TechTerms

- ▶ **battery:** series of electrochemical cells connected together
- ▶ **electrochemical cell:** device that changes chemical energy into electrical energy
- ▶ **electrode:** positive or negative pole of an electrochemical cell
- ▶ **electrolyte:** substance that is an electrical conductor

Battery A **battery** is a source of direct current. When you connect the negative and positive poles of a battery to a conductor, electric current flows through the conductor. A battery is a series of **electrochemical cells** that are connected together. An electrochemical cell changes chemical energy into electrical energy.

▶**Describe:** How does a battery produce electricity? It changes chemical energy into electrical energy.

Wet Cell The simplest type of electrochemical cell is called a wet cell. A wet cell has three parts. They are a negative pole, a positive pole, and an **electrolyte.** An electrolyte is a substance that conducts electricity. The negative and positive poles of a wet cell are called **electrodes.** The negative electrode is made of zinc. The positive electrode is made of copper. Sulfuric acid is often used as an electrolyte.

A chemical reaction in the wet cell causes electrons to build up on the zinc electrode. When the two electrodes are connected by a wire, electrons flow from the zinc to the copper electrode. A car battery is made up of several wet cells.

In a car's storage battery, two sets of metal plates act as electrodes. One set of plates is made of lead. The other set is made of lead dioxide. The electrolyte is sulfuric acid.

▶**Identify:** What are the three parts of a wet cell? negative electrode, positive electrode, and an electrolyte

Dry Cell The most familiar type of electrochemical cell is called a dry cell. Dry cells are used in flashlights and radios. A dry cell works the same way as a wet cell. It has a positive electrode, a negative electrode, and an electrolyte. The electrolyte is a moist paste inside the cell. The outside case of the battery is usually made of zinc. It is the negative electrode. The positive electrode is sometimes made of carbon. It is inside the dry cell.

Another type of dry cell is the nickel–cadmium cell. The positive electrode is made of nickel oxide. The negative electrode is made of cadmium.

▶**Contrast:** How is the electrolyte in a dry cell different from that in a wet cell? The electrolyte is a moist paste, not a liquid.

182

LESSON SUMMARY

▶ A battery is made up of a series of electro-chemical cells.

▶ The simplest type of electrochemical cell is a wet cell.

▶ The most familiar type of electrochemical cell is a dry cell.

CHECK *Complete the following.*

1. A _____ cell is used in a flashlight. dry

2. A car battery is made up of _____ cells. wet

3. The negative electrode of a wet cell is usually made of _____ . zinc

4. An _____ is a substance that conducts electricity. electrolyte

5. An electrochemical cell converts _____ into electrical energy. chemical energy

6. In a dry cell, the electrolyte is found _____ the cell. inside

7. A battery is a source of _____ current. direct

8. The positive and negative poles of a battery are called _____ . electrodes

APPLY *Complete the following.*

9. **Hypothesize:** Why do you think a car battery is sometimes called a storage battery?

10. **Infer:** Some batteries last longer if they are kept in a cold place, such as a freezer. What effect does the cold have on the battery?

Designing an Experiment...........

Design an experiment to solve the problem.

PROBLEM: How can you show that a battery will last longer if it is kept in a freezer before it is used?

Your experiment should:

1. List the materials you would need.

2. Identify safety precautions that should be followed.

3. List a step-by-step procedure.

4. Describe how you would record your data.

ACTIVITY

MAKING A LEMON WET CELL

You will need a lemon, a strip of copper, a strip of zinc, two wires, and a voltmeter.

1. Attach a wire to one end of each metal strip.

2. Insert the free end of each strip of metal at opposite ends of the lemon.

3. Attach the free ends of the wires to the terminals of the voltmeter. The voltmeter will indicate if electricity is flowing.

Questions

1. What is the positive electrode of your wet cell?

2. What is the negative electrode?

3. What is the electrolyte?

10-5 What is a series circuit?

Objectives
▶ Explain how electricity flows through a closed circuit. ▶ Describe a series electric circuit.

TechTerms

▶ **electric circuit:** path that an electric current follows

▶ **series circuit:** circuit in which electric current follows only one path

Circuits An **electric circuit** is the path that an electric current follows. When wires are connected to a battery and to a lamp, the lamp will light. The wires, lamp, and battery form an electric circuit. If the wires are disconnected, current cannot flow through the circuit.

All electric circuits have four parts. A circuit needs a source of electric current. This can be a battery or a wall outlet. The load is the device that uses the electric current. The load can be a light bulb, a motor, or some other electric device. Wires connect the battery to the load. All electric circuits also have a switch.

▷▶ *Describe:* What are the parts of an electric circuit?
a source of electric current, a load, wires, and a switch

Open and Closed Circuits Electric current cannot flow through an open circuit. Electric current can flow only through a closed circuit. A switch is used to open and close an electric circuit. When a switch is off, the circuit is called an open circuit. When the switch is on, the circuit is called a closed circuit.

▷▶ *Explain:* Why is a switch used in an electric circuit? to turn a circuit on or off

Series Circuit The simplest type of electric circuit is a **series circuit.** In a series circuit, the current follows only one path. Figure 1 shows a series circuit. The path of the current is from the negative electrode of the battery, to the lamp, to the positive electrode of the battery. Figure 2 also shows a series circuit. In this circuit, the battery is connected to two lamps. The current first goes to one lamp, and then to the other lamp.

▷▶ *Define:* What is a series circuit?
a circuit in which the current follows only one path

Figure 2

Light bulb (load)

Wire

Switch

Source of electric current

Figure 1

LESSON SUMMARY

▶ An electric circuit is the path an electric current follows.

▶ All electric circuits have four parts: a source of electric current, a load, wires, and a switch.

▶ Current flows only through a closed circuit.

▶ In a series circuit, the current follows only one path.

CHECK *Complete the following.*

1. A lamp that has been switched on is an example of a _____ circuit. closed

2. A battery connected to one lamp is an example of a _____ circuit. series

3. A circuit needs a source of current, a load, a _____ , and wires. switch

4. Current flows through one path in a _____ circuit. series

5. Current cannot flow in an _____ circuit. open

APPLY *Complete the following.*

6. Why does an electric circuit need a switch?

▲ 7. **Model:** Think of an electric circuit as water flowing through pipes. What represents the electric current in this model?

Skill Builder

Diagraming You know that all electric circuits have four parts. You can use this information to draw a circuit diagram. In order for someone else to understand your diagram, it is helpful to use standard symbols for each part of the circuit. Here is a list of some of the most common circuit diagram symbols.

This is the source of electric current.

This is a lamp.

This is the wire.

This is the switch.

The diagram shows an open series circuit with one lamp connected to the current source.

Use this information to draw a circuit diagram for a closed series circuit with two lamps.

▼▼▼ LOOKING BACK IN SCIENCE ▼▼▼▼▼▼▼▼▼▼▼▼

COMPUTERS

The idea for computers dates back to the 17th century. However, the first real computer was not built until the 20th century. In 1946, two American scientists were the first to build a working computer.

Early computers were very large. They had only a limited amount of memory, or ability to store information. Over the years, the technology for building computers improved. Computers could be made much smaller but with more speed and memory. In the 1950s and 1960s, computers were important in space exploration.

In the 1970s and 1980s, small, powerful computers were developed for home use. Microcomputers, or personal computers, are used by schools and businesses. Computers are now used for bookkeeping, drawing illustrations, word processing, and telecommunications.

Computers still are changing. They are being made smaller and faster. In the future, artificial intelligence research may result in computers that can "think" like humans.

10-6 What is a parallel circuit?

Objectives ▶ Describe a parallel electric circuit. ▶ Compare a parallel circuit and a series circuit.

TechTerm

▶ **parallel circuit:** circuit in which electric current can follow more than one path

Parallel Circuits In a **parallel circuit,** the electric current can follow more than one path. Look at Figure 1. Three lamps are connected to one battery. Notice how the wires are connected to each lamp. The current flows along three separate paths, one for each lamp. If one of the lamps goes out, the other two lamps will still remain lit.

Figure 1

▶ *Define:* What is a parallel circuit?
a circuit in which the current can follow more than one path

Series vs Parallel Circuits Suppose a battery, wires, and three lamps are connected in a series circuit. What happens if one of the lamps goes out? The current flowing to that lamp will be stopped. When the current in a series circuit is stopped at any point, the whole circuit becomes open. None of the lights will work. In a parallel circuit, the current can follow more than one path. The lamps are on different branches. If one lamp goes out, current can still reach the other lamps.

▶ *Describe:* What happens to the current in a series circuit when a lamp goes out?
The current stops flowing.

Uses for Parallel Circuits The electric circuits in your home are parallel circuits. When appliances are plugged into wall outlets, they are connected to parallel circuits. If one appliance stops working, electric current keeps flowing in the circuit. The other appliances keep working. Most schools and office buildings also use parallel electric circuits.

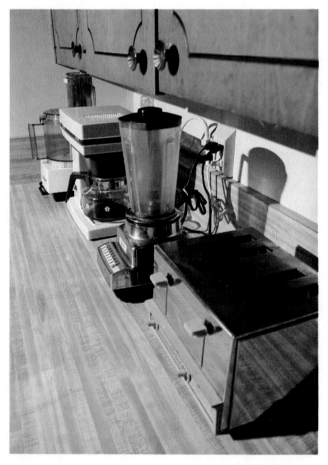

Figure 2

▶ **Infer:** Why are parallel circuits used in homes, schools, and office buildings?
so that current will keep flowing if one part of the circuit is broken

LESSON SUMMARY

▶ In a parallel circuit, the electric current can follow more than one path.

▶ If several appliances are connected in a parallel circuit, current can still reach the other appliances if one of them stops working.

▶ Parallel circuits are used in homes, schools, and office buildings.

CHECK *Complete the following.*

1. In a _____ circuit, the current follows only one path. series

2. In a _____ circuit, the current can follow more than one path. parallel

3. If two lamps are connected in a series circuit and one of them goes out, the circuit becomes an _____ circuit. open

4. Homes, offices, and schools use _____ electric circuits. parallel

APPLY *Complete the following.*

5. A string of Christmas tree lights goes out when one bulb stops working. Are the lights connected in series or parallel?

6. **Interpret:** Look at the circuit diagram. Which lamps are connected in parallel? Which are in series?

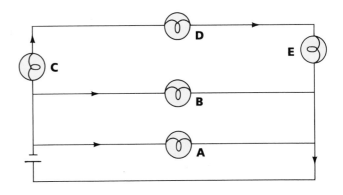

7. **Infer:** Are the circuits in a car parallel or series? How do you know?

Ideas in Action ....

IDEA: Parallel circuits are used more often than series circuits.

ACTION: Look at the manual for a portable radio and find the circuit diagram. How many series circuits does it have? How many parallel circuits?

CAREER IN PHYSICAL SCIENCE

COMPUTER PROGRAMMER

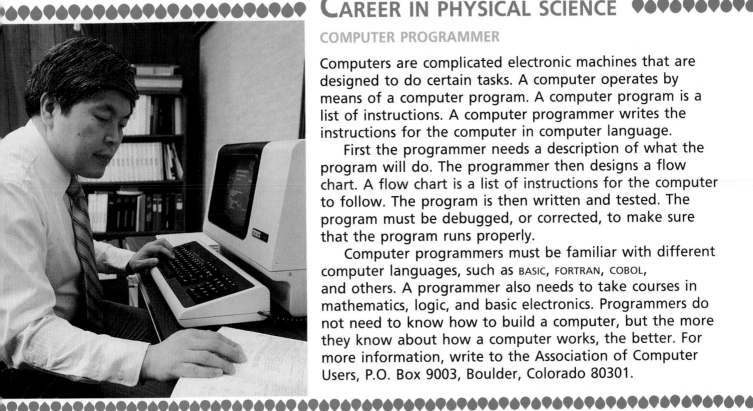

Computers are complicated electronic machines that are designed to do certain tasks. A computer operates by means of a computer program. A computer program is a list of instructions. A computer programmer writes the instructions for the computer in computer language.

First the programmer needs a description of what the program will do. The programmer then designs a flow chart. A flow chart is a list of instructions for the computer to follow. The program is then written and tested. The program must be debugged, or corrected, to make sure that the program runs properly.

Computer programmers must be familiar with different computer languages, such as BASIC, FORTRAN, COBOL, and others. A programmer also needs to take courses in mathematics, logic, and basic electronics. Programmers do not need to know how to build a computer, but the more they know about how a computer works, the better. For more information, write to the Association of Computer Users, P.O. Box 9003, Boulder, Colorado 80301.

Objective ► Use the correct units to measure voltage, current, and resistance.

TechTerms

- ► **ampere:** unit used to measure electric current
- ► **ohm:** unit used to measure resistance
- ► **resistance:** opposition to the flow of electric current
- ► **volt:** unit used to measure voltage
- ► **voltage:** force that makes electrons move

Voltage Electric current is made up of electrons flowing through an electric conductor. The electrons are pushed from place to place by an electric force. The force needed to make the electrons move along the conductor is the **voltage.** The unit used to measure voltage is the **volt.** Voltage is measured with a device called a voltmeter.

▐▐▐▶ *Define:* What is voltage?

the force that makes electrons move

Current The amount of electric current depends on the number of electrons flowing through a wire. The unit for measuring electric current is the **ampere,** or amp. The ampere is a measure of the number of electrons flowing in a circuit in a given amount of time. One ampere is the amount of current passing a point each second. Current is measured with a device called an ammeter.

▐▐▐▶ *Identify:* What does an ampere measure?

current

Resistance When electric current flows through a wire, the electrons are slowed down a bit by the wire. The electric current meets a certain **resistance** from the wire. Resistance opposes the flow of electrons in a circuit. Resistance is similar to friction. The unit for measuring resistance is the **ohm.** Four things affect the resistance of a wire.

- ► **Length** The longer a wire is, the more resistance it has.
- ► **Width** The thinner a wire is, the more resistance it has.
- ► **Material** Wires made of poor conductors have more resistance than wires made of good conductors.
- ► **Temperature** As a wire gets hotter, its resistance increases.

▐▐▐▶ *Name:* In what units is resistance measured?

ohms

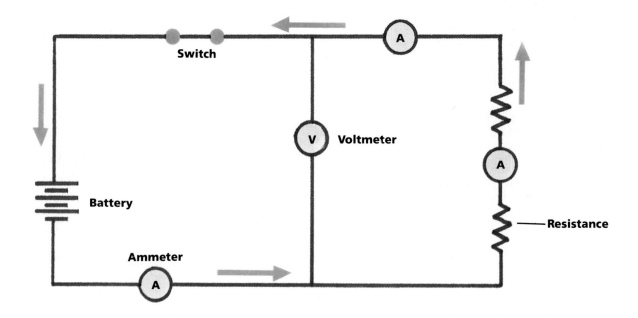

LESSON SUMMARY

▶ Voltage is the force that moves electrons in a circuit.

▶ The amount of electric current depends on the number of electrons flowing in a circuit.

▶ As current flows through a wire, it meets resistance from the wire.

▶ Resistance of a wire depends on length, thickness, material, and temperature.

CHECK *Write true if the statement is true. If the statement is false, change the underlined term to make the statement true.*

1. A thick wire will have <u>less</u> resistance than a thin wire. true

2. Current is measured in <u>volts</u>. amperes, or amps

3. A cool wire will have <u>more</u> resistance than a hot wire. less

4. The force that makes electrons move is called <u>voltage</u>. true

5. The resistance in a wire <u>helps</u> the flow of electric current. opposes

6. A long wire has <u>less</u> resistance than a short wire. more

7. The unit for voltage is the <u>volt</u>. true

APPLY *Complete the following.*

8. **Hypothesize:** Which has more resistance, a conductor or an insulator? Why?

InfoSearch

Read the passage. Ask two questions about the topic that you cannot answer from the information in the passage.

Volts, Amps, Ohms Many units of measurement in science are taken from the names of famous scientists. Volts are named after Alessandro Volta. Volta was an Italian professor of physics. Amperes are named in honor of Andre Marie Ampere. He was a French physicist and mathematician. Ohms are named for Georg Ohm, a German physicist.

SEARCH: Use library references to find answers to your questions.

TECHNOLOGY AND SOCIETY

SUPERCONDUCTORS

Electric current needs a conductor in which to flow. Usually, this conductor is a metal wire. If the wire is cooled, its resistance decreases. As the wire gets colder and colder, the resistance keeps decreasing. The lower the resistance, the more current can flow in the wire.

The coldest possible temperature is −273 °C, or absolute zero. At this temperature, a wire will have zero resistance. A conductor with near-zero resistance is known as a superconductor. Superconductors can improve many devices that use electricity. Superconducting wires could carry electricity great distances with no loss due to resistance.

To be really useful, materials that become superconductors at room temperature must be developed. In the late 1980s, some high-temperature superconductors were discovered. However, much research still needs to be done.

10-8 What is Ohm's law?

Objective ▶ Relate electric current, voltage, and resistance.

TechTerm

▶ **Ohm's law:** current in a wire is equal to the voltage divided by the resistance

I, V, and R Every closed circuit has an electric current (I), voltage (V), and resistance (R). Current, voltage, and resistance vary from circuit to circuit. Different power sources have different amounts of voltage. For example, a 9-volt radio battery has less voltage than a 12-volt car battery. The resistance varies depending on the type of wires used. The current is affected by the voltage and the resistance.

▶**Name:** What three things does every closed circuit have? electric current, voltage, and resistance

Ohm's Law Even though the current, voltage, and resistance vary from circuit to circuit, there is a simple relationship among them. This relationship is called **Ohm's law.** Ohm's law states that the current is equal to the voltage divided by the resistance.

$$I = V/R$$

Suppose a 12-volt battery is connected to a circuit with a resistance of 6 ohms. What is the current?

$$I = V/R$$
$$I = 12 \text{ volts}/6 \text{ ohms}$$
$$I = 2 \text{ amps}$$

▶**Describe:** What does Ohm's law state about the current, voltage, and resistance in an electric circuit?
The current is equal to the voltage divided by the resistance, or I = V/R.

12 volts **6 ohms** **Ammeter**

Figure 1

Using Ohm's Law As long as you know any two values, you can use Ohm's law to find the remaining value. Look at Figure 2. Cover the value you want to find. Then divide or multiply the other two values as shown to find the correct answer.

Voltage (volts)
Divide   Divide
Current (amperes)   Resistance (ohms)
Multiply

Figure 2

To find the voltage, multiply the current times the resistance.

$$V = I \times R$$

To find the resistance, divide the voltage by the current.

$$R = V/I$$

▶**Calculate:** What is the resistance of a toaster that uses 5 amps of current when it is plugged into a 110-volt outlet?
R = V/I, R = 110 volts/5 amps, R = 22 ohms

190

LESSON SUMMARY

▶ Every closed circuit has current (I), voltage (V), and resistance (R).

▶ Ohm's law states that the current is equal to the voltage divided by the resistance, or I = V/R.

▶ Ohm's law can be used to find the value for I, V, or R if any two of the values are known.

CHECK *Complete the following.*

1. According to Ohm's law, I = _____ /R. ᵛ

2. If the resistance (R) in a circuit increases, the current (I) _____ . decreases.

3. Every closed circuit has current, resistance, and _____ . voltage

4. If the voltage in a circuit decreases and the resistance stays the same, the current _____ . ^decreases

APPLY *Copy the table at the top of the next column on a sheet of paper. Use Ohm's law (I = V/R) to complete the table.*

I (amps)	V (volts)	R (ohms)
	10	70
15	250	
5		25
5	110	

Skill Builder

◼︎ *Analyzing* You can use the equation for Ohm's law to analyze circuit diagrams. The circuit diagram shows a series circuit with three resistors. The total resistance is the sum of all the resistances in the circuit. The power source is a 20-volt battery. What is the current in the circuit?

R₁ = 2 ohms

V = 20 volts

R₂ = 3 ohms

I = ?

R₃ = 5 ohms

CAREER IN PHYSICAL SCIENCE

ELECTRICIAN

An electrician is a person who is trained to install and repair electrical equipment. An electrician must know how much electricity a building needs. A house with basic appliances such as a refrigerator and a television needs a certain amount of electricity. A high-rise apartment building needs much more.

Most homes have wall outlets that supply 120 volts. Appliances made in Europe are sometimes meant for a different voltage. An electrician needs to make sure that the correct amount of voltage is being used. An electrician must be familiar with different types of wires. Most houses use copper wiring, but some use aluminum. Older homes may need to have all the wiring replaced. An electrician needs to make sure that the correct type of wiring is used.

People interested in a career as an electrician should take courses in mathematics, physics, and chemistry in high school. For more information, write to the International Brotherhood of Electrical Workers, 1125 15th Street NW, Washington, DC 20005.

10-9 How can you use electricity safely?

Objective ▶ Describe ways to use electricity safely.

TechTerms

▶ **circuit breaker:** switch that opens a circuit if too much current is flowing

▶ **fuse:** wire that melts and breaks a circuit if too much current is flowing

Electrical Safety Electricity is a very useful source of energy, but it can also be dangerous. Too many appliances connected to one outlet may cause a short circuit. In a short circuit, too much current flows through a wire. The wire may overheat or cause sparks. The heat and sparks could result in a fire.

▶ *Identify:* What happens in a short circuit? Too much current flows in a wire.

Fuses The electric current that comes into your home comes through several power lines. The power lines lead to a breaker box. From the breaker box, separate wires go to different parts of the house. These wires are connected to the wall outlets that provide electric current.

Some breaker boxes, especially in older homes, have **fuses.** A fuse is a thin piece of metal that melts when too much current flows through it. By melting, the fuse stops current from flowing and prevents a short circuit. When you replace the fuse, the current begins flowing again. The photograph shows several types of fuses.

▶ *Describe:* How does a fuse prevent a short circuit? A piece of metal melts and stops the flow of current.

Circuit Breakers Most newer homes have **circuit breakers.** A circuit breaker is a switch that turns off when there is too much current. A circuit

breaker is another way of preventing a short circuit. To start the current flowing again, the switch must be turned on. The photograph also shows a breaker box with circuit breakers.

▶ *Describe:* How does a circuit breaker prevent a short circuit? A switch is turned off when too much current is flowing.

Electrical Safety Rules

▶ Never touch an electrical appliance when your hands are wet; do not use electrical appliances near water.

▶ Do not connect several electrical extension cords together.

▶ Do not plug too many appliances into one outlet.

▶ Do not allow electric cords to become worn.

▶ Do not run electric cords under carpets.

LESSON SUMMARY

▶ Too much current in a circuit can cause a short circuit.

▶ Electric current comes into a house from power lines to a breaker box.

▶ A fuse is a thin piece of metal that melts and opens a circuit before a short circuit can occur.

▶ A circuit breaker is a switch that turns a circuit off before a short circuit can occur.

CHECK *Complete the following.*

1. If too many appliances are plugged into an outlet, there may be a _____ circuit. short

2. A _____ is a thin piece of metal that melts when there is too much current in a circuit. fuse

3. The _____ is where the current from the power plant comes into a house. breaker box

4. A circuit breaker switches off in order to prevent a _____ . short circuit

5. You should never use electrical appliances when your hands are _____ . wet

6. Electric cords should not be allowed to become _____ . worn

APPLY *Complete the following.*

▶ 7. **Infer:** When too many appliances are connected to one wall outlet, a circuit breaker may switch off. None of the appliances connected to this outlet will work until the circuit breaker is switched on. What kind of circuit do the appliances form?

8. **Hypothesize:** Resistance in a wire increases with the length of the wire. Why do you think it is not a good idea to connect several extension cords?

State the Problem ....

Study the illustration below.

V =
120 volts

A

R =
4 ohms

State the problem for this experiment.

SCIENCE CONNECTION ◆○◆○◆○◆○◆○◆○◆○◆○◆○◆○◆○◆○◆○◆○◆○◆○◆

HIGH-VOLTAGE WIRES

An electric power plant can generate enough electricity for an entire city. The electricity must be transported to homes, schools, and office buildings.

Electricity is like a reservoir of water that must be pumped throughout a city. Large, strong pipes are used to carry the water. High-pressure pumps are used to raise the water to the tops of tall buildings.

How do electric power plants transport electricity? Remember that voltage is the force that pushes electrons along a wire. The greater the voltage, the more electrons can be pushed along a wire. High-voltage wires are used to carry electric current long distances. You should stay away from high-voltage wires on power poles. The very high voltages used can be deadly.

DANGER
HIGH VOLTAGE

UNIT 10 Challenges

STUDY HINT Before you begin the Unit Challenges, review the TechTerms and Lesson Summary for each lesson in this unit.

TechTerms

alternating current (180)	electricity (176)	ohm (188)
ampere (188)	electrochemical cell (182)	Ohm's law (190)
battery (182)	electrode (182)	parallel circuit (186)
circuit breaker (192)	electrolyte (182)	proton (176)
conductors (178)	electron (176)	resistance (188)
direct current (180)	fuse (192)	series circuit (184)
electric circuit (184)	insulators (178)	volt (188)
electric current (180)	neutron (176)	voltage (188)

TechTerm Challenges

Matching *Write the TechTerm that matches each description.*

1. unit for measuring resistance ohm
2. series of electrochemical cells battery
3. circuit in which the current follows one path series circuit
4. negatively charged atomic particle electron
5. electric current that changes direction at a regular rate alternating current
6. unit for measuring electric current ampere, or amp
7. switch that opens a circuit if too much current is flowing circuit breaker
8. substance that conducts electric current electrolyte
9. positively charged atomic particle proton
10. material that prevents electric current from flowing insulator
11. unit used to measure voltage volt
12. flow of electrons electric current
13. atomic particle with neither a positive nor a negative charge neutron
14. opposition to the flow of electric current resistance

Fill In *Write the TechTerm that best completes each statement.*

1. A battery is a series of _____ . electrochemical cells
2. A car battery is a source of _____ . direct current
3. In a _____ , the current can follow more than one path. parallel circuit
4. According to _____ , the current in a closed circuit is equal to the voltage divided by the resistance. Ohm's law
5. The force that pushes the electrons in an electric circuit is called the _____ . voltage
6. A _____ allows electric current to flow. conductor
7. The form of energy associated with moving electrons is _____ . electricity
8. The positive or negative pole of a battery is also called an _____ . electrode
9. A _____ is a wire that melts and breaks a circuit if too much current is flowing. fuse
10. An _____ is the path that an electric current follows. electric circuit

Content Challenges

Multiple Choice *Write the letter of the term or phrase that best completes each statement.*

1. A car battery is an example of
   a. wet cells. **b.** dry cells. **c.** alternating current. **d.** a parallel circuit. a

2. According to Ohm's law, current is equal to
   a. resistance divided by voltage. **b.** voltage divided by resistance.
   c. resistance times voltage. **d.** voltage plus resistance. b

3. One of the best conductors of electric current is a wire made of
   a. brass. b. magnesium. c. copper. d. rubber. c
4. Electric current can follow more than one path in
   a. a series circuit. b. an alternating circuit. c. a direct circuit. d. a parallel circuit. d
5. The most common type of current used in homes, offices, and schools is
   a. direct current. b. alternating current.
   c. both alternating and direct current. d. neither alternating nor direct current. b
6. The resistance in a wire will increase if
   a. the thickness of the wire is decreased. b. the temperature of the wire is increased. c. the length of the wire is increased. d. all of the above. d
7. If several appliances are connected in series and one of the appliances stops working
   a. none of the appliances will work. b. the rest of the appliances will continue to work.
   c. there may be a short circuit. d. the current needs to be increased. a
8. A circuit breaker is useful in case there is
   a. a short circuit. b. not enough current. c. not enough voltage. d. too much resistance. a
9. Electric current is measured in
   a. volts. b. ohms. c. amps. d. newtons. c
10. Most of the circuits in the home are
    a. series circuits. b. parallel circuits. c. electronic circuits. d. short circuits. b

True/False *Write true if the statement is true. If the statement is false, change the underlined term to make the statement true.*

1. Electricity is the energy associated with moving <u>protons</u>. electrons
2. Electric current moves from a <u>negative</u> pole to a positive pole. true
3. <u>A conductor</u> prevents electric current from flowing. An insulator
4. As current flows in a wire, it meets a certain <u>voltage</u> from the wire. resistance
5. Current can flow only in <u>an open</u> circuit. a closed
6. A <u>switch</u> is used to turn a circuit on or off. true
7. In a <u>parallel</u> circuit, the current follows only one path. series
8. The advantage of a <u>parallel</u> circuit is that current will reach all appliances even if one of them stops working. true
9. The current in a circuit is found by using the equation I = <u>V/R</u>. true
10. It is <u>not safe</u> to connect several extension cords together. true

Understanding the Features ....

Reading Critically *Use the feature reading selections to answer the following. Page numbers for the features are in parentheses.*

1. **Compare:** How are superconducting circuits different from ordinary electric circuits? (189)
2. What is a computer program? (187)
3. What courses should you take in high school if you decide you want to become an electrician? (191)
4. In what year was the first successful computer invented? (185)
5. Do model trains use alternating current or direct current? (181)
6. What is electronics? (179)
7. **Infer:** Why do electric power plants use high-voltage wires to carry electricity? (193)

Concept Challenges

Critical Thinking *Answer each of the following in complete sentences.*

1. **Hypothesize:** During a storm, a power line falls from its pole. Is it safe to touch the wires? Explain.

2. **Infer:** Some appliances can use batteries or a wall outlet. What is the difference between the two power sources? What does this tell you about the current in an electric circuit?

3. **Calculate:** Which circuit has more current? **a.** voltage = 12 volts, resistance = 2 ohms **b.** voltage = 36 volts, resistance = 6 ohms

4. In a parallel circuit, the current follows more than one path. Does this mean that parallel circuits can use only alternating current? Why or why not?

Interpreting a Diagram *Use the circuit diagrams to complete the following.*

1. What kind of circuit is diagram A?
2. What kind of circuit is diagram B?
3. What does I stand for in an electric circuit?
4. What does V stand for in an electric circuit?
5. What are R_1, R_2, and R_3 in diagram A?
6. What is the total resistance in diagram A?
7. What is the current, in amperes, in diagram A?
8. Find I_1 and I_2 in diagram B, in amperes.

Finding Out More

1. **Classify:** Make a list different appliances and electric devices. These can include portable radios, televisions, flashlights, and so on. Arrange your list into three separate groups: appliances that run only on direct current; appliances that run only on alternating current; and appliances that run on either direct or alternating current.

2. Electric power plants use electric generators to produce electric energy. Electric generators convert energy from another form into electric energy. Use library reference to find out what kinds of energy are used by power plants to make electricity.

3. Look at a copy of your family's electric bill. The bill will show the number of kilowatts used in your home in one month. A kilowatt is 1000 watts. A watt is 1 amp/sec. Use the information on the electric bill to find out how many amperes of current your house uses in one month. To do this, multiply the number of kilowatts by 1000. Then multiply the number of watts by the number of seconds in one month. Try to come up with a list of suggestions for cutting down the number of kilowatts used.

MAGNETISM

CONTENTS

STUDY HINT Before beginning Unit 11, scan through the lessons in the unit looking for words that you do not know. On a sheet of paper, list these words. Work with a class-mate to try to define each word on your list.

What is a magnet?

Objective ▸ Describe the properties of a magnet.

TechTerms

▸ **magnetism:** force of attraction or repulsion
▸ **poles:** two ends of a magnet

Magnetism When you hold a magnet close to certain types of metal, the metal will move toward the magnet. The metal is pulled toward the magnet by a force called **magnetism.** Magnetism is a force of attraction or repulsion.

Not all substances are affected by the force of magnetism. For example, if you hold a magnet near a piece of wood, the wood will not be attracted to the magnet. Glass, plastic, and paper also are not affected by magnetism. Common metals affected by magnetism are iron, nickel, and cobalt.

▸**Predict:** What will happen if you hold a magnet near a wooden toothpick?
The toothpick will not be attracted to the magnet.

Magnetic Poles Each end of a magnet is called a magnetic **pole.** Every magnet has two poles. They

are called a north pole (N) and a south pole (S). The north pole of a magnet points North. For this reason, the north pole of a magnet is sometimes called the "North-seeking" pole. A magnet always has two poles. Even if you break a magnet in half, each half will have a north pole and a south pole.

▸*Identify:* What are the two ends of a magnet called? north pole and south pole

Properties of a Magnet When two magnets are brought near each other, a magnetic force acts on each of them. If a north pole and a south pole are brought together, they will attract each other. The unlike poles of any two magnets always attract each other.

If the north pole of one magnet is brought near the north pole of another magnet, they will repel each other. Two magnetic south poles also will repel each other. The like poles of any two magnets always repel each other.

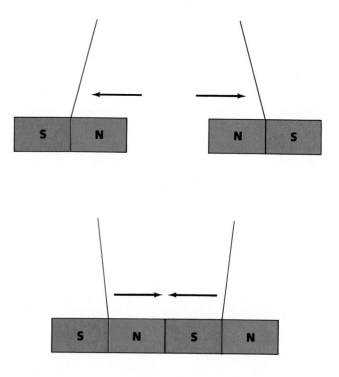

▸*Describe:* How do the poles of a magnet affect each other? Like poles repel; unlike poles attract.

LESSON SUMMARY

▶ Magnetism is a force of attraction or repulsion.

▶ Not all substances are affected by magnetism.

▶ Every magnet has two poles, called a north pole and a south pole.

▶ Like magnetic poles repel each other; unlike poles attract each other.

CHECK *Complete the following.*

1. The force that pulls an iron nail toward a magnet is called _____ . magnetism

2. The south pole of a magnet will _____ the south pole of another magnet. repel

3. The north pole of a magnet points _____ . North

4. The north pole of a magnet will _____ the south pole of another magnet. attract

5. The only metals attracted by a magnet are iron, _____ , and cobalt. nickel

6. The force of magnetism is a force of _____ or repulsion. attraction

7. If you cut a magnet in half, each half will have a north pole and a _____ . south pole

APPLY *Complete the following.*

▶ 8. **Predict:** If you keep cutting a magnet into smaller and smaller pieces, will you ever reach a point when you have only a magnetic north pole? Explain.

▶ 9. **Infer:** Why could the south pole of a magnet be called the "South-seeking" pole?

10. Describe what will happen if you bring the south pole of one magnet near the south pole of another magnet.

Designing an Experiment............

Design an experiment to solve the problem.

PROBLEM: How can you identify the north and south magnetic poles of a magnet?

Your experiment should:

1. List the materials you would need.

2. Identify safety precautions that should be followed.

3. List a step-by-step procedure.

4. Describe how you would record your data.

LOOKING BACK IN SCIENCE

MAGNETIC COMPASSES

Magnets can be found in nature. Naturally occurring substances with magnetic properties are called natural magnets. The Greeks were the first to discover natural deposits of a magnetic iron ore. They discovered this ore in a region of Turkey known as Magnesia. The ore was called magnetite (MAG-nuh-tyt). Magnetite also can be found in many other parts of the world.

Magnetite was used by sailors and navigators to find directions at sea. The sailors found that a piece of magnetite always pointed toward the North Star. The North Star was also called the lodestar, or leading star. So the sailors called a piece of magnetite a "lodestone." The sailors made magnetic compasses with a pointer made of lodestone.

Today, magnetic compasses are often used by hikers and backpackers. The pointer of a magnetic compass always points North. Once you know where North is, you can easily locate the other three directions, or compass points.

11-2 What is a magnetic field?

Objective ▶ Demonstrate the shape of a magnetic field.

TechTerms

- ▶ **magnetic field:** area around a magnet where magnetic forces can act
- ▶ **magnetic lines of force:** lines that show the shape of a magnetic field

Magnetic Field If you bring a magnet close to a piece of iron, the magnetic force will attract the iron. The magnet does not need to touch the piece of iron in order for the magnetic force to attract the iron. Every magnet has a **magnetic field** around it. The magnetic field is the area around a magnet where magnetic forces can act. Figure 1 shows the magnetic field around a bar magnet. The shape of the magnetic field was shown by sprinkling iron filings around the magnet.

▐▐▐▶*Define:* What is a magnetic field?
area around a magnet where magnetic forces can act

Magnetic Lines of Force A magnetic field is made up of **magnetic lines of force.** The magnetic lines of force show the shape of a magnetic field. The magnetic lines of force reach from the magnet's north pole to the south pole. When iron filings are sprinkled around a magnet, they will line up along the magnetic lines of force.

The magnetic lines of force are closest together at the poles of a magnet. This is because the magnetic field is strongest at the poles. Figure 2 shows the magnetic lines of force between two like poles and two unlike poles of a magnet. The lines of force show that two like poles repel each other. Two unlike poles attract each other. The lines of force come together for two like poles. They bend apart for two unlike poles.

▐▐▐▶*Describe:* Where is the magnetic field of a magnet strongest? at the poles

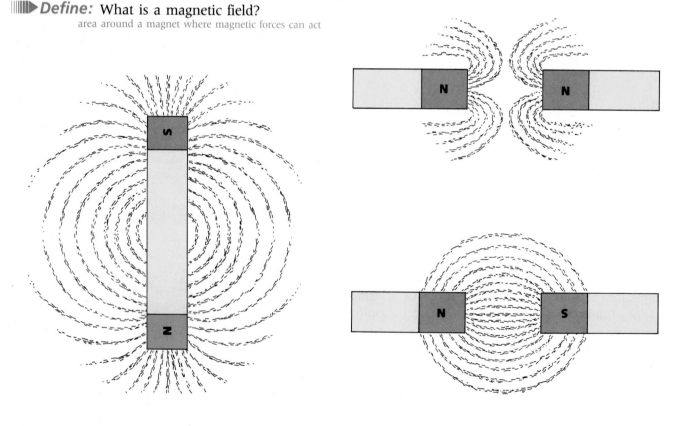

Figure 1

Figure 2

LESSON SUMMARY

▶ A magnetic field is the area around a magnet where magnetic forces can act.

▶ A magnetic field is made up of magnetic lines of force that show the shape of the magnetic field.

▶ The magnetic lines of force are closest together at the poles of a magnet.

CHECK *Complete the following.*

1. What is a magnetic field?

2. What happens when you bring a magnet near a piece of iron?

3. What are magnetic lines of force?

4. How can iron filings be used to show magnetic lines of force?

5. Why are magnetic lines of force closest together at the poles of a magnet?

APPLY *Complete the following.*

6. **Predict:** What will happen to the magnetic lines of force when the north poles of two magnets are placed next to each other?

7. **Analyze:** Look at the diagram. Is pole A a north pole or a south pole? How do you know?

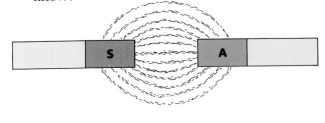

InfoSearch

Read the passage. Ask two questions about the topic that you cannot answer from the information in the passage.

Magnetic Domains Matter is made up of atoms. Atoms contain electrically charged particles called electrons. Each electron has a magnetic field around it. In most atoms, electrons are found in pairs. The magnetic fields of the two electrons cancel each other. Some metals have unpaired electrons. The magnetic fields of the unpaired electrons may be lined up in the same direction. When this happens, the metal is magnetized.

SEARCH: Use library references to find answers to your questions.

ACTIVITY

OBSERVING MAGNETIC LINES OF FORCE

You will need two bar magnets, a sheet of construction paper, and iron filings. **Caution: Wear your safety goggles.**

1. Place the sheet of paper over one of the magnets.

2. Sprinkle the iron filings over the part of the paper that covers the magnet. Try to spread the filings evenly. Observe the shape of the magnetic field.

3. Repeat steps 1 and 2 using two magnets. Have the north poles of the magnets face each other.

4. Repeat step 3 with a north pole and a south pole facing each other. Do not let the magnets touch.

Questions

👁 1. **Observe:** Where are the iron filings closest together?

👁 2. **Observe:** What do the lines of force look like when two north poles are facing each other?

👁 3. **Observe:** What do the lines of force look like when a north pole and a south pole are facing each other?

▲ 4. **Model:** Draw a diagram showing the magnetic fields.

11-3 How can you make a magnet?

Objectives ▶ Explain how materials can be magnetized. ▶ Compare permanent magnets and temporary magnets.

TechTerm

▶ **magnetic induction:** process by which a material can be made into a magnet

Natural Magnets Some magnets occur in nature. These magnets are called natural magnets. Lodestone, or magnetite (MAG-nuh-tyt), is an example of a natural magnet. Materials that are not natural magnets can be made into magnets, or magnetized (MAG-nuh-tyzed). For example, rubbing an iron nail in the same direction with a magnet will magnetize the nail. The magnetized nail will have the same properties as a natural magnet.

▶*Name:* What is an example of a natural magnet? lodestone, or magnetite

Magnetic Induction The process by which a material is magnetized, or made into a magnet, is called **magnetic induction.** Some materials are easier to magnetize than others. Iron, nickel, and cobalt are very easy to magnetize. Placing any of these materials near a magnet will immediately magnetize them. A material such as aluminum is harder to magnetize. If a piece of aluminum is placed in a magnetic field, it may become magnetized. However, the induced magnetic field of aluminum is weaker than the induced magnetic field of iron.

▶*Define:* What is magnetic induction? process by which a material is magnetized

Temporary and Permanent Magnets Once a material is magnetized, it may or may not remain magnetized. Iron is easily magnetized, but the iron loses its magnetism just as easily. A material that is easily magnetized tends to lose its magnetism quickly. A magnet made of this kind of material is called a temporary magnet.

Materials that are hard to magnetize will also stay magnetized for a long time. A piece of aluminum is harder to magnetize than a piece of iron. However, the aluminum will keep its magnetic properties much longer than a piece of iron. A magnet that is hard to magnetize but tends to keep its magnetism is called a permanent magnet.

▶*Classify:* Is an iron magnet a temporary magnet or a permanent magnet? temporary

Figure 1 Natural magnet

Figure 2 Magnetic induction

LESSON SUMMARY

▶ Materials that are not natural magnets can be magnetized.

▶ The process by which a material is made into a magnet is called magnetic induction.

▶ Temporary magnets are easy to magnetize, but lose their magnetism quickly.

▶ Permanent magnets are hard to magnetize, but tend to keep their magnetism.

CHECK *Write true if the statement is true. If the statement is false, change the underlined term to make the statement true.*

1. A piece of iron will be magnetized <u>slowly</u>. *quickly*

2. A magnetic field is induced when a material becomes <u>magnetized</u>. *true*

3. A <u>temporary magnet</u> is hard to magnetize. *permanent magnet*

4. A <u>permanent magnet</u> will keep its magnetic properties for a long time. *true*

5. A temporary magnet is <u>hard</u> to magnetize. *easy*

6. A piece of aluminum that has been magnetized is an example of a <u>permanent magnet</u>. *true*

APPLY *Complete the following.*

7. **Infer:** An iron nail has been magnetized. Does the magnetized nail have a north pole and a south pole? Explain.

8. **Contrast:** What is the difference between a temporary magnet and a permanent magnet?

9. **Predict:** Soft iron is very easy to magnetize. Will soft iron make a good permanent magnet? Why or why not?

TECHNOLOGY AND SOCIETY

MAGNETIC LEVITATION

Magnetism is a force of attraction or repulsion. The like poles of any two magnets will repel each other. This force of repulsion can be used to lift objects. The process by which an object can be kept floating above a magnet is called magnetic levitation (mag-NET-ik lev-uh-TAY-shun). "Maglev" is short for magnetic levitation.

Scientists are studying ways to use magnetic levitation. Several companies are experimenting with maglev trains. A train fitted with special magnets could float 10 to 15 cm in the air. There would be no friction with the tracks to slow down the train. Maglev trains could reach very high speeds.

Maglev trains are still being researched. Many questions about the technology involved still need to be answered. Sometime in the near future, however, maglev trains may be used worldwide.

11-4 How is the earth like a magnet?

Objective ▶ Explain how the earth acts like a magnet.

TechTerm

▶ **magnetosphere** (mag-NEET-uh-sfir): region of the earth's magnetic field

William Gilbert William Gilbert was a British scientist. He was interested in magnetism. Gilbert observed that if you hang a bar magnet from a string, one pole of the magnet always points North. The other pole points South. Gilbert was the first person to explain this observation. He said that the earth has north and south poles like a bar magnet.

▶ *State:* What did Gilbert say about the earth's magnetic properties?
The earth has north and south poles like a bar magnet.

Earth's Magnetic Poles The earth acts as if a bar magnet was buried inside it. The earth has two magnetic poles. One of the earth's magnetic poles is in Canada. This pole is called the North Magnetic Pole. It is about 2000 km from the geographic North Pole. The other magnetic pole is near the geographic South Pole. This pole is called the South Magnetic Pole.

Remember that the north pole of a magnet always points North. However, like magnetic poles repel each other. This means that the earth's North Magnetic Pole is really like the south pole of a bar magnet. It is called the "North" Magnetic Pole because it is near the geographic North Pole. The earth's South Magnetic Pole is really a north pole.

▶ *Infer:* Why is the South Magnetic Pole called a "south" pole? It is near the geographic South Pole.

Magnetosphere Like a bar magnet, the earth is surrounded by a magnetic field. The earth's magnetic field extends far into space. The magnetic field of the earth is also called the **magnetosphere** (mag-NEET-uh-sfir). The magnetosphere traps charged particles from the sun. When these particles hit particles in the atmosphere, an aurora (ow-ROWR-uh) is formed. Auroras are also called the northern or southern lights.

▶ *Define:* What is the magnetosphere?
the earth's magnetic field

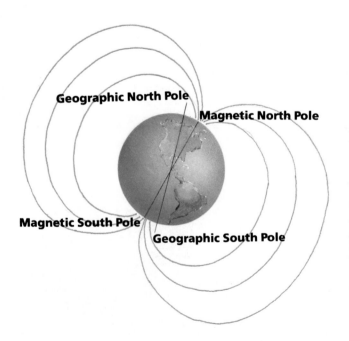

Geographic North Pole
Magnetic North Pole
Magnetic South Pole
Geographic South Pole

LESSON SUMMARY

▶ William Gilbert was a British scientist who said that the earth has poles like a bar magnet.

▶ The earth has a North Magnetic Pole and a South Magnetic Pole.

▶ The North Magnetic Pole is really a south pole, and the South Magnetic Pole is a north pole.

▶ The earth has a magnetic field that is called the magnetosphere.

CHECK *Write true if the statement is true. If the statement is false, correct the underlined term to make the statement true.*

1. William Gilbert was a British scientist who studied <u>magnetism</u>. true

2. When a compass points North, it is pointing to the earth's Magnetic <u>South</u> Pole. North

3. The South Magnetic Pole is like the <u>south</u> pole of a bar magnet. north

4. The earth's magnetic field is similar to that of a <u>horseshoe</u> magnet. bar

5. Another name for the northern lights is the <u>aurora</u>. true

6. The <u>magnetosphere</u> traps charged particles from the sun. true

APPLY *Complete the following.*

7. **Classify:** Is the earth a permanent magnet or a temporary magnet? How do you know?

8. **Hypothesize:** Two forces affect objects on the earth. They are the force of gravity and the force of the earth's magnetic field. Suppose you drop an iron bar. Which force would have a greater effect on the bar? Explain.

InfoSearch

Read the passage. Ask two questions about the topic that you cannot answer from the information in the passage.

Nikola Tesla One of the most famous researchers in the area of magnetism was Nikola Tesla. Tesla was born in 1856 in Croatia. Croatia is today part of Yugoslavia. He later became a citizen of the United States. One of Tesla's greatest inventions was the Tesla coil. Tesla had so many ideas for inventions that he recorded them all in notebooks. Today, scientists and engineers still study Tesla's notebooks to test some of his ideas.

SEARCH: Use library references to find answers to your questions.

ACTIVITY

MAKING A MAGNETIC COMPASS

You will need a sewing needle, a bar magnet, a thumbtack, a cork, and a small container of water.

1. Rub the magnet against the needle about 10 times. Rub in only one direction.

2. Stick the thumbtack into the bottom of the cork.

3. Float the cork with the thumbtack pointing down in a container of water.

4. Place the needle on top of the cork. Observe in which direction the needle points when it stops turning. If you have a real magnetic compass, compare the direction of the needle with the direction of the compass pointer.

Questions

1. What effect does rubbing with a magnet have on the needle?

2. **Observe:** In which direction does the floating needle point?

11-5 How are electricity and magnetism related?

Objective ▶ Relate electricity and magnetism.

TechTerms

▶ **electromagnetic induction:** process by which an electric current is produced by moving a wire in a magnetic field

▶ **electromagnetism:** relationship between electricity and magnetism

Oersted's Discovery Hans Christian Oersted (UR-sted) was a Danish scientist. He lived more than 200 years ago. By accident, he discovered an important property of electric current. Oersted connected a simple series circuit made up of a battery and a wire. As current flowed through the wire, Oersted noticed that a nearby compass needle moved. When the circuit was disconnected, the compass needle returned to its original position.

▶ *Describe:* What happens to a compass needle when current flows in a wire?
The compass needle moves.

Electromagnetism Oersted discovered that an electric current causes a magnetic field. Magnet-ism can be made from electricity. When an electric current passes through coils of wire, the wire acts like a magnet. The relationship between magnet-ism and electricity is called **electromagnetism.**

▶ *Define:* What is electromagnetism?
the relationship between magnetism and electricity

Electromagnetic Induction An electric current causes a magnetic field. Does a magnetic field cause an electric current? A British scientist named Michael Faraday tried to answer this question. Faraday moved a wire through the magnetic field of a strong magnet. He found that an electric current was produced in the wire. The American scientist Joseph Henry made the same discovery at about the same time. The process by which an electric current is produced by moving a wire in a magnetic field is called **electromagnetic induction.** The current is induced, or caused to flow, in the wire when the wire cuts across the magnetic lines of force.

Direction of current flow

▶ *Explain:* How is an electric current induced in a wire? by moving the wire across magnetic lines of force

206

LESSON SUMMARY

▶ Oersted found that an electric current flowing in a wire causes a compass needle to move.

▶ The relationship between electricity and magnetism is called electromagnetism.

▶ When a wire is moved across a magnetic field, an electric current is induced in the wire.

CHECK *Complete the following.*

1. Oersted discovered that an _____ caused a magnetic field. electric current

2. When a _____ flows in a wire, a compass needle near the wire will move. current

3. Electromagnetism is the relationship between _____ and magnetism. electricity

4. Michael Faraday discovered electromagnetic _____ . induction

5. A current is induced when a _____ cuts magnetic lines of force. wire

6. The scientist who made the same discovery about electromagnetism as Faraday was _____ . Joseph Henry

APPLY *Complete the following.*

7. **Infer:** An electric current produces a weaker magnetic field in a straight wire than in a coil of wire. Would adding more coils to the wire increase or decrease the strength of the magnetic field around the wire? Explain.

8. **Hypothesize:** Do you think an electric current could be induced by moving a magnet in and out of coils of wire? Why or why not?

Skill Builder ....

Researching Many scientists from different countries were involved in experiments on electromagnetism. These scientists include Hans Christian Oersted, Andre Marie Ampere, Nikola Tesla, Joseph Henry, and Michael Faraday. Units of measurement have been named after each of these scientists. Use library references to find out what the units of measurement are, and what properties they are used to measure. Write a report of your findings.

▼▼▼ LOOKING BACK IN SCIENCE ▼▼▼▼▼▼▼▼▼▼▼▼▼▼▼▼▼▼▼▼▼

INVENTION OF THE TELEGRAPH

Before telephones were invented, people used the telegraph to send messages. Telegraph messages could be sent over long distances. At one time, the United States was crisscrossed with telegraph wires.

Samuel Morse, an American inventor, developed a telegraph that was widely used throughout this country. Morse's telegraph was made up of a circuit connected by long wires. A telegraph operator used a key to close and open the circuit. Morse also invented a code for sending messages by telegraph. By pressing the key in different combinations of dots and dashes, an operator could send a coded signal. A person at the other end of the telegraph line could translate the signal and decode the message.

Telegraphs are still used today, but the technology has changed. Now satellites, underground cables, or radio transmitters are used to send messages all over the world.

What is an electromagnet?

Objective ▶ Describe how to make an electromagnet.

TechTerm

▶ **electromagnet:** temporary magnet made by wrapping a current-carrying wire around an iron core

Magnetic Field Strength A wire carrying an electric current always has a magnetic field around it. The magnetic field around a straight wire is not very strong. If the wire is wound into a coil, the magnetic field becomes much stronger. The magnetic fields around the coils of wire add together. The more coils you add to the wire, the stronger the magnetic field will be.

▶**Predict:** Which will have a stronger magnetic field, a wire with two coils or a wire with four coils? four coils

Electromagnets Wrapping coils of wire around a piece of iron will produce a strong magnet. This type of magnet is called an **electromagnet.** The iron in the center of an electromagnet is called a core. An electromagnet is a temporary magnet. As long as current is flowing through the wire, the electromagnet has a magnetic field. When the current is turned off, there is no longer a magnetic field.

The strength of an electromagnet can be increased in two ways. Adding more coils of wire around the iron core makes an electromagnet stronger. Increasing the current through the wire also makes an electromagnet stronger.

▶*Infer:* Why is an electromagnet a temporary magnet? It can be turned on and off.

Uses of Electromagnets Electromagnets have several important uses. They are used in many devices, including radios, telephones, and computers. Large electromagnets are used to lift heavy pieces of metal. Electromagnets are very useful because they can be turned on and off.

Iron core   Coil of wire

+ − Flow of electrons

Battery

▶*Explain:* Why are electromagnets so useful? They can be turned on and off.

LESSON SUMMARY

▶ A coil of wire has a stronger magnetic field than a straight wire.

▶ An electromagnet is made by wrapping coils of wire around an iron core.

▶ An electromagnet can be made stronger by adding more coils of wire or by increasing the current in the wire.

▶ Electromagnets are useful because they can be turned on and off.

CHECK *Complete the following.*

1. What is an electromagnet?

2. Why is an electromagnet considered to be a temporary magnet?

3. What are two ways to increase the strength of an electromagnet?

4. What is the piece of iron in the center of an electromagnet called?

5. Name three devices that use electromagnets.

APPLY *Complete the following.*

6. **Contrast:** How is an electromagnet different from a permanent magnet?

7. **Compare:** Which electromagnet in each of the following pairs is stronger?
   a. an electromagnet with 20 coils of wire, or an electromagnet with 40 coils of wire
   b. an electromagnet with 20 coils of wire that uses 1 amp of current, or an electromagnet with 20 coils of wire that uses 3 amps of current

InfoSearch

Read the passage. Ask two questions about the topic that you cannot answer from the information in the passage.

Alnico The core of an electromagnet can be made from any magnetic material. Iron, nickel, and cobalt can be used to make cores for electromagnets. A material called alnico (AL-ni-koh) is made from iron, nickel, cobalt, aluminum, and copper. Two of these materials are not magnetic. However, alnico is used to make very strong permanent magnets. An alnico magnet is many times stronger than an iron magnet of the same size.

SEARCH: Use library references to find answers to your questions.

ACTIVITY

MAKING AN ELECTROMAGNET

You will need paper clips, two 1.5-volt batteries, insulated wire, and an iron nail.

1. Wrap 20 turns of wire around an iron nail. Connect the ends of the wire to the terminals of a 1.5-volt battery.

2. Try to pick up some paper clips.

3. Now wrap 40 turns of wire around the nail. Connect the ends of the wire to the battery. See how many paper clips you can pick up.

4. Connect two batteries in series. Connect the electromagnet with 40 turns of wire to the two batteries. See how many paper clips you can pick up.

Questions

1. **Measure:** How many paper clips did the electromagnet with 20 turns of wire pick up? With 40 turns?

2. **Compare:** Were you able to pick up more paper clips when the electromagnet was connected to one battery or to two batteries? Explain.

11-7 What is a transformer?

Objectives ▶ Explain how a transformer works.
▶ Compare a step-up and a step-down transformer.

TechTerm

▶ **transformer:** device in which alternating current in one coil of wire induces a current in a second coil

Transformers Suppose you wrap two coils of wire around a nail. Attach one coil of wire to a dry cell and a switch. This coil of wire is called the primary coil. The second coil of wire is called the secondary coil. When you close the switch, a current is induced in the secondary coil. A current is also induced when you open the switch. Whenever the current in the primary coil is turned on or off, a current is induced in the secondary coil.

Primary coil

Secondary coil

A **transformer** is a device that uses alternating current in the primary coil to induce a current in the secondary coil. The primary coil is connected to a source of alternating current. In alternating current, the flow of electrons changes direction regularly. Alternating current in the primary coil induces a current in the secondary coil.

▶*Predict:* What will happen in the secondary coil when the primary coil is attached to a source of alternating current?
A current will be induced.

Step-up Transformers A transformer can be used to increase voltage. This kind of transformer is called a step-up transformer. A step-up transformer has more turns of wire in the secondary coil than in the primary coil. The more turns of wire there are in the secondary coil, the higher the voltage will be. Power companies use step-up transformers to send high-voltage electricity over long distances.

▶*Describe:* What does a step-up transformer do? increases voltage

Step-down Transformers A transformer can also decrease voltage. A transformer that decreases voltage is called a step-down transformer. A step-down transformer has fewer turns of wire in the secondary coil than in the primary coil. The fewer the turns of wire in the secondary coil, the lower the voltage will be. Step-down transformers lower the high voltage carried by power lines so that it can be used in homes.

▶*Explain:* Why are step-down transformers used? to lower voltage so that it can be used in homes

STEP-UP TRANSFORMER

STEP-DOWN TRANSFORMER

210

LESSON SUMMARY

▶ Changing the current in one coil of wire can induce a current in a second coil of wire.

▶ A transformer uses alternating current in the primary coil to induce a current in the secondary coil.

▶ A step-up transformer increases voltage.

▶ A step-down transformer decreases voltage.

CHECK *Complete the following.*

1. What is a transformer?

2. Which coil of a transformer is attached to a source of current?

3. What happens in the secondary coil of wire whenever a current in the primary coil is turned on or off?

4. What kind of transformer can be used to increase voltage?

5. What kind of transformer can be used to decrease voltage?

APPLY *Complete the following.*

6. **Contrast:** What is the difference between a step-up transformer and a step-down transformer?

7. **Infer:** Is the induced current in the secondary coil of a transformer direct current or alternating current? How do you know?

8. The electricity that comes into your home is carried by high-voltage wires. The voltage in these wires can be as high as 20,000 volts. The voltage from a wall outlet in your home is 120 volts. What happened to the voltage?

9. **Classify:** Identify each of the following as a step-up or a step-down transformer.

   a. A transformer with fewer turns of wire in the primary coil than in the secondary coil.

   b. A transformer with more turns of wire in the primary coil than in the secondary coil.

   c. A transformer that changes high voltage to low.

Ideas in Action ....

IDEA: Transformers are used to change the voltage in an alternating current.

ACTION: The voltage from a wall outlet in your home is 120 volts. Look at several appliances around your house. How much voltage do they need to operate? The voltage is usually given on the appliance or in the owner's manual. Which appliances need less than 120 volts? Which need more than 120 volts? Do the appliances use step-up or step-down transformers?

PEOPLE IN SCIENCE

JOSEPH HENRY (1797–1878)

Joseph Henry was an American scientist. He studied electricity and magnetism. He and the British scientist Michael Faraday made similar discoveries at around the same time. Joseph Henry discovered that a changing magnetic field will induce a current in a conductor. Faraday is given credit for making the same discovery. Both scientists discovered electromagnetic induction at around the same time.

Henry was born in Albany, New York, on December 17, 1797. He helped Samuel Morse develop the telegraph. In 1831, Henry built a telegraph of his own. He also built one of the most powerful electromagnets known at the time. Henry's electromagnet could lift more than a ton.

In 1846, Henry was placed in charge of the Smithsonian Institution in Washington, D.C. He was the main organizer of the National Academy of Science. He also started the U.S. Weather Service. In honor of Joseph Henry, the unit for electromagnetic induction is called the henry.

What is a motor?

Objective ► Explain how an electric motor works.

TechTerm

► **electric motor:** device that changes electrical energy into mechanical energy

Reversing Magnetic Poles Changing the direction of an electric current will reverse the poles of an electromagnet. Suppose you connect an electromagnet to a dry cell. As the current flows, one end of the electromagnet becomes a north pole. The other end becomes a south pole. You can use a compass to identify the poles. What will happen if you switch the connections to the dry cell? The direction of the current will be reversed. The compass will show that the magnetic poles also are reversed.

❯**Predict:** What will happen to the poles of an electromagnet if you change the direction of the electric current? The poles will be reversed.

Electric Motors The ability to reverse the poles of an electromagnet explains how an **electric motor** works. An electric motor changes electrical energy into mechanical energy. An electric motor is made up of an electromagnet and a permanent magnet. The electromagnet is free to turn. It is attached to a source of alternating current. As the direction of the current changes, the poles of the electromagnet are reversed. Attraction and repulsion between the electromagnet and the per-

manent magnet cause the electromagnet to spin. An electric motor also can run on direct current. A motor that runs on direct current has a special switch. The switch reverses the direction of the current.

▷**Define:** What is an electric motor?
device that changes electrical energy into mechanical energy

Uses of Electric Motors Almost all electrical appliances have motors. Electric fans, food processors, and refrigerators all use electric motors. There are many kinds of electric motors that are designed for specific uses. All electric motors have one thing in common. They all use the force of magnetism to change electrical energy into useful mechanical energy.

▷**Explain:** Why does an electrical appliance need a motor?
to change electrical energy into mechanical energy

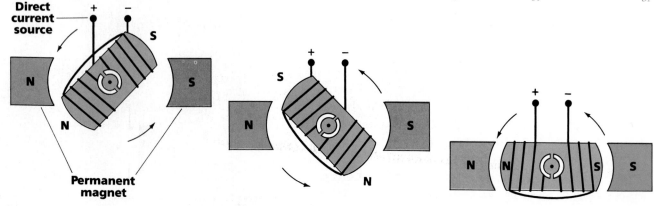

LESSON SUMMARY

▶ Changing the direction of an electric current causes the poles of an electromagnet to reverse.

▶ Electric motors change electrical energy into mechanical energy.

▶ Most electrical appliances use electric motors.

CHECK *Write true if the statement is true. If the statement is false, change the underlined term to make the statement true.*

1. The poles of an electromagnet are reversed when the <u>direction</u> of an electric current is changed. true

2. An electric motor changes electrical energy into <u>chemical</u> energy. mechanical

3. An electric motor contains an electromagnet and a <u>temporary</u> magnet. permanent

4. Without a <u>motor</u>, an electrical appliance would not be able to change electrical energy into mechanical energy. true

5. The <u>permanent magnet</u> in an electric motor is free to turn. electromagnet

6. A motor that uses <u>alternating</u> current has a special switch to change the direction of the current. direct

APPLY *Complete the following.*

▶ 7. **Infer:** Why does a motor that uses direct current need a special switch to change the direction of the current?

8. **Hypothesize:** What do you think would happen to an electric motor if the amount of current in the motor was reduced? Why?

9. An electric motor could also be called a magnetic motor. Explain.

Skill Builder

▲ *Modeling* Draw a diagram of an electric motor. Use the following terms to label the parts of the motor: armature (AHR-muh-chur), field magnet, brushes, and commutator (KAHM-yuh-tayt-ur). Use library references to identify these parts of a motor.

◆◆◆◆◆ CAREER IN PHYSICAL SCIENCE ◆◆◆◆◆◆◆◆

AUTOMOBILE MECHANIC

An automobile engine is a complex machine. It is made up of a starting motor, a battery, a generator, and many other electrical parts. Many jobs for automobile mechanics are available in automobile service stations or in the service departments of automobile dealers.

Automobile mechanics are trained to make repairs on automobile engines. Beginning mechanics get on-the-job training. They may work with an experienced mechanic. Special training is also available from factory service instructors employed by automobile manufacturers. Specialty mechanics may work on only one part of a car's engine.

To be an automobile mechanic, you should graduate from high school. Courses in automobile engine repair are available at most technical and vocational schools. For more information about jobs in automobile servicing, write to the Motor Vehicle Manufacturers Association, Inc., 320 New Center Building, Detroit, MI 48202.

Objective ▶ Explain how a generator works.

TechTerm

▶ **generator:** device that changes mechanical energy into electrical energy

Induced Current A current can be induced in a loop of wire by spinning the loop inside a magnetic field. Remember that an electric current is induced when a wire cuts across magnetic lines of force. Figure 1 shows a loop of wire in a magnetic field. The loop is spinning clockwise. As the wire moves across the magnetic field, it cuts the magnetic lines of force. As a result, current flows through the wire. When the loop moves down through the magnetic field, the current flows in one direction. When the loop moves up, the current flows in the opposite direction. The change in direction produces an alternating current in the wire.

▥▶*Identify:* What kind of current is induced when a loop of wire spins in a magnetic field?
alternating current

Figure 1

214

Electric Generators An electric **generator** is a device that changes mechanical energy into electrical energy. An electric generator is made up of an insulated loop of wire and a U-shaped magnet. The loop of wire is attached to a power source that causes the loop to turn. Spinning the loop of wire in the magnetic field of the magnet produces an electric current. The mechanical energy used to spin the loop of wire is changed into electrical energy.

▥▶*Define:* What is an electric generator?
a device that changes mechanical energy into electrical energy

Figure 2

Uses of Generators Most of the electricity you use every day comes from generators. Power plants use large generators to supply electricity for homes, offices, and other buildings. The mechanical energy for these generators is supplied by turbines (TUR-bins). A turbine is a large wheel that is turned by moving steam or water. Moving water to turn a turbine may come from a dam or river. Steam is produced by burning fuels, such as coal or oil, or from nuclear energy.

▥▶*Name:* What are two fuels that are burned to make steam for turbines in generators?
oil and coal

LESSON SUMMARY

▶ When a loop of wire spins in a magnetic field, an alternating current is induced in the wire.

▶ An electric generator is a device that changes mechanical energy into electrical energy.

▶ Most of the electricity you use every day comes from generators.

CHECK *Complete the following.*

1. What happens when a loop of wire cuts magnetic lines of force?

2. What kind of electric current is induced when a loop of wire turns in a magnetic field?

3. What is a generator?

4. What supplies the mechanical energy in a generator?

5. What causes a turbine to turn?

6. Name three sources of energy that are used to make steam for turbines.

APPLY *Complete the following.*

7. **Contrast:** A generator is the opposite of an electric motor. Explain.

▶ 8. **Infer:** What would happen if the wire loop in a generator stopped spinning?

9. Some homeowners have emergency generators that they can use if there is a blackout. These generators use gasoline as a fuel. Describe the energy changes needed to produce electricity with this type of generator.

InfoSearch

Read the passage. Ask two questions about the topic that you cannot answer from the information in the passage.

Geothermal Energy In some parts of the world, heat inside the earth produces steam and boiling water. This heat is called geothermal (jee-oh-THUR-mul) energy. The steam and boiling water may reach the surface through geysers (GY-zurs). Geothermal power plants can use the steam and boiling water to spin turbines and generate electricity.

SEARCH: Use library references to find answers to your questions.

SCIENCE CONNECTION

ELECTRIC POWER PLANTS

Electric power plants provide electricity for large numbers of people. Towns and cities in all parts of the country need large amounts of electricity. Electric power plants use electric generators to produce electricity.

The design of a commercial generator is very similar in most power plants. Instead of a simple wire loop spinning in the magnetic field of a magnet, commercial generators use many coils of wire and strong electromagnets. The spinning wires are connected to a turbine. The energy to spin the turbine comes from steam or moving water.

Some generators use the energy from waterfalls to spin the turbine. These generators are known as hydroelectric (hy-druh-i-LEK-trik) plants. The gravitational energy from the falling water is changed into mechanical energy to spin the turbine. Hydroelectric plants are built near waterfalls. For this reason, hydroelectric plants can provide only a limited supply of electrical energy.

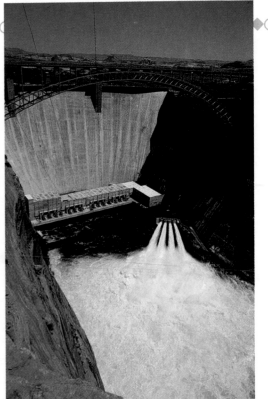

UNIT 11 Challenges

STUDY HINT Before you being the Unit Challenges, review the TechTerms and Lesson Summary for each lesson in this unit.

TechTerms

electric motor (212)
electromagnet (208)
electromagnetic induction (206)
electromagnetism (206)

generator (214)
magnetic field (200)
magnetic induction (202)
magnetic lines of force (200)

magnetism (198)
magnetosphere (204)
poles (198)
transformer (210)

TechTerm Challenges

Matching *Write the TechTerm that matches each description.*

1. force of a magnet magnetism
2. two ends of a magnet poles
3. device that changes the voltage of an alternating current transformer
4. shape of a magnetic field magnetic lines of force
5. temporary magnet formed by wrapping a coil of wire around an iron core electromagnet
6. process of making an object into a magnet
magnetic induction

Fill In *Write the TechTerm that best completes each statement.*

1. Because of _____ , a current will flow in a wire moving across a magnetic field.
electromagnetic induction
2. A _____ converts mechanical energy into electrical energy. generator
3. The magnetic field around the earth is also called the _____ . magnetosphere
4. The area around a magnet is called the _____ . magnetic field
5. The study of the relationship between electricity and magnetism is called electromagnetism
6. An _____ converts electrical energy into mechanical energy. electric motor

Content Challenges

Multiple Choice *Write the letter of the term or phrase that best completes each statement.*

1. In a step-down transformer
   a. the secondary coil has more turns. **b.** the primary coil has fewer turns. **c.** the secondary coil has fewer turns. **d.** the primary and secondary coils have the same number of turns. c

2. Oersted discovered
   a. electromagnetism. **b.** magnetic induction. **c.** electromagnetic induction.
   d. transformers. a

3. A temporary magnet
   a. keeps its magnetic properties for a long time. **b.** is hard to magnetize.
   c. loses its magnetic properties quickly. **d.** is a natural magnet. c

4. When the north and south poles of two magnets are facing each other, they
   a. attract each other. **b.** repel each other. **c.** have no effect on each other.
   d. magnetize each other. a

5. One way to increase the strength of an electromagnet is to
   a. increase the resistance in the wire. **b.** decrease the voltage in the wire.
   c. decrease the current in the wire. **d.** increase the current in the wire. d

216

6. In order for a 9-volt appliance to use the 120 volts from a wall outlet, you need a
   a. step-up transformer. b. step-down transformer. c. Tesla coil. d. electric motor. b
7. William Gilbert was one of the first scientists to study
   a. electric motors. b. generators. c. electromagnetism. d. magnetism. d
8. The earth's Magnetic North Pole is located
   a. near the geographic North Pole. b. near the geographic South Pole.
   c. at the geographic North Pole. d. at the geographic South Pole. a
9. An iron bar becoming magnetized is an example of
   a. magnetic induction. b. electromagnetic induction. c. a transformer.
   d. an electric motor. a
10. Electromagnetic induction was discovered by
    a. Tesla. b. Oersted. c. Faraday. d. Ampere. c

True/False *Write true if the statement is true. If the statement is false, change the underlined term to make the statement true.*

1. In a step-up transformer, the <u>primary</u> coil has fewer turns of wire. true
2. The magnetic lines of force for any magnetic field point from <u>north to south</u>. true
3. An electric <u>generator</u> converts electrical energy into mechanical energy. motor
4. To increase the strength of an electromagnet, <u>decrease</u> the number of turns of wire around the iron core. increase
5. Every magnet has <u>one magnetic pole</u>. two magnetic poles
6. Before the current from high-voltage wires reaches your home, a <u>step-up</u> transformer decreases the voltage to 120 volts. step-down
7. An electric motor can run on <u>direct</u> current. true
8. The mechanical energy in a generator is supplied by a <u>turbine</u>. true
9. The earth's South Magnetic Pole is the same as the <u>south</u> pole of a bar magnet. north
10. The magnetosphere traps charged particles from <u>the sun</u>. true

Understanding the Features . . . .

Reading Critically *Use the feature reading selections to answer the following. Page numbers for the features are in parentheses.*

1. What is lodestone? Why was it used in early magnetic compasses? (199)
2. **Infer:** Why could a maglev train reach very high speeds? (203)
3. What unit of measurement is named in honor of Joseph Henry? (211)
4. What kind of power plants use the energy from waterfalls to spin turbines? (215)
5. Where are jobs available for automobile mechanics? (213)
6. Who invented the telegraph that was widely used before the invention of the telephone? (207)

Concept Challenges . . . .

Critical Thinking *Answer each of the following in complete sentences.*

1. Explain the difference between magnetic induction and electromagnetic induction.
2. Describe the energy changes that take place in an electric motor and in an electric generator.
3. Describe how you could use a compass to identify the north and south poles of a magnet.

4. Why does spinning a loop of wire in a magnetic field produce alternating current instead of direct current?

5. Hypothesize: In most transformers, some energy is lost between the primary coil and the secondary coil. What do you think might cause this loss of energy?

Interpreting a Diagram *Use the diagram showing two transformers to complete the following.*

1. How many turns of wire are there in the primary coil in Diagram A? In the secondary coil?
2. How many turns of wire are there in the primary coil in Diagram B? In the secondary coil?
3. Which diagram shows a step-up transformer? Which shows a step-down transformer?
4. Which transformer would you use to change the voltage from 10 volts to 50 volts? Explain.
5. Which transformer would you use to change the voltage from 50 volts to 10 volts? Explain.
6. Which kind of transformer must be used before the electricity from power lines can be used in your home? Explain.

Finding Out More . . . .

1. Collect pictures from magazines of different types of appliances that use electric motors. Use the pictures to create a bulletin board display illustrating "Uses of Electric Motors."

2. Use library references, or visit a local hospital, to find out about a medical technology called magnetic resonance imaging, or MRI. Using MRI, doctors can study three-dimensional images of a patient's internal organs. Find out how MRI works. Share your findings with the class in an oral report.

3. Visit your local electric company. Find out how large a generator the company uses to produce electricity for your community. Ask what kind of energy is used to run the generator, for example, coal, oil, nuclear energy, or another source of energy. Write a report of your findings.

4. Use library references to learn more about the northern lights, or aurora borealis (bowr-ee-AL-is). Find out exactly what role the magnetosphere plays in forming the aurora. Where can the aurora be seen? What effects do the weather or the time of year have on the aurora?

MATTER

CONTENTS

STUDY HINT After you read each lesson in Unit 12, write a brief summary on a sheet of paper explaining how the information in each lesson applies to your everyday life.

12-1 What are the properties of matter?

Objective ▶ Identify four basic properties of matter.

TechTerms

- **chemistry** (KEM-is-tree): study of matter and its reactions
- **matter:** anything that has mass and takes up space
- **properties** (PROP-ur-tees): characteristics used to describe a substance

Matter Look around you. What do all the objects you see around you have in common? They are all made up of **matter.** Matter is anything that has mass and takes up space. Mass is the amount of matter an object contains. The amount of space an object takes up is its volume.

Water is matter. A glass filled with water is heavier than an empty glass. A filled glass is heavier because water has mass. If you kept adding water to a filled glass, the water would overflow. It would overflow because water takes up space.

Air is matter, too. A balloon filled with air is heavier than a balloon that is not blown up because air has mass. When you blow air into a balloon, the balloon gets larger as air takes up space.

▶**Define:** What is matter?
anything that has mass and takes up space.

Properties of Matter How would you describe an apple? You might say that an apple is red, round, and hard. Color, shape, and hardness are three **properties** (PROP-ur-tees) of matter. Properties are characteristics used to describe an object.

Mass and volume are two basic properties of matter. Weight and density also are basic properties of matter. Weight is a measure of the pull of gravity on an object. Density tells you how much matter is in a certain volume.

▶**List:** What are the four basic properties of matter? mass, volume, weight, and density.

Studying Matter The study of matter and the reactions of matter is called **chemistry** (CHEM-is-tree). Scientists who study matter are called chemists. Chemists study what different substances are made of. They do experiments to learn how different kinds of matter can change and combine.

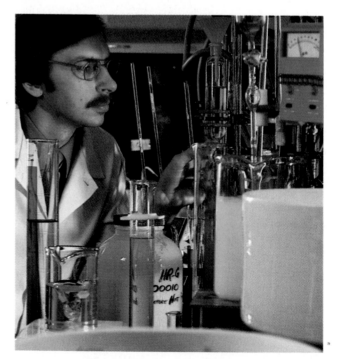

▶**Define:** What is chemistry?
the study of matter and its reactions

LESSON SUMMARY

▶ Matter is anything that has mass and takes up space.

▶ Water is matter.

▶ Air has mass and takes up space.

▶ Properties are characteristics used to describe an object.

▶ Mass, volume, weight, and density are the four basic properties of matter.

▶ Chemistry is the study of matter and the reactions of matter.

CHECK *Complete the following.*

1. All the objects you see around you are made up of _____ . matter

2. Matter is anything that has mass and takes up _____ . space

3. Mass is a basic _____ of matter. property

4. Weight is a measure of the pull of _____ on an object. gravity

5. The amount of space taken up by matter is its _____ . volume

6. The amount of matter per unit volume is called _____ . density

APPLY *Complete the following.*

7. **Compare:** What is the difference between mass and weight?

8. **Explain:** Why do you think scientists can use the basic properties of matter to help identify an unknown substance?

Skill Builder ....

Calculating You can find the density of an object by dividing its mass by its volume. Mass is measured in grams. Volume is measured in milliliters or cubic centimeters. The units for density are g/mL or g/cm³. Find the density of each of the following objects: a wooden block with a volume of 1 cm³ and a mass of 0.8 g; a 10-cm³ piece of lead that has a mass of 113 g; an ice cube that has a volume of 2 cm³ and a mass of 1.8 g. Organize your data in a table.

Ideas in Action ....

IDEA: Some properties can be observed using your five senses.

ACTION: Choose three common objects. Describe 10 properties of each object you choose. Which of your senses helped you describe the objects?

ACTIVITY

OBSERVING THAT AIR IS MATTER

You will need a glass, a marking pen, a tissue, a pail, and water.

1. Stuff a tissue into the bottom of a glass. Fill a pail with water.

2. Turn the glass upside down and push it straight down into the pail of water.

3. Pull the glass straight out of the water and feel the tissue. Record your observations.

Questions

1. **Observe:** Did the tissue feel wet?

2. Why did water not enter the glass?

3. **Relate:** How does this activity show that air is matter?

12-2 What are three phases of matter?

Objective ▶ Recognize three phases of matter.

TechTerms

- **gas:** phase of matter that has no definite shape or volume
- **liquid:** phase of matter with a definite volume, but no definite shape
- **phase** (FAYZ): form of matter
- **solid:** phase of matter with a definite shape and volume

Phases of Matter You cool drinks with ice cubes. You wash your hands in liquid water. You iron the wrinkles out of your clothes with steam. Ice, liquid water, and steam all are made up of particles of water. Different forms of the same substance are called **phases.** A phase (FAYZ) is a form of matter. There are three main phases of matter. They are solids, liquids, and gases.

▷ *Identify:* In how many main phases can matter exist? three

Solids Most of the objects that surround you are made of **solids.** A solid is a phase of matter that has a definite shape and volume. In a solid, particles of matter are tightly packed together. The par-

ticles cannot change position easily. They can only vibrate (VI-brayt), or move back and forth in place.

▷ *Define:* What is a solid?
phase of matter that has a definite shape and volume

Liquids Milk is a **liquid.** A liquid has a definite volume, but no definite shape. Liquids are able to change shape because the particles of a liquid can change position. They can slide past one another. If you pour a liter of milk into different containers, the milk always takes the shape of the containers. However, the volume of the milk stays the same. You cannot make a liter of milk fit into a half-liter bottle.

▷ *Explain:* Why can liquids change shape?
Particles in a liquid can change position.

Gases A **gas** is a phase of matter that has no definite shape or volume. Air is a gas. Air has different shapes in a basketball, a football, and a bicycle tire. If you fill a balloon with air, the air completely fills the balloon. A container of a gas is always completely full. The particles of a gas are in constant motion. They are much farther apart than the particles in solids or liquids. They can move freely to all parts of a container.

▷ *Define:* What is a gas?
phase of matter that has no definite shape or volume

Solid

Liquid

Gas

LESSON SUMMARY

▶ A phase is a form of matter.

▶ A solid is a phase of matter with a definite shape and volume.

▶ A liquid is a phase of matter with a definite volume, but no definite shape.

▶ A gas is a phase of matter that has no definite shape or volume.

CHECK *Complete the following.*

1. What are the three main phases of matter?

2. In what phase of matter do particles vibrate in place?

3. What happens to the shape of a liquid when you pour it into a container?

4. What determines the volume of a gas?

5. What phase of matter is air?

APPLY *Complete the following.*

6. **Explain:** What will happen to the particles of a gas if the gas is transferred from a small container into a much larger container?

7. **Classify:** Classify the following substances as a solid, a liquid, or a gas:

a. cotton cloth   f. sugar
b. seltzer   g. helium
c. rain   h. orange juice
d. hydrogen   i. salt
e. carbon dioxide   j. bricks

InfoSearch

Read the passage. Ask two questions about the topic that you cannot answer from the information in the passage.

Plasma Matter can exist in a fourth phase called plasma (PLAZ-muh). Plasma is very rare. It can exist only at very high temperatures and pressures. Plasma has been found in stars, where the temperatures and pressures are very high. Scientists also have been able to make plasma in the laboratory.

SEARCH: Use library references to find answers to your questions.

SCIENCE CONNECTION

THE EARTH'S MANTLE

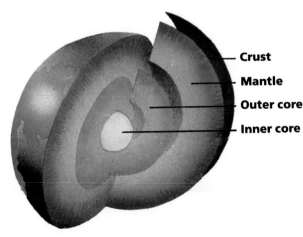

- Crust
- Mantle
- Outer core
- Inner core

The outer layer of the earth is called the crust. The crust is 8–32 km deep. Beneath the crust is the mantle. The mantle is about 2900 km deep. The mantle has two parts. The upper part is solid rock. Below the solid rock, the mantle rock behaves like a very thick liquid, such as molasses. The mantle rock behaves like a liquid because of very high temperatures and pressures. Like all liquids, the rock can flow. The rock also has some properties of a solid. It is an in-between phase of matter called the plastic phase. A plastic material is neither a solid nor a liquid. It has properties of both solids and liquids.

Scientists would like to learn more about the plastic material in the mantle. So far, they have not been able to drill that far into the earth. What they know has been learned indirectly, by studying shock waves from earthquakes. Someday, scientists hope to find a way to drill deep enough to reach the mantle. Then they will be able to study the properties of plastic rock directly.

How does matter change phase?

Objective ▶ Identify ways in which matter can change from one phase to another.

TechTerms

- **condensation** (kon-duhn-SAY-shun): change from a gas to a liquid
- **evaporation** (ih-vap-uh-RAY-shun): change from a liquid to a gas at the surface of the liquid
- **freezing:** change from a liquid to a solid
- **melting:** change from a solid to a liquid

Phase Changes Matter can change from one phase to another. For example, water can change from a solid to a liquid. A change in matter from one phase to another is called a phase change. There are four main kinds of phase changes. During a phase change, there is a change in heat energy. A substance either gains or loses heat as it changes from one phase to another.

▶ **Describe:** What is a phase change?

change in matter from one phase to another

Freezing and Melting If you fill an ice cube tray with water and place it in the freezer, the water will change to ice. Water changing into ice is an example of **freezing.** Freezing is a change from a liquid to a solid. Freezing occurs when the temperature of a liquid reaches its freezing point. At its freezing point, a liquid loses enough heat to change to a solid.

When the temperature rises above the freezing point of water, ice changes to a liquid. A change from solid to liquid is called **melting.** Melting occurs when a solid gains enough heat to change into a liquid.

▶ **Explain:** What causes a liquid to freeze?

loss of heat

Evaporation and Condensation Before you go to bed tonight, fill a plastic container with water. Mark the level of the water. Place the container in a warm, dry place. When you get up tomorrow, see what has happened to the water level. You will find that some of the water has "disappeared." Particles at the surface of the water gained enough heat energy to change into the gas phase. **Evaporation** (ih-vap-uh-RAY-shun) is a change from a liquid to a gas at the surface of the liquid.

You probably have noticed drops of water on your bathroom mirror after taking a hot shower. Hot water from the shower causes the temperature in the bathroom to rise. Some water particles gain enough heat energy to change to water vapor. Water vapor is the gas phase of water. As particles of water vapor hit the cool surface of a mirror, they lose heat energy and change back into liquid water. This process is called **condensation** (KON-den-SAY-shun). Condensation is a change from a gas to a liquid.

▶ **Explain:** What causes condensation?

gas particles lose heat

LESSON SUMMARY

▶ Matter can change phase.

▶ Freezing is a change from liquid to solid.

▶ Melting is a change from solid to liquid.

▶ Evaporation is a change from liquid to gas at the surface of the liquid.

▶ Condensation is a change from gas to liquid.

CHECK *Write the letter of the term that best completes each statement.*

1. Melting is a change from a solid to a **a.** gas. **b.** liquid. **c.** plasma. **d.** plastic. b

2. Water changing into ice is an example of **a.** freezing. **b.** melting. **c.** evaporation. **d.** condensation. a

3. A change in matter from one phase to another is called **a.** condensation. **b.** evaporation. **c.** a phase change. **d.** freezing. c

4. A change from a liquid to a gas at the surface of the liquid is **a.** freezing. **b.** melting. **c.** evaporation. **d.** condensation. c

5. Water vapor changing to liquid water is an example of **a.** freezing. **b.** melting. **c.** evaporation. **d.** condensation. d

APPLY *Complete the following.*

▶ 6. **Infer:** What happens to the particles of a liquid as the liquid freezes?

▶ 7. **Infer:** What happens to the particles of a liquid as the liquid evaporates?

📁 8. **Classify:** Identify the phase change taking place in each of the following situations.
   a. Water droplets form on the inside of your window on a chilly winter night.
   b. A full perfume bottle left open for several days is now half empty.
   c. A block of baking chocolate is heated until it can be poured into a measuring cup.

Ideas in Action

IDEA: Many appliances in your home cause phase changes.

ACTION: Look around your home. Identify three appliances that can cause phase changes. Describe the phase changes that each of the appliances can produce.

CAREER IN PHYSICAL SCIENCE

REFRIGERATION TECHNICIAN

Imagine a hot day in July. You are a restaurant owner and you discover that your largest refrigerator has stopped working. Luckily, you know a good refrigeration technician to call. A refrigeration technician installs and repairs cooling systems in homes, offices, and other buildings. Cooling systems include refrigerators, freezers, and air conditioners.

The principles of cooling are basically the same in all cooling systems. Cooling occurs when heat is drawn away from a substance. A refrigeration technician must understand the principles of heat transfer.

To be a refrigeration technician, you need a high school diploma. Technical training also may be helpful. Some employers provide on-the-job training. If you think you might be interested in a career as a refrigeration technician, write to the Air-Conditioning and Refrigeration Institute, 1501 Wilson Boulevard, Suite 600, Arlington, VA 22209.

What are physical and chemical changes?

Objective ▸ Distinguish between physical and chemical changes in matter.

TechTerms

- **chemical change:** change that produces new substances
- **physical change:** change that does not produce new substances

Physical Properties So far, all of the properties of matter that you have studied are physical properties. Mass, weight, volume, and density are physical properties. The phases of matter also are physical properties. Some other physical properties of matter include shape, size, taste, color, smell, and texture. Physical properties are characteristics that can be observed or measured without changing the makeup of a substance.

▸**List:** What are some physical properties of matter? mass, weight, phase, density, volume, color, shape, smell, taste, texture

Physical Changes If you cut an apple in half and share it with a friend, it is still an apple. If you change water to ice, it is still water. If you crumple a piece of paper into a ball, it is still paper. All of these changes are examples of **physical changes.** A physical change does not produce new substances. A physical change only changes some of the physical properties of a substance. Cutting an apple in half changes its size. Freezing liquid water changes its phase. Crumpling up a piece of paper changes its size and shape.

▸**Define:** What is a physical change?
A chemical change produces new substances.

Chemical Changes If you take a crumpled piece of paper and smooth it out, you can still write on it. It is still paper. Suppose you burn a piece of paper. When substances burn, they combine with oxygen. Burning is an example of a **chemical change.** A chemical change forms new substances. When paper burns, ashes, soot, heat, and light are produced. You no longer have paper. Some other examples of chemical changes include the rusting of iron, the digestion of food, and the burning of gasoline in a car engine.

Physical change

Chemical change

▸**Contrast:** How is a chemical change different from a physical change?
change that does not produce new substances

LESSON SUMMARY

▶ Physical properties can be observed or measured without changing the makeup of a substance.

▶ A physical change does not produce any new substances.

▶ A chemical change produces new substances.

CHECK *Write true if the statement is true. If the statement is false, change the underlined term to make the statement true.*

1. Volume and density are examples of <u>chemical</u> properties of matter. physical

2. Phase is a <u>physical</u> property of matter. true

3. Changing water to ice is an example of a <u>chemical</u> change. physical

4. New substances are produced by a <u>physical</u> change. chemical

5. Iron rusting is an example of a <u>chemical</u> change. true

6. Color, shape, and taste are <u>physical</u> properties. true

APPLY *Complete the following.*

7. **Analyze:** Describe the changes that take place when a match burns. Are these changes physical or chemical changes?

8. **Interpret:** Mixing vinegar with baking soda produces carbon dioxide. Is this a physical change or a chemical change? How do you know?

Skill Builder ....

Classifying Classify each of the following examples as a physical change or a chemical change. Explain your answers.

a. match burns
b. glass breaks
c. rubber band is stretched
d. iron rusts
e. ice melts
f. sugar cube is crushed

ACTIVITY

OBSERVING PHYSICAL CHANGES

You will need a glass bottle or jar with a narrow neck, ice cubes, and hot water.

1. Put a small amount of hot water into a bottle or jar. Place an ice cube over the top of the bottle so that the ice cube will not fall in.

2. Observe what you see coming from the surface of the hot water.

3. Observe the bottle for a few minutes and note what you see happening near the top of the bottle.

4. Watch for another minute or two. Note if you see anything fall from the top of the bottle.

Questions

1. **a. Observe:** What did you observe coming from the surface of the hot water in Step 2? **b.** What caused this to happen?

2. **a.** What did you see near the top of the bottle in Step 3? **b.** What caused this to happen?

3. **a.** Did you see anything falling inside the bottle in Step 4? **b.** If so, explain what you saw and how it was produced.

What are elements?

Objective ▶ Identify elements as simple substances that cannot be broken down.

TechTerm

▶ **elements** (EL-uh-munts): simple substances that cannot be broken down into simpler substances

Elements Some substances can be broken down into other substances. Water can broken down into hydrogen and oxygen. Sugar can be broken down into carbon, hydrogen, and oxygen. Salt can be broken down into sodium and chlorine. However, hydrogen, oxygen, carbon, sodium, and chlorine cannot be broken down. All of these substances are **elements** (EL-uh-munts). An element is a simple substance that cannot be broken down into simpler substances.

▶**Define:** What is an element?
a substance that cannot be broken down into a simpler substance.

The Known Elements There are 109 known elements. Ninety-two elements are found in nature. The other 17 elements have been made by scientists under special laboratory conditions. Most elements are solids at room temperature.

Some common examples of solid elements are iron, tin, lead, silver, gold, calcium, and copper. A few elements, such as mercury and bromine, are liquids. Other elements, such as oxygen, hydrogen, and nitrogen, are gases.

▶**Identify:** How many elements are found in nature? 92

Elements and Matter All matter is made up of elements. Some types of matter are made up of only one element. An iron nail contains only the element iron. Aluminum foil is made up of only the element aluminum. Gold and silver are other familiar substances that are made up of only one element.

Other substances are made up of more than one element. Water is made up of hydrogen and oxygen. Table salt is made up of sodium and chlorine. Sugar is made up of carbon, oxygen, and hydrogen. Brass is made up of copper and zinc. In the laboratory, a chemist can break down a substance such as sugar into the elements that make it up.

▶**Name:** What three elements make up sugar?
carbon, hydrogen, oxygen

LESSON SUMMARY

▶ Elements are simple substances that cannot be broken down into simpler substances.

▶ Most of the 109 known elements are solids at room temperature.

▶ All matter is made up of elements.

▶ Some substances are made up of more than one element.

CHECK *Complete the following.*

1. An _____ is a substance that cannot be broken into simpler substances. element

2. There are _____ known elements. 109

3. Mercury and bromine are _____ at room temperature. liquids

4. Substances made up of more than _____ element can be broken down into simpler substances. one

5. Most of the known elements are _____ at room temperature. solids

6. Water is made up of hydrogen and __oxygen__ .

APPLY *Complete the following.*

7. **Infer:** Mercuric oxide is made up of the elements mercury and oxygen. Could a chemist break down mercuric oxide into simpler substances? Explain.

8. **Classify:** Which of the following substances are elements?

   a. carbon   **f.** iron
   b. water   **g.** hydrogen
   c. gold   **h.** salt
   d. silver   **i.** brass
   e. sugar   **j.** tin

....
Health and Safety Tip ....

Some elements, such as mercury and chlorine, are interesting to study. They also are poisonous. You must always use caution when handling chemicals. Review the rules for chemical safety in Appendix B. Make a poster showing chemical safety rules.

PEOPLE IN SCIENCE ▶◼◀◇▶◼◀◇▶◼◀◇▶◼◀◇▶◼◀◇▶◼◀

MARIE CURIE (1867–1934)

Marie Sklodowska Curie was a Polish chemist. She worked and taught in Paris at the School of Physics and Chemistry and at the Sorbonne. While teaching in Paris, Marie Curie met her husband, the French scientist Pierre Curie.

Marie and Pierre Curie worked together. They became interested in the work of another French scientist named Henri Becquerel (bek-uh-REL). Becquerel had discovered that the element uranium is radioactive (ray-dee-oh-AK-tiv). Radioactive elements give off a form of energy that is known as radiation (ray-dee-AY-shun). The Curies discovered the radioactive elements polonium and radium. In 1903, the Curies were awarded a Nobel Prize in physics.

After the death of her husband in 1906, Marie Curie continued to study radioactive elements. In 1911, she was awarded a Nobel Prize in chemistry. Marie Curie is the only person ever to receive two Nobel Prizes in science.

12-6 What are chemical symbols?

Objective ▶ Explain how chemical symbols are used to write the names of the elements.

TechTerm

▶ **chemical symbols:** shorthand way of writing the names of the elements.

Chemical Shorthand Chemists have a special shorthand for writing the names of the elements. These shortened names for the elements are called **chemical symbols.** Chemical symbols are used by scientists all over the world. Chemical symbols are a universal (yoo-nuh-VUR-sul) language that all chemists understand.

▌▌▌▶*Define:* What are chemical symbols?
shorthand way of writing the names of the elements

Chemical Names and Symbols Chemical symbols come from the names of elements. The chemical symbol of many elements is made up of the first letter of the elements' names. For example, the symbol for oxygen is O. The symbol for carbon is C. The symbol for hydrogen is H. When more than one element has the same first letter, a second letter is added. The symbol for helium is He. Some chemical symbols come from the Latin name of the element. The Latin name for silver is *argentum.* The symbol for silver is Ag.

👁*Observe:* Use Table 1. What is the Latin name for lead? plumbum

Writing Chemical Symbols In order for everyone to understand chemical symbols, the symbols always must be written the same way. You should use the following rules when writing chemical symbols.

▶ A chemical symbol is always either one letter or two letters.

▶ The first letter is always a capital letter.

▶ If the symbol is two letters, the second letter is a lower case letter.

Table 1 Symbols of Some Common Elements	
ELEMENT	SYMBOL
Aluminum	Al
Calcium	Ca
Carbon	C
Gold (aurum)	Au
Iodine	I
Iron (ferrum)	Fe
Lead (plumbum)	Pb
Mercury (hydrargyrum)	Hg
Nitrogen	N
Oxygen	O
Silver (argentum)	Ag

▌▌▌▶*State:* What are the three rules for writing chemical symbols? Students should restate the rules listed above.

LESSON SUMMARY

▶ Chemical symbols are a shorthand way of writing the names of the elements.

▶ Chemical symbols are taken from the names of the elements.

▶ All chemical symbols consist of one or two letters; the first letter or single letter is always a capital letter, and the second letter is a lower case letter.

CHECK *Complete the following.*

1. Chemical symbols are a shorthand way to write the names of the _____ . elements

2. Each chemical symbol consists of one or two _____ . letters

3. The symbol for gold is Au, from its _____ name *aurum*. Latin

4. The first letter of a chemical symbol is always a _____ letter. capital

5. The symbol for the element _____ silver is Ag.

6. The chemical symbol for hydrogen is _____ H .

APPLY *Use Table 1 on page 230 to answer the following questions.*

7. **a.** What is the chemical symbol for calcium?
   b. For iodine?

8. What element does the symbol Fe represent?

9. What is the Latin name for iron?

InfoSearch ....

Read the passage. Ask two questions about the topic that you cannot answer from the information in the passage.

Ancient Chemistry Many ancient Greeks thought that all matter on Earth was made up of four basic "elements." These elements were earth, fire, air, and water. Ancient Greeks had picture symbols to represent each of these elements. The Greeks thought that these basic substances could be combined to make other substances. For example, they thought that steam could be made by combining fire and water.

SEARCH: Use library references to find answers to your questions.

PEOPLE IN SCIENCE

JONS JAKOB BERZELIUS (1779–1848)

Scientists today owe a lot to Jons Jakob Berzelius (buhr-ZEE-lee-us). He developed the modern system of writing chemical symbols and chemical formulas. Chemical formulas use chemical symbols to describe substances that are made up of two or more elements.

Berzelius was a Swedish chemist. He taught medicine, pharmacy, and chemistry at the University of Stockholm. In addition to using chemical symbols, Berzelius made many other contributions to science. He analyzed many chemical compounds to find out what elements they were made of. Berzelius discovered the elements selenium, cerium, and thorium. He was the first to identify the elements calcium and silicon. He also classified minerals according to their chemical composition. For all his work in the field of chemistry, the Swedish government honored Berzelius by making him a baron.

12-7 What is the periodic table?

Objective ▶ Trace the development of the modern periodic table of the elements.

TechTerms

▶ **group:** vertical column of elements in the periodic table

▶ **period:** horizontal row of elements in the periodic table

▶ **periodic** (pir-ee-AHD-ik): repeating pattern

Arranging the Elements By the 1800s, scientists had discovered many elements. Scientists began to search for ways to organize these elements. In 1869, a Russian chemist named Dmitri Mendeleev (men-duh-LAY-uf) listed the elements in order of increasing mass.

Mendeleev noticed that elements with similar properties occurred periodically. The word **"periodic"** (pir-ee-AHD-ik) means to repeat in a certain pattern. Based on the pattern he observed, Mendeleev arranged the elements in rows in a chart. Elements with similar properties were in the same columns of his chart, one under the other. Mendeleev's chart was the first periodic table of the elements. The diagram shows what Mendeleev's periodic table looked like.

▶ **Describe:** How were elements arranged in Mendeleev's periodic table of the elements?
according to mass

The Modern Periodic Table Mendeleev's periodic table was very useful, but it had some problems. Some elements did not seem to fit the pattern. Scientists tried to improve the periodic table. They arranged the elements by atomic (uh-TOM-ik) number, or number of protons, instead of mass. When the elements were arranged by atomic number, all the elements fell into place in the table. Atomic numbers are used to arrange the elements in the modern periodic table.

▶ **Compare:** How is the modern periodic table different from Mendeleev's periodic table?
Elements are arranged according to increasing atomic number, rather than increasing mass.

Using the Periodic Table Look at the periodic table on pages 234–235. Notice that there are 18 vertical columns in the table. These columns are called **groups,** or families. Elements in the same group have similar physical and chemical properties. Now look at the horizontal rows of the table. These rows are called **periods.** You can see that period 1 has only two elements, hydrogen and helium.

▶ **Define:** What is a group?
vertical column in the periodic table

			Ti = 50	Zr = 90	? = 180
			V = 51	Nb = 94	Ta = 182
			Cr = 52	Mo = 96	W = 186
			Mn = 55	Rh = 104,4	Pt = 197,4
			Fe = 56	Rn = 104,4	Ir = 198
		Ni =	Co = 59	Pd = 106,6	Os = 199
H = 1			Cu = 63,4	Ag = 108	Hg = 200
	Be = 9,4	Mg = 24	Zn = 65,2	Cd = 112	Au = 197?
	B = 11	Al = 27,4	? = 68	Ur = 116	Bi = 210?
	C = 12	Si = 28	? = 70	Sn = 118	Ti = 204
	N = 14	P = 31	As = 75	Sb = 122	Pb = 207
	O = 16	S = 32	Se = 79,4	Te = 128?	
	F = 19	Cl = 35,5	Br = 80	J = 127	
Li = 7	Na = 23	K = 39	Rb = 85,4	Cs = 133	
		Ca = 40	Sr = 87,6	Ba = 137	
		? = 45	Ce = 92		
		?Er = 56	La = 94		
		?Yt = 60	Di = 95		
		?In = 75,6	Th = 118?		

LESSON SUMMARY

▶ Mendeleev arranged the elements according to increasing mass.

▶ Mendeleev constructed the first periodic table of the elements.

▶ In the modern periodic table, elements are arranged in order of increasing atomic number.

▶ Vertical columns of elements in the periodic table are called groups; horizontal rows of elements are called periods.

CHECK *Complete the following.*

1. How are the elements arranged in the modern periodic table?

2. What are the horizontal rows of the periodic table called?

3. How are elements in the same group similar?

4. How did Mendeleev arrange the elements in his periodic table?

5. Why did scientists try to improve Mendeleev's table?

APPLY *Use the periodic table on pages 234–235 to answer the following questions.*

6. **Identify:** What elements are found in group 17 of the periodic table?

7. **Identify:** What elements are found in period 2 of the periodic table?

8. **Explain:** Do you think calcium and magnesium have similar properties? Explain.

9. What is the atomic number of carbon?

10. What is the atomic mass of sodium?

Skill Builder

Analyzing Many of the elements listed in the periodic table are familiar to you. Find 10 elements in the periodic table whose names are familiar to you. Make a list of the names and chemical symbols for these elements. Describe what you know about each element or how you have used it. For example, you might say that fluorine is added to toothpaste or to the water supply of the city or town in which you live.

LEISURE ACTIVITY

AMATEUR CHEMISTRY

Would you like a fun way to learn about the basic principles and processes of chemistry? Amateur chemistry sets let people explore chemistry as a hobby. Different kinds of amateur chemistry sets are available. Beginner chemistry sets introduce simple ideas such as the properties of the elements. More advanced sets explore complex topics, such as water purification and pollution.

Chemistry sets also can help a person enjoy other hobbies. For example, gardening and raising tropical fish both use a basic knowledge of chemistry. There are demonstrations in chemistry sets that show you how to check the pH of aquarium water or potting soil.

If you are interested in buying a chemistry set, visit your local hobby shop. Ask the sales people to help you choose the chemistry set that would be best for you.

Safety Tip: When using a chemistry set, always follow all safety rules. Remember that chemistry experiments can be dangerous if not carried out properly. Never try something in the science laboratory on your own before first checking with your science teacher.

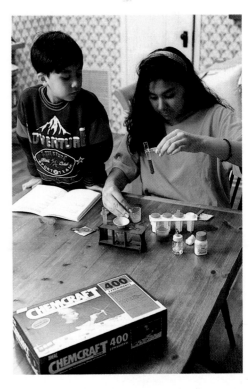

PERIODIC TABLE OF THE ELEMENTS

Atomic number — 6
Element's symbol — C
Element's name — Carbon
Atomic mass — 12.011

1

1
1
H
Hydrogen
1.00794

2

2	
3	4
Li	**Be**
Lithium	Beryllium
6.941	9.0122

Transition Metals

		3	4	5	6	7	8	9
11	12							
Na	**Mg**							
Sodium	Magnesium							
22.990	24.305							
19	20	21	22	23	24	25	26	27
K	**Ca**	**Sc**	**Ti**	**V**	**Cr**	**Mn**	**Fe**	**Co**
Potassium	Calcium	Scandium	Titanium	Vanadium	Chromium	Manganese	Iron	Cobalt
39.098	40.08	44.956	47.88	50.94	51.996	54.938	55.847	58.9332
37	38	39	40	41	42	43	44	45
Rb	**Sr**	**Y**	**Zr**	**Nb**	**Mo**	**Tc**	**Ru**	**Rh**
Rubidium	Strontium	Yttrium	Zirconium	Niobium	Molybdenum	Technetium	Ruthenium	Rhodium
85.468	87.62	88.9059	91.224	92.91	95.94	(98)	101.07	102.906
55	56	57 to 71	72	73	74	75	76	77
Cs	**Ba**		**Hf**	**Ta**	**W**	**Re**	**Os**	**Ir**
Cesium	Barium		Hafnium	Tantalum	Tungsten	Rhenium	Osmium	Iridium
132.91	137.33		178.49	180.95	183.85	186.207	190.2	192.22
87	88	89 to 103	104	105	106	107	108	109
Fr	**Ra**		**Unq**	**Unp**	**Unh**	**Uns**	**Uno**	**Une**
Francium	Radium		Unnilquadium	Unnilpentium	Unnilhexium	Unnilseptium	Unniloctium	Unnilennium
(223)	226.025		(261)	(262)	(263)	(262)	(265)	(266)

Row labels: 1, 2, 3, 4, 5, 6, 7

Rare-Earth Elements

Lanthanoid Series

57	58	59	60	61	62
La	**Ce**	**Pr**	**Nd**	**Pm**	**Sm**
Lanthanum	Cerium	Praseodymium	Neodymium	Promethium	Samarium
138.906	140.12	140.908	144.24	(145)	150.36

Actinoid Series

89	90	91	92	93	94
Ac	**Th**	**Pa**	**U**	**Np**	**Pu**
Actinium	Thorium	Protactinium	Uranium	Neptunium	Plutonium
227.028	232.038	231.036	238.029	237.048	(244)

C	solid	
Br	liquid	
H	gas	

Nonmetals

18

					2 **He** Helium 4.003
13	**14**	**15**	**16**	**17**	
5 **B** Boron 10.81	6 **C** Carbon 12.011	7 **N** Nitrogen 14.007	8 **O** Oxygen 15.999	9 **F** Fluorine 18.998	10 **Ne** Neon 20.179
13 **Al** Aluminum 26.98	14 **Si** Silicon 28.086	15 **P** Phosphorus 30.974	16 **S** Sulfur 32.06	17 **Cl** Chlorine 35.453	18 **Ar** Argon 39.948

10	**11**	**12**						
28 **Ni** Nickel 58.69	29 **Cu** Copper 63.546	30 **Zn** Zinc 65.39	31 **Ga** Gallium 69.72	32 **Ge** Germanium 72.59	33 **As** Arsenic 74.922	34 **Se** Selenium 78.96	35 **Br** Bromine 79.904	36 **Kr** Krypton 83.80
46 **Pd** Palladium 106.42	47 **Ag** Silver 107.868	48 **Cd** Cadmium 112.41	49 **In** Indium 114.82	50 **Sn** Tin 118.71	51 **Sb** Antimony 121.75	52 **Te** Tellurium 127.60	53 **I** Iodine 126.905	54 **Xe** Xenon 131.29
78 **Pt** Platinum 195.08	79 **Au** Gold 196.967	80 **Hg** Mercury 200.59	81 **Tl** Thallium 204.383	82 **Pb** Lead 207.2	83 **Bi** Bismuth 208.98	84 **Po** Polonium (209)	85 **At** Astatine (210)	86 **Rn** Radon (222)

Metals

63 **Eu** Europium 151.96	64 **Gd** Gadolinium 157.25	65 **Tb** Terbium 158.925	66 **Dy** Dysprosium 162.50	67 **Ho** Holmium 164.93	68 **Er** Erbium 167.26	69 **Tm** Thulium 168.934	70 **Yb** Ytterbium 173.04	71 **Lu** Lutetium 174.967
95 **Am** Americium (243)	96 **Cm** Curium (247)	97 **Bk** Berkelium (247)	98 **Cf** Californium (251)	99 **Es** Einsteinium (252)	100 **Fm** Fermium (257)	101 **Md** Mendelevium (258)	102 **No** Nobelium (259)	103 **Lr** Lawrencium (260)

12-8 What are metals and nonmetals?

Objectives ▶ Locate metals and nonmetals on the periodic table. ▶ Identify the properties of metals and nonmetals.

TechTerms

- **ductile** (DUK-tul): able to be drawn into thin wires
- **luster** (LUS-tur): shine
- **malleable** (MAL-ee-uh-bul): able to be hammered into different shapes
- **metals** (MET-uls): elements that have the properties of luster, ductility, and malleability
- **nonmetals:** elements that have none of the properties of metals

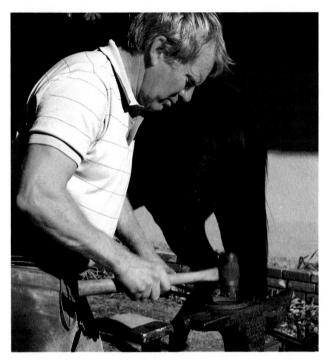

Metals and Nonmetals If you look at the periodic table, you will see a dark zigzag line running from the top of Group 13 to the bottom of Group 16. This line separates elements that are metals from elements that are nonmetals. Metals are all of the elements to the left of the zigzag line. Nonmetals are all of the elements to the right of the line.

◉ *Observe:* Use the periodic table. Are there more metals or nonmetals? metals

Properties of Metals Gold, silver, copper, aluminum, and tin are some **metals** (MET-uls). Metals have certain properties.

- All metals, except mercury, are solids at room temperature. Mercury is a liquid.
- Metals are shiny. For example, a gold ring is shiny. Aluminum foil is shiny. The shine of a metal is called its **luster** (LUS-tur). A metal has luster because it can reflect light.
- A sheet of aluminum foil and an aluminum pot have different shapes even though they are both made of aluminum. Aluminum and other metals are **malleable** (MAL-ee-uh-

bul), or able to be hammered into different shapes.

- Metals are **ductile** (DUK-tul). They can be made into thin wires. Most of the wires in electrical appliances are made of metals.
- Metals are good conductors of electricity. Wires made of a metal such as copper are able to carry an electric current.
- Metals also are good conductors of heat. Many pots and pans are made of metals because metals are good heat conductors.

▶ *List:* What are three properties of metals?
Students should list three of the properties given above.

Properties of Nonmetals Unlike metals, **nonmetals** do not have luster, so they look dull. Solid nonmetals are brittle. They are easily broken. They cannot be pounded into different shapes or drawn into thin wires the way metals can. Nonmetals are poor conductors of electricity and heat. Nonmetal elements may exist at room temperature as solids, liquids, or gases.

▶ *Relate:* Why do nonmetals look dull?
They do not have luster.

LESSON SUMMARY

▶ A zigzag line on the periodic table separates the metals from the nonmetals.

▶ Metals are shiny, ductile, malleable, and good conductors of electricity and heat.

▶ Nonmetals do not have the properties of metals.

CHECK *Complete the following.*

1. Metals can be hammered into thin sheets because they are _____ . malleable

2. Elements to the right of the zigzag line in the periodic table are _____ . nonmetals

3. It is possible to make silver wire because silver is _____ . ductile

4. All _metals_ are good conductors of heat.

5. Poor conductors of heat and electricity are _____ . nonmetals

6. Metals look shiny because of their _luster_ .

APPLY *Complete the following.*

7. **Identify:** What two properties of metals make them useful materials for the electrical wiring in your home?

8. **Infer:** What property of metals allows a jeweler to hammer a piece of silver to make jewelry?

9. **Hypothesize:** Why do you think nonmetals are used as insulators in thermos bottles?

10. **Classify:** Use the periodic table on pages 234–235 to identify the following elements as metals or nonmetals.
   a. zinc
   b. sulfur
   c. boron
   d. potassium
   e. silicon
   f. cobalt
   g. magnesium
   h. platinum
   i. antimony
   j. arsenic

Designing an Experiment ....

Design an experiment to solve the problem.

PROBLEM: Is an unidentified element a metal or a nonmetal?

Your experiment should:

1. List the materials you would need.

2. Identify safety precautions that should be followed.

3. List a step-by-step procedure.

4. Describe how you would record your data.

TECHNOLOGY AND SOCIETY

PLASTIC AND GRAPHITE

Imagine a car engine so light that you can easily carry it under your arm. Imagine a car that never gets dented. The car just bounces back into shape like a rubber ball. These ideas may sound impossible, but the materials for cars like this already have been invented. In fact, lightweight plastic automobile engines actually have been used to power racing cars.

Scientists are developing materials to replace metals for many reasons. Metals tend to be heavy. They are not very flexible. Metals are not renewable. Once the supply of metals is used up, there will be no more on the earth.

Some of the materials being developed to replace metals are made from plastics and graphite (GRAF-yt). In the 1970s, scientists developed a material made out of graphite and plastic to replace steel in airplane wings. Steel wings tend to break off at high speeds. Wings made of graphite and plastic are stronger and more flexible. What things can you think of that might someday be made of materials other than metals?

12-9 What are the inert gases?

Objectives ▶ Locate the inert gases on the periodic table. ▶ Identify the properties of the inert gases.

TechTerm

▶ **inert** (in-URT) **gases:** six elements that make up the last group in the periodic table of the elements

The Inert Gases Look at Group 18 of the periodic table. What do the elements in group 18 have in common? All of the elements in group 18 are gases. They are called the **inert** (in-URT) **gases.** Inert gases are sometimes called the noble gases. The word "inert" means "inactive." The inert gases are not chemically active. They do not combine readily with the other elements in chemical reactions.

▶ *Identify:* What are the elements in group 18 called? inert gases

Common Inert Gases The names of some of the inert gases may be familiar to you. You probably have heard of helium and neon. The names and chemical symbols of the six inert gases are listed in Table 1. All the inert gases are found in small amounts in the earth's atmosphere. The most common inert gas is argon. Argon makes up about 1% of the atmosphere.

Table 1 Inert Gases	
ELEMENT	SYMBOL
Helium	He
Neon	Ne
Argon	Ar
Krypton	Kr
Xenon	Xe
Radon	Rn

▶ *Name:* What are the six inert gases?
helium, neon, argon, krypton, xenon, radon

Uses of Inert Gases The inert gases have important uses. You may have heard of "neon lights." Neon is used for lighting because it gives off a bright red light when electricity passes through it. By mixing neon with different gases, many colors can be produced. Many of the signs that light up theaters, restaurants, and other businesses are neon lights. Argon and krypton also are used for lighting. They keep ordinary light bulbs from burning out. Helium is used to fill balloons and airships. Xenon is used in photographic lamps. Radon, which is radioactive, is used to treat cancer.

▶ *Describe:* What happens when electricity passes through neon gas?
Neon gives off a bright red light.

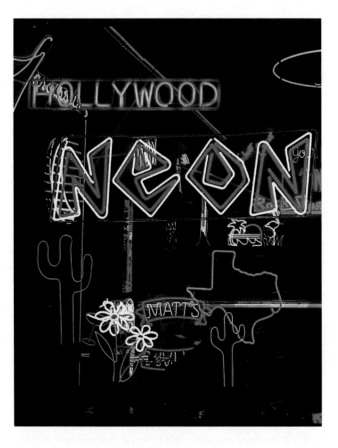

LESSON SUMMARY

▶ The six elements in the last group of the periodic table are called the inert gases.

▶ The six inert gases are helium, neon, argon, krypton, xenon, and radon.

▶ The inert gases have important uses.

CHECK *Complete the following.*

1. Where are the inert gases located in the periodic table?

2. What is the meaning of the word "inert"?

3. How do the inert gases behave in chemical reactions?

4. Which inert gas makes up about 1% of the earth's atmosphere?

5. Which inert gas is used to make balloons float in air?

APPLY *Complete the following.*

6. **Analyze:** Based on the position of helium in the periodic table, why does a helium balloon float in air?

7. **Hypothesize:** Why do you think the inert gases are not chemically active?

8. Why do you think radon is useful in fighting cancer?

InfoSearch

Read the passage. Ask two questions about the topic that cannot answer from the information in the passage.

Xenon Tetraflouride Inert gases do not combine chemically in nature. In 1962, a scientist named Neils Bartlett combined an inert gas with another element in a chemical reaction for the first time. He combined the inert gas xenon with the element fluorine to make xenon tetrafluoride (TET-ruh-floor-yd). Since 1962, other inert gases have been combined in chemical reactions.

SEARCH: Use library references to find answers to your questions.

◆◆◆◆ CAREER IN PHYSICAL SCIENCE ◆◆◆◆◆◆◆

ENVIRONMENTAL CHEMIST

You probably have read or heard about the pollution problem called acid rain. Acid rain results when certain pollutants (puh-LOOT-ents), or harmful materials, released into the air by industry combine with water and form drops of acid. The drops fall to the earth as acid rain. Acid rain is harmful to many living things. Environmental chemists are scientists who study problems such as acid rain.

Environmental chemists may analyze air, soil, or water to find out what substances are harming the environment. They also may determine how pollutants combine with one another. Some environmental chemists suggest solutions to pollution problems. They also may manage the clean-up of polluted areas.

To be an environmental chemist, you need a college degree in chemistry or environmental science and an advanced degree in environmental science. If you are interested in this career, write to the Environmental Protection Agency, Public Information Center, Mail Code PM-211B, 401 M Street SW, Washington, DC 20460.

UNIT 12 Challenges

STUDY HINT Before you begin the Unit Challenges, review the TechTerms and Lesson Summary for each lesson in this unit.

TechTerms

chemical change (226)
chemical symbols (230)
chemistry (220)
condensation (224)
ductile (236)
elements (228)
evaporation (224)
freezing (224)

gas (222)
group (232)
inert gases (238)
liquid (222)
luster (236)
malleable (236)
matter (220)
melting (224)

metals (236)
nonmetals (236)
period (232)
periodic (232)
phase (222)
physical change (226)
properties (220)
solid (222)

TechTerm Challenges

Matching *Write the TechTerm that matches each description.*

1. able to be hammered into different shapes malleable
2. form of matter phase
3. study of matter and its reactions chemistry
4. change from a gas to a liquid condensation
5. repeating pattern periodic
6. shine luster
7. simplest substances that cannot be broken down into simpler substances elements
8. elements that have the properties of luster, ductility, and malleability metals
9. horizontal row of elements in the periodic table period
10. change that does not produce a new substance physical change
11. change from a solid to a liquid melting
12. phase of matter that has no definite shape or volume gas

Fill In *Write the TechTerm that best completes each statement.*

1. Elements in the same _____ in the periodic table have similar properties. group
2. A shorthand way of writing the names of the elements are _____ . chemical symbols
3. Elements that have none of the properties of metals are _____ . nonmetals
4. New substances are produced during a _____ . chemical change
5. Water changing into ice is an example of _____ . freezing
6. Helium and xenon are two of the six _____ . inert gases
7. During _____ , liquids gain enough heat energy to change into the gas phase. evaporation
8. A _____ is a phase of matter with a definite shape and volume. solid
9. Anything that has mass and takes up space is _____ . matter
10. Mass and volume are two basic _____ of matter. properties
11. Because copper is , it can be drawn into thin wires. ductile
12. Milk is an example of a _____ . liquid

Content Challenges

Multiple Choice *Write the letter of the term that best completes each statement.*

1. In the modern periodic table, elements are arranged by
   a. density. **b.** atomic number. **c.** volume. **d.** weight. b

2. A metal can be hammered into different shapes because it is
   a. ductile. **b.** inert. **c.** malleable. **d.** lustrous. c

3. Water vapor changes to liquid water in a process called
   a. condensation. **b.** evaporation. **c.** melting. **d.** freezing. a

4. Elements found on the left side of the periodic table are
   a. gases. **b.** nonmetals. **c.** liquids. **d.** metals. d

5. The three main phases of matter are solids, liquids, and
   a. metals. **b.** nonmetals. **c.** plasma. **d.** gases. d

6. Burning is an example of a
   a. phase change. **b.** chemical change. **c.** physical change. **d.** physical property. b

7. Sugar is made up of the elements carbon, hydrogen, and
   a. chlorine. **b.** zinc. **c.** oxygen. **d.** sodium. c

8. The first letter of a chemical symbol is always
   a. a capital letter. **b.** a lower case letter. **c.** underlined. **d.** script. a

9. At room temperature, all metals, except mercury, are
   a. solids. **b.** liquids. **c.** gases. **d.** dull. a

10. An inert gas used to treat cancer is
    a. argon. **b.** helium. **c.** neon. **d.** radon. d

11. Particles of matter are tightly packed together in
    a. a solid. **b.** a liquid. **c.** a gas. **d.** plasma. a

12. The four basic properties of matter are mass, volume, weight, and
    a. luster. **b.** temperature. **c.** density. **d.** shape. c

13. Some common examples of solid elements are iron, tin, and
    a. mercury. **b.** oxygen. **c.** gold. **d.** nitrogen. c

14. The chemical symbol for calcium is
    a. Hg. **b.** Al. **c.** Fe. **d.** Ca. d

True/False *Write true if the statement is true. If the statement is false, change the underlined term to make the statement true.*

1. Substances that cannot be broken down into simpler substances are <u>elements</u>. true

2. Iron rusting is an example of a <u>physical</u> change. chemical

3. <u>Weight</u> is a measure of the pull of gravity on an object. true

4. There are <u>eight</u> groups of elements in the periodic table. eighteen

5. There are 109 known <u>metals</u>. elements

6. When a solid melts, its particles <u>lose</u> heat energy. gain

7. The most common inert gas is <u>argon</u>. true

8. The particles of a <u>liquid</u> can only vibrate in place. solid

9. <u>Physical</u> properties can be observed without changing the makeup of a substance. true

10. All matter takes up space and has <u>luster</u>. mass

11. <u>Metals</u> are good conductors of heat and electricity. true

Understanding the Features

Reading Critically *Use the feature reading selections to answer the following. Page number for the features are in parentheses.*

1. How does acid rain form? (238)
2. What two elements did Marie and Pierre Curie discover? (228)
3. **List:** What are three contributions to science made by Jons Jakob Berzelius? (230)
4. What are some hobbies that involve a basic knowledge of chemistry? (232)

5. **Infer:** Why does fixing an air conditioner require the same basic skills as fixing a refrigerator? (224)
6. Why are scientists developing materials to replace metals? (236)
7. What is an in-between state of matter that is neither solid nor liquid called? (222)

Concept Challenges

Critical Thinking *Answer each of the following in complete sentences.*

1. **Compare:** What are the differences between physical properties and chemical properties?
2. **Describe:** What happens to the particles of a substance as it changes from a solid to a liquid to a gas?

3. What properties of an element can be determined from the periodic table?
4. Why does evaporation require that heat energy be added to a substance?
5. What is the relationship between the mass and volume of an object?

Interpreting a Chart *Use the section of the periodic table to complete the following.*

1. Which element has a greater atomic mass, silver or copper?
2. Name three elements that have properties similar to sulfur.
3. What elements are found in group 13?
4. In which group and period is aluminum found?
5. What is the atomic number of tin?
6. Name two metals that are in the same group as carbon.
7. **Infer:** Do you think zinc is a good conductor of electricity? Explain.

Finding Out More

1. Make a list of at least 10 items in your home or school that are made of metal. In an oral report, describe what properties of metals make each item you listed useful or attractive.
2. In certain parts of the country, radon gas is a serious threat to public health. Research the problems concerning radon gas, and what has caused them. Find out what is being done to correct the problems. Write your findings in a report.
3. Collect pictures that show various elements. Label each picture with the name and chemical symbol of the element that is shown. Display all your pictures on a poster.
4. Use library references to find out more about the work of Dmitri Mendeleev. Write a short biography of his life.
5. Find out about the process of sublimation. Write your findings in a brief report.

DENSITY

CONTENTS

STUDY HINT As you read each lesson in Unit 13, write the topic sentence for each paragraph in the lesson on a sheet of paper. After you complete each lesson, compare your list of topic sentences to the Lesson Summary.

What is density?

Objective ▶ Define density.

TechTerm

▶ **density** (DEN-suh-tee): mass per unit volume

Density Which do you think is heavier, a kilogram of feathers or a kilogram of lead? You may already know the answer to this riddle. They both weigh the same amount. However, a kilogram of feathers takes up a large amount of space, or volume. A kilogram of lead is small enough to hold in your hand. A kilogram of lead takes up less space because lead has a much greater **density** (DEN-suh-tee) than feathers. Density is the mass per unit volume of a substance. Substances that are very heavy for their volume are called dense (DENS) substances. A large mass of a dense substance fits into a small volume.

▶**Define:** What is density? mass per unit volume of a substance

Units of Density You can find the density of a substance by finding the mass of a certain volume of the substance. Units of density include units of

mass and volume. Mass is measured in grams. The volume of solids is measured in cubic centimeters. The volume of liquids is measured in milliliters. One milliliter is equal to one cubic centimeter. Therefore, the density of any substance can be given in grams per cubic centimeter, or g/cm^3. For example, water has a density of 1 g/cm^3. There is 1 g of mass in 1 cm^3 of water. The densities of other common substances are listed in Table 1.

Table 1 Densities of Some Common Substances	
SUBSTANCE	DENSITY (g/cm³)
Air	0.0013
Alcohol	0.8
Aluminum	2.7
Cork	0.2
Gold	19.3
Iron	7.9
Lead	11.3
Mercury	13.6
Silver	10.5
Steel	7.8
Water	1.0

▶**Identify:** In what units is density measured? g/cm^3

Using Density Density is a basic physical property of all matter. Every substance has a density that can be measured. The density of a substance is always the same. The density of lead is always 11.3 g/cm^3. The density of mercury is always 13.6 g/cm^3. Density does not depend on the size or shape of the substance.

People can use density to help identify different kinds of matter. Suppose you have a metal and want to know what the metal is. You could identify the metal by finding its density. If the density is 10.5 g/cm^3, the metal is silver. If the density is 2.7 g/cm^3, the metal is aluminum.

▶**Identify:** What kind of property is density? physical property

LESSON SUMMARY

▶ Density is the mass per unit volume of a substance.

▶ Density is measured in grams per cubic centimeter, or g/cm³.

▶ Density is a basic property of all matter.

▶ Density can be used to identify different substances.

CHECK *Complete the following.*

1. Density is the _____ per unit volume of a substance. mass

2. When a substance has a high density, a large mass fits into a _____ volume. small

3. The units of _____ are grams per cubic centimeter. density

4. Density is a physical _property_ of all matter.

5. The density of a substance is always the _____ . same

APPLY *Complete the following.*

6. **Calculate:** What is the density of a metal block that has a mass of 750 g and a volume of 55 cm³?

Use Table 1 on page 244 to answer the following questions.

7. **a.** What is the density of iron?
   b. Of steel?
   c. Of mercury?

8. How large a container would be needed to hold 800 g of water?

9. **Sequence:** List the following substances in order from lowest density to highest density: iron, gold, steel, water, air, silver, aluminum, gasoline.

Designing an Experiment............

Design an experiment to solve the problem.

PROBLEM: What is the density of chalk?

Your experiment should:

1. List the materials you need.

2. Identify safety precautions that should be followed.

3. List a step-by-step procedure.

4. Describe how you would record your data.

SCIENCE CONNECTION ◆○◆○◆○◆○◆○◆○

NEUTRON STARS

Stars are not alive, but they have life cycles. The life cycle of a star is billions of years. Stars go through different stages during their life cycles. When a star first forms, it is made up mostly of hydrogen. As time passes, the hydrogen is changed into helium. When all the hydrogen is used up, the star becomes larger. It forms a large, cool star called a red giant.

After becoming a red giant, a very massive star may blow off its outer layers in a huge explosion. The star blows itself apart and becomes a supernova. All the matter of a supernova may be squeezed into a very small volume. The star shrinks into a dense ball. This dense ball is called a neutron (NOO-trahn) star. The density of a neutron star is so great that it is hard to imagine. One cubic centimeter of the material in a neutron star would have a mass of 1,000,000,000,000,000,000 kg.

Star

Red giant

Supernova

Neutron star

How is density measured?

Objective ▶ Explain how to find the density of a solid or a liquid.

Finding Density In order to find the density of a substance, you must measure both mass and volume. Once you have made these measurements, you can find density by dividing the mass by the volume. Remember that mass is measured in grams. Volume is measured in cubic centimeters.

density = mass/volume

▐▐▐▶ *Identify:* What measurements must you make before you can calculate the density of a substance? mass and volume

Density of a Liquid You can find the density of a liquid using a graduated cylinder and a balance. First, find the mass of the graduated cylinder alone. Record your measurement. Next, pour some of the liquid you want to measure into the graduated cylinder. Write down the volume of the liquid. Place the graduated cylinder with the liquid on the balance. Record the mass. Then find the mass of the liquid. To find the mass of the liquid, subtract the mass of the empty graduated cylinder from the mass of the graduated cylinder with liquid.

Now you are ready to calculate the density of the liquid. Look at the example shown. The mass of the liquid is 10 g. The volume is 10 mL. To find the liquid's density, divide its mass by its volume.

$$density = mass/volume$$
$$density = 10 \text{ g}/10 \text{ mL}$$
$$density = 1 \text{ g/mL}$$

Notice that in this example, density is measured in grams per milliliter. One milliliter is equal to one cubic centimeter. The density of a liquid can be measured in g/mL or g/cm^3.

▐▐▐▶ *Explain:* Why can density be measured either in g/cm^3 or in g/mL? because 1 mL is equal to 1 cm^3

Density of a Solid You can find the density of any solid if you know its mass and volume. You can use a balance to find the mass of a solid. You can find the volume of a solid with a regular shape by multiplying its length by its width by its height. Look at the aluminum bar. Its mass is equal to 270 g. Its volume is equal to 10 cm × 5 cm × 2 cm, or 100 cm^3. To find the density of the aluminum bar, divide its mass by its volume.

$$density = mass/volume$$
$$density = 270 \text{ g}/100 \text{ cm}^3$$
$$density = 2.7 \text{ g/cm}^3$$

10 mL   40 g

30 g

5 cm

10 cm

2 cm

270 g

▐▐▐▶ *Describe:* How can you find the volume of a solid with a regular shape? multiply its length by its width by its height

LESSON SUMMARY

▶ Density is equal to mass divided by volume.

▶ The density of a liquid can be measured in g/mL or g/cm³.

▶ To find the density of a solid with a regular shape, measure its mass and find its volume by multiplying its length by its width by its height.

Use the diagram to answer the questions.

CHECK *Complete the following.*

1. What measurements must be known in order to find the density of a substance?

2. What are the units of density for a liquid?

3. What equipment do you need to find the density of a liquid?

4. What three measurements must you make when finding the density of a liquid?

5. How can you find the density of a solid with a regular shape?

APPLY *Complete the following.*

6. **Calculate:** If 5 mL of a liquid have a mass of 10 g, what is the density of the liquid?

7. When finding the density of a liquid, why must you first find the mass of a graduated cylinder alone?

8. What is the volume of the bar?

9. If the bar has a mass of 5 grams, what is its density?

Skill Builder

Analyzing Gold was discovered in California in 1848. Many people went West to look for gold and become rich. One problem they had in hunting for gold was a substance known as fool's gold. Fool's gold looks like gold, but it is really a compound of iron and sulfur and sometimes copper. The density of real gold is 19.3 g/cm³. The density of fool's gold is about 5 g/cm³. If you were searching for gold, how could you tell whether a material was real gold or fool's gold?

ACTIVITY

FINDING THE DENSITIES OF DIFFERENT LIQUIDS

You will need a graduated cylinder, water, corn syrup, vegetable oil, and glycerine.

1. One at a time, slowly pour the water, corn syrup, vegetable oil, and glycerine into the graduated cylinder.

2. Observe the liquids as they form separate layers.

▲ 3. **Model:** Make a sketch showing the order in which the liquids have settled in the graduated cylinder.

Questions

1. **Relate:** What caused the liquids to separate into layers?

2. **a.** Which liquid is the most dense? **b.** Which liquid is the least dense?

3. List the four liquids in order from least to most dense.

13-3 What is specific gravity?

Objective ▶ Explain what is meant by specific gravity.

TechTerms

▶ **hydrometer** (hy-DRAHM-uh-tuhr): device used to measure specific gravity

▶ **specific** (spi-SIF-ik) **gravity:** density of a substance compared with the density of water

Specific Gravity It is often useful to compare the density of a substance with the density of water. Water is used as the standard for comparison because its density is 1 g/cm³. **Specific** (spi-SIF-ik) **gravity** is the density of a substance compared with the density of water. You can find the specific gravity of a substance by dividing its density by the density of water.

Suppose you want to find the specific gravity of copper. The density of copper is 8.9 g/cm³. The density of water is 1.0 g/cm³. To find the specific gravity of copper, divide the density of copper by the density of water. The specific gravity of copper is 8.9.

Notice that specific gravity has no units. The density units cancel each other out. The specific gravities of some common substances are listed in Table 1.

Table 1 Specific Gravities	
SUBSTANCE	SPECIFIC GRAVITY
Aluminum	2.7
Corn syrup	1.38
Diamond	3.5
Gasoline	0.7
Glycerine	1.26
Gold	19.3
Ice	0.92
Marble	2.7
Rubber	1.34
Water	1.00

▶**Define:** What is specific gravity? density of a substance compared with the density of water

Uses of Specific Gravity Specific gravity has many practical uses. It can be used to identify substances because each substance always has the same specific gravity. Specific gravity often is used to check the battery acid in a car. The amount of battery acid increases the specific gravity of liquid in the battery. Specific gravity also can be used to check the chemical purity of substances. Industries use specific gravity to check the quality of many of their products. For example, specific gravity is used to check the amount of cane sugar in solution. It also is used to check the purity of milk.

▶**Name:** What are two uses of specific gravity? Students should restate two of the uses given above.

Measuring Specific Gravity The specific gravity of a liquid can be measured with a device called a **hydrometer** (hy-DRAHM-uh-tuhr). A hydrometer is a sealed tube with a weight in the bottom and markings along one side. When a hydrometer is placed in a liquid, it floats. The height at which the hydrometer floats is determined by the specific gravity of the liquid. The higher the specific gravity of a liquid, the higher the hydrometer floats. You can tell the specific gravity of the liquid by reading the marking at the surface of the liquid.

▶**Explain:** What determines the height at which a hydrometer floats in a liquid? density of the liquid

248

LESSON SUMMARY

▶ Specific gravity is the density of a substance compared with the density of water.

▶ Specific gravity has no units.

▶ Specific gravity can be used to check chemical purity.

▶ A hydrometer is a device used to measure the specific gravity of a liquid.

CHECK *Complete the following.*

1. Specific gravity compares the density of a substance with the density of _____ . water

2. Specific gravity has no _____ because the density units cancel out. units

3. Specific gravity can be used to check the battery acid in a _____ . car

4. A hydrometer is a device that can be used to measure the specific gravity of a _____ . liquid

5. The _____ at which the hydrometer floats depends on the density of the liquid. height

APPLY *Use Table 1 on page 248 to answer the following questions.*

▶ 6. **Predict:** In which liquid would a hydrometer float lower, gasoline or corn syrup? Explain.

7. What substance has a specific gravity of 1.34?

Complete the following.

8. **Calculate:** Silver has a density of 10.5 g/cm³. What is the specific gravity of silver?

9. Why does specific gravity have no units?

Skill Builder

Researching In a brief report, explain how specific gravity could be useful in each of the following situations:
a. determining the purity of a diamond
b. determining whether a rock is real gold or fool's gold
c. separating corn oil from corn syrup
d. determining whether gasoline has been contaminated with water

CAREER IN PHYSICAL SCIENCE
MINERALOGIST

Minerals are natural substances found in soil and rock. Many products are made from minerals. For example, quartz is a mineral that is used to make watches. Diamonds are precious stones. Sulfur is a mineral that is used to make medicines.

To obtain useful minerals, the minerals must be mined, or taken from the earth. When a mineral deposit is found, a mining company needs to know how much of the mineral is present and what form the mineral is in. A mineralogist (min-uh-RAHL-uh-gist) finds answers to these questions. Mineralogists take samples from mineral deposits. They perform tests to analyze the properties of minerals. One of the properties mineralogists study is specific gravity. Specific gravity is useful in identifying minerals and determining their purity.

Mineralogists are employed by private industry, research laboratories, and the government. To be a mineralogist, a person needs a college degree. Many mineralogists also have advanced degrees. If you are interested in this career, you should have a good background in science and mathematics.

13-4 What is displacement?

Objectives ► Define displacement. ► Find the volume of an irregular solid.

TechTerm

► **displacement** (dis-PLAYS-muhnt): amount of water an object replaces

Displacement About 2000 years ago, a Greek scientist named Archimedes (ahr-kuh-MEE-deez) made an interesting observation. He stepped into a bathtub full of water and noticed that the water level rose. Some of the water spilled over the edge of the tub.

What Archimedes observed occurs whenever an object is placed in water. When objects are placed in water, they make the water level rise. The water level rises because water is pushed out of the way by the object. The amount of water that an object replaces is called its **displacement** (dis-PLAYS-muhnt).

▐▐▐▶ *Define:* What is displacement? amount of water that an object replaces

Displacement and Volume When an object is placed in water, the volume of the water that the object displaces is equal to the volume of the object. Many objects, such as rocks, do not have a regular shape. You can use displacement to find the volume of an irregularly shaped object.

To find the volume of an irregularly shaped object, pour some water into a graduated cylinder or a beaker that is marked to show volume. Record the volume of the water. Place the object into the water. Notice that the water level rises. Record the new volume of the water. The volume of the object is equal to the amount that the water level rose. Suppose a rock displaces 40 mL of water. The volume of the rock is 40 mL.

60 mL

20 mL

▐▐▐▶ *Explain:* How can you find the volume of an irregularly shaped object? displacement

250

LESSON SUMMARY

► Archimedes observed that when an object is placed in water, it causes the water level to rise.

► The amount of water that an object replaces is called displacement.

► The volume of water that an object displaces is equal to the volume of the object.

► The volume of an irregularly shaped solid can be found by placing the object in water and measuring the volume of water that the object displaces.

CHECK *Write true if the statement is true. If the statement is false, changed the underlined term to make the statement true.*

1. If an object displaces 50 mL of water, the object's <u>volume</u> is 50 mL. true

2. When an object is placed in water, the water level <u>falls</u>. rises

3. The amount of water that an object replaces is called <u>displacement</u>. true

4. A rock is an <u>irregularly</u> shaped object. true

5. The <u>mass</u> of an irregular object is equal to the volume of water it displaces. volume

APPLY *Complete the following.*

6. **Analyze:** Does the amount of water displaced by an object depend on the object's mass? Explain your answer.

7. Why is displacement useful in finding the density of an irregularly shaped object?

Use the diagram to answer the following.

10 mL   30 mL

8. **Observe:** How much water was displaced?

9. What is the volume of the rock?

Ideas in Action

IDEA: You probably observe displacement everyday. When you add ice to a drink, the ice displaces some of the liquid.

ACTION: Keep a record for the next three days of all the activities you observe that involve displacement of a liquid.

ACTIVITY

MEASURING DISPLACEMENT

You will need a small rock, a golf ball, a graduated cylinder, water, and a balance.

1. Use a balance to find the masses of a small rock and a golf ball. Record your measurements in a data table.

2. Fill the graduated cylinder with water to the 50-mL mark.

3. Gently place the rock sample in the water. Notice how much the water level rises. This is equal to the volume of water displaced. Record the volume of the water in your data table.

4. Repeat Step 3 with the golf ball.

50 mL

Questions

1. **a.** What is the volume of the rock? **b.** Of the golf ball?

2. **Analyze:** Does the amount of water displaced by an object depend on its mass? How do you know?

What is buoyancy?

Objective ▶ Explain Archimedes' principle in terms of buoyancy and displacement.

TechTerm

▶ **buoyancy** (BOI-uhn-see): upward force exerted by a gas or liquid

Archimedes' Principle Archimedes observed water rising when he stepped into a tub. He also noticed that his body seemed to feel lighter in water. Archimedes thought that the rising of the water in the tub and his feeling of weight loss must be related. He found that the loss of weight of an object in water is equal to the weight of the displaced water. This is called Archimedes' principle (PRIN-suh-puhl).

State: What does Archimedes' principle state? The weight lost by an object in water is equal to the weight of the displaced water.

Buoyancy When an object is placed in water, it seems to weigh less. The water exerts an upward force on the object. The upward force decreases the weight. It pushes up against the downward force of gravity. This upward force is called **buoyancy** (BOI-uhn-see). Buoyancy is the upward force exerted by a gas or a liquid.

You can observe buoyancy in action when you watch an airplane fly or a cork bob up and down in the ocean. You can experience buoyancy yourself by standing in the shallow end of a swimming pool and lifting your leg. Your leg will seem very light. Your leg feels light because the buoyant force of the water is helping to hold your leg up.

Define: What is buoyancy? the upward force exerted by a gas or a liquid

Buoyancy and Archimedes' Principle Archimedes' principle states that the amount of weight lost by an object in water is equal to the weight of the water that the object displaces. Buoyancy also is related to displacement. The buoyant, or upward, force on an object is equal to the weight of the water that the object displaces. If a rock weighing 4 N displaces an amount of water weighing 1 N, the buoyant force on the rock is 1 N. The rock's weight in the water is 4 N − 1 N, or 3 N.

Apply: If a buoyant force of 6 N acts on a block placed in water, what is the weight of the water that the block displaces? 6 N

Floating Buoyancy explains why an object sinks or floats. Suppose that an object displaces enough water so that the weight of the displaced water is equal to its own weight. The buoyant force on the object is equal to the object's weight. As a result, the weight of the object in water is zero. The object floats. An object also floats if it displaces a weight of water greater than its own weight. When a ship is placed in water, the weight of the water it displaces is equal to or greater than the ship's weight. That is why big, heavy ships can float.

Recognize: When will an object float in water? When it displaces a weight of water that is equal to or greater than its own weight.

LESSON SUMMARY

▶ Archimedes' principle states that the loss of weight of an object in water is equal to the weight of water the object displaces.

▶ Buoyancy is the upward force exerted by a gas or a liquid.

▶ Buoyancy helps hold objects in water or in the air.

▶ The buoyant force on an object is equal to the weight of the water that the object displaces.

▶ An object floats if the weight of the water it displaces is equal to or greater than its own weight.

CHECK *Complete the following.*

1. Buoyancy is the _____ force exerted by a gas or a liquid. upward

2. The buoyant force on an object is equal to the weight of the water it _____ . displaces

3. When the buoyant force on a object is equal to or greater than its weight, the object _____ . floats

4. Buoyancy decreases the downward pull of _____ on an object. gravity

5. A ship floats because the weight of water it displaces is equal to or greater than its own _____ . weight

APPLY *Complete the following.*

6. **a.** A wood block is 10 cm long, 5 cm high, and 3 cm wide. How much water will the block displace? **b.** If the density of water is 1 g/cm³, what will be the buoyant force on the block?

7. How are displacement and buoyant force related?

Health and Safety Tip

Always wear a life jacket if you go sailing or canoeing. If you fall into the water, the air in the jacket will decrease your density and help you to float, even if you cannot swim. Visit a local swimming pool. Ask the swimming instructor to describe how people are taught to float.

LEISURE ACTIVITY

BALLOONING

A hot-air balloon can float because the hot air in the balloon is less dense than the surrounding air. People have risen to very high altitudes in balloons. In 1978, a balloonist crossed the Atlantic Ocean for the first time in history.

Ballooning is a sport that involves competition among balloonists. Some of the events in balloon competitions include long-distance races and spot-landing matches. In a long-distance race, the winning balloon is the one that travels the farthest and stays in the air the longest. In a spot-landing match, a prize is given to the balloonist who is able to land closest to a certain point.

In the United States, balloon pilots must be licensed by the Federal Aviation Agency. Competitions are governed by the National Aeronautic Association (NAA) and the Balloon Federation of America. To find out more about ballooning, contact the Hot Air Balloon Club of America in Concord, California, or the Balloon Club of America in Newtown, Pennsylvania.

UNIT 13 Challenges

STUDY HINT Before you begin the Unit Challenges, review the TechTerms and Lesson Summary for each lesson in this unit.

TechTerms . . . .

buoyancy (250)   displacement (248)   specific gravity (246)
density (242)   hydrometer (246)

TechTerm Challenges . . . .

Matching *Write the TechTerm that matches each description.*

1. device used to measure specific gravity hydrometer
2. upward force exerted by a gas or a liquid buoyancy
3. density of a substance compared with the density of water specific gravity
4. amount of water an object replaces displacement
5. mass per unit volume density

Fill In *Write the TechTerm that best completes each statement.*

1. To find the volume of an irregular solid, measure its _____ . displacement
2. A _____ floats high in a liquid with a high specific gravity. hydrometer
3. The _____ of a substance tells the amount of mass in a certain volume. density
4. A ship can float in water because of _____ . buoyancy
5. If the _____ of a substance is greater than 1.0, the substance is more dense than water. specific gravity

Content Challenges . . . .

Multiple Choice *Write the letter of the term or phrase that best completes each statement.*

1. Density is measured in
   a. mL/cm^3. b. cm^3/g. c. g/cm^3. d. mL/g. c

2. Density is a basic physical property of
   a. gases. b. all matter. c. solids. d. liquids. b

3. A hydrometer is used to measure
   a. mass. b. volume. c. length. d. specific gravity. d

4. Archimedes discovered that objects weigh less when they
   a. have a larger volume. b. are placed in water. c. are suspended in air.
   d. are irregularly shaped. b

5. The amount of water that a floating object displaces is equal to the object's
   a. mass. b. weight. c. volume. d. density. c

6. If the buoyant force on an object is equal to the object's weight, then the object will
   a. sink. b. float. c. become less dense. d. increase in mass. b

7. A ship floats because it
   a. is made of dense materials. b. has a high specific gravity. c. is less dense than air.
   d. displaces a weight of water equal to its own weight. d

8. One milliliter is equal to one
   a. gram. b. cubic centimeter. c. meter. d. centimeter. b

9. To find the density of a substance, you must measure
   a. mass and length.   b. mass and specific gravity.   c. mass and volume.
   d. volume and specific gravity. c
10. If a rock displaces 50 mL, the volume of the rock is
   a. 50 cm.   b. 50 g/cm³.   c. 5 mL.   d. 50 mL. d

True/False *Write true if the statement is true. If the statement is false, change the underlined term to make the statement true.*

1. Specific gravity compares the density of a substance to the <u>weight</u> of water. density
2. The units of <u>density</u> include units of mass and volume. true
3. <u>Displacement</u> is a force exerted upward by a gas or a liquid. buoyancy
4. When an object is placed in water, it will lose all or some of its <u>mass</u>. weight
5. The buoyant force of an object is equal to the <u>volume</u> of the water it displaces. weight
6. When a substance has a high density, a large amount of mass takes up a <u>large</u> volume. small
7. Specific gravity is a ratio between two <u>weights</u>. densities
8. The density of a liquid can be measured in grams per cubic centimeter or grams per <u>milliliter</u>. true

Understanding the Features ....

Reading Critically *Use the feature reading selections to answer the following. Page numbers for the features are in parenthesis.*

1. What causes a neutron star to form? (243)
2. What kinds of events make up a ballooning competition? (251)
3. **Describe:** How do mineralogists use specific gravity in their work? (247)

Concept Challenges ....

Interpreting a Table *Use the table of specific gravities to answer the following questions.*

1. What is the specific gravity of gold?
2. Which substance has a higher specific gravity, diamond or silver?
3. What are three substances that will float in water?
4. **Predict:** What will happen if water and gasoline are mixed together?
5. **Analyze:** Which will take up a greater volume, a kilogram of marble or a kilogram of rubber?
6. Will a hydrometer float higher in glycerine or in gasoline? Explain.

Table 1 Specific Gravities	
SUBSTANCE	SPECIFIC GRAVITY
Aluminum	2.7
Corn syrup	1.38
Diamond	3.5
Gasoline	0.7
Glycerine	1.26
Gold	19.3
Ice	0.92
Marble	2.7
Rubber	1.34
Silver	10.5
Water	1.00

255

Critical Thinking *Answer each of the following in complete sentences.*

1. Why is 1 cm³ of wood lighter than 1 cm³ of iron?
2. **Analyze:** Suppose two objects look alike, but one is made of marble and the other is made of plastic. How could you use specific gravity to identify the objects?
3. Why is it possible to measure the volume of a liquid in milliliters when you are finding its density?
4. **Relate:** What is the relationship between the density of a substance and its specific gravity?
5. **Relate:** What is the relationship between the buoyant force on an object and the amount of water that the object displaces?

Finding Out More . . . .

1. **Sequence:** Collect pictures showing different common substances. Find out the density of each substance. Arrange the pictures in order of increasing density on a poster.
2. Specific gravity is an important property of gems. Gems are precious or semi-precious stones that have been cut and polished. Research the specific gravities of ten precious and semi-precious stones. List the information in a chart.
3. Archimedes made many contributions to the fields of science and mathematics. Use library references to find out more about the life and work of Archimedes. Present your findings in an oral report.
4. A submarine uses the principle of buoyancy to rise or sink in water. Find out how a submarine is constructed and how it operates. Write your findings in a report. Include a diagram showing the parts of a submarine.
5. Mathematical formulas make it possible to calculate the volumes of regular solids. Once the volumes are known, the densities of the solids can be found. Using library references, find out the formulas for calculating the volume of a cylinder, a cone, a pyramid, and a sphere. Organize your findings in a table.

ATOMS

CONTENTS

STUDY HINT Before beginning Unit 14, scan through the lessons in the unit looking for words that you do not know. On a sheet of paper, list these words. Work with a classmate to try to define each word on your list.

14-1   What are atoms?

Objectives ▶ Identify an atom as the smallest part of an element. ▶ List the parts of Dalton's atomic theory.

TechTerm

▶ **atom:** smallest part of an element that can be identified as that element

Atoms An **atom** is the smallest part of an element that can be identified as that element. Elements are simple substances that cannot be broken down into simpler substances. What happens if you keep cutting an element into smaller and smaller pieces? There is a smallest piece of an element that cannot be divided any further. This smallest piece is called an atom.

Atom of hydrogen   **Atom of oxygen**   **Atom of hydrogen**

▶ **Define:** What is an atom? smallest part of an element that can be identified as that element

Democritus The first person to suggest the idea of atoms was the Greek philosopher Democritus (di-MAHK-ruh-tus). More than 2400 years ago, Democritus asked whether it is possible to divide a sample of matter forever into smaller and smaller pieces. After much thought, Democritus came to the conclusion that it is not possible to divide matter forever. At some point, a smallest piece would be reached. Democritus named this smallest piece an atom. The word "atom" comes from a Greek word that means "cannot be divided."

Democritus and his students did not know what scientists today know about atoms. However, they hypothesized that atoms were small, hard particles that were all made out of the same material. They also thought that atoms were infinite in number, that they were always moving, and that they could be joined together.

▶ **Identify:** What does the word "atom" mean? cannot be divided

Dalton's Atomic Theory In the early 1800s, an English chemist named John Dalton did some experiments. Based on his observations, Dalton stated an atomic theory of matter. The main parts of Dalton's atomic theory are as follows:

▶ All elements are composed of atoms. Atoms cannot be divided or destroyed.

▶ Atoms of the same element are exactly alike.

▶ Atoms of different elements are different.

▶ The atoms of two or more elements can join together to form compounds.

Like Democritus, Dalton had some ideas about atoms that scientists no longer agree with. However, Dalton's atomic theory was the beginning of the modern theory of atoms.

▶ **List:** What are the parts of Dalton's atomic theory? Students should restate the list given above.

LESSON SUMMARY

▶ An atom is the smallest part of an element that can be identified as that element.

▶ The first person to suggest the existence of atoms was the Greek philosopher Democritus.

▶ The Greeks believed that atoms were small, hard particles that were infinite in number, always moving, and could be joined together.

▶ In the early 1800s, the English chemist John Dalton proposed an atomic theory of matter.

CHECK *Complete the following.*

1. An atom is the smallest part of an __element__ .

2. Dalton stated that atoms can join together to form _____ . compounds

3. The first person to use the word ''atom'' was _____ . Democritus

4. The _____ believed that atoms were hard particles that were always moving. Greeks

5. Dalton based his atomic theory on experiments and _____ . observations

APPLY *Complete the following.*

6. **Compare:** How were the ideas of Democritus and Dalton similar?

7. Do you agree with Democritus that atoms are ''small, hard particles''? Why or why not?

8. **Hypothesize:** What kind of information might be available to scientists today that would lead them to disagree with Dalton's atomic theory?

InfoSearch

Read the passage. Ask two questions about the topic that you cannot answer from the information in the passage.

STM Atoms are too small to be seen except with the most powerful microscopes. These microscopes are called scanning-tunneling microscopes, or STMs. This type of microscope uses electrons instead of light to observe atoms. STMs have also been used to move individual atoms from place to place on a surface.

SEARCH: Use library references to find answers to your questions.

TECHNOLOGY AND SOCIETY

PARTICLE ACCELERATORS

Have you ever heard or read about an ''atom smasher''? Yes, there are machines that can smash atoms. These machines are called particle accelerators. Particle accelerators speed up atoms so that the atoms gain huge amounts of kinetic energy. Then the atoms are smashed into one another or into a target. The atoms collide in much the same way as speeding cars smash into each other in an auto accident. After an atomic collision, scientists examine the pieces that are left. By studying these pieces, scientists can learn more about the basic structure of matter.

Particle accelerators are among the largest scientific instruments ever built. In Batavia, Illinois, a particle accelerator called the Tevatron is built around a 6.4-kilometer circular track. A particle accelerator in Switzerland is built in a circular tunnel that is 27 kilometers long. Plans are currently underway to build a Superconducting Supercollider that will circle an entire Texas town. The collider's tunnel will be almost 70 kilometers long.

14-2 What are the parts of an atom?

Objective ▸ Name the three basic parts of an atom.

TechTerms

- **electron:** negatively charged particle
- **neutron:** neutral particle
- **nucleus:** center, or core, of an atom
- **proton:** positively charged particle

Structure of an Atom According to modern atomic theory, an atom has a center, or core, called the **nucleus.** In the nucleus are **protons** and **neutrons.** Protons are positively charged particles. Neutrons are neutral particles. Surrounding the nucleus is a cloud of very small particles called **electrons.** Electrons are negatively charged particles.

▸**List:** What are the three types of particles in an atom? protons, neutrons, electrons

Thomson's Model The first scientist to suggest that atoms contain smaller particles was J. J. Thomson of England. In 1897, Thomson passed an electric current through a gas. He found that the gas gave off rays made of negatively charged particles. Today, these particles are known as electrons. Because atoms are neutral, Thomson reasoned that there must also be positively charged particles in an atom. Thomson hypothesized that an atom was made up of a positively charged material with electrons scattered evenly throughout.

▸**Identify:** What type of particles did Thomson discover in atoms? electrons

Rutherford's Model In 1908, a scientist from New Zealand named Ernest Rutherford performed an experiment to test Thomson's atomic model. Rutherford discovered that an atom is mostly empty space. He concluded that the protons are contained in a small central core. Rutherford called this core the nucleus.

▸**Describe:** What did Rutherford discover about an atom? An atom is mostly empty space.

Bohr's Model Rutherford's model of the atom did not explain the arrangement of electrons. In 1913, the Danish scientist Neils Bohr proposed that electrons in an atom are found in energy levels. Each energy level is at a certain distance from the nucleus. Electrons in different energy levels move around the nucleus in different orbits, much as the planets move in orbits around the sun.

Scientists now know that atoms are more complex than Bohr's model. The exact location of an electron cannot be predicted. Instead, energy levels are used to predict the place where an electron is most likely to be found outside the nucleus.

▸**Locate:** Where did Bohr say that electrons are found in an atom? in energy levels

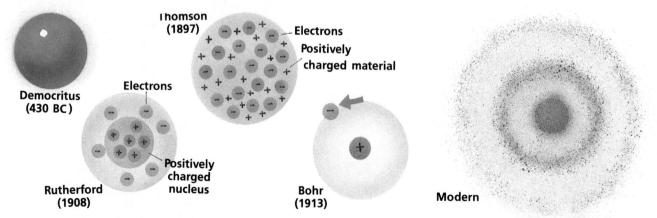

Democritus (430 BC)

Thomson (1897) — Electrons, Positively charged material

Electrons — Rutherford (1908), Positively charged nucleus

Bohr (1913)

Modern

LESSON SUMMARY

▶ Atoms are made up of protons, neutrons, and electrons.

▶ The first scientist to suggest that atoms contain smaller particles was J. J. Thomson.

▶ Ernest Rutherford proposed that an atom is mostly empty space with a small nucleus at the center.

▶ Neils Bohr proposed that electrons occupy specific energy levels as they orbit the nucleus of an atom.

CHECK *Complete the following.*

1. What are electrons?

2. Where are protons found in an atom?

3. What are the neutral particles in an atom called?

4. Who first suggested that atoms are made up of smaller particles?

5. Who discovered that an atom is mostly empty space?

6. Who said that electrons are found in energy levels around the nucleus?

APPLY *Complete the following.*

7. **Contrast:** How did Rutherford's model of the atom differ from Thomson's model?

8. Suppose that you could look at the electrons in a hydrogen atom, an oxygen atom, and a carbon atom. What would the electrons look like?

Skill Builder

▬ *Analyzing* In his experiment to test Thomson's model of the atom, Ernest Rutherford shot positively charged particles at a sheet of gold foil. The foil was only a few atoms thick. Rutherford observed that most of the particles went right through the foil. A few particles were deflected, or bent away, from the foil. A few particles bounced straight back from the foil. How did these results help Rutherford reach the following conclusions? (Hint: Remember that like charges repel each other.)

a. Most of an atom is empty space.

b. An atom has a dense nucleus in the center.

c. The nucleus is positively charged.

SCIENCE CONNECTION

QUARKS AND LEPTONS

In recent years, scientists have discovered that protons, neutrons, and electrons are made up of even smaller particles. Based on experiments using particle accelerators, scientists have identified two groups of subatomic particles. All matter is made up of these particles. These groups of particles are known as quarks (KWORKZ) and leptons (LEP-tahnz). The word "quark" was first used as the name of a subatomic particle by the American physicist Murray Gell-Mann. The word was invented by the writer James Joyce in his book *Finnegans Wake.* The word "lepton" comes from a Greek word that means "small" or "thin."

Quarks make up protons, neutrons, and other particles found in the nucleus of an atom. Leptons make up electrons. They also make up other particles called neutrinos (noo-TREE-nohz) and muons (MYOO-ahnz). There are six types of quarks and six types of leptons. Scientists have named the six types of quarks. Their names are strange, charm, up, down, top, and bottom. The basic particles in an atom are made up of combinations of two or three different quarks or leptons.

14-3 What is atomic number?

Objective ▶ Explain what is meant by the atomic number of an element.

TechTerm

▶ **atomic number:** number of protons in the nucleus of an atom

Elements and Atomic Number Atoms of different elements have different numbers of protons. The number of protons found in the nucleus of an atom is called the **atomic number.** The atoms of each element are different because each element has a different atomic number.

▷ **Define:** What is atomic number? the number of protons in the nucleus of an atom

Importance of Atomic Number The atomic number of an element is very important, because it identifies that element. No two elements have the same atomic number. Recall that the elements are arranged in order of increasing atomic number in the Periodic Table. The element with the smallest atomic number is hydrogen. Hydrogen has an atomic number of 1. This means that a hydrogen atom has one proton in its nucleus. Oxygen has an atomic number of 8. Gold has much larger atoms than either hydrogen or oxygen. Gold has an atomic number of 79. Table 1 lists the atomic numbers of some common elements.

▶ **Infer:** How many protons are there in the nucleus of a gold atom? 79

Table 1	Atomic Numbers of Common Elements	
ELEMENT	SYMBOL	ATOMIC NUMBER
Hydrogen	H	1
Helium	He	2
Carbon	C	6
Nitrogen	N	7
Oxygen	O	8
Sodium	Na	11
Aluminum	Al	13
Sulfur	S	16
Chlorine	Cl	17
Calcium	Ca	20
Iron	Fe	26
Copper	Cu	29
Silver	Ag	47
Gold	Au	79
Lead	Pb	82

Atomic Number and Electrons If you know the atomic number of an element, you can find the number of electrons in an atom of that element. An atom is neutral. It has neither a positive nor a negative charge. In order for an atom to be neutral, the number of electrons must equal the number of protons. The positive and negative charges cancel each other. So the number of electrons is always equal to the atomic number, or the number of protons.

▷ **Calculate:** How many electrons are there in an atom of an element with atomic number 14? 14

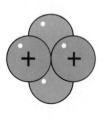

Helium
Atomic number = 2

Beryllium
Atomic number = 4

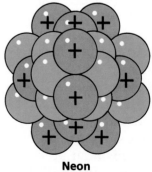

Neon
Atomic number = 10

LESSON SUMMARY

▶ The atomic number is the number of protons in the nucleus of an atom.

▶ Every element has its own atomic number which identifies that element.

▶ Because an atom is neutral, the number of electrons is equal to the number of protons.

CHECK *Complete the following.*

1. The atomic number is the number of _____ in the nucleus of an atom. protons

2. Every _element_ has its own atomic number.

3. Elements are arranged in order of increasing atomic number in the _____ . Periodic Table

4. If an atom of an element contains 12 protons and 12 electrons, the atomic number of that element is _____ . 12

5. An atom is _neutral_ because the negative charges and the positive charges cancel.

6. The number of protons in an atom is equal to the number of _____ . electrons

APPLY *Use Table 1 on page 262 to identify each of the following elements.*

7. Atoms of this element contain 17 protons.

8. This element has an atomic number of 20.

9. The atoms of this element have 8 protons and 8 electrons.

10. An atom of this element has 10 more protons than an atom of sulfur.

Complete the following.

11. **Analyze:** What is wrong with this statement? An atom of iron contains 13 protons and 13 electrons.

Skill Builder

Interpreting a Chart Use the Periodic Table on pages 234–235 to answer the following questions.

1. What is the atomic number of copper?

2. How many protons are there in an atom of chromium?

3. How many electrons are there in an atom of silicon?

4. What are the atomic numbers of the first three elements in Group 1?

5. In which group is the element with the highest known atomic number found? What is the atomic number of this element?

PEOPLE IN SCIENCE

CHIEN-SHIUNG WU (1912–PRESENT)

Chien-Shiung Wu was born in Liu Ho, China. She is a theoretical physicist. Chien-Shiung Wu came to America in 1936 to study for a doctorate in physics. She received her Ph.D. from the University of California at Berkeley. During World War II, she taught physics at Smith College and at Princeton University. After the war, she went to Columbia University to do research in nuclear physics. She became a professor of physics at Columbia in 1957.

Chien-Shiung Wu's area of specialization is beta decay. Beta decay is a form of radioactivity. In beta decay, the nucleus of an atom gives off electrons. The atom changes into another element. Chien-Shiung Wu has made many important contributions to scientists' present knowledge of the atom. Her experiments on beta decay confirmed a theory proposed by two other scientists. These scientists, Tsung Dao Lee and Chen Ning Yang, later won a Nobel Prize for their theory. Chien-Shiung Wu was the first woman to receive the Comstock Prize from the National Academy of Sciences.

14-4 What is atomic mass?

Objective ▶ Explain how to find the atomic mass and mass number of an atom.

TechTerms

▶ **atomic mass:** total mass of the protons and neutrons in an atom, measured in atomic mass units

▶ **mass number:** number of protons and neutrons in the nucleus of an atom

Mass of an Atom The mass of an atom is very small. Scientists cannot measure the mass of an atom in grams. In order to measure the mass of an atom, scientist have developed a special unit. This unit is called the atomic mass unit, or amu. One amu is equal to the mass of one proton. Neutrons and protons have the same mass. Therefore, one amu is also equal to the mass of one neutron. The mass of an electron is equal to 1/1836 amu. Because electrons are so small, only the masses of protons and neutrons are used to find the mass of an atom.

▶ *Infer:* What is the mass, in amu, of an atom with one proton and two neutrons? ₃ amu

Atomic Mass Because atoms of different elements have different numbers of protons and neutrons, they also have different masses. The total mass of the protons and neutrons in an atom is called the **atomic mass.** Atomic mass is measured in atomic mass units.

▶ *Define:* What is atomic mass? total mass of the protons and neutrons in an atom

Mass Number The total number of protons and neutrons in the nucleus of an atom is called the **mass number.** Each element has its own mass number. The mass number is equal to the atomic mass rounded off to the nearest whole number. You can find the number of neutrons in an atom by using this formula:

neutrons = mass number (protons + neutrons) − atomic number (protons)

Table 1 lists the atomic numbers and mass numbers of some common elements. You can use this information to determine the number of protons, neutrons, and electrons in an atom of an element.

LITHIUM

Mass number 7   Atomic number 3

protons and neutrons   protons   neutrons

Mass number   Atomic number   Number of neutrons

Table 1 Atomic Number and Mass Number			
ELEMENT	SYMBOL	ATOMIC NUMBER	MASS NUMBER
Hydrogen	H	1	1
Helium	He	2	4
Carbon	C	6	12
Nitrogen	N	7	14
Oxygen	O	8	16
Sodium	Na	11	23
Aluminum	Al	13	27
Sulfur	S	16	32
Chlorine	Cl	17	35
Calcium	Ca	20	40
Iron	Fe	26	56
Copper	Cu	29	64
Silver	Ag	47	108
Gold	Au	79	197
Lead	Pb	82	207

▶ *Analyze:* How many neutrons are in the nucleus of an atom of chlorine? ₁₈

LESSON SUMMARY

▶ The mass of an atom is measured in atomic mass units, or amu.

▶ The total mass of the protons and neutrons in an atom is the atomic mass.

▶ The mass number of an element is equal to the number of protons and neutrons in the nucleus of an atom of that element.

▶ The number of neutrons in an atom can be found by subtracting the atomic number from the mass number.

CHECK *Write true if the statement is true. If the statement is false, change the underlined term to make the statement true.*

1. The mass of a <u>neutron</u> is the same as the mass of a proton. true

2. Because they are so small, <u>neutrons</u> are not counted when measuring the mass of an atom. electrons

3. The <u>gram</u> is the unit used by scientists to measure the mass of an atom. atomic mass unit

4. The <u>mass number</u> tells the number of protons and neutrons in the nucleus of an atom. true

5. If the atomic number of an element is 8 and the mass number is 16, the number of neutrons in an atom of that element is <u>24</u>. 8

APPLY *Complete the following.*

6. **Calculate:** The atomic number of element X is 30 and the mass number is 65. Find the number of protons, neutrons, and electrons in an atom of element X.

7. **Hypothesize:** A few of the heavier elements have the same mass number. How is this possible, if no two elements have the same atomic number?

Skill Builder

Interpreting a Table Use Table 1 on page 264 to answer the following questions.

a. Which element has no neutrons in the nuclei of its atoms?

b. Which element has atoms containing 20 electrons and 20 neutrons?

c. Which element has 30 neutrons in its atoms?

d. Which elements have atoms with the same number of neutrons as protons? How can you tell?

e. Lead has a higher density than the other metals listed in the table. How can you explain this observation?

◼◀◇▶◼◀ PEOPLE IN SCIENCE ▶◼◀◇▶◼◀◇▶◼◀◇▶◼◀◇▶◼◀◇▶◼◀◇▶◼◀◇▶◼◀◇▶◼◀◇▶◼◀◇▶◼◀◇▶

DMITRI MENDELEEV (1834–1907)

Dmitri Mendeleev (men-duh-LAY-uf) was a Russian chemist. He was a professor of chemistry at St. Petersburg University. Mendeleev also held the important position of Director of the Bureau of Weights and Measures.

Mendeleev is best known for developing the first periodic table of the elements. He wrote a book called *Principles of Chemistry.* For this book, Mendeleev collected thousands of facts about the 63 elements that were known at that time. He tried to find a way to organize this information. Mendeleev thought that a certain pattern, or order, of the elements must exist. Mendeleev decided to test his hypothesis. He wrote the name of each element and the properties of that element on an index card. Then he tried different arrangements of the cards for the 63 elements. When he arranged the cards in order of increasing atomic mass, the elements fell into groups with similar properties.

What are isotopes?

▶ Explain what an isotope of an element is. ▶ Compare the three isotopes of hydrogen.

TechTerm

▶ **isotope** (Y-suh-tohp): atom of an element with the same number of protons but a different number of neutrons

Different Atomic Masses The atomic number of an element never changes. All atoms of the same element have the same atomic number. Thus, all atoms of an element have the same number of protons in the nucleus. However, all atoms of the same element may not have the same atomic mass. The difference in atomic mass is caused by a different number of neutrons in the nuclei of the atoms.

▶ **State:** What causes atoms of the same element to have different atomic masses? different numbers of neutrons in the nucleus

Isotopes Atoms of the same element that have different atomic masses are called **isotopes** (Y-suh-tohps). An isotope is an atom of an element with the same number of protons but a different number of neutrons. The Periodic Table of the Elements gives the mass number of the most common isotope of an element. The atomic mass of an element is an average of the atomic masses of all the isotopes of the element. That is why an element's atomic mass is not a whole number.

▶ **Define:** What is an isotope? atom of an element with the same number of protons but a different number of neutrons

Isotopes of Common Elements Most elements have two or more isotopes. An example of an element with three isotopes is hydrogen. The three isotopes of hydrogen are known as hydrogen-1, hydrogen-2, and hydrogen-3. The numbers 1, 2, and 3 represent the mass numbers of the isotopes. Sometimes the three isotopes of hydrogen are called protium (PROHT-ee-um), deuterium (doo-TIR-ee-um), and tritium (TRIT-ee-um). An atom of hydrogen-1 has only one proton and no neutrons in its nucleus. An atom of hydrogen-2 has one proton and one neutron. An atom of hydrogen-3 has one proton and two neutrons. The only difference among the three isotopes is the number of neutrons.

Other familiar elements that have isotopes include carbon, nitrogen, and uranium. Two isotopes of carbon are carbon-12 and carbon-14, or C-12 and C-14. Nitrogen-15 is an isotope of nitrogen. Two important isotopes of uranium are U-235 and U-238.

▶ **List:** What are the three isotopes of hydrogen called? hydrogen-1, hydrogen-2, hydrogen-3

Protium (H-1) nucleus

Deuterium (H-2) nucleus

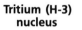

Tritium (H-3) nucleus

LESSON SUMMARY

▶ Different atoms of the same element may have different atomic masses.

▶ Atoms of the same element that have different numbers of neutrons are called isotopes.

▶ Most elements have two or more isotopes.

CHECK *Find the sentence in the lesson that answers each question. Then write the sentence.*

1. What causes some atoms of the same element to have different atomic masses?

2. What is an isotope?

3. How many isotopes do most elements have?

4. What are some familiar elements that have isotopes?

APPLY *Complete the following.*

5. How many neutrons are there in an atom of hydrogen-1? Of hydrogen-2? Of hydrogen-3?

6. Why is the atomic mass of an element not a whole number?

7. Calculate: The atomic number of carbon is 6. How many protons and neutrons are there in an atom of carbon-12? Of carbon-14?

8. Analyze: The mass number of oxygen is 16. Its atomic mass is 15.999. The atomic number of oxygen is 8. Which of the following statements about the isotopes of oxygen are true? Why?

a. All of the isotopes have 8 neutrons.

b. All of the isotopes have 8 or more neutrons.

c. Some of the isotopes have fewer than 8 neutrons.

d. All of the isotopes have fewer than 8 neutrons.

InfoSearch

Read the passage. Ask two questions about the topic that you cannot answer from the information in the passage.

Radioisotopes Some isotopes of certain elements give off particles or rays. These isotopes are called radioactive isotopes, or radioisotopes. Radioisotopes are very useful in medicine. They are helpful in diagnosing cancer. They are also used to treat cancer with radiation therapy.

SEARCH: Use library references to find answers to your questions.

CAREER IN PHYSICAL SCIENCE

RADIATION THERAPIST

Radiation therapy is an important form of cancer treatment. People who give radiation therapy to cancer patients are called radiation therapists. Radiation therapists prepare patients for treatment, position the patient properly under radiation equipment, and then operate the equipment. Radiation therapy can produce harmful side effects. For this reason, radiation therapists must watch a patient's reactions closely and keep doctors informed of the patient's condition. Radiation therapists must also follow important safety measures. Radioactive substances must be handled with great care.

Most radiation therapists work in hospitals. Others work in doctors' offices, clinics, and laboratories. Many part-time jobs are also available in this field. To become a radiation therapist, a person must finish high school and complete a training program in radiation therapy. Training programs are offered by hospitals, vocational or technical schools, and colleges and universities. If you would like to learn more about this career, write to the Society of Nuclear Medicine, 136 Madison Avenue, New York, NY 10016.

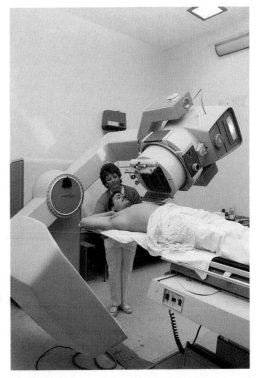

14-6 How are electrons arranged in an atom?

Objective ► Describe how the electrons in an atom are arranged in energy levels.

TechTerm

► **energy level:** place in an atom where an electron is most likely to be found

Electron Cloud Model For many years, scientists thought that electrons circled the nucleus of an atom in much the same way as planets orbit the sun. Scientists now know that it is not possible to predict the exact path of an electron. The area in an atom where electrons are likely to be found is often called the electron cloud. Scientists use the word "cloud" because they know that they cannot predict the exact location of electrons at any given time. The electron cloud is often compared to bees buzzing around a beehive.

▌▌▌► **Describe:** What is the electron cloud? area in an atom where electrons are likely to be found

Energy Levels In the modern atomic theory, electrons are arranged in **energy levels.** An energy level is the place in the electron cloud where an electron is most likely to be found. Each energy level is a different distance from the nucleus. The lowest, or first, energy level is closest to the nucleus. Electrons with higher energy are found in energy levels farther away from the nucleus.

Each energy level can hold only a certain number of electrons. The first energy level can hold only 2 electrons. The second energy level can hold 8 electrons. Energy levels beyond the second

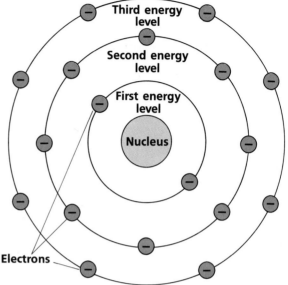

level can hold up to 32 electrons. For the first 20 elements, the electrons in an atom of an element fill up the energy levels in order, beginning with the lowest. An atom of helium has 2 electrons. These 2 electrons fill the first energy level. An atom of lithium has 3 electrons. Two of these electrons fill the first energy level. The third electron occupies the second energy level.

► **Predict:** Where would you expect to find the 6 electrons in an atom of carbon? 2 in the first energy level and 4 in the second

Changing Energy Levels Electrons can move from one energy level to another. If an electron gains enough energy, it jumps to a higher energy level. If an electron loses enough energy, it drops back to a lower energy level.

◢ **Analyze:** What causes an electron to change energy levels? gaining or losing energy

Electron gains energy

Electron jumps to higher energy level

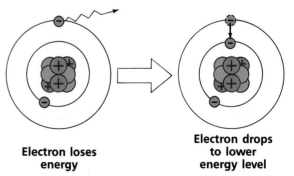

Electron loses energy

Electron drops to lower energy level

LESSON SUMMARY

▶ The area in an atom where electrons are likely to be found is called the electron cloud.

▶ An energy level is the place in the electron cloud where an electron is most likely to be found.

▶ Each energy level in an atom can hold a certain number of electrons.

▶ An electron can change energy levels if it gains or loses a certain amount of energy.

CHECK *Complete the following.*

1. The term _____ refers to the area in an atom where electrons are likely to be found. electron cloud

2. An energy level is the place where an _____ is most likely to be found in an atom. electron

3. The _____ energy level is located closest to the nucleus of an atom. lowest or first

4. The second energy level can hold _____ electrons. 8

5. Some energy levels far from the nucleus can hold up to _____ electrons. 32

6. An electron will drop to a lower energy level when it _____ energy. loses

APPLY *Complete the following.*

7. **Analyze:** The atoms of a certain element have the first and second energy levels filled with electrons. What is the atomic number of this element? How can you tell?

8. **Infer:** What is the relationship between the amount of energy that an electron has and its distance from the nucleus of an atom?

InfoSearch

Read the passage. Ask two questions about the topic that you cannot answer from the information in the passage.

Orbitals Energy levels of an atom are divided into sublevels called orbitals. Orbitals can be thought of as more specific areas in which electrons are likely to be found. The first energy level has only one orbital, called an s orbital. The shape of an s orbital is a sphere. The second and third energy levels have one s orbital and three p orbitals. The p orbitals are dumbbell-shaped. Each orbital can hold a maximum of two electrons. These electrons must have opposite spins.

SEARCH: Use library references to find answers to your questions.

◼◁◇▷◼ PEOPLE IN SCIENCE ▷◼◁◇▷◼◁◇▷◼◁◇▷◼◁◇▷◼◁◇▷◼◁◇▷◼◁◇▷◼◁◇▷◼◁◇▷◼◁◇▷◼◁◇▷◼◁◇▷

MARIA GOEPPERT MAYER (1906–1972)

Maria Goeppert Mayer came from a German family with seven generations of university professors. She received her doctorate at the age of 24. Goeppert Mayer had to study for the exams by herself, because the only school that prepared women for the university had closed.

Maria Goeppert Mayer worked with some of the most important physicists and chemists of the twentieth century, including the famous German physicist Max Born. Born's work helped describe the structure of atoms and the motion of atomic particles. Goeppert Mayer was interested in atomic structure. She proposed a shell model for the nucleus of the atom. For this work, she won a Nobel Prize in 1963. Maria Goeppert Mayer was the second woman in history to win a Nobel Prize in physics. At the time of her death in 1972, Maria Goeppert Mayer was on the faculty of the University of California at San Diego.

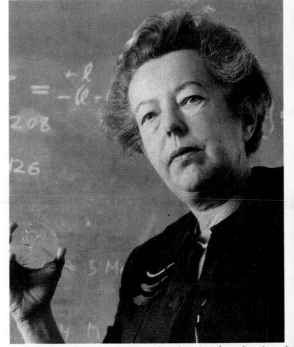

Challenges

STUDY HINT Before you begin the Unit Challenges, review the TechTerms and Lesson Summary for each lesson in this unit.

TechTerms

atom (258)
atomic mass (264)
atomic number (262)
electron (260)

energy level (268)
isotope (266)
mass number (264)
neutron (260)

nucleus (260)
proton (260)

TechTerm Challenges

Matching *Write the TechTerm that matches each description.*

1. number of protons and neutrons in the nucleus of an atom mass number
2. negatively charged particle in an atom electron
3. number of protons in the nucleus of an atom atomic number
4. positively charged particle in an atom proton
5. place where an electron is most likely to be found energy level
6. smallest part of an element atom
7. neutral particle in the nucleus of an atom neutron
8. has the same number of protons but a different number of neutrons isotope

Fill In *Write the TechTerm that best completes each statement.*

1. The small, dense core of an atom is called the _____ . nucleus
2. A proton and a _neutron_ have the same mass.
3. The _____ of an element is an average of all the isotopes of that element. atomic mass
4. You can find the number of neutrons in an atom by subtracting the atomic number from the _____ . mass number
5. The mass of an _____ is only 1/1836 amu. electron
6. Every neutral atom has the same number of _____ and electrons. protons

Content Challenges

Multiple Choice *Write the letter of the term or phrase that best completes each statement.*

1. The first person to suggest that matter is made up of atoms was
   a. Dalton. **b.** Rutherford. **c.** Democritus. **d.** Bohr. c

2. The Greeks believed all of the following about atoms <u>except</u> that
   a. atoms are small. **b.** atoms are always moving. **c.** atoms can be joined together.
   d. atoms contain smaller particles. d

3. In the early 1800s, an atomic theory of matter was developed by
   a. Democritus. **b.** Dalton. **c.** Rutherford. **d.** Thomson. b

4. The first scientist to discover that atoms contain smaller particles was
   a. Thomson. **b.** Rutherford. **c.** Bohr. **d.** Dalton. a

5. J. J. Thomson pictured atoms as being made up mostly of
   a. empty space. **b.** positively charged material. **c.** electrons. **d.** the nucleus. b

6. Rutherford's model of the atom included
   a. a small, dense nucleus. **b.** a positive material studded with electrons.
   c. energy levels for electrons. **d.** neutrons. a

7. Neils Bohr proposed that electrons
   a. orbit the nucleus.   b. are inside the nucleus.   c. do not exist.
   d. cannot be located exactly in an atom. a

8. The atomic number of an atom is equal to
   a. the number of protons and neutrons.   b. the number of electrons and neutrons.
   c. the number of protons.   d. the number of protons and electrons. c

9. The letters "amu" stand for
   a. atomic measuring unit.   b. alternate mass unit.   c. atomic mass unit.
   d. atomic matter unit. c

10. Tritium is an
    a. electron energy level.   b. isotope of hydrogen.   c. element with atomic number 102.
    d. isotope of carbon. b

True/False *Write true if the statement is true. If the statement is false, change the underlined term to make the statement true.*

1. An isotope of an element has the same number of protons but a different number of <u>electrons</u>. neutrons

2. Electrons with the lowest energy are <u>farthest from</u> the nucleus of an atom. closest to

3. The second energy level can hold <u>8</u> electrons. true

4. Scientists <u>can</u> predict the exact location of an electron in an atom. cannot

5. When an electron <u>gains</u> enough energy, it moves to a higher energy level. true

6. Democritus concluded that it <u>is</u> possible to divide matter into smaller and smaller pieces forever. is not

7. Dalton stated that atoms of the same element are <u>alike</u>. true

8. Scientists today think that atoms <u>cannot</u> be divided into smaller particles. can

9. <u>Neutrons</u> are positively charged particles. Protons

10. The number of protons and <u>electrons</u> in a neutral atom must be equal. true

11. All atoms of the same element have the same atomic <u>mass</u>. number

12. To find the number of <u>electrons</u> in an atom, subtract the atomic number from the mass number. neutrons

Understanding the Features ....

Reading Critically *Use the feature reading selections to answer the following. Page numbers for the features are in parentheses.*

1. What does a particle accelerator do to atoms? (259)

2. What was Mendeleev's hypothesis about the elements? (265)

3. What smaller particles make up protons, neutrons, and electrons? (261)

4. What is Chien-Shiung Wu's area of specialization in science? (263)

5. Infer: Why must radiation therapists follow important safety measures? (267)

6. Infer: What was the status of education for women in Germany during the years when Maria Goeppert Mayer was growing up? (269)

Concept Challenges ·

Critical Thinking *Answer each of the following in complete sentences.*

1. Explain how Democritus came to the conclusion that matter is made up of atoms.
2. **Compare:** In what ways did Dalton, Thomson, and Rutherford have similar ideas about atoms? In what ways did their ideas differ?
3. Is it possible for the atoms of two different elements to have the same number of electrons? Explain your answer.

4. Why is the atomic mass of an element not a whole number, while the mass number is always a whole number?
5. **Predict:** How many electrons would be found in the second energy level of an atom of nitrogen, which has an atomic number of 7?
6. **Hypothesize:** Would you expect isotopes of the same element to have the same properties or different properties? Why?

Interpreting a Diagram *Use the diagrams showing the atomic structure of elements A and B to complete the following.*

Element A   Element B

1. Which element, A or B, would appear first in the Periodic Table? How can you tell?
2. An isotope of element B has two additional neutrons. What is the mass number of this isotope? What is the atomic number?
3. What is the mass number of element A? What is the atomic number?
4. In which element is the second energy level filled with electrons?
5. In elements A and B, which electrons have the most energy? Which of the two element has more high-energy electrons?

Finding Out More ·

1. **Model:** Choose four elements from the Periodic Table of the Elements. Draw diagrams to show an atom of each of the elements. Clearly show the number of protons, neutrons, and electrons in each atom. Also show the energy levels in which the electrons are located.
2. A radioactive isotope of iodine is important in medicine. Find out the name, atomic number, and mass number of this isotope.

Find out how the isotope is used in medicine. Write a few paragraphs to describe what you have learned.
3. Use library references to identify the four forces that exist within an atom. Write a brief description of these forces. Include a description of how each force affects the behavior of subatomic particles, and how they hold an atom together.

COMPOUNDS AND MIXTURES

CONTENTS

STUDY HINT Before beginning each lesson in Unit 15, review the TechTerms for each lesson and read the Lesson Summary.

15-1 What is a compound?

Objective ▸ Recognize that compounds can be broken down into simpler substances.

TechTerm

▸ **compound:** substance made up of two or more elements chemically combined

Combining Elements Hydrogen and oxygen are elements. They are both gases at room temperature. These two elements can combine to form a liquid. When hydrogen combines chemically with oxygen, water is formed. Water is a **compound.** A compound is a substance made up of two or more elements that are chemically combined.

▸**Define:** What is a compound? a substance made up of more than one element

Compounds Most of the matter making up the earth is composed of compounds. Sugar is another compound. It is made up of the elements carbon, hydrogen, and oxygen. Table salt is a compound. It is made of the elements sodium and chlorine.

Sugar = Carbon + Hydrogen + Oxygen

▸**Name:** What are three examples of common compounds? water, sugar, and salt

Properties of Compounds The properties of a compound are very different from the properties of the elements that make it up. Sodium is a metal that burns very easily. Chlorine is a poisonous gas. Sodium and chlorine combine to form the compound sodium chloride, or table salt. Sodium chloride is neither a metal nor a poisonous gas. Salt is a white solid. It has its own properties.

▸**Contrast:** How do the properties of salt differ from the properties of sodium and chlorine? Salt is a white solid. Sodium is a metal, and chlorine is a poisonous gas.

Forming Compounds A compound is formed as a result of a chemical change. A chemical change causes elements to lose their original properties. The elements combine to form a new substance with different properties. A chemical change also can cause a compound to break down into the elements that make it up. When sugar is heated, it melts into a liquid. If the liquid is heated long enough, hydrogen and oxygen gas are released into the air. Finally, only a black solid remains. This solid is carbon. Heating the sugar caused it to break down into the elements that formed it.

Water

Sugar

Carbon

▸**Describe:** How can a compound be broken down into the elements that formed it? by a chemical change

LESSON SUMMARY

▶ A compound is a substance made up of more than one element.

▶ Most of the matter making up the earth is composed of compounds.

▶ The properties of a compound are very different from the properties of the elements that form it.

▶ Compounds are formed as a result of a chemical change.

CHECK *Complete the following.*

1. A compound is made up of more than one _____ . element

2. The compound _____ is made up of the elements hydrogen and oxygen. water

3. The properties of a compound are _____ from the properties of the elements that form it. different

4. When sodium and chlorine combine chemically, the compound _____ forms.
   sodium chloride, or salt

5. Compounds are formed as a result of a _____ . chemical change

6. Heating _____ will cause it to break down into hydrogen, oxygen, and carbon. sugar

APPLY *Complete the following.*

7. **Infer:** A recipe calls for a cup of sugar to be heated slowly over a low flame. What could happen if the sugar is heated slowly over a high flame?

8. When an unknown solid is heated, two different gases and a liquid are formed. Is the solid a compound or an element? How do you know?

Ideas in Action

IDEA: Rust is a compound that forms when iron is exposed to moist air.

ACTION: Identify different examples of rusted iron around your home. What are some ways to protect a bicycle from rusting?

ACTIVITY

BREAKING DOWN HYDROGEN PEROXIDE

You will need a 150-mL beaker, hydrogen peroxide, a raw potato, and a knife.

1. Carefully cut a slice from the center of the potato about 0.5 cm thick. **Caution: Be careful when using a knife.**

2. Pour hydrogen peroxide into the beaker up to the 75-mL mark.

3. Place the potato slice into the beaker. Observe what happens.

Questions

1. **Observe:** What happened when you put the potato slice into the beaker of hydrogen peroxide?

2. **Analyze:** Hydrogen peroxide is a compound made up of the elements hydrogen and oxygen. What do you think caused the results you observed?

15-2 What are molecules?

Objective ▶ Identify a molecule as the smallest part of a compound.

TechTerm

▶ **molecule:** smallest part of a compound that has all the properties of the compound

Parts of a Compound Suppose you could look at a sugar cube under a microscope. You would see that it is made up of many small grains. A single grain of sugar can be broken down into millions of smaller particles called **molecules.** A molecule is the smallest part of a compound that has all the properties of the compound. A molecule of sugar is the smallest form of sugar that has the properties of sugar.

◀▶**Define:** What is a molecule? smallest part of a compound that has all the properties of the compound

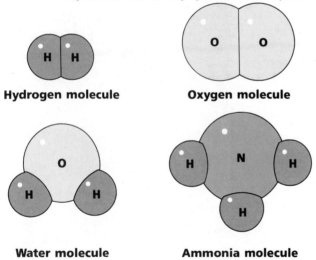

Hydrogen molecule   **Oxygen molecule**

Water molecule   **Ammonia molecule**

Diatomic Molecules Some elements always occur as diatomic (dy-uh-TAHM-ik) molecules. A diatomic molecule contains two atoms of the same element joined together. A molecule of oxygen contains two oxygen atoms joined together. Hydrogen also is always found in diatomic molecules. In fact, most gaseous elements form diatomic molecules.

◀▶**Identify:** How many atoms are there in a molecule of hydrogen? two

Combining Atoms Water is a compound made up of the elements hydrogen and oxygen. One molecule of water contains two atoms of hydrogen and one atom of oxygen. Any molecule of water contains the same kinds of atoms joined together. Whether the water exists as a solid, a liquid, or a gas, one molecule of water always contains two atoms of hydrogen joined to one atom of oxygen.

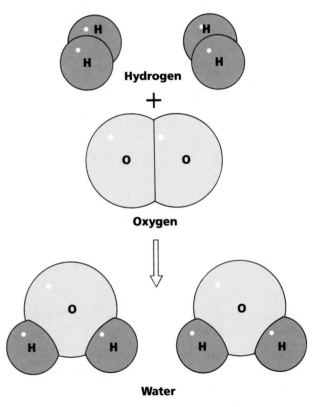

Hydrogen

+

Oxygen

Water

Different types of molecules are made up of different combinations of atoms. The compound salt contains the elements sodium and chlorine. A molecule of salt is made up of one atom of sodium joined with one atom of chlorine. The elements carbon, hydrogen, and oxygen are found in the compound sugar. A molecule of sugar contains 12 atoms of carbon, 22 atoms of hydrogen, and 11 atoms of oxygen. Every sugar molecule always contains the same combination of atoms.

◀▶**Identify:** What is one molecule of water made of? two atoms of hydrogen and one atom of oxygen

LESSON SUMMARY

▶ A molecule is the smallest part of a compound that has all the properties of the compound.

▶ Some elements occur as diatomic molecules.

▶ Any molecule of water contains two atoms of hydrogen joined to one atom of oxygen.

▶ All molecules of a particular compound contain the same combination of atoms.

CHECK *Complete the following.*

1. What is the smallest part of a compound that has all the properties of the compound?

2. What is a molecule made up of?

3. How many atoms are there in a molecule of water?

4. What molecule is made up of atoms of sodium and chlorine?

5. What kinds of atoms are there in a molecule of sugar?

6. What is a diatomic molecule?

APPLY *Use the diagram of an alcohol molecule to answer the following.*

👁 **7. Observe:** How many atoms does one molecule of alcohol contain?

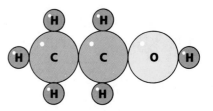

8. Identify: List the number and type of each atom found in a molecule of alcohol.

InfoSearch ....

Read the passage. Ask two questions about the topic that you cannot answer from the information in the passage.

Combinations of Atoms Two different compounds may contain the same elements. However, different numbers of atoms of each element make up different compounds. For example, water and hydrogen peroxide both contain the elements hydrogen and oxygen. In a molecule of water, two atoms of hydrogen are joined to one atom of oxygen. In a molecule of hydrogen peroxide, two atoms of hydrogen are joined to two atoms of oxygen. The different combinations of atoms of the same elements give these two compounds their different properties.

SEARCH: Use library references to find answers to your questions.

SCIENCE CONNECTION

MASS SPECTROSCOPY

Have you ever wondered how scientists identify the atoms that make up a particular molecule? Because atoms are so tiny, scientists must use special instruments to study them. One such tool used to study atoms is called a mass spectrometer (spek-TRAHM-uh-tur).

In a mass spectrometer, a sample of a substance is bombarded with electrons. The substance takes in the electrons and forms electrically charged particles. The charged particles are then put into a magnetic field. The magnetic field separates the charged particles according to their mass. A pattern called a mass spectrum is formed. Scientists can identify elements and molecules by the kind of mass spectrum they form. Mass spectroscopy also can be used to separate different isotopes of an element, and to find impurities in samples of different elements.

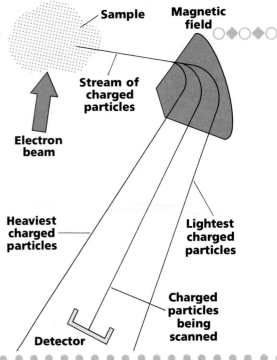

What are mixtures?

Objective ▶ Describe the physical properties of a mixture.

TechTerm

▶ **mixture:** two or more substances that have been combined, but not chemically changed

Mixtures Have you ever swallowed a mouthful of water while swimming in the ocean? If you have, you know that the earth's oceans are made up of salt water. Salt water contains particles of salt and other substances mixed in water. The substances in salt water are not chemically combined. The salt and the water keep their own properties. Salt water is an example of a **mixture.** A mixture contains two or more substances that have been combined, but not chemically changed. The molecules in a mixture are not all alike. Salt water contains molecules of salt and molecules of water.

▶**Define:** What is a mixture? a substance that contains two or more substances that have been combined, but not chemically changed

Making a Mixture The substances in a mixture can be present in any amount. A mixture of salt water may contain equal numbers of salt molecules and water molecules. It also may have twice as many water molecules as salt molecules. The amount of each type of substance present in a mixture can change. However, the substances always keep their own properties. All salt–water mixtures have the properties of both salt and water.

▶*Infer:* Why do the substances in a mixture keep their own properties? because they have not been chemically changed

Separating a Mixture The properties of the substances in a mixture can be used to separate the mixture. The substances in a mixture are not chemically combined. Therefore, they can be separated by physical means. Water evaporates when it is heated. If salt water is heated, the water will evaporate out of the mixture. Salt and other impurities will be left behind. Another way of separating a mixture is by filtering the mixture. Suppose you wanted to separate a mixture of sand and water. You could pour the mixture through a piece of filter paper. The water would pass through the paper. The sand would not pass through the paper. It would collect on the paper.

▶*Name:* What are two possible ways of separating a mixture? evaporation and filtering

Water vapor

Salt water

Salt crystals

Sand and water

Sand

Filter paper

Funnel

Water

LESSON SUMMARY

► A mixture contains two or more substances that have been combined, but not chemically changed.

► The substances in a mixture keep their original properties.

► The substances in a mixture can be present in any amount.

► A mixture can be separated by physical means.

CHECK *Write true if the statement is true. If the statement is false, change the underlined term to make the statement true.*

1. Salt water is an example of a <u>compound</u>. ^(mixture)

2. The substances in salt water <u>are not</u> chemically combined. ^(true)

3. All the molecules in a <u>mixture</u> are alike. ^(compound)

4. The substances in a <u>mixture</u> keep their original properties. ^(true)

5. A mixture can be separated by <u>chemical</u> means. ^(physical)

6. Salt water can be separated by <u>filtering</u>. ^(evaporation)

APPLY *Complete the following.*

⟩ 7. **Predict:** How could a mixture of iron and sulfur be separated? (Hint: Iron is magnetic.)

Sulfur and iron filings

8. **Classify:** A teaspoon of instant coffee is placed in a cup of water. Is this a mixture or a compound? Explain your answer.

....

Designing an Experiment............

Design an experiment to solve the problem.

PROBLEM: Is gelatin a compound or a mixture?

Your experiment should:

1. List the materials you would need.

2. Identify safety precautions that should be followed.

3. List a step-by-step procedure.

4. Describe how you would record your data.

LEISURE ACTIVITY

COOKING

Many people enjoy cooking as a hobby. In many ways, the principles used in cooking are similar to those of chemistry. Food scientists have been able to make many different types of prepared foods available to millions of people. By analyzing the senses of smell and taste, scientists can combine seasonings to improve the flavor of foods.

Amateur cooks may specialize in certain kinds of cooking, such as baking, or in regional cooking, such as southern cooking. Good cooks must be able to follow a recipe, but they also experiment with different procedures and ingredients. Cooking is both an art and a science.

Some people teach themselves to cook by reading cookbooks. You also could watch cooking programs on television. Most communities have cooking schools.

15-4 How are compounds and mixtures different?

Objective ▶ Contrast the properties of compounds with the properties of mixtures.

Making a Mixture The substances in a mixture have been combined, but not chemically changed. You can make a mixture of iron filings and sulfur by simply mixing the substances together. Each substance in the mixture will keep its own properties after being mixed. Iron filings are magnetic slivers of grey metal. Sulfur is a nonmetallic yellow powder. You can see the grains of yellow powder and slivers of grey metal in a mixture of these substances.

Sulfur

Iron filings

Sulfur and iron filings

Iron sulfide

▶*Infer:* How could a mixture of iron filings and sulfur be separated? A magnet will attract the iron filings and leave the sulfur behind.

Making a Compound A compound is made up of more than one element. The elements in a compound have been chemically combined. Iron sulfide is a compound. It contains the elements iron and sulfur. If you heat a mixture of iron filings and sulfur, you will produce the compound iron sulfide. Heating the mixture causes the iron filings and sulfur to combine chemically.

▷*Explain:* What happens when a mixture of iron filings and sulfur is heated? The compound iron sulfide is produced.

Comparing Mixtures and Compounds Mixtures and compounds are different in several ways. A mixture of iron and sulfur does not have a definite chemical composition. The mixture might contain equal parts of each element. It also might have twice as much iron as sulfur. Each substance in a mixture of iron and sulfur keeps its own properties. A mixture of iron and sulfur can be separated by simple physical means.

The compound iron sulfide always has a definite chemical composition. A molecule of iron sulfide always contains one part iron and one part sulfur. This is because the elements in the compound have been joined chemically. When elements are combined chemically, each element loses its properties. The iron and sulfur in iron sulfide cannot be separated by physical means.

Table 1 Comparing Mixtures and Compounds	
MIXTURES	COMPOUNDS
Made of two or more substances mixed together	Made of two or more substances chemically combined
Substances keep their own properties	Substances lose their own properties
Can be separated by physical means	Can be separated only by chemical means
Have no definite chemical composition	Have a definite chemical composition

▷*Contrast:* How are mixtures and compounds different? Students should restate the information in Table 1.

280

LESSON SUMMARY

▶ The substances in a mixture keep their own properties.

▶ Heating a mixture of iron and sulfur will produce the compound iron sulfide.

▶ A compound has a definite chemical composition, while a mixture does not.

▶ A mixture can be separated by physical means, while a compound can be separated only by chemical means.

CHECK *Complete the following.*

1. The elements in a _____ are combined chemically. compound

2. Each substance in a _____ keeps its own properties. mixture

3. A _____ does not have a definite chemical composition. mixture

4. A _____ can only be separated by chemical means. compound

5. A _____ always has the same chemical composition. compound

6. A _____ can be separated by simple physical means. mixture

APPLY *Complete the following.*

▶ 7. **Infer:** When a certain poisonous gas is combined with a flammable metal, a fine white powder results. The powder is neither flammable nor poisonous. Is the powder a mixture or a compound? How do you know?

8. **Hypothesize:** Would all samples of salt water taste the same? Explain.

Skill Builder ....

Building Vocabulary The suffix "-ide" is often used in chemistry to form the names of compounds. Iron sulfide is a compound made up of the elements iron and sulfur. Sodium chloride, or table salt, is a compound made up of sodium and chlorine. Look up the words "oxide," "bromide," and "carbide" in a dictionary. Identify the elements found in compounds that have these words in their names.

SCIENCE CONNECTION
CHEMISTRY OF COOKING

Have you ever made pancakes for breakfast? If so, you actually made both a mixture and then a compound. The pancake mix, water, and eggs formed a mixture. Each of the substances in the batter kept their original properties. The batter could be separated into the substances that formed it by physical means such as filtering and evaporation.

When you poured the batter onto a hot skillet, you made a compound. Heat from the stove caused the substances in the mixture to combine chemically. If you were able to observe the pancake molecules, you would see that each molecule was made up of the same elements in the same combination.

When cooking, it is important to follow recipe directions very carefully. Adding too much or too little of one ingredient can change the taste and texture of the final product. Adding too much water to the pancake mixture might prevent it from cooking properly and changing into a compound.

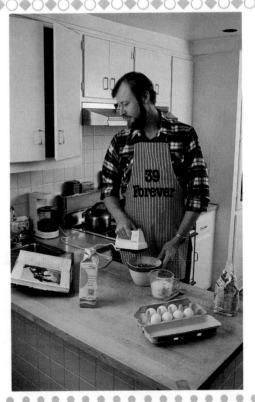

15-5 What are ionic bonds?

Objective ▶ Describe how atoms combine in ionic bonds.

TechTerms

- ▶ **crystal lattice** (LAT-is): ions arranged in a regular pattern
- ▶ **ion** (Y-un): charged particle
- ▶ **ionic bond:** bond formed when atoms gain and lose electrons

Neutral Atoms All matter is made up of atoms. Atoms contain smaller particles called protons, neutrons, and electrons. A proton has a positive charge. An electron has a negative charge. A neutron has neither a positive nor a negative charge. In an atom, the number of protons equals the number of electrons. Since their electrical charges are balanced, the atom is neutral.

▶**Explain:** Why are atoms neutral? The electrical charges of the protons and electrons are balanced.

Charged Particles Sometimes an atom gains or loses electrons. When the number of electrons is different from the number of protons in an atom, the atom has an electrical charge. An atom with an electrical charge is called an **ion** (Y-un). An ion is a charged particle. If a neutral atom gains electrons, it becomes a negative ion. If a neutral atom loses electrons, it becomes a positive ion.

▶**Compare:** Does a negative ion have more protons or more electrons? A negative ion has more electrons than protons.

Ionic Bonds Particles of matter are held together by atomic bonds. The bond that is formed when two atoms trade electrons is called an **ionic bond.** One atom gains electrons and becomes a negative ion. The other atom loses electrons and becomes a positive ion. The two ions have opposite electrical charges. As a result, they are attracted to each other. This force of attraction is what holds atoms together in an ionic bond.

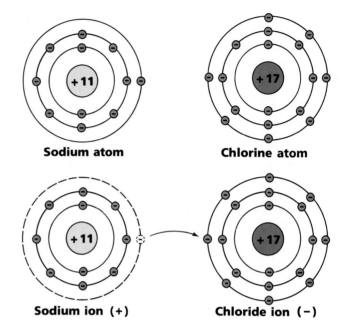

Sodium atom   **Chlorine atom**

Sodium ion (+)   **Chloride ion (−)**

▶**Explain:** How does an ionic bond form?
It forms when two atoms trade electrons.

Crystals The force of attraction in an ionic bond is very strong. Many compounds that contain ionic bonds are solids. A crystal is a solid that contains ions arranged in a regular pattern. The pattern of positive and negative ions forms a **crystal lattice** (LAT-is). The shape of a crystal is determined by its crystal lattice. Scientists have found that there are six basic types of crystal lattices.

▶**Identify:** What is a crystal lattice? a pattern of positive and negative ions

282

LESSON SUMMARY

▶ Neutral atoms have equal numbers of protons and electrons.

▶ When an atom gains or loses electrons, it becomes an ion.

▶ An ionic bond forms when two atoms trade electrons.

▶ Ions arranged in a regular pattern form a crystal lattice.

CHECK *Complete the following.*

1. Electrons have a ___negative___ electrical charge.

2. When an atom loses electrons, it becomes a _____ ion. positive

3. An _____ forms when two atoms trade electrons. ionic bond

4. Particles with opposite electrical charges _____ each other. attract

5. A _____ is a solid that contains ions arranged in a regular pattern. crystal

6. The patterns that form a crystal are called the _____ . crystal lattice

APPLY *Complete the following.*

☛ 7. **Infer:** Could an atom ever lose an electron without another atom gaining the electron? Explain.

❯ 8. **Predict:** Two different crystals are found to have identical shapes. What does this tell you about the pattern of ions in the crystals?

Skill Builder

▲ *Modeling* Table salt, or sodium chloride, is an example of a crystal. Since sodium chloride is a crystal, its ions are arranged in a repeating pattern. Research the type of crystal lattice found in table salt. Draw a diagram of the crystal lattice of sodium chloride and display it to the class. Label the sodium and chloride ions in the lattice.

LOOKING BACK IN SCIENCE

CRYSTALLOGRAPHY

Crystals have been the object of scientific study for hundreds of years. Early mineralogists classified crystals according to observable properties such as shape and color. Around 1800, mineralogists began measuring the angles found on a crystal's surface. The mineralogists thought that the size of a crystal's angles was related to the type of substances that make up the crystal. However, they had no way of looking at the internal structure of a crystal.

In 1895, X rays were discovered. Using X rays, scientists could examine the structure of crystals. They discovered that crystal angles are caused by common structural patterns inside the crystal. As a result of X-ray crystallography (kris-tuh-LAHG-ruh-fee), scientists were able to identify six basic types of crystal lattices.

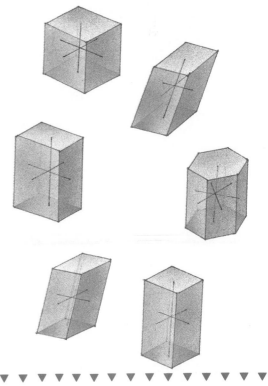

15-6 What are covalent bonds?

Objective ▶ Describe how atoms combine in covalent bonds.

TechTerm

▶ **covalent bond:** bond formed when atoms share electrons

Outer Energy Levels Most atoms do not have complete outer energy levels. Atoms can complete their outer energy levels by either gaining, losing, or sharing electrons. Ionic bonds form when two atoms trade electrons. The result is an ionic compound. Another way elements can form compounds is by sharing electrons. In a **covalent bond**, two atoms share electrons. The covalent compound has a complete outer energy level.

▶ *Identify:* How can atoms complete their outer energy levels? by gaining and losing electrons, or by sharing electrons

Covalent Compounds Water is an example of a covalent compound. A molecule of water forms as a result of a covalent bond between an atom of oxygen and two atoms of hydrogen. The oxygen atom has six electrons in its outer energy level. It needs two more electrons to complete the energy level. A hydrogen atom has one electron in its outer energy level. This energy level is complete when it has two electrons. So a hydrogen atom needs only one more electron to complete its energy level.

The diagram shows the covalent bonds in a molecule of water. Each hydrogen atom shares its one electron with the oxygen atom. The oxygen atom shares one of its electrons with each hydrogen atom. Each atom has a complete outer energy level.

Classify: Is water an ionic compound or a covalent compound? covalent

Ionic and Covalent Compounds Atoms joined by a covalent bond do not lose or gain electrons. The atoms remain electrically neutral. They do not become positively or negatively charged ions. Table 1 lists the main points to remember about compounds formed by ionic and covalent bonds.

Table 1 Ionic and Covalent Compounds	
IONIC COMPOUNDS	COVALENT COMPOUNDS
Atoms complete outer energy level	Atoms complete outer energy level
Electrons are lost and gained	Electrons are shared
Atoms form ions	Atoms remain neutral

▶ *Compare:* How are ionic compounds and covalent compounds similar? The atoms in each type of compound have a complete outer energy level.

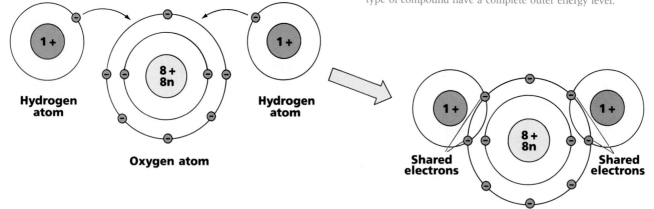

Hydrogen atom   **Oxygen atom**   **Hydrogen atom**

Shared electrons   **Shared electrons**

Water molecule

284

LESSON SUMMARY

▶ A covalent bond forms when atoms share electrons.

▶ Both ionic bonds and covalent bonds result in atoms with complete outer energy levels.

▶ A molecule of water contains two atoms of hydrogen that are joined to an atom of oxygen by covalent bonds.

▶ Atoms joined by a covalent bond remain electrically neutral.

CHECK *Identify whether each of the following statements describes an ionic bond, a covalent bond, or both.*

1. Atoms are joined together. both
2. Atoms become electrically charged. ionic bond
3. Electrons are shared. covalent bond
4. Atoms remain neutral. covalent bond
5. Atoms complete their outer energy levels. both
6. Electrons are lost or gained. ionic bond

APPLY *Complete the following.*

7. **Infer:** A hydrogen atom contains one electron. A molecule of hydrogen gas contains two hydrogen atoms bonded together. The atoms in hydrogen gas are neutral. Are the hydrogen atoms joined by an ionic or a covalent bond? Explain.

8. A carbon atom has four electrons in its outer energy level. A chlorine atom has seven electrons in its outer energy level. In the compound carbon tetrachloride, covalent bonds join chlorine atoms to one atom of carbon. How many chlorine atoms are in a molecule of this compound?

Skill Builder

▶ ***Predicting*** Electricity is the flow of electrons. A conductor is a substance that electricity flows through easily. Predict whether ionic compounds or covalent compounds are good conductors of electricity. Research the properties of each type of compound to check the accuracy of your prediction.

ACTIVITY

MAKING A MOLECULAR MODEL

You will need white modeling clay, red modeling clay, and toothpicks.

1. Using the red clay, make four round balls that are the same size.
2. Make one round ball of white clay that is the same size as the red balls.
3. Use the toothpicks to connect each of the four red balls to the white ball. Space the red balls equally around the white ball. You have just made a model of a methane molecule. A molecule of methane contains four carbon atoms joined to one hydrogen atom.

Questions

1. What element is represented by the red balls?
2. What element is represented by the white ball?
3. What type of bond joins the atoms?
4. **Analyze:** Do the atoms in a methane molecule have an electrical charge? Why or why not?

What are organic compounds?

Objective ▶ Identify some organic compounds.

TechTerms

▶ **organic chemistry:** study of organic compounds

▶ **organic compound:** compound containing carbon

▶ **structural formula:** molecular model of an organic compound

Classifying Compounds Scientists place compounds into two different groups. **Organic compounds** contain the element carbon. Scientists once thought that only living things contained organic compounds. Today, scientists know that some nonliving substances, such as plastic, contain organic compounds. The other group of compounds are inorganic compounds. Most inorganic compounds do not contain carbon. The compound carbon dioxide contains carbon. However, it is classified as an inorganic compound. Water is an example of an inorganic compound.

▶ **Compare:** What is the difference between organic and inorganic compounds? Organic compounds contain carbon. Most inorganic compounds do not contain carbon.

Organic Chemistry About 95% of all known substances are organic compounds. The study of organic compounds is called **organic chemistry.** Scientists have learned that a molecule of an organic compound can contain large numbers of atoms.

A carbon atom has four electrons in its outer energy level. As a result, a carbon atom can form covalent bonds with four other atoms. A carbon atom can form three different kinds of covalent bonds. In a single bond, a carbon atom shares one pair of electrons with another atom. In a double bond, two pairs of electrons are shared between atoms. In a triple bond, three pairs of electrons are shared between atoms. These three types of bonds allow many different organic compounds to be formed.

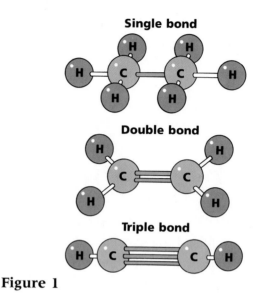

Single bond

Double bond

Triple bond

Figure 1

▶ **Identify:** What is organic chemistry?
study of organic compounds

Structural Formulas When carbon atoms join together, they can form many different structures. The atoms can join in a straight chain or a branched chain, or curve around in a ring. These arrangements of atoms can be shown in a **structural formula.** A structural formula is a molecular model of an organic compound. Figure 2 shows the structural formulas of some organic compounds.

▶ **Identify:** What is a structural formula?
a molecular model of an organic compound

Methane

Ethane

Propane

Butane

Figure 2

LESSON SUMMARY

► Organic compounds contain the element carbon.

► Carbon dioxide is an inorganic compound that contains carbon.

► The study of organic compounds is called organic chemistry.

► Carbon atoms can join other atoms in single, double, or triple covalent bonds.

► A structural formula is a molecular model of an organic compound.

CHECK *Complete the following.*

1. All organic compounds contain the element _____ . carbon

2. Water is an example of an __inorganic__ compound.

3. A carbon atom can form covalent bonds with _____ other atoms. four

4. In a _____ bond, two pairs of electrons are shared between atoms. double

5. The study of organic compounds is called _____ . organic chemistry

6. A _____ shows the arrangement of atoms in a molecule of an organic compound. structural formula

APPLY *Complete the following.*

► 7. **Infer:** An atom of hydrogen contains one electron in its outer energy level. Could a hydrogen atom form a triple bond with another atom? Explain.

8. Are the bonds in a methane molecule single, double, or triple bonds? Explain.

InfoSearch

Read the passage. Ask two questions about the topic that you cannot answer from the information in the passage.

Isomers Certain organic compounds have the same chemical formula but different structural formulas. These compounds are called isomers (Y-suh-murz). Butane is an example of an isomer. All molecules of butane contain four atoms of carbon bonded to 10 atoms of hydrogen. These atoms can join together to form two different structures. Some molecules of butane form a straight-chain structure. Other molecules of butane form a branched-chain structure. As a result, butane is said to have two isomers.

SEARCH: Use library references to find the answers to your questions.

CAREER IN PHYSICAL SCIENCE

PHARMACIST

Do you enjoy studying math and science? Are you able to talk with new people easily? Can you keep accurate, detailed records? If you answered yes to these questions, then you may enjoy a career as a pharmacist.

Most pharmacists work in community pharmacies or drug stores. The pharmacists prepare medications according to the directions of a doctor. Pharmacists also explain to patients how a medication is to be taken. Some pharmacists work for companies that make medicines. They work in laboratories developing new or better drugs. Other pharmacists work in hospitals. Hospital pharmacists work closely with doctors and nurses.

To become a pharmacist, a person needs a college degree in pharmacy. Most states require passing a pharmacy examination, followed by a one-year internship. The student then receives a license to practice pharmacy. For more information, write to the American Council on Pharmaceutical Education, One East Wacker Drive, Chicago, IL 60601.

15-8 What compounds are needed by living things?

Objective ► Identify organic compounds needed by living things.

TechTerms

► **amino acids:** building blocks of proteins
► **carbohydrates** (kahr-buh-HY-drayts): sugars and starches
► **lipids:** fats and oils
► **proteins:** compounds needed to build and repair the body

Needs of Living Things All living things need certain organic compounds to stay alive. An organism gets the organic compounds it needs from the food it eats. Most food is made up of carbohydrates (kahr-buh-HY-drayts), lipids, and proteins.

▶*Identify:* How do organisms obtain the organic compounds they need? from the food they eat

Carbohydrates Organic compounds made up of carbon, hydrogen, and oxygen are called carbohydrates. Sugars and starches are carbohydrates. These organic compounds are the body's main source of energy. Foods such as cereals, grains, vegetables, and fruits are good sources of carbohydrates.

▶*Explain:* Why do all living things need carbohydrates? They are the body's main source of energy.

Lipids Organic compounds made up mostly of carbon and hydrogen are called **lipids.** Fats and oils are lipids. These compounds are another energy source for the body. Lipids can be stored in the body for use at a later time. For this reason, lipids are often called the body's "stored energy supply." Foods such as butter, meat, cheese, and nuts are good sources of lipids.

Cholesterol (kuh-LES-tuh-rohl) is a kind of lipid. Animal fat contains cholesterol. Eating foods high in cholesterol can be harmful to the body. Too much cholesterol may build up in the body. Cholesterol forms fatty deposits on the walls of blood vessels. These fatty deposits can interfere with the flow of blood throughout the body.

▶*Describe:* Why do living things need lipids? Lipids provide an organism with energy and act as a stored energy supply.

Proteins Organic compounds that are used to build and repair the body are called **proteins.** Proteins are made up of substances called **amino acids.** Amino acids contain carbon, hydrogen, oxygen, and nitrogen. Amino acids join together to form long chains of proteins. For this reason, amino acids are called the building blocks of proteins. Milk, fish, eggs, and beans are good sources of protein. Certain foods such as fish and soybeans provide the body with all the amino acids it needs.

▶*Relate:* What is the relationship between amino acids and proteins? Amino acids join in long chains to form proteins.

LESSON SUMMARY

▶ All living things need certain organic compounds to stay alive.

▶ Carbohydrates are an organism's main source of energy.

▶ Fats and oils are lipids that store energy for an organism to use at a later time.

▶ Cholesterol is a harmful form of lipid.

▶ Amino acids are called the building blocks of proteins.

▶ An organism uses proteins to build and repair its body parts.

CHECK *Identify whether each of the following examples describes a carbohydrate, a lipid, or a protein.*

1. stores energy lipid

2. made up of amino acids protein

3. sugars and starches carbohydrate

4. used to build and repair body parts protein

5. fats and oils lipid

6. body's main source of energy carbohydrate

APPLY *Complete the following.*

▶ **7. Infer:** Why do many long-distance runners eat a meal of spaghetti before running a race?

8. Which is a better source of lipids, a cheeseburger or a peanut butter sandwich? Explain.

Skill Builder

Organizing Information A table is one way of organizing information. Make a table with the following headings: "Carbohydrates," "Lipids," and "Proteins." Under each heading, identify five types of foods that you enjoy eating which contain each type of organic compound.

SCIENCE CONNECTION

A BALANCED DIET

In many ways, your body is like a car. Cars need fuel to operate. Your body also needs fuel in the form of food to work properly. Putting the wrong type of fuel into a car can seriously damage the machine. Eating the wrong types of food can seriously injure your body.

One way of keeping your body healthy is by eating a balanced diet. A balanced diet provides the body with different types of foods. Foods are usually divided into four different food groups. These groups are the milk group, the meat group, the fruit and vegetable group, and the grain group. A balanced diet includes daily servings from each of these four food groups. By eating different types of foods, you are making sure your body receives the organic compounds it needs to work properly.

Challenges

STUDY HINT Before you begin the Unit Challenges, review the TechTerms and Lesson Summary for each lesson in this unit.

TechTerms

amino acids (288)
carbohydrates (288)
compound (274)
covalent bond (284)
crystal lattice (282)

ion (282)
ionic bond (282)
lipids (288)
mixture (278)
molecule (276)

organic chemistry (286)
organic compound (286)
proteins (288)
structural formula (286)

TechTerm Challenges

Matching *Write the TechTerm that matches each description.*

1. ions arranged in a regular pattern crystal lattice
2. study of organic compounds organic chemistry
3. compounds containing carbon organic compounds
4. model of an organic compound structural formula
5. fats and oils lipids
6. sugars and starches carbohydrates

Applying Definitions *Explain the difference between the words in each pair. Write your answers in complete sentences.* Answers will vary. Accept all logical responses.

1. compound, mixture
2. ionic bond, covalent bond
3. ion, molecule
4. amino acids, proteins

Content Challenges

Multiple Choice *Write the letter of the term or phrase that best completes each statement.*

1. The compound formed when hydrogen and oxygen combine chemically is
   a. salt. **b.** water. **c.** sugar. **d.** hydrogen dioxide. b

2. Compounds are formed as a result of
   a. evaporation. **b.** filtration. **c.** a physical change. **d.** a chemical change. d

3. A molecule is the smallest part of
   a. an element. **b.** a compound. **c.** an atom. **d.** a mixture. b

4. Molecules are made up of different combinations of
   a. atoms. **b.** elements. **c.** mixtures. **d.** compounds. a

5. A mixture of salt and water can be separated by
   a. freezing. **b.** filtering. **c.** melting. **d.** evaporating. d

6. The compound iron sulfide can be formed by
   a. heating. **b.** filtering. **c.** evaporating. **d.** mixing. a

7. Diatomic molecules contain
   a. one atom. **b.** two or more atoms. **c.** two atoms of the same element.
   d. two atoms of different elements. c

8. Because they have the same number of protons and neutrons, atoms are
   a. positively charged. **b.** negatively charged. **c.** ions. **d.** neutral. d

9. An ionic bond is formed when atoms
   a. share electrons. **b.** gain electrons. **c.** lose electrons. **d.** trade electrons. d

10. Most compounds that have ionic bonds are
   a. solids. b. liquids. c. gases. d. covalent compounds. a
11. In a covalent bond, two atoms
   a. trade electrons. b. share electrons. c. gain electrons. d. lose electrons. b
12. Water is an example of
   a. an ionic compound. b. an element. c. a covalent compound. d. a crystal. c
13. All organic compounds contain
   a. hydrogen b. oxygen. c. nitrogen. d. carbon. d

Completion *Write the term or phrase that best completes each statement.*

1. Hydrogen and oxygen are _____ that are gases at room temperature. elements
2. Compounds are formed as a result of _____ changes. chemical
3. Most gases exist as _____ molecules. diatomic
4. Water is always made up of _____ atoms of hydrogen and one atom of oxygen. two
5. The substances in a mixture have not been _____ combined. chemically
6. You could separate a mixture of sand and water by _____ the mixture. filtering
7. A compound is made up of two or more _____ . elements
8. The substances in a compound _____ their own properties. lose
9. An ionic bond is formed when two atoms _____ electrons. trade
10. Atoms can complete their _____ energy levels by gaining, losing, or sharing electrons. outer
11. The atoms in a covalent bond are electrically _____ . neutral
12. Most inorganic compounds do not contain the element _____ . carbon
13. Most food is made up of organic compounds called carbohydrates, _____ , and proteins. lipids
14. Eating foods with a lot of _____ can be harmful to the body. cholesterol

Understanding the Features

Reading Critically *Use the feature reading selections to answer the following. Page numbers for the features are shown in parentheses.*

1. What is the pattern formed by a mass spectrometer called? (277)
2. **Predict:** What might happen if you added too much or too little of an ingredient to a recipe? (281)
3. What are three ways you could learn to cook as a hobby? (279)
4. What are the four different food groups called? (289)
5. Where do most pharmacists work? (287)
6. **Analyze:** How did the discovery of X rays help scientists study crystals? (283)

Concept Challenges

Critical Thinking *Answer each of the following in complete sentences.*

1. Describe a method of separating each of the following mixtures:
   a. sand and sugar b. sugar and water
   c. sawdust and iron filings
   d. nickels and dimes.
2. **Compare:** How are ionic and covalent compounds the same? How are they different?
3. How are the properties of table salt different from the properties of the elements that make it up?
4. **Predict:** What might happen if you did not get enough carbohydrates in your diet? Explain your answer.
5. **Analyze:** Why can carbon form so many different kinds of organic compounds?

Interpreting a Diagram *Use the diagrams of atomic bonds to answer the following.*

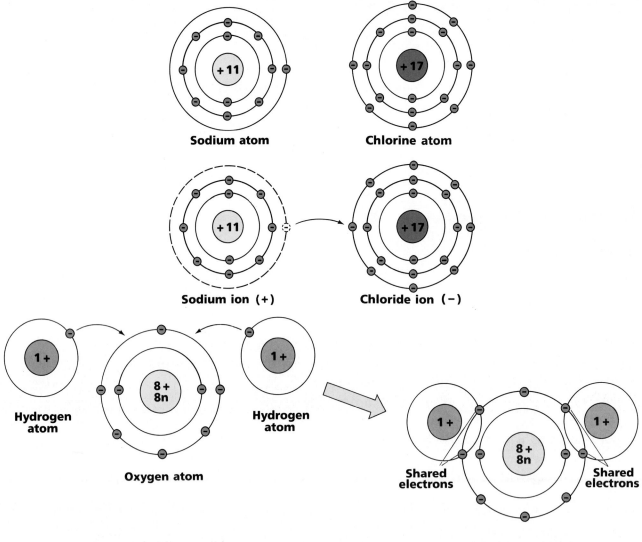

Sodium atom

Chlorine atom

Sodium ion (+)

Chloride ion (−)

Hydrogen atom

Oxygen atom

Hydrogen atom

Shared electrons

Shared electrons

Water molecule

1. Which diagram shows an ionic bond? Which shows a covalent bond?
2. How many electrons are there in the outer energy level of a sodium atom? Of a chlorine atom?
3. Which element forms a positive ion? Which forms a negative ion?
4. How many electrons are there in the outer energy level of a hydrogen atom? Of an oxygen atom?
5. How many electrons does each hydrogen atom share with an oxygen atom to form a water molecule?

Finding Out More . . . .

1. A device called the Hoffman apparatus can be used to separate water into the elements hydrogen and oxygen. Use library references to find out how the Hoffman apparatus works. Draw a diagram of the Hoffman apparatus, and describe its operation to the class.

2. Carbon dioxide is an inorganic compound that contains carbon. You can make carbon dioxide gas by putting some baking soda in a cup and adding some vinegar. Try this experiment at home. Observe the bubbles of carbon dioxide gas that form.

3. **Classify:** Make a list of at least 10 items in your home that are elements, 10 that are mixtures, and 10 that are compounds.

UNIT **16**

CHEMICAL FORMULAS

CONTENTS

STUDY HINT As you read each lesson in Unit 16, write the lesson title and lesson objective on a sheet of paper. After you complete each lesson, write the sentence or sentences that answer each objective.

16-1 What is a chemical formula?

Objective ▶ Write chemical formulas for some simple compounds.

TechTerms

- ▶ **chemical formula:** way of writing the name of a compound using chemical symbols
- ▶ **subscript:** number written to the lower right of a chemical symbol

Chemical Symbols Scientists use symbols to represent elements. Each element has its own symbol. Most symbols are made up of one letter. H is the symbol for hydrogen. O is the symbol for oxygen. Some symbols are made up of two letters. Fe is the symbol for iron.

▶**Identify:** What is the symbol for iron? Fe

Chemical Formulas Compounds consist of two or more atoms joined together. Scientists use a **chemical formula** to represent one molecule of a compound. A chemical formula is a way of writing the name of a compound using chemical symbols. The compound water contains the elements hydrogen and oxygen. Each molecule of water contains two atoms of hydrogen and one atom of oxygen. The chemical formula for water is H_2O. The chemical formula includes the symbols for each element in the compound. Table 1 shows the chemical formulas of some compounds.

▶**Describe:** What is a chemical formula?
way of writing the name of a compound using chemical symbols

Subscripts Chemical formulas also indicate how many atoms of each type of element are in a compound. Numbers of atoms are shown by a **subscript.** A subscript is a number written to the lower right of a chemical symbol. The 2 in the chemical formula H_2O is a subscript. It indicates that there are two atoms of hydrogen in a molecule of water. There is also one atom of oxygen in a molecule of water. No subscript is written after the O in H_2O. This is because the subscript 1 is never written in a chemical formula. When no subscript appears after a symbol in a formula, you know that there is one atom of that element.

▶**Infer:** How many atoms of hydrogen are there in a molecule of sugar if the chemical formula for sugar is $C_{12}H_{22}O_{11}$? 22

Writing Chemical Formulas In many compounds, a metallic element is combined with a nonmetallic element. Sodium chloride is made up of the metal sodium and the nonmetal chlorine. In a chemical formula, the symbol for the metallic element is always written first. Table 2 shows some compounds and the metallic and nonmetallic elements that make them up.

▶**Calculate:** How many atoms are there in a molecule of aluminum chloride? 4

Table 1 Chemical Formulas of Compounds		
COMPOUND	FORMULA	ELEMENTS
Sodium chloride	NaCl	sodium chlorine
Hydrochloric acid	HCl	hydrogen chlorine
Sodium hydroxide	NaOH	sodium oxygen hydrogen
Calcium oxide	CaO	calcium oxygen

Table 2 Metallic and Nonmetallic Elements			
COMPOUND AND FORMULA		METALLIC ELEMENT	NONMETALLIC ELEMENT
Sodium chloride	NaCl	Na (sodium)	Cl (chlorine)
Aluminum chloride	$AlCl_3$	Al (aluminum)	Cl (chlorine)
Silver sulfide	Ag_2S	Ag (silver)	S (sulfur)
Aluminum oxide	Al_2O_3	Al (aluminum)	O (oxygen)

LESSON SUMMARY

▶ Every element has its own chemical symbol.

▶ A chemical formula identifies the elements in a compound.

▶ Subscripts show how many atoms of an element are found in a molecule of a compound.

▶ The symbol for a metallic element is always written first in a chemical formula.

CHECK *Complete the following.*

1. The chemical symbol for hydrogen is ___H___ .

2. The chemical formula for water is ___H_2O___ .

3. A _____ shows how many atoms of an element are contained in one molecule of a compound. subscript

4. The nonmetallic element in sodium chloride is _____ . chlorine

5. In a chemical formula, the symbol for a _____ element is always written first. metallic

APPLY *Complete the following.*

6. **Analyze:** Identify the elements and the number of each kind of atom in one molecule of the following compounds: ammonia (NH_3), barium chloride ($BaCl_2$).

7. **Contrast:** The chemical formula for water is H_2O. The chemical formula for hydrogen peroxide is H_2O_2. How does a molecule of water differ from a molecule of hydrogen peroxide?

8. What two things does a chemical formula tell you about a compound?

Skill Builder

Interpreting Chemical Formulas On a sheet of paper, list the names of the elements that make up each of the following compounds.

a. carbon dioxide (CO_2)

b. ammonia (NH_3)

c. calcite ($CaCO_3$)

d. potassium hydroxide (KOH)

e. baking soda ($NaHCO_3$)

SCIENCE CONNECTION

DESTRUCTION OF THE OZONE LAYER

Ozone is a molecule made of three atoms of oxygen. Ozone forms in the atmosphere when ultraviolent rays from the sun strike oxygen molecules in the air. A layer of ozone in the atmosphere absorbs about 99% of the sun's ultraviolet radiation.

Scientists have found that certain chemicals called CFCs destroy ozone molecules. CFCs are found in gases that are used in spray cans. CFCs also are used in some factories, air conditioners, refrigerators, and aircraft engines. The use of these chemicals releases CFCs into the atmosphere. When the CFCs reach the ozone layer, ozone molecules are destroyed. As a result, more harmful ultraviolet radiation reaches the earth's surface. Exposure to increased levels of ultraviolet radiation can cause skin cancer in humans.

In the 1970s, the United States government banned the use of CFCs in spray cans. However, these chemicals are still used in some other products.

NIMBUS-7 : TOMS OZONE

DAY:237   AUG 25

16-2 What is an oxidation number?

Objective ▶ Describe how to use oxidation numbers to write the chemical formula of a compound.

TechTerms

- ▶ **oxidation number:** number of electrons an atom gains, loses, or shares when it forms a chemical bond
- ▶ **valence electrons:** electrons in an atom's outer energy level

Valence Electrons Electrons move around the nucleus of an atom in energy levels. When the outer energy level of an atom holds eight electrons, it is complete. Atoms tend to complete their outer energy levels when they bond with other atoms. The electrons in an atom's outer energy level are called **valence electrons.**

▶ *Identify:* How many electrons are in a complete outer energy level? eight

Oxidation Number An **oxidation number** shows how many electrons an atoms gains, loses, or shares when it forms a chemical bond. A chlorine atom has seven valence electrons. In order to complete its outer energy level, the atom must gain one electron from another atom. An atom that gains electrons has a negative oxidation number. When it gains an electron, the chlorine atom has a negative charge. Therefore, the oxidation number of chlorine is 1−.

An atom that loses electrons has a positive oxidation number. The outer energy level of a sodium atom contains one electron. If sodium loses this electron, then the next energy level becomes its outer level. This level already has eight electrons, so it is complete. A sodium atom has one valence electron. When it loses this electron, sodium has a positive charge. Therefore, the oxidation number of sodium is 1+.

▶ *Infer:* What is the oxidation number of an atom with six valence electrons? 2−

Writing Chemical Formulas The oxidation number of an element tells you if it loses (+) or gains (−) electrons. Knowing the oxidation numbers of elements can help you predict the formula for a compound. To write the formula for a compound, remember this rule: The oxidation numbers of the elements in a compound must add up to zero. For example, sodium chloride is a compound made up of sodium and chlorine. The oxidation number of sodium is 1+. The oxidation number of chlorine is 1−. The oxidation numbers of the elements cancel one another: (1+) + (1−) = 0. In the compound sodium chloride, one atom of sodium combines with one atom of chlorine. The chemical formula for sodium chloride is NaCl.

▶ *Predict:* The oxidation number of hydrogen is 1+ and of sulfur is 2−. What is the formula for hydrogen sulfide? H_2S

Sodium atom
1 valence electron

Chlorine atom
7 valence electrons

LESSON SUMMARY

▶ The number of electrons in an atom's outer energy level determines how the atom will combine with other atoms.

▶ Valence electrons are the number of electrons in an atom's outer energy level.

▶ Atoms that lose electrons have positive oxidation numbers.

▶ Atoms that gain electrons have negative oxidation numbers.

▶ Elements with positive oxidation numbers are written first in a chemical formula.

CHECK *Complete the following.*

1. What is the number of electrons an atom gains, loses, or shares when it forms a chemical bond called?

2. What determines how an atom will combine with other atoms?

3. Is the oxidation number of an atom that gains electrons positive or negative?

4. How many valence electrons does an atom of chlorine have?

5. What is the oxidation number of chlorine?

6. What is true about the oxidation numbers of the elements in a compound?

APPLY *Complete the following.*

7. **Infer:** The chemical formula for aluminum chloride is $AlCl_3$. Which atom in a molecule of aluminum chloride has a positive oxidation number? How do you know?

8. **Analyze:** What is the oxidation number of aluminum in $AlCl_3$? What is the oxidation number of chlorine? Explain how you arrived at your answers.

Ideas in Action

IDEA: Many household products are compounds. Each compound has its own chemical formula.

ACTION: Make a list of five compounds found in products used in your home. Write the chemical formula of each compound and identify the elements it contains.

ACTIVITY

WRITING CHEMICAL FORMULAS

You will need a pencil and a sheet of paper.

1. Copy the table on a separate sheet of paper.
2. Use the oxidation numbers to find the formulas of the compounds that will be formed when each metal combines with each nonmetal. The first one has been done for you.

Questions

1. Which type of element, a metal or a nonmetal, did you write first in your chemical formulas?
2. Why is it possible to use oxidation numbers to write chemical formulas for compounds?

NONMETALS					
		Cl^{1-}	Br^{1-}	I^{1-}	O^{2-}	S^{2-}
	H^{1+}	HCl				
M	Na^{1+}					
E	Au^{1+}					
T						
A	K^{1+}					
L	Fe^{3+}					
S	Al^{3+}					

16-3 How are chemical compounds named?

Objective ► Explain how chemical compounds are named.

TechTerm

► **binary** (BY-nur-ee) **compound:** compound containing two elements

Binary Compounds A **binary** (BY-nur-ee) **compound** contains two different elements bonded together. The name of a binary compound tells which elements are found in a molecule of the compound. Sodium chloride is an example of a binary compound. It is formed from the elements sodium and chlorine. The name of the element with a positive oxidation number is written first. The name of the second element is changed to end in -ide and is written last. In sodium chloride, chlorine is changed to chloride.

► **Predict:** What elements would you expect to find in the compound hydrogen chloride?
hydrogen and chlorine

Different Oxidation Numbers Some elements have more than one oxidation number. For example, the oxidation number of iron can be either 2+ or 3+. Scientists indicate different oxidation numbers of an element by including a Roman number in parentheses in the name of the element. Iron (II) shows that an atom of iron has an oxidation number of 2+. Iron (III) shows that an atom of iron has an oxidation number of 3+.

Iron and chlorine can combine to form the compound $FeCl_2$. Iron and chlorine can also combine to form the compound $FeCl_3$. $FeCl_2$ and $FeCl_3$ are two different compounds. In one compound, the oxidation number of iron is 2+. In the other compound, the oxidation number of iron is 3+. Another way of identifying these compounds is iron (II) chloride and iron (III) chloride.

► **Infer:** What is the oxidation number of an atom of gold (III)? 3+

Other Elements Most elements have more than one oxidation number. Tin can have an oxidation number of 4+ or 2+. Gold can have an oxidation number of 3+ or 1+. Copper can have an oxidation number of 2+ or 1+. The Roman number in the name of the compound shows the oxidation number of the element in the compound.

► **Identify:** What is the oxidation number of copper in the compound copper (II) chloride? 2+

Table 1	Oxidation Numbers of Some Metals	
METAL	HIGHER OXIDATION NUMBER	LOWER OXIDATION NUMBER
Iron	3+, iron (III)	2+, iron (II)
Mercury	2+, mercury (II)	1+, mercury (I)
Copper	2+, copper (II)	1+, copper (I)
Tin	4+, tin (IV)	2+, tin (II)
Nickel	3+, nickel (III)	2+, nickel (II)
Gold	3+, gold (III)	1+, gold (I)

iron (II) chloride

iron (III) chloride

LESSON SUMMARY

► A binary compound contains atoms of two different elements.

► An element can have more than one oxidation number.

► A Roman number in the name of a compound shows the oxidation number of an element in the compound.

CHECK *Complete the following.*

1. The oxidation number of iron (III) is _____ . 3+

2. Iron (II) chloride and iron (III) chloride are different _____ . compounds

3. In the compound hydrogen chloride (HCl) _____ has a positive oxidation number. hydrogen

4. Since the compound water contains the elements hydrogen and oxygen, it is a _____ compound. binary

5. The name mercury (II) bromide indicates that the oxidation number of _____ is 2+. mercury

APPLY *Complete the following.*

► 6. **Infer:** The chemical formula for oxygen gas is O_2. Is oxygen gas a binary compound?

7. Mercury can have two oxidation numbers, 1+ or 2+. Are mercury (I) chloride and mercury (II) chloride the same compound? Explain.

InfoSearch

Read the passage. Ask two questions about the topic that you cannot answer from the information in the passage.

Naming Compounds Some compounds contain two nonmetals. In naming compounds made up of two nonmetals, prefixes are used. For example, the compound carbon monoxide contains one atom of carbon and one atom of oxygen. The compound carbon dioxide contains one atom of carbon and two atoms of oxygen. The prefix "mono-" means "one," while the prefix "di-" means "two."

SEARCH: Use library reference to find answers to your questions.

CAREER IN PHYSICAL SCIENCE

CHEMICAL TECHNICIAN

Do you enjoy working in a science laboratory? Can you keep detailed records? Do you like putting complicated things together? If so, you may enjoy a career as a chemical technician. Chemical technicians help develop, test, and manufacture chemical products.

Chemical technicians work in many different settings. Some chemical technicians perform tests to check the quality of chemical products. Others assist in the design and production of food or drug products. Still others aid in the research and development of new products in the aerospace and automobile industries.

To become a chemical technician, you should attend a two-year college or technical school. You should take courses in mathematics, chemistry, and other sciences. If you would like more information, write to the American Chemical Society, Education Department, 1155 16th Street NW, Washington, DC 20036.

What is a polyatomic ion?

Objective ▶ Recognize the chemical formula for a polyatomic ion.

TechTerm

▶ **polyatomic** (PAHL-i-uh-tahm-ik) **ion:** group of atoms that acts as a single atom when combining with other atoms

Polyatomic Ions Sometimes, a group of atoms stays together when combining with other atoms. The group of atoms acts as one single atom. A group of atoms that acts as a single atom when combining with other atoms is called a **polyatomic** (PAHL-i-uh-tahm-ik) **ion.** The prefix "poly-" means "more than one." The group of atoms is called an ion because it has an electrical charge. The atoms in a polyatomic ion are joined by covalent bonds.

▶*Define:* What is a polyatomic ion? group of atoms that acts as a single atom when combining with other atoms

Hydroxide Ions An example of a polyatomic ion is the hydroxide ion. A hydroxide ion is made up of one oxygen atom and one hydrogen atom. The oxygen atom and the hydrogen atom are joined by a covalent bond. A hydroxide ion has a negative electrical charge. The chemical formula for a hydroxide ion is OH^-.

▶*Identify:* What is the formula for a hydroxide ion? OH^-

Ammonium Ions The ammonium ion is made up of one nitrogen atom and four hydrogen atoms. The ammonium ion has a positive electrical charge. The chemical formula for the ammonium ion is NH_4^+. The ammonium ion is the only common polyatomic ion with a positive charge. Table 1 shows some other polyatomic ions and their chemical formulas.

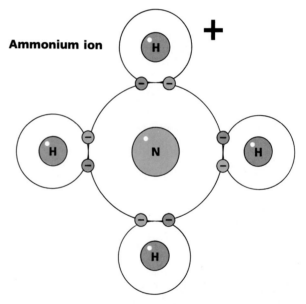

Ammonium ion

Hydroxide ion

Table 1 Common Polyatomic Ions	
POLYATOMIC ION	CHEMICAL FORMULA
Sulfate ion	SO_4^{-2}
Carbonate ion	CO_3^{-2}
Phosphate ion	PO_3^{-3}
Nitrate ion	NO_3^{-1}
Chlorate ion	ClO_3^{-1}

▬*Analyze:* What is the formula for a nitrate ion? NO_3^{-1}

LESSON SUMMARY

▶ A group of atoms that acts as a single atom when combining with other atoms is called a polyatomic ion.

▶ The atoms in a polyatomic ion are joined by covalent bonds.

▶ A hydroxide ion has a negative charge.

▶ An ammonium ion has a positive charge.

CHECK *Write true if the statement is true. If the statement is false, change the underlined term to make the statement true.*

1. A polyatomic ion contains <u>only</u> one atom. *more than*

2. The polyatomic ion made up of oxygen and hydrogen atoms is the <u>ammonium</u> ion. *hydroxide*

3. The atoms in a polyatomic ion are joined by <u>an ionic</u> bond. *a covalent*

4. A polyatomic ion <u>always</u> has an electrical charge. *true*

5. An ammonium ion contains one hydrogen atom and <u>four</u> nitrogen atoms. *true*

APPLY *Complete the following.*

6. **Classify:** Compounds that contain a hydroxide ion (OH^-) combined with a metal are called bases. Look at the following chemical formulas. Which are bases? Which are not? Explain your answers.
   a. limewater, $Ca(OH)_2$
   b. milk of magnesia, $Mg(OH)_2$
   c. table salt, NaCl
   d. water, H_2O
   e. lye, NaOH

7. **Infer:** When polyatomic ions combine with other elements to form compounds, do they form covalent bonds or ionic bonds? Explain.

Skill Builder ....

Building Vocabulary The prefix "poly-" means "more than one." Use a dictionary to find the definitions of the words "polygon," "polymer," and "polysyllabic." Write the definitions on a sheet of paper.

SCIENCE CONNECTION ◆○◆○◆○◆○◆○◆○◆○◆○◆○◆○◆○◆○◆○◆○◆

NITRATES

A nitrate ion is a polyatomic ion made up of one atom of nitrogen and three atoms of oxygen. Different compounds containing this ion are often used in the food industry. Sodium nitrate is added to fresh meat. The compound prevents the growth of harmful organisms in the meat. Other nitrates are used to give meat a particular flavor and color. The flavor of cured meats such as ham and bacon is due in part to the addition of nitrates.

The safety of adding these compounds to meat has been questioned by some nutrition experts. Nitrates may combine with other chemical compounds in the body to form a compound called nitrosamine (ny-TROH-suh-meen). Experiments have shown that nitrosamine causes cancer in animals.

More research is needed on the effect of nitrates on the body. However, you should limit the amount of nitrate-containing food you eat. In addition to ham and bacon, other foods that contain nitrates include hot dogs and bologna.

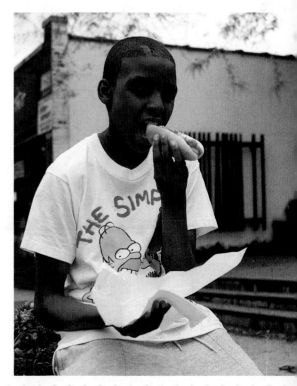

16-5 What are diatomic molecules?

Objective ▶ Identify elements that are made up of diatomic molecules.

TechTerm

▶ **diatomic molecule:** molecule made up of two atoms of the same element

Hydrogen Molecules An atom of hydrogen has one electron in its outer energy level. A hydrogen atom needs another electron in order to complete its outer energy level. It can receive that extra electron from another hydrogen atom. By sharing two electrons, the two hydrogen atoms complete their outer energy levels. The two hydrogen atoms combine to form a hydrogen molecule, H_2. Hydrogen gas is always made up of pairs of hydrogen atoms.

Hydrogen atom   **Hydrogen atom**

Diatomic Molecule of Hydrogen   **Shared electrons**

▶ **Infer:** What kind of bond is formed between the two atoms in a hydrogen molecule?
covalent

Diatomic Molecules The molecule formed by the covalent bonding of two hydrogen atoms is called a **diatomic molecule.** A diatomic molecule is made up of two atoms of the same element.

Most gaseous elements form diatomic molecules. Atoms in a diatomic molecule are always joined by a covalent bond.

▷ **Define:** What is a diatomic molecule?
a molecule made up of two atoms of the same element

Oxygen Molecules Oxygen (O_2) is another element that occurs naturally as diatomic molecules. Most oxygen gas in the air is formed from two oxygen atoms joined by a covalent bond. However, some oxygen molecules contain three atoms of oxygen. The molecule that contains three atoms of oxygen is called ozone (O_3). A layer of ozone is found in the earth's atmosphere from about 10 km to about 45 km above the earth. Table 1 shows some other diatomic molecules and their chemical formulas.

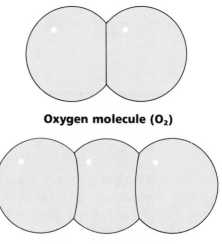

Oxygen molecule (O_2)

Ozone molecule (O_3)

Table 1 Some Diatomic Molecules	
MOLECULE	CHEMICAL FORMULA
Iodine	I_2
Nitrogen	N_2
Chlorine	Cl_2
Fluorine	F_2
Bromine	Br_2

▷ **Contrast:** How do oxygen gas and ozone differ? *Oxygen gas contains two atoms of oxygen. Ozone contains three atoms of oxygen.*

LESSON SUMMARY

▶ Hydrogen is always found as two atoms joined by a covalent bond.

▶ A diatomic molecule contains two atoms of the same element.

▶ Atoms in a diatomic molecule are always joined by a covalent bond.

▶ A diatomic molecule of oxygen contains two atoms of oxygen joined by a covalent bond.

CHECK *Complete the following.*

1. A hydrogen atom needs _____ electrons to complete its outer energy level. two

2. A molecule that contains two atoms of the same element is called a _____ molecule. diatomic

3. Most _____ elements form diatomic molecules. gaseous

4. A molecule of _____ contains three atoms of oxygen. ozone

5. The atoms in a diatomic molecule are joined by a _____ bond. covalent

APPLY *Complete the following.*

6. **Classify:** Which of the following molecules are diatomic molecules?
   a. Br_2
   b. H_2O
   c. CO_2
   d. Cl_2

7. The elements fluorine (Fl), chlorine (Cl), bromine (Br), iodine (I), and astatine (At) are called halogens (HAL-uh-junz). As gases, the halogens exist as diatomic molecules. Write the formula for one molecule of each of these elements.

Skill Builder

Building Vocabulary Use a dictionary to find the meaning of the prefix "di-." Look up the definitions of the terms "diagonal," "dicotyledon," "dibromide," and "dichloride." Write the definitions on a sheet of paper.

SCIENCE CONNECTION

THE NITROGEN CYCLE

Almost all living things need nitrogen to survive. Organisms use nitrogen to build proteins. The proteins are then used to build and repair body parts.

About 79% of the earth's atmosphere is made up of nitrogen. However, most organisms cannot use nitrogen directly from the air. Nitrogen must first be changed into a form the organisms can use. This is done by the nitrogen cycle.

The nitrogen cycle is a path by which nitrogen is removed from the air, used by organisms, and returned to the air. Some kinds of bacteria can take in nitrogen directly from the air. These bacteria then combine the nitrogen with oxygen to make nitrates. Plants remove nitrates from the soil. Animals eat the plants and obtain the nitrates. When plants or animals die, other types of bacteria break down the nitrates in their bodies. Nitrogen gas is released into the air. The nitrogen cycle passes nitrogen back and forth between living things and the environment.

Green plants

Nitrogen gas

Animals

Bacteria

Nitrogen compounds

16-6 What is formula mass?

Objective ▶ Explain how to find the formula mass of a molecule.

TechTerm

▶ **formula mass:** sum of the mass numbers of all the atoms in a molecule

Formula Mass A molecule contains atoms joined together. Each type of atom in a molecule has its own mass number. Remember that the mass number of an atom is the number of protons and neutrons in the nucleus. It is equal to the atomic mass rounded off to the nearest whole number. The sum of the mass numbers of all the atoms in a molecule is called the **formula mass** of the molecule. Formula mass is also sometimes called molecular mass.

Mass number of carbon = 12

12 hydrogen atoms

Mass number of hydrogen = 1

▶ **Define:** What is formula mass? sum of the mass numbers of all the atoms in a molecule

Finding Formula Mass A molecule of mercury contains one atom of mercury. The mass number of mercury is 201. Because there is only one atom in a molecule of mercury, the formula mass of mercury is also 201.

Most molecules contain more than one atom. To find the formula mass of molecules that contain more than one atom, follow these steps:

▶ Write the chemical formula of the compound.

▶ Use the Periodic Table to find the atomic mass of each element in the compound. Round off the atomic mass to find the mass number.

▶ Multiply the mass number of each element by its subscript. The subscript shows how many atoms of each element are present in the molecule. If there is no subscript, multiply the mass number by 1.

▶ Add the total masses of all the elements in the compound. The total is the formula mass of one molecule of the compound.

▶ **Explain:** Why is the mass number of each element in a molecule multiplied by the subscript of the element to find the formula mass? The subscript shows the number of atoms of each element in the molecule.

Formula Mass of Ethyl Chloride You can find the formula mass of the compound ethyl chloride by following the steps listed. First, write the chemical formula for ethyl chloride. The formula for ethyl chloride is C_2H_5Cl. This represents one molecule of ethyl chloride. Next, find the mass number for each element in the compound. The mass number of C = 12, H = 1, and Cl = 35. Then multiply the mass number of each element by its subscript: $12 \times 2 = 24$; $1 \times 5 = 5$; $35 \times 1 = 35$. Finally, add the total masses of all the elements in the compound: $24 + 5 + 35 = 64$. The formula mass of ethyl chloride is 64.

Ethyl chloride (C_2H_5Cl)

▶ **Calculate:** If the mass number of hydrogen is 1 and the mass number of oxygen is 16, what is the formula mass of water (H_2O)? 18

LESSON SUMMARY

▸ Each atom in a molecule has its own mass number.

▸ The sum of the mass numbers of all the atoms in a molecule is equal to the formula mass of the molecule.

▸ In a molecule containing only one atom, the formula mass is the same as the mass number.

CHECK *Write true if the statement is true. If the statement is false, change the underlined term to make the statement true.*

1. Each type of atom in a molecule has its own <u>mass number</u>. true

2. The mass number of an atom of mercury is the same as the <u>formula mass</u> of a molecule of mercury. true

3. In the chemical formula of a compound, a <u>superscript</u> shows how many atoms of each element are present in one molecule of the compound. subscript

4. The formula mass is the sum of the <u>atomic</u> numbers of all the atoms in a molecule. mass

5. Formula mass is also sometimes called <u>atomic</u> mass. molecular

APPLY *Complete the following.*

▸ 6. **Infer:** The mass number of a certain element is the same as the formula mass of the element. What does this tell you about a molecule of the element?

7. Could the formula mass of an element ever be less than the mass number of the element?

8. **Calculate:** What is the formula mass of a molecule of sugar, $C_{12}H_{22}O_{11}$?

Skill Builder ....

❯ *Predicting* The chemical formula of the compound carbon monoxide is CO. The chemical formula of the compound carbon dioxide is CO_2. Predict how the formula mass of a molecule of carbon dioxide differs from the formula mass of a molecule of carbon monoxide. Check the accuracy of your prediction by calculating the formula mass of each molecule.

ACTIVITY

CALCULATING FORMULA MASS

You will need a pencil and a sheet of paper.

1. Review the steps to follow when calculating formula mass.

2. Copy the following table on a sheet of paper. Use the periodic table on pages 234–235 to complete the table.

Questions

1. What is the difference between the atomic mass and the mass number of an element? Which is used to calculate formula mass?

2. What is the atomic mass of sodium? What is its mass number?

COMPOUND	CHEMICAL FORMULA	FORMULA MASS
Hydrogen peroxide	H_2O_2	
Chloroform	$CHCl_3$	
Nitrogen dioxide	NO_2	
Sodium carbonate	Na_2CO_3	
Calcium oxide	CaO	
Magnesium chloride	$MgCl_2$	
Calcium carbonate	$CaCO_3$	

STUDY HINT Before you begin the Unit Challenges, review the TechTerms and Lesson Summary for each lesson in this unit.

TechTerms . . . .

binary compound (298)
chemical formula (294)
diatomic molecule (302)

formula mass (304)
oxidation number (296)
polyatomic ion (300)

subscript (294)
valence electrons (296)

TechTerm Challenges . . . .

Matching *Write the term that matches each description.*

1. number written to the lower right of a chemical symbol subscript
2. electrons in an atom's outer energy level valence electrons
3. compound containing two elements binary compound
4. molecule made up of two atoms of the same element diatomic molecule
5. sum of the mass numbers of the atoms in a molecule formula mass

Fill In *Write the TechTerm that best completes each statement.*

1. An example of a _____ is NaCl. chemical formula
2. An atom that loses one electron to another atom has an _____ of 1+. oxidation number
3. The ammonium ion is an example of a _____ . polyatomic ion
4. Another term for _____ is molecular mass. formula mass
5. A molecule of hydrogen gas is an example of a _____ . diatomic molecule
6. The 2 in the formula H_2O is a _____ . subscript

Content Challenges . . . .

Multiple Choice *Write the letter of the term that best completes each statement.*

1. Scientists represent elements by using
   a. formulas. **b.** subscripts. **c.** symbols. **d.** words. c

2. The compound NaOH contains the elements
   a. sodium and hydrogen. **b.** sodium and oxygen. **c.** oxygen and hydrogen.
   d. sodium, oxygen, and hydrogen. d

3. The oxidation number of chlorine is
   a. 1−. **b.** 1+. **c.** 7−. **d.** 7+. a

4. The oxidation numbers of the elements in a compound must add up to
   a. 8. **b.** 0. **c.** 1. **d.** 7. b

5. The oxidation number of iron in iron (II) chloride is
   a. 2+. **b.** 2−. **c.** 3+. **d.** 0. a

6. An example of a binary compound is
   a. NaOH. **b.** O_2. **c.** NaCl. **d.** Na. c

7. A polyatomic ion always has
   a. no electrical charge. **b.** a positive charge. **c.** a negative charge. **d.** an electrical charge. d

8. The only common polyatomic ion with a positive charge is the
   a. nitrate ion. **b.** ammonium ion. **c.** sulfate ion. **d.** phosphate ion. b

9. Of the following molecules, the one that is <u>not</u> a diatomic molecule is
   a. ozone. **b.** oxygen. **c.** hydrogen. **d.** iodine. _a

10. Most diatomic molecules are
    a. solids. **b.** liquids. **c.** gases. **d.** crystals. _c

11. The formula mass of a molecule is found by adding all the
    a. molecular masses. **b.** atomic masses. **c.** mass numbers. **d.** atomic numbers. _c

True/False *Write true if the statement is true. If the statement is false, change the underlined term to make the statement true.*

1. Formula mass is also called <u>atomic</u> mass. _{molecular}
2. The mass number of an atom is equal to the <u>atomic mass</u> rounded off to the nearest whole number. _{true}
3. <u>All</u> symbols for elements are made up of one letter. _{Most} _{elements}
4. The chemical formula of a compound tells you what <u>molecules</u> the compound contains.
5. In a chemical formula, the element with a positive oxidation number is written <u>last</u>. _{first}
6. Chlorine has seven valence electrons, so it must <u>lose</u> one electron to complete its outer energy level. _{gain}
7. Sodium has a <u>positive</u> charge after it loses its valence electron. _{true}
8. Two hydrogen atoms form a hydrogen molecule by <u>gaining</u> electrons. _{sharing}
9. An ammonium ion has a <u>positive</u> electrical charge. _{true}
10. The oxidation number of copper in copper (II) chloride is <u>2−</u>. ₂₊
11. H_2, O_2, and Cl_2 are all examples of <u>polyatomic ions</u>. _{diatomic molecules}
12. The chemical formula for the <u>hydroxide</u> ion is OH^-. _{true}

Understanding the Features ....

Reading Critically *Use the feature reading selections to answer the following. Page numbers for the features are in parentheses.*

1. Why is ultraviolet radiation from the sun dangerous to humans? (295)
2. What is the nitrogen cycle? (303)
3. What are nitrates used for? (301)
4. What do chemical technicians do? (299)

Concept Challenges ....

Critical Thinking *Answer each of the following in complete sentences.*

1. **Hypothesize:** Why do you think chemical symbols and formulas are an important part of communication among scientists?
2. **Calculate:** Find the formula mass for each of the following compounds. **a.** NaOH **b.** HCl **c.** $AgCl_3$ **d.** Al_2O_3 **e.** $NaHCO_3$.
3. What is the relationship between an atom's oxidation number and the number of valence electrons the atom has?
4. **Hypothesize:** Why do you think carbon can have an oxidation number of 4+ or 4−? Hint: Carbon has four valence electrons.

Interpreting a Table *Use the table showing the metallic and nonmetallic elements in certain compounds to answer the following.*

COMPOUND	FORMULA	METALLIC ELEMENT	NONMETALLIC ELEMENT
Sodium chloride	NaCl	Na	Cl
Aluminum chloride	$AlCl_3$	Al	Cl
Silver sulfide	Ag_2S	Ag	S
Aluminum oxide	Al_2O_3	Al	O

1. What are the metallic elements listed in this table? What are the nonmetallic elements?
2. What kind of element is written first in the chemical formulas of the compounds?
3. What is the oxidation number of aluminum in aluminum chloride? In aluminum oxide?
4. What is the oxidation number of chlorine in sodium chloride? In aluminum chloride?

Finding Out More ....

1. Find out how to use the "crisscross" method of writing chemical formulas. Demonstrate this method of writing chemical formulas to the class.
2. What is the chemical formula for each of the following compounds: sodium nitrate; calcium carbonate; copper (II) sulfate; potassium chlorate? Notice that the names of all these compounds end in "-ate." What element do all of these compounds have in common?

▲ 3. **Model:** Choose four compounds discussed in this unit. Use modeling clay and toothpicks to make a model of one molecule of each compound. Display your models to the class.
4. The number 4 in the chemical formula SO_4^{-2} is a subscript. The number −2 is called a superscript. Look up each of these words in a dictionary. What does the prefix "sub-" mean? What does the prefix "super-" mean?

CHEMICAL REACTIONS

CONTENTS

STUDY HINT Before beginning Unit 17, scan through the lessons in the unit looking for words that you do not know. On a sheet of paper, list these words. Work with a classmate to try to define each word on your list.

17-1 What is conservation of matter?

Objective ▶ Describe what happens in a chemical reaction.

TechTerms

- ▶ **chemical reaction:** process in which new substances with new chemical and physical properties are formed
- ▶ **product:** substance that is formed in a chemical reaction
- ▶ **reactant:** substance that is changed in a chemical reaction

Chemical Reactions New substances are formed as a result of a chemical change. The process by which a chemical change takes place is called a **chemical reaction.** In a chemical reaction, new substances with new chemical and physical properties are formed.

Chemical reactions are taking place all around you. You may notice that rust has formed on an iron fence or on your bicycle. A series of chemical reactions starts a car's engine and keeps it running. If you could look inside your body, you would see thousands of chemical reactions taking place as your body digests the food you eat.

▸▸▶ **Describe:** What always happens in a chemical reaction?
    New substances with new physical and chemical properties are formed.

Reactants and Products In any chemical reaction, certain substances are present at the start of the reaction and different substances are present at the end of the reaction. A substance that is present at the start of a chemical reaction is called a **reactant.** A reactant is a substance that is changed in a chemical reaction. A substance that is present at the end of a chemical reaction is called a **product.** A product is a substance that is formed in a chemical reaction.

▸▸▶ **Contrast:** What is the difference between a reactant and a product? A reactant is present at the start of a chemical reaction; a product is present at the end of a chemical reaction.

Conservation of Matter When a log burns, a chemical reaction is taking place. The wood disappears and ashes are left. The ashes seem much smaller than the wooden log. What has happened? Has matter disappeared? Can matter be lost during a chemical reaction? The answer is no. In addition to the ashes, water and carbon dioxide are produced when a log burns. If you could find the mass of these products and add them together, you would find that they equal the mass of the original log. This illustrates an important scientific law called the law of conservation of matter. The law of conservation of matter states that matter cannot be created or destroyed by a chemical change. This means that in a chemical reaction, the amount of matter present in the products must always equal the amount of matter present in the reactants.

▸▸▶ **State:** What is the law of conservation of matter? Matter cannot be created or destroyed by a chemical change.

LESSON SUMMARY

▶ A chemical reaction is a process in which new substances with new physical and chemical properties are formed.

▶ Some examples of chemical reactions include rusting, the burning of gasoline in a car engine, and digestion.

▶ Substances present at the start of a chemical reaction are called reactants, while substances present at the end of a chemical reaction are called products.

▶ The law of conservation of matter states that matter cannot be created or destroyed in a chemical change.

CHECK *Complete the following.*

1. A substance that is changed during a chemical reaction is called a _____ . reactant

2. A chemical reaction is the process by which a _____ takes place. chemical change

3. In a chemical reaction, the mass of the products must always _____ the mass of the reactants. equal

4. According to the law of conservation of matter, matter cannot be created or _____ in a chemical reaction. destroyed

5. Digestion is an example of a _____ . chemical reaction

APPLY *Complete the following.*

▰ 6. **Analyze:** How do you know that a chemical reaction produces a chemical rather than a physical change in the substances that are reacting?

7. When rust forms on a piece of iron, what evidence do you have that a chemical reaction has taken place?

InfoSearch

Read the passage. Ask two questions about the topic that you cannot answer from the information in the passage.

Reactions and Energy All chemical reactions involve a change in energy. Some chemical reactions give off energy. This energy may be in the form of heat, light, or mechanical energy. Burning is an example of a chemical reaction that gives off energy in the form of heat and light. Reactions that give off energy are called exothermic (ek-suh-THUR-mik) reactions. Other chemical reactions absorb energy. They need a steady supply of energy in order to take place. Reactions that absorb energy are called endothermic (en-duh-THUR-mik) reactions.

SEARCH: Use library references to find answers to your questions.

ACTIVITY

OBSERVING A CHEMICAL REACTION

You will need a sheet of paper, aluminum foil, matches, and a balance.

1. Crumple the paper into a ball. Place it on the aluminum foil. Use the balance to find the mass of the combination. Record the mass.

2. Use a match to set fire to the paper ball. Do not add the match to the fire. **Caution:** Be careful when using matches. Discard used matches in a container of water.

3. Measure and record the mass of the ashes.

Questions

1. Compare the mass of the paper with the mass of the ashes.

▰ 2. **Hypothesize:** How can you account for the difference in mass?

17-2 What are chemical equations?

Objectives ▶ Explain how a chemical equation describes a chemical reaction. ▶ Write balanced chemical equations.

TechTerms

▶ **chemical equation:** statement in which chemical formulas are used to describe a chemical reaction

▶ **coefficient** (koh-uh-FISH-unt): number that shows how many molecules of a substance are involved in a chemical reaction

Chemical Equations To describe chemical reactions in a simple way that people all over the world can understand, scientists use chemical equations. A **chemical equation** is a statement in which chemical formulas are used to describe a chemical reaction.

Na + Cl → NaCl

sodium and chlorine yield sodium chloride

Reactants   Product

When you write a chemical equation, you must first write the correct chemical formulas for the reactants and products. Write the reactants on the left side of the equation with a + sign between them. The + sign means "and." Write the products on the right side of the equation with a + sign between them. Draw an arrow between the reactants and the products. The arrow means "yields." An arrow is used in place of an equal sign in a chemical equation. The following example shows the chemical equation for the reaction between sodium and chlorine.

$$2Na + Cl_2 \longrightarrow 2NaCl$$

Read the equation as "sodium and chlorine yield sodium chloride."

▶ **Define:** What is a chemical equation? a statement in which chemical formulas are used to describe a chemical reaction

Balanced Chemical Equations A chemical equation shows the atoms and molecules that are involved in the reaction. Because matter cannot be created or destroyed during a chemical reaction, the number of atoms of each element must be the same on both sides of the equation. When the number of atoms of each element is the same on both sides of the equation, the equation is said to be balanced.

$$2\,Na + Cl_2 \longrightarrow 2\,NaCl$$

To balance a chemical equation, you place numbers called coefficients in front of chemical formulas. A **coefficient** (koh-uh-FISH-unt) is a number that shows how many molecules or atoms of a substance are involved in the reaction. Look at the reaction between sodium and chlorine. If the equation were written with just the formulas for the reactants and products, it would read:

$$Na + Cl_2 \longrightarrow NaCl$$

There are two chlorine atoms on the left side of the equation and only one chlorine atom on the right side. Write a 2 before the product:

$$Na + Cl_2 \longrightarrow 2NaCl$$

Now the chlorine atoms are balanced, but the sodium atoms are not. Write another 2 on the left side of the equation:

$$2Na + Cl_2 \longrightarrow 2NaCl$$

Now the equation is balanced.

▶ **Define:** What is a coefficient? a number that shows how many molecules of a substance are involved in a chemical reaction

LESSON SUMMARY

▸ A chemical equation is a statement in which chemical formulas are used to describe a chemical reaction.

▸ A chemical equation is made up of the formulas for the reactants and the products with an arrow between them.

▸ When the number of atoms of each element are the same on both sides of the equation, the equation is balanced.

▸ To balance chemical equations, numbers called coefficients are placed in front of chemical formulas.

CHECK *Complete the following.*

1. A chemical equation describes a _____ . chemical reaction.

2. The arrow in a chemical equation means _____ . yields

3. A + sign in a chemical equation means _____ . and

4. The products in a chemical equation are written on the _____ side of the equation. right

5. A balanced chemical equation must show the same number of _____ of each element on both sides of the equation. atoms

APPLY *Complete the following.*

6. **Identify:** Which of the following equations are balanced?
   a. $H_2O_2 \longrightarrow H_2O + O_2$
   b. $CO + O_2 \longrightarrow CO_2$
   c. $Si + O_2 \longrightarrow SiO_2$
   d. $KClO_3 \longrightarrow KClO + O_2$
   e. $N_2 + H_2 \longrightarrow NH_3$

7. Write balanced equations for the unbalanced equations in question 6.

8. When there is no number in front of a chemical formula in a chemical equation, what number is understood?

State the Problem

A student balanced the following equations as shown. What mistakes did this student make?

a. $Cl_2 + NaBr \longrightarrow Br_2 + NaCl$
$2Cl + 2NaBr \longrightarrow 2Br + 2NaCl$

b. $Mg + O_2 \longrightarrow MgO$
$Mg + O \longrightarrow MgO$

c. $Na + O_2 \longrightarrow Na2O$
$2Na + 2O_2 \longrightarrow 2Na_2O$

ACTIVITY

BALANCING CHEMICAL EQUATIONS

You will need a pencil and paper.

1. Copy the equations on a sheet of paper.

2. Count the number of atoms of each element on both sides of the equation.

3. If the number of atoms of an element is not the same on both sides, balance the equation by using coefficients. Write a coefficient in front of a symbol or formula so that the number of atoms of that element is the same on both sides of the equation. Continue doing this until you have balanced the atoms of all the elements in the equation.

4. Check your work by counting the number of atoms of each element on both sides of the equation.

$C + Br_2 \longrightarrow CBr_4$
$Fe + O_2 \longrightarrow FeO$
$B + Cl_2 \longrightarrow BCl_3$
$Mg + O_2 \longrightarrow 2MgO$
$BaCl_2 + H_2SO_4 \longrightarrow BaSO_4 + HCl$
$P + O_2 \longrightarrow P_4O_{10}$
$Zn + HCl \longrightarrow ZnCl + H_2$
$PbO2 \longrightarrow PbO + O2$
$Ch_4 + O_2 \longrightarrow CO_2 + H_2O$
$Na + H_2O \longrightarrow NaOH + H_2$

Questions

1. What is the difference between a balanced and an unbalanced equation?

2. Why is it necessary to write balanced equations?

What are oxidation and reduction?

Objective ▶ Compare oxidation and reduction reactions.

TechTerms

- ▶ **oxidation** (ahk-suh-DAY-shun): chemical reaction in which electrons are lost
- ▶ **reduction** (ri-DUK-shun): chemical reaction in which electrons are gained

Oxidation The rusting of iron is a chemical reaction. In this reaction, iron combines with oxygen in air to form iron oxide. Iron oxide is a reddish-brown compound that is commonly known as rust. The burning of a match also is a chemical reaction. In this reaction, the hydrogen and carbon in the wood of the match react with oxygen in air to form carbon dioxide and water.

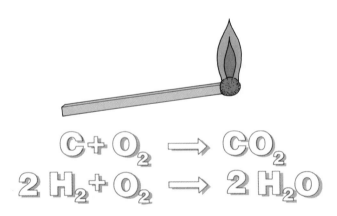

When a substance combines chemically with oxygen, the chemical reaction is called **oxidation.** The rusting of iron is an example of oxidation. Oxidation is a chemical reaction in which electrons are lost.

▶*Describe:* What happens during an oxidation reaction? A substance combines chemically with oxygen and electrons are lost.

Reduction When atoms of different elements react chemically, electrons are transferred or shared between atoms. Some atoms gain electrons in the reaction. Other atoms lose electrons. When the atoms of an element lose electrons, the reac-

tion is oxidation. When the atoms of an element gain electrons, the reaction is **reduction** (ri-DUK-shun). Oxidation and reduction always take place together The electrons lost by one element are gained by another element. Reactions involving oxidation and reduction are often referred to as "redox" reactions. In a redox reaction, some elements are oxidized, and other elements are reduced.

◼*Analyze:* What is the relationship between oxidation and reduction reactions? The atoms of some elements lose electrons, while the atoms of other elements gain electrons.

Examples of Oxidation and Reduction The rusting of iron is one of the most common examples of oxidation. The chemical equation for this reaction is

$$4Fe + 3O_2 \longrightarrow 2Fe_2O_3$$

The reaction between sodium and chlorine is an example of an oxidation-reduction reaction. In this reaction, sodium loses an electron and chlorine gains an electron.

$$2Na + Cl_2 \longrightarrow 2NaCl$$

When the flashbulb in a camera goes off, magnesium metal combines with oxygen to produce magnesium oxide. In this reaction, magnesium is oxidized because it loses two electrons.

$$2Mg + O_2 \longrightarrow 2MgO$$

◼*Analyze:* In the reaction between sodium and chlorine, which element is reduced? chlorine

LESSON SUMMARY

▶ The rusting of iron and the burning of wood are chemical reactions in which substances combine with oxygen.

▶ When a substance combines chemically with oxygen, the reaction is called oxidation.

▶ When atoms of an element gain electrons, the reaction is reduction.

▶ Examples of oxidation-reduction reactions are the rusting of iron, the combination of sodium and chlorine to form sodium chloride, and the reaction of magnesium with oxygen to form magnesium oxide.

CHECK *Complete the following.*

1. When iron rusts, iron combines chemically with _____ . oxygen

2. In an oxidation reaction, electrons are _____ . lost

3. In a reduction reaction, electrons are __gained__ .

4. Oxidation and _____ reactions always occur together. reduction

5. When sodium reacts with chlorine, _____ is oxidized. sodium

APPLY *Complete the following.*

▶ 6. **Infer:** Why must oxidation and reduction always happen together in a chemical reaction?

▶ 7. **Hypothesize:** Is oxygen always needed in order for an oxidation reaction to occur? Explain your answer.

InfoSearch ....

Read the passage. Ask two questions about the topic that you cannot answer from the information in the passage.

Reducing Metal Ores The first meaning that chemists gave to the term "reduction" referred to the process of reducing metal ore to the pure metal. For example, copper occurs in the ore chalcopyrite (kal-kuh-PY-ryt) as copper sulfide. The copper can be removed from the copper sulfide by heating the ore in the presence of air. In the chemical reaction that takes place, free copper is released and sulfur dioxide gas is formed.

SEARCH: Use library references to find answers to your questions.

SCIENCE CONNECTION ◆○◆○◆○◆○◆○◆○◆○◆○◆○◆○◆○◆○

RESPIRATION

Respiration (res-puh-RAY-shun) is an important process in all living things. During respiration, oxygen combines with sugars, starches, and fats to release energy. The energy is then used by the organism to carry out life activities. In birds and mammals, energy from respiration is also used to maintain a constant body temperature.

Respiration takes place inside body cells. As the hydrogen and carbon in foods are oxidized, carbon dioxide, water, and energy are produced. The carbon dioxide and water are given off as waste products. In humans and other animals, oxygen is taken in as the organism inhales. Carbon dioxide is given off as the organism exhales. Water is released as urine and perspiration. Plants absorb oxygen through tiny openings in their leaves. Carbon dioxide and water are released through these openings in the leaves.

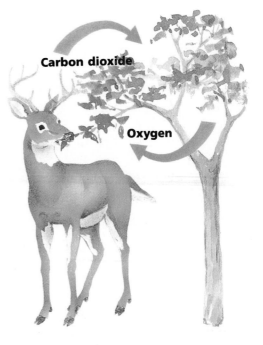

Carbon dioxide

Oxygen

What is a synthesis reaction?

Objective ▶ Describe what happens in a synthesis reaction.

TechTerm

▶ **synthesis** (SIN-thuh-sis) **reaction:** reaction in which substances combine to form a more complex substance

Synthesis Reactions There are many types of chemical reactions. One type of chemical reaction is called a **synthesis** (SIN-thuh-sis) **reaction.** In a synthesis reaction, two or more substances combine to form a more complex substance. The reactants in a synthesis reaction can be elements, compounds, or both. The product of a synthesis reaction is always a compound.

▰▰▰▶ **Define:** What is a synthesis reaction? reaction in which two or more substances combine to form a more complex substance

Examples of Synthesis Reactions Many oxidation reactions are synthesis reactions. For example, the burning of carbon to form carbon dioxide is a synthesis reaction.

$$C + O_2 \longrightarrow CO_2$$

The rusting of iron and the oxidation of magnesium are also synthesis reactions:

$$4Fe + 3O_2 \longrightarrow 2Fe_2O_3$$
$$2Mg + O_2 \longrightarrow 2MgO$$

Many synthesis reactions involve the combination of two elements to produce a compound. Two familiar examples are the formation of sodium chloride from sodium and chlorine, and the combination of hydrogen and oxygen to produce water.

$$2Na + Cl_2 \longrightarrow 2NaCl$$
$$2H_2 + O_2 \longrightarrow 2H_2O$$

Some synthesis reactions involve the combination of two compounds or the combination of an element and a compound. When carbon monoxide burns, a compound combines with an element to produce a more complex compound.

$$2CO + O_2 \longrightarrow 2CO_2$$

Two compounds react when calcium oxide combines with water to produce calcium hydroxide.

$$CaO + H_2O \longrightarrow Ca(OH)_2$$

▶ **Predict:** When carbon burns, what elements combine and what compound is produced? Carbon and oxygen combine and carbon dioxide is produced.

Figure 1 Burning carbon

Figure 2 Burning magnesium

LESSON SUMMARY

▶ In a synthesis reaction, two or more substances combine to form a more complex substance.

▶ Common examples of synthesis reactions include the burning of carbon, the rusting of iron, the formation of table salt, and the formation of water.

CHECK *Write true if the statement is true. If the statement is false, change the underlined term to make the statement true.*

1. In a synthesis reaction, the reactants may be two <u>elements</u>. true

2. In a synthesis reaction, the product formed is <u>sometimes</u> a compound. always

3. In a synthesis reaction, two or more substances produce a <u>simpler</u> substance. more complex

4. In a synthesis reaction, the product <u>can</u> be an element. cannot

5. The oxidation of magnesium <u>is</u> an example of a synthesis reaction. true

APPLY *Complete the following.*

6. **Identify:** Which of the following reactions are synthesis reactions?
   a. $2H_2O \longrightarrow 2H_2 + O_2$
   b. $N_2 + 2O_2 \longrightarrow 2NO_2$
   c. $4Al + 3O_2 \longrightarrow 2Al_2O_3$
   d. $Zn + 2HCl \longrightarrow ZnCl_2 + H_2$
   e. $CaO + H_2O \longrightarrow Ca(OH)_2$
   f. $Ti + N_2 \longrightarrow Ti_3N_4$

7. Explain why the following general equation describes a synthesis reaction:
$$A + B \longrightarrow C$$

Ideas in Action

IDEA: Many common substances are chemical combinations of two or more substances.

ACTION: List as many common substances as you can think of that are chemical combinations of two or more elements or compounds. Use your science book or other reference materials to help you write equations to show how these substances are formed.

LOOKING BACK IN SCIENCE

DEVELOPMENT OF SYNTHETIC FABRICS

Do you own a rayon blouse or a polyester shirt? Can some of your clothes be washed and then worn with little or no ironing? Are your bathing suit or exercise clothes made of a special material that stretches for a good fit? If you answered "yes" to any of these questions, you are familiar with synthetic (sin-THET-ik) fabrics.

Synthetic fabrics are materials that were invented by chemists to take the place of natural fibers such as silk, cotton, and wool. Synthetic fabrics are made from very large molecules called polymers (PAHL-i-murz). Polymers are made up of many smaller molecules joined together in long chains.

The first synthetic fiber to be developed was rayon. Rayon was invented in the early part of this century. Rayon is manufactured from cellulose. Cellulose comes from wood pulp. This synthetic fiber can be treated to resemble wool, cotton, linen, or silk. Nylon was introduced in 1938 as a substitute for silk. It soon became popular in the manufacture of "silk" stockings.

17-5 What is a decomposition reaction?

Objective ▶ Describe what happens in a decomposition reaction.

TechTerms

- ▶ **decomposition** (dee-kahm-puh-ZISH-un) **reaction:** reaction in which a complex substance is broken down into two or more simpler substances
- ▶ **electrolysis** (i-lek-TRAHL-uh-sis): process by which water is decomposed using an electric current

Decomposition Reactions When a substance breaks down into simpler substances, the reaction is called a **decomposition** (dee-kahm-puh-ZISH-un) **reaction.** Decomposition reactions are the opposite of synthesis reactions. In a decomposition reaction, a complex substance is broken down into two or more simpler substances. The products of a decomposition reaction can be elements, compounds, or both.

▐▐▐▶*Define:* What is a decomposition reaction?
reaction in which complex substance is broken down into two or more simpler substances.

Electrolysis of Water An example of a decomposition reaction is the **electrolysis** (i-lek-TRAHL-uh-sis) of water. In electrolysis, an electric current is passed through water. The energy from the electric current causes a chemical reaction to take place. When an electric current is passed through water, the water breaks down, or decomposes, into oxygen and hydrogen.

$$2H_2O \longrightarrow 2H_2 + O_2$$

Electrolysis can be used to break down other substances as well. If table salt is melted, it can be decomposed by passing an electric current through it.

$$2NaCl \longrightarrow 2Na + Cl_2$$

▐▐▐▶*Describe:* What happens during the electrolysis of water? An electric current is passed through water, and the water decomposes into hydrogen and oxygen.

Oxygen

Hydrogen

Decomposition and Energy Most decomposition reactions need energy in order to take place. The energy is usually in the form of heat or electricity. Chemists often show the type of energy that is needed for a decomposition reaction by writing a symbol above the arrow in the chemical equation. A triangle drawn above the arrow means that heat is needed. You can see how the heat symbol is used in the following equation. This equation shows the decomposition of mercuric oxide:

$$2HgO \xrightarrow{\Delta} 2Hg + O_2$$

▐▐▐▶*Explain:* What role does energy play in most decomposition reactions? Most decomposition reactions need energy in the form of heat or electricity.

318

LESSON SUMMARY

► In a decomposition reaction, a complex substance is broken down into two or more simpler substances.

► Water is decomposed using an electric current in a process called electrolysis.

► Most decomposition reactions require energy in order to take place.

CHECK *Complete the following.*

1. What is the opposite of a decomposition reaction called?

2. What are the products formed in decomposition reactions?

3. How many products are formed as a result of a decomposition reaction?

4. What is used to decompose water in an electrolysis reaction?

5. What is needed for most decomposition reactions to occur?

APPLY *Complete the following.*

◢ 6. **Analyze:** The bubbles in a carbonated beverage come from the decomposition of carbonic acid. The products of this reaction are carbon dioxide and water. Use the partial equation shown below to determine the chemical formula for carbonic acid. Assume that the equation is balanced.

$$carbonic\ acid \longrightarrow CO_2 + H_2O$$

◢ 7. **Infer:** Why can the reactant in a decomposition reaction never be an element?

·····························
Designing an Experiment···········

Design an experiment to solve the problem.

PROBLEM: What are the products when ordinary sugar, $C_6H_{12}O_6$, is decomposed by heating?

Your experiment should:

1. List the materials you would need.

2. Identify safety precautions that should be followed.

3. List a step-by-step procedure.

4. Describe how you would record your data.

TECHNOLOGY AND SOCIETY

DISPOSAL OF CHEMICAL WASTES

Almost two-thirds of the hazardous wastes produced in the United States each year are produced by the chemical industry. Chemical wastes are dangerous because they contain chemicals that can harm or kill living things. These chemicals also tend to stay in the environment for a long time.

Getting rid of chemical wastes is one of the major environmental problems at the present time. About 22,000 to 25,000 chemical waste dumping sites exist in the United States. Dangerous chemicals tend to leak out of them and into the surrounding land, air, and water.

What can be done to reduce the problems caused by chemical wastes? The Federal government is trying to help by providing funds to clean up existing waste sites. The government is also trying to regulate the wastes that are dumped by various industries. In addition, some industries are experimenting with ways to recycle chemical wastes so that they can be recovered and used again.

What is a single-replacement reaction?

Objective ▶ Describe what happens in a single-replacement reaction.

TechTerm

▶ **single-replacement reaction:** reaction in which one element replaces another element in a compound.

Single-Replacement Reactions In certain types of chemical reactions, one element replaces another element in a compound. This type of chemical reaction is called a **single-replacement reaction.** All single-replacement reactions begin with a compound and a free element. When the element and the compound react, the free element replaces one of the elements in the compound. The element that is replaced then becomes a free element. The reactants in a single-replacement reaction are always a compound and a free element. The products are always a new compound and a new free element. This general equation will help you to remember what happens in a single-replacement reaction.

$$A + BC \longrightarrow AC + B$$

Define: What type of reaction is a single-replacement reaction?
reaction in which one element replaces another element in a compound

Examples Many metals react with acids to produce hydrogen gas and a salt. In this type of reaction, the metal is the free element and the acid is the compound. When the metal and the acid react, the metal replaces hydrogen in the acid. When the reaction is over, hydrogen is the free element and a salt is the new compound. You can see how this works by looking at the equation for the reaction between zinc and hydrochloric acid.

$$Zn + 2HCl \longrightarrow ZnCl_2 + H_2$$

Some other types of single-replacement reactions include the reaction between sodium and water, and the reaction between chlorine and sodium bromide.

$$2Na + 2H_2O \longrightarrow 2NaOH + H_2$$
$$Cl_2 + 2NaBr \longrightarrow 2NaCl + Br_2$$

▶ **Infer:** What is the compound produced in the reaction between chlorine and sodium bromide? sodium chloride

Collected hydrogen

Hydrogen

Hydrogen

Hydrochloric acid

Water

Zinc

LESSON SUMMARY

▶ In a single-replacement reaction, one element replaces another element in a compound.

▶ The reactants in a single-replacement reaction are always a compound and a free element, and the products are always a new compound and a new free element.

▶ An example of a single-replacement reaction is the reaction between a metal and an acid to produce a salt and hydrogen gas.

CHECK *Complete the following.*

1. A single-replacement reaction always takes place between an element and a __compound__ .

2. In a single-replacement reaction, one element _____ another element. replaces

3. The products of a single-replacement reaction are always a compound and an __element__ .

4. When zinc reacts with hydrochloric acid, hydrogen is replaced by _____ . zinc

5. When the reaction between zinc and hydrochloric acid is complete, _____ is the free element. hydrogen

APPLY *Complete the following.*

6. **Analyze:** When an iron nail is placed in a solution of copper sulfate for several minutes, a coating of copper forms on the nail. What type of chemical reaction has taken place? Write a word equation to describe this reaction.

7. Explain in words what is happening in the following reaction:

$$Cu + AgNO_3 \longrightarrow CuNO_3 + Ag$$

InfoSearch ....

Read the passage. Ask two questions about the topic that you cannot answer from the information in the passage.

Reaction Speed Some chemical reactions occur slowly. Other reactions occur very quickly. The factors that determine the speed of chemical reactions include temperature, pressure, and particle size. The single-replacement reaction between sodium and water is an example of a fast reaction. In this reaction, sodium and water combine and release energy in such a short time that an explosion results.

SEARCH: Use library references to find answers to your questions.

ACTIVITY

REPLACING METALS

You will need copper sulfate, silver nitrate, water, an iron nail, a copper strip, two 100-mL beakers, 2 test tubes, a graduated cylinder, a clock or timer, and a balance.

1. In a beaker, make a solution of copper sulfate by adding 1 g of copper sulfate to 100 mL of water. Fill a test tube half full with the solution.

2. Place the iron nail in the solution for 5 min. Remove the nail and examine it.

3. In another beaker, prepare a solution of silver nitrate by dissolving 1 g of silver nitrate in 100 mL of water. Fill a clean test tube half full with the solution.

4. Place the copper strip in the solution for 5 min. Remove the copper strip and examine it.

Questions

👁 1. **Observe:** What happened to the iron nail?

👁 2. **Observe:** What happened to the copper strip?

Copper sulfate

Silver nitrate

Iron nail

Copper strip

17-7 What is a double-replacement reaction?

Objective ▶ Describe what happens in a double-replacement reaction.

TechTerms

- ▶ **double-replacement reaction:** reaction in which elements from two different compounds replace each other, forming two new compounds
- ▶ **precipitate** (pri-SIP-uh-tayt): solid that settles to the bottom of a mixture

Double-Replacement Reactions In certain types of chemical reactions, the elements from two different compounds replace each other, forming two new compounds. This type of reaction is called a **double-replacement reaction.** A double-replacement reaction can be shown by the following general equation.

$$AB + CD \longrightarrow AD + CB$$

▬▶ *Define:* What is a double-replacement reaction? reaction in which elements from two different compounds replace each other, forming two new compounds

Formation of a Precipitate In many double-replacement reactions, a substance called a **precipitate** (pri-SIP-uh-tayt) is formed. A precipitate is a solid that settles to the bottom of a mixture. When a precipitate forms in a chemical reaction, it sinks to the bottom of the container in which the reaction is taking place. The precipitate silver chlo-ride forms in the reaction between silver nitrate and sodium chloride.

$$AgNO_3 + NaCl \longrightarrow AgCl + NaNO_3$$

In this double-replacement reaction, the nitrate polyatomic ion, NO_3^-, trades places with the chlorine atom. The following equations show two other reactions that form precipitates.

$$BaCl_2 + Ca(OH)_2 \longrightarrow 2NaOH + CaCO_3$$
$$Al_2(SO_4)_3 + 3BaCl_2 \longrightarrow 2AlCl_3 + 3BaSO_4$$

▬▶ *Define:* What is a precipitate? solid that settles to the bottom of a mixture

Acid-Base Reactions Another common type of double-replacement reaction is the reaction that takes place between an acid and a base. This type of reaction is called a neutralization (NOO-truh-liz-ay-shun) reaction. In a neutralization reaction, an acid and a base combine to form a salt and water. The following equation describes the reaction between the base sodium hydroxide and hydrochloric acid.

$$NaOH + HCl \longrightarrow NaCl + HOH$$

The formula HOH is another way to write the formula for water. You could also write this equation as follows:

$$NaOH + HCl \longrightarrow NaCl + H_2O$$

▬▶ *Describe:* What happens in a neutralization reaction? An acid and a base react to form a salt and water.

LESSON SUMMARY

▶ In a double-replacement reaction, elements from two different compounds replace each other, forming two new compounds.

▶ In many double-replacement reactions, a solid called a precipitate is formed.

▶ A neutralization reaction is a double-replacement reaction in which an acid and a base react to form a salt and water.

CHECK *Complete the following.*

1. In a double-replacement reaction, two new _____ are formed. compounds

2. A precipitate is a _____ that settles to the bottom of a mixture. solid

3. When silver nitrate and sodium chloride react, the precipitate formed is _____ . silver chloride

4. In this reaction, the nitrate ion trades places with a _____ atom. chlorine

5. When a _____ and an acid react, water and a salt are formed. base

APPLY *Complete the following.*

▶ 6. **Predict:** What products are formed in the following reaction?

$$KOH + HCl \longrightarrow ? + ?$$

7. Which of the following reactions are double-replacement reactions?
   a. $2Al_2O_3 \longrightarrow 4Al + 3O_2$
   b. $Ca + 2H_2O \longrightarrow Ca(OH)_2 + H_2$
   c. $AgNO_3 + NaBr \longrightarrow AgBr + NaNO_3$
   d. $2Na + Br_2 \longrightarrow 2NaBr$
   e. $2NaOH + H_2SO_4 \longrightarrow Na_2SO_4 + 2H_2O$

InfoSearch

Read the passage, Ask two questions about the topic that you cannot answer from the information in the passage.

Colored Precipitates Certain precipitates have distinct colors. For example, silver chloride forms a white precipitate, silver iodide forms a yellow precipitate, and silver sulfide forms a black precipitate. Precipitates can often be used to identify unknown substances. If a solution of silver nitrate is mixed with a solution containing an unknown sodium salt, the color of the precipitate formed can indicate if the salt is sodium chloride, sodium iodide, or neither.

SEARCH: Use library references to find answers to your questions.

ACTIVITY

FORMATION OF A PRECIPITATE

You will need potassium carbonate, calcium chloride, two 100-mL beakers, a graduated cylinder, a balance, and a stirring rod.

1. Measure 2 g of potassium carbonate and 2 g of calcium chloride. Place each compound into a beaker.

2. Using the graduated cylinder, add 50 mL of water to the beaker with the potassium carbonate. Stir.

3. Use the graduated cylinder to add 50 mL of water to the beaker with the calcium chloride. Stir.

4. Mix both solutions together and stir.

Questions

👁 1. **Observe:** What did each solution look like before you mixed them together?

▶ 2. **Infer:** How do you explain the appearance of the mixture?

Potassium carbonate

Calcium chloride

STUDY HINT Before you begin the Unit Challenges, review the TechTerms and Lesson Summary for each lesson in this unit.

TechTerms . . . .

chemical equation (312)
chemical reaction (310)
coefficient (312)
decomposition reaction (318)
double-replacement reaction (322)

electrolysis (318)
oxidation (314)
precipitate (322)
product (310)
reactant (310)

reduction (314)
single-replacement reaction (320)
synthesis reaction (316)

TechTerm Challenges . . . .

Matching *Write the TechTerm that matches each description.*

1. substance formed in a chemical reaction ^product
2. chemical reaction in which electrons are gained reduction
3. decomposition using an electric current ^electrolysis
4. chemical reaction in which electrons are lost oxidation
5. substance changed in a chemical reaction ^reactant
6. one element replaces another element in a compound single-replacement reaction
7. solid that settles out of a mixture precipitate

Fill in *Write the TechTerm that best completes each statement.*

1. In a _____ two new compounds are formed. double-replacement reaction
2. A _____ describes a chemical reaction. ^chemical equation
3. When new substances with new physical and chemical properties are formed, a _____ has occurred. chemical reaction
4. The opposite of a synthesis reaction is a _____ . decomposition reaction
5. A more complex substance is formed from two simpler substances in a synthesis reaction .
6. During _____ , water is broken down into hydrogen and oxygen. electrolysis
7. One element replaces another element in a compound in a single-replacement reaction .
8. To balance a chemical equation you would use a _____ . coefficient

Content Challenges . . . .

Multiple Choice *Write the letter of the term or phrase that best completes each statement.*

1. The substance produced when iron is oxidized is
   a. water. **b.** oxygen. **c.** iron precipitate. **d.** rust. d
2. The reactants in the equation $2H_2 + O_2 \longrightarrow 2H_2O$ are
   a. hydrogen and energy. **b.** hydrogen and oxygen. **c.** water and energy.
   d. oxygen and water. b
3. In a chemical reaction, matter cannot be
   a. changed. **b.** conserved. **c.** lost. **d.** measured. c
4. The arrow in a chemical equation means
   a. yields. **b.** and. **c.** balances. **d.** changes. a
5. The number of atoms of each element on both sides of a chemical equation must always be
   a. greater than one. **b.** less than two. **c.** different. **d.** equal. d

6. The chemical formula 3H₂O means
   a. 3 atoms of hydrogen and 3 atoms of oxygen. **b.** 6 atoms of hydrogen and 3 atoms of oxygen. **c.** 3 atoms of water. **d.** 3 atoms of hydrogen and 2 atoms of oxygen. b

7. The burning of wood is an example of
   a. rusting. **b.** oxidation. **c.** synthesis. **d.** respiration. b

8. In a chemical reaction, atoms gain and lose
   a. oxygen. **b.** energy. **c.** electrons. **d.** matter. c

9. The product of a synthesis reaction is always
   a. an element. **b.** a solid. **c.** a precipitate. **d.** a compound. d

10. In electrolysis, an electric current produces a chemical reaction by providing
    a. reactants. **b.** energy. **c.** oxygen. **d.** water. b

True/False *Write true if the statement is true. If the statement is false, change the underlined term to make the statement true.*

1. In a synthesis reaction, the product is <u>sometimes</u> an element. never
2. The electrolysis of water is a <u>single-replacement</u> reaction. decomposition
3. Many metals replace <u>hydrogen</u> when they react with acids. true
4. A double-replacement reaction involves the reaction of two <u>elements</u>. compounds
5. A <u>neutralization</u> reaction involves an acid and a base. true
6. The mass of matter present at the end of a chemical reaction <u>is not</u> the same as the mass of matter present at the beginning of the reaction. is
7. Oxidation and reduction <u>always</u> occur together in a chemical reaction. true
8. Two or more substances are <u>broken down</u> in a synthesis reaction. combined
9. When a substance is reduced, it <u>loses</u> electrons. gains
10. A free element is <u>always</u> one of the products in a single-replacement reaction. true
11. Oxidation-reduction involves gaining and losing <u>atoms</u>. electrons
12. You can recognize a precipitate because it forms at the <u>surface</u> of a mixture. bottom

Understanding the Selections . . . .

Reading Critically *Use the feature reading selections to answer the following. Page numbers for the features are in parentheses.*

1. **Classify:** What are the reactants in respiration? What are the products? (315)
2. **Infer:** Why are chemical wastes considered dangerous to the environment? (319)
3. What was the first synthetic fabric to be developed? (317)

Concept Challenges . . . .

Critical Thinking *Answer each of the following in complete sentences.*

1. Explain how balanced chemical equations illustrate the law of conservation of matter.
2. How can you tell from a chemical equation that a chemical change is taking place?
3. Discuss the two different meanings of the term "oxidation."
4. Explain the difference between a single-replacement reaction and a double-replacement reaction.

5. Describe the physical and chemical changes that take place when silver nitrate reacts with sodium chloride.

6. **Classify:** Identify each of the following reactions as synthesis, decomposition, single-replacement, or double-replacement.
   a. $2Mg + O_2 \longrightarrow 2MgO$
   b. $BaCl_2 + H_2SO_4 \longrightarrow BaSO_4 + 2HCl$
   c. $Si + O_2 \longrightarrow SiO_2$
   d. $Zn + 2HCl \longrightarrow ZnCl_2 + H_2$
   e. $H_2CO_3 \longrightarrow CO_2 + H_2O$
   f. $NaOH + HBr \longrightarrow NaBr + H_2O$
   g. $2CO + O_2 \longrightarrow 2CO_2$
   h. $Ca + 2H_2O \longrightarrow Ca(OH)_2 + H_2$
   i. $2H_2O_2 \longrightarrow 2H_2O + O_2$
   j. $BaCl_2 + Na_2SO_4 \longrightarrow BaSO_4 + 2NaCl$

Interpreting a Diagram *Use the diagram to answer the following questions.*

1. What kind of a reaction is shown in the diagram?
2. What is this reaction called?
3. What are the reactants in this reaction?
4. What are the products?
5. Write a balanced equation for this reaction.

Oxygen

Hydrogen

Finding out More ·

1. **Model:** Make a model to show the four types of chemical reactions. One way to do this is to cut circles of different color paper to represent the atoms that are involved in chemical reactions. Then use the circles to show how atoms combine and separate during each type of reaction.

2. The explosion of fireworks is a chemical reaction that releases energy. Use library references to find out what substances are used to make fireworks, and what chemical reactions take place when they explode. Write a report of your findings.

3. Design an experiment to show that matter is neither created nor destroyed in a chemical reaction.

4. The eruption of a volcano is the result of a chemical reaction. Use library references to find out what type of chemical reaction takes place and what substances are involved. Find out where the energy for this reaction comes from.

5. Many of the colors in paints are produced by double-replacement reactions. Go to the library and find a book on paint chemistry. Find out what compounds give various paints their colors. Make a chart showing each color and the name and formula of the compound that produces that color. If you can, write an equation to show the double-replacement reaction that produces each compound.

METALS

CONTENTS

STUDY HINT Before beginning Unit 18, scan through the lessons in the unit looking for words that you do not know. On a sheet of paper, list these words. Work with a class-mate to try to define each word on your list.

18-1 What is an ore?

Objective ▶ Name some common ores and the metals that are removed from them.

TechTerm

▶ **ore:** rock or mineral from which a useful metal can be removed

Useful Metals Metals are a useful part of your life. Some of the most useful metals are iron, copper, aluminum, and lead. These metals are used to make cars, appliances, electrical wires, jewelry, as well as many other objects.

Metals are found in nature. However, they are usually found combined with other elements. The useful metal has to be removed before it can be used. An **ore** is a rock or mineral from which a useful metal can be removed.

▶**Define:** What is an ore?
rock or mineral from which useful metals can be removed

Common Ores Many types of ores are found in nature. Each type of ore contains a metal that can be removed from the ore. Pyrite is an ore of iron. It is a combination of iron and sulfur. Pyrite is called iron disulfide. Its chemical symbol is FeS_2. Pyrite has a yellow color and a shiny surface. Pure iron can be removed from the pyrite. Table 1 shows some common ores and the metals that can be removed from them.

Table 1	Metallic Ores	
ORE	CHEMICAL FORMULA	METAL
Hematite	Fe_2O_3	iron
Bauxite	Al_2O_3	aluminum
Galena	PbS	lead
Litharge	PbO	lead
Cuprite	Cu_2O	copper
Sphalerite	ZnS	zinc
Magnesite	$MgCO_3$	magnesium

▰**Analyze:** What metal is removed from the ore galena? lead

Type of Ores The metals in an ore are usually combined with oxygen, sulfur, or carbon. A metal that combines with oxygen is called an oxide. A metal that combines with sulfur is called a sulfide. A metal that combines with carbon is called a carbonate.

▶**Identify:** Which of the ores in Table 1 is a carbonate? magnesite

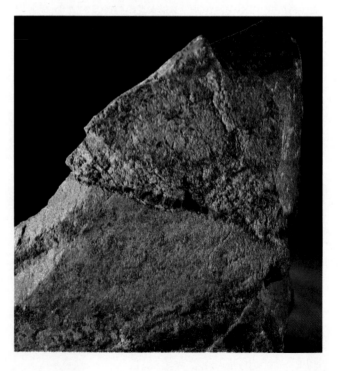

LESSON SUMMARY

▶ Many useful metals are found in nature.

▶ Metals are usually found combined with other elements.

▶ An ore is a rock or mineral from which a useful metal can be removed.

▶ Three types of ores are oxides, sulfides, and carbonates.

CHECK *Complete the following.*

1. An ore is made of a _____ and other elements. metal

2. Three types of ores are sulfides, _____ , and oxides. carbonates

3. Pyrite is a combination of _____ and sulfur. iron

4. Useful _____ are usually removed from rocks and minerals. metals

APPLY *Complete the following.*

5. What element must be removed from an oxide ore in order to obtain the useful metal?

6. What element must be removed from magnesite in order to obtain magnesium?

7. **Infer:** Gold and silver are not found combined with other elements. Are gold and silver examples of oxides, sulfides, or carbonates? How do you know?

8. Write the chemical formula for each of the following ores:

   a. pyrite   **e.** hematite
   b. cuprite   **f.** litharge
   c. magnesite   **g.** galena
   d. sphalerite   **h.** bauxite

Ideas in Action

IDEA: Metals are a useful part of your life.
ACTION: Look at the objects you use around the house. Make a list of all the objects that are made of metal. How many objects have metal parts? Do most of the objects have some parts that are metal? What kinds of metals are used?

LEISURE ACTIVITY

DESIGNING AND MAKING JEWELRY

Jewelry making goes back to prehistoric times. The first jewelry was made from objects found in nature, such as shells, stones, and feathers. Jewelry has always involved interesting designs and colors. As people began to use metals for tools, they also used metals to make jewelry.

Today, jewelry is usually made of metals such as gold and silver. However, you can make attractive jewelry without using gold and silver. For example, you can use colored beads, shells, or any small colorful objects.

The first step in making jewelry is creating a design. Usually this is a simple sketch of what the piece of jewelry will look like. You should indicate what colors will be used. Once you have made a sketch, you can begin making the piece of jewelry. Decide what materials you will need for each piece. For example, if you are making a simple necklace, you will need a string or thin chain to hold the beads or shells.

18-2 How are metals removed from their ores?

Objective ▶ Describe how some metals are removed from their ores.

TechTerms

- ▶ **reduction:** process of removing oxygen from an ore
- ▶ **roasting:** process in which an ore is heated in air to produce an oxide

Removing Metals from Ores Metals are removed from their ores through chemical processes. The three common types of ores are oxides, sulfides, and carbonates. The different types of ores require different processes to remove the metals they contain. Of the three ores, metals are most easily removed from oxides.

▶**Predict:** Would it be easier to remove a metal from an oxide or a sulfide? from an oxide

Metals from Oxides Suppose you want to separate copper from cuprite (Cu_2O). You could combine two parts of cuprite to one part of carbon and heat the mixture. This process is shown in the following chemical equation:

$$C + 2Cu_2O \longrightarrow 2Cu + CO_2$$

One atom of carbon and two atoms of oxygen combined to form one molecule of carbon dioxide. The oxygen was separated from the copper. When the carbon dioxide was formed, two atoms of copper were left over. For any oxide ore, the metal is obtained by removing the oxygen. The process of removing oxygen from an ore is called **reduction.**

▶**Describe:** How is a metal removed from an oxide? by removing the oxygen

Metals from Sulfides and Carbonates Two steps are needed to remove the metal from a sul-

fide or a carbonate ore. The first step is called **roasting.** In roasting, an ore is heated in air to produce an oxide. For example, suppose galena (PbS) is roasted. Galena is a sulfide, but after roasting it becomes an oxide. This process is shown in the following chemical equation:

$$2PbS + 2O_2 \longrightarrow 2PbO + 2SO_2$$

Lead combined with oxygen to form the oxide litharge (PbO).

Once the ore has been changed into an oxide, the oxygen can be removed. This second step requires the process of reduction. The oxygen is removed from the litharge as shown in the following equation:

$$2PbO + C \longrightarrow 2Pb + CO_2$$

The same two-step process is used to remove metals from carbonates.

▶**Identify:** How are metals removed from sulfides and carbonates? by roasting and reduction

LESSON SUMMARY

▶ Metals are removed from their ores through chemical processes.

▶ For oxides, the metal is obtained through the process of reduction.

▶ For sulfides and carbonates, the metal is obtained by first roasting and then reducing the ore.

CHECK *Complete the following.*

1. Obtaining metals from _____ ores is a two-step process. sulfide and carbonate

2. Metals are most easily removed from _____ ores. oxide

3. The reaction for removing copper from cuprite is $C +$ _____ $\longrightarrow 2Cu + CO_2$. Cu_2O

4. In the process of _____ , a sulfide or carbonate is made into an oxide. roasting

5. Both roasting and reduction require energy in the form of _____ . heat

6. In the process of _____ , oxygen is removed from the ore. reduction

APPLY *Complete the following.*

7. **Analyze:** Look at the chemical equation for removing aluminum from bauxite:
$$3C + 2Al_2O_3 \longrightarrow 2Al + 3CO_2$$
Why are three carbon atoms needed? How many atoms of aluminum are produced?

8. **Sequence:** The two chemical equations show the two steps involved in removing zinc from the ore sphalerite. Which equation shows step 1? Which equation shows step 2?
$$2ZnO + C \longrightarrow 2Zn + CO_2$$
$$2ZnS + 2O_2 \longrightarrow 2ZnO + 2SO_2$$

9. **Analyze:** How many molecules of sphalerite were used in step 1? How many atoms of zinc were produced in step 2?

Skill Builder

Balancing Equations Write balanced equations for the following reduction reactions.
$$C + CuO \longrightarrow Cu + CO_2$$
$$C + FeO \longrightarrow Fe + CO_2$$
$$C + ZnO \longrightarrow Zn + CO_2$$

◆○◆○◆ SCIENCE CONNECTION ◆○◆○◆○◆○◆○◆○◆

METALLURGY

Metallurgy (MET-uh-lur-jee) is the science of removing metals from their ores. Metallurgy began in about 4000 BC in the Middle East. Copper was the first metal to be used. The remains of copper mines have been found throughout parts of the Middle East. Over the centuries, such metals as tin, iron, silver, and gold were successfully mined and separated from their ores.

Ores must be removed from the earth by mining before the metals they contain can be removed. For example, the main source of the iron ore hematite was at one time the area around Lake Superior. In some places, open-pit mines have damaged the environment.

There are two steps to separating a metal from its ore. First, the ore is broken up and crushed. The smaller particles are then separated either by hand or using machines. Next, the metal is removed from the remaining bits of ore. This process can involve roasting and reduction, or simply heating the ore in a furnace. The metal that results from the second step is almost, but not quite, pure. Further refining of the metal is necessary to obtain a completely pure metal.

18-3 What are alloys?

Objective ▶ Identify some alloys and their uses.

TechTerm

▶ **alloy:** substance made up of two or more metals

Pure Metals Through roasting and reduction, a pure metal can be removed from its ore. Pure metals have certain properties.

▶ They are good conductors of both heat and electricity.

▶ They are malleable. They can be hammered into different shapes.

▶ They are ductile. They can be made into wires.

▶ They are solids at room temperature. The only exception is mercury, which is a liquid at room temperature.

▶ **List:** What are the properties of pure metals?
Students should restate the properties listed above.

Alloys Several pure metals can be combined to form a substance with different properties from the properties of the pure metal. A substance that is made up of two or more metals is called an alloy. For example, brass is an alloy of copper and zinc. The advantage of brass is that it is harder than either copper or zinc. Brass is used in water pipes because it lasts longer than copper or zinc. Table 1 shows some other common alloys and the metals that make them up.

Table 1	Alloys	
ALLOY	ELEMENTS	USES
Steel	iron, chromium, carbon	knives, forks, bridges
Brass	copper, zinc	plumbing
Bronze	copper, tin	machine parts
Pewter	tin, copper, antimony	dishes, cups
Sterling silver	silver, copper	jewelry
Alnico	iron, aluminum, nickel, cobalt	magnets
Nichrome	nickel, iron, chromium, manganese	electrical wire

▶ **Define:** What is an alloy?
mixture of two or more metals

Figure 1 Brass

Figure 2 Pewter

LESSON SUMMARY

▶ The metals removed from ores are pure metals.

▶ An alloy is a substance that is made up of two or more metals.

▶ Alloys have different properties from the metals that make them up.

CHECK *Write true if the statement is true. If the statement is false, change the underlined term to make the statement true.*

1. <u>Steel</u> is an alloy of iron, chromium, and nickel. true

2. Most metals are <u>liquid</u> at room temperature. solid

3. Pewter is a combination of tin, <u>nickel</u>, and antimony. copper

4. <u>Brass</u> is harder than the metals that are used to make it. true

5. All <u>alloys</u> are malleable. metals

6. <u>Mercury</u> is the only metal that is a liquid at room temperature. true

7. <u>Steel</u> is used in making cooking utensils. true

8. Pure metals <u>are not</u> good conductors of heat and electricity. are

9. Because metals are <u>malleable</u>, they can be made into wires. ductile

10. <u>Sterling silver</u> is made of silver and copper. true

APPLY *Complete the following.*

▶ 11. **Infer:** Electrical wires are usually made of copper. Why is a pure metal better than an alloy for electrical wires?

InfoSearch

Read the passage. Ask two questions about the topic that you cannot answer from the information in the passage.

Amalgams An alloy of mercury is called an amalgam (uh-MAL-gum). One of the most used alloys of mercury is an amalgam of silver, tin, copper, and zinc. This amalgam is used by dentists for filling teeth. The advantage of this alloy is that it will fill a cavity completely. It also withstands the temperature changes in the mouth from hot and cold foods.

SEARCH: Use library references to find answers to your questions.

TECHNOLOGY AND SOCIETY

ALLOYS OF STEEL

Steel is one of the most useful alloys ever developed. Iron is the main element used to make steel. By combining iron with different amounts of chromium, nickel, and carbon, different alloys of steel can be made.

An alloy of steel commonly used today is called low-alloy steel. Low-alloy steel contains small amounts of nickel, chromium, molybdenum, tungsten, titanium, niobium, and vanadium. Low-alloy steel is very strong and sturdy. It is used to make machine parts, and the metal supports on bridges and buildings. Stainless steel is a type of low-alloy steel.

Another commonly used alloy of steel is called high-alloy steel. High-alloy steel contains larger amounts of nickel, chromium, and the other metals found in low-alloy steel. High-alloy steel is very shiny. It is used to make cooking utensils, cutting tools, and jet-engine parts.

Why are some metals more active than others?

Objective ▶ List the most active metals in the periodic table.

TechTerms

▶ **alkali metals:** metals in Group 1 of the periodic table

▶ **alkaline earth metals:** metals in Group 2 of the periodic table

Group 1 Metals There are many types of metals. Some metals are more chemically active than others. One group of very active metals is called the **alkali metals.** They include lithium, sodium, potassium, rubidium, cesium, and francium. The most common alkali metals are sodium and potassium. The least common alkali metals are rubidium, cesium, and francium. The alkali metals are also called the sodium family.

In the periodic table, the alkali metals make up Group 1. The Group 1 metals are so chemically active that they are always found combined with other elements. These metals are rarely found as pure metals in nature. The structure of Group 1 metals is similar. Each alkali metal has one valence electron.

▦▶*Identify:* What are the six alkali metals?
lithium, sodium, potassium, rubidium, cesium, and francium

Group 2 Metals The metals next to the Group 1 metals in the periodic table are the Group 2 metals. Group 2 metals are also called the **alkaline earth metals.** They include beryllium, magnesium, calcium, strontium, barium, and radium. Like the Group 1 metals, the Group 2 metals are chemically active. They readily combine with other elements. The most common Group 2 metals are calcium, strontium, and barium. Another name for the Group 2 metals is the calcium family.

The structure of the Group 2 metals is similar. Each metal has two valence electrons. This is one more electron than in the Group 1 metals. The extra electron makes the Group 2 metals harder and stronger than the Group 1 metals.

▦▶*Identify:* What are the six Group 2 metals?
beryllium, magnesium, calcium, strontium, barium, and radium

Electromotive Series Some metals are more active than others. In a chemical reaction, some metals will replace other metals in a compound. For example, when iron reacts with copper sulfate, the iron will replace the copper to form iron sulfate.

Table 1 shows the electromotive (i-lek-truh-MOHT-iv) series of metals. The most active metals are at the top of the table. The least active metals are at the bottom of the table. A metal will replace those metals below it on the list. For example, lithium will replace sodium, but sodium will not replace calcium.

Table 1 Electromotive Series of Metals	
Lithium (Li)	most active
Potassium (K)	
Barium (Ba)	
Calcium (Ca)	
Sodium (Na)	
Magnesium (Mg)	
Aluminum (Al)	
Zinc (Zn)	
Iron (Fe)	
Tin (Sn)	
Lead (Pb)	
Hydrogen (H)	
Copper (Cu)	
Silver (Ag)	
Platinum (Pt)	
Gold (Au)	least active

▬*Analyze:* Which metals will replace magnesium in a chemical reaction?
lithium, potassium, barium, calcium, sodium

334

LESSON SUMMARY

▶ The Group 1 metals, or alkali metals, include lithium, sodium, potassium, rubidium, cesium, and francium.

▶ The alkali metals have one valence electron.

▶ The alkaline earth metals, or Group 2 metals, include beryllium, magnesium, calcium, strontium, barium, and radium.

▶ The alkaline earth metals have two valence electrons.

▶ Some metals can replace others in a chemical reaction.

CHECK *Complete the following.*

1. Another name for the Group 1 metals is the _____ family. sodium

2. Another name for the Group 2 metals is the _____ family. calcium

3. In the electromotive series, a metal can be replaced by a metal _____ it on the list. above

4. Another name for the Group 1 metals is the _____ metals. alkali

5. Another name for the Group 2 metals is the _____ metals. alkaline earth

6. The most active metal in the electromotive series is _____ . lithium

7. The least active metal in the electromotive series is _____ . gold

APPLY *Complete the following.*

8. Look at the following chemical equation:
$$Fe + CuSO_4 \longrightarrow Cu + FeSO_4$$
Why was iron able to replace the copper in copper sulfate?

▶ 9. **Infer:** Why is gold usually found as a pure metal in nature, unlike most of the other metals?

Skill Builder ....

Sequencing When you sequence, you place things in order. Arrange the following metals in order from most active to least active:

aluminum	calcium	gold
lithium	silver	zinc
barium	copper	
magnesium	sodium	

◆◆◆◆◆ CAREER IN PHYSICAL SCIENCE ◆◆◆◆◆◆◆

WELDER

Welding is a process used to join two pieces of metal. The process usually involves heating a metal and hammering it into shape. Several layers of metals can be heated and hammered together. This process is known as forge welding. A forge is a furnace in which the metal is heated for welding.

Today, most welders use a technique known as arc welding. In arc welding, heat and electric current are used to weld together two pieces of metal. Another type of welding uses lasers. The intense light and heat from a laser is used to weld two pieces of metal. High-frequency, or ultrasonic, sounds can also be used in welding. The vibrations from the sound waves are transferred to the metals as heat. The heat is used to weld the metals.

Welders need to take many safety precautions. Special equipment is needed to protect the welders from the intense heat produced. Welders may work for automobile manufacturers, construction companies, or in other industries.

18-5 What is corrosion?

Objective ▶ Explain what causes corrosion.

TechTerm

▶ **corrosion:** wearing away of a metal

Corrosion Metals are chemically active. One result of this chemical activity is **corrosion.** Corrosion is a chemical change in metals. It is the wearing away of metals. When a metal becomes corroded, it changes color and loses its shine.

▶▶**Describe:** What happens when a metal becomes corroded? It changes color and loses its shine.

Types of Corrosion The most common example of corrosion is rust. A piece of iron that is exposed to air and moisture will become rusted. The rusted iron has a reddish brown color and no shine. Tarnished metals are another example of corrosion. A tarnished metal has lost its shine and has a dull color. For example, tarnished silver looks dark and dull.

Corrosion forms layers on the surface of a metal. When you polish silver, you remove the layers of tarnish. Once the corrosion is removed, the metal will have its usual color and shine.

▶▶**Identify:** Name two types of corrosion.
rust and tarnish

Cause of Corrosion Corrosion is caused by oxidation. Remember that oxidation is a loss of electrons. Rust is formed when iron combines with oxygen in the air to form iron oxide.

$$4Fe + 3O_2 \longrightarrow 2Fe_2O_3$$

Iron has three valence electrons. Oxygen needs to gain two electrons. The four atoms of iron will lose 12 electrons. The six atoms of oxygen will gain the 12 electrons.

▶▶**Describe:** What happens to a metal during oxidation? The metal loses electrons.

Preventing Corrosion Corrosion wears away and weakens metals. Therefore, it is important to prevent metals from corroding. There are several ways to prevent corrosion:

▶ Paint is a good way to prevent rust. A layer of paint over a piece of iron will keep air and moisture away from the surface of the metal.

▶ Coating the surface of a metal with a layer of oil protects the metal from corrosion.

▶ Using alloys is a way of preventing rust. For example, steel is an alloy of iron. Unlike iron, steel will not rust.

▶▶**List:** What are three ways to prevent corrosion in metals? painting the metal or coating it with oil, and using alloys

LESSON SUMMARY

- Corrosion is a chemical change in metals.
- Two examples of corrosion are rust and tarnish.
- Corrosion is caused by oxidation.
- Three ways to prevent corrosion are painting or coating the surface of the metal, or using alloys.

CHECK *Complete the following.*

1. What is corrosion?
2. What is the chemical formula for rust?
3. What causes corrosion?
4. Why is it important to prevent metals from corroding?
5. How does painting a metal prevent corrosion?
6. What happens when iron combines with oxygen in the air?

APPLY *Complete the following.*

7. **Predict:** Will corrosion take place if a metal is kept in a vacuum? Why or why not?
8. **Infer:** Why do ships always have more rust along the bottom?

ACTIVITY

OBSERVING THE TARNISHING OF SILVER

You will need a silver spoon and a hard-boiled egg.

1. Peel the hard-boiled egg and remove the yolk.
2. Place the silver spoon into the egg yolk. Leave it in the yolk for several minutes.
3. After several minutes, remove the spoon from the yolk. Observe the appearance of the spoon.

Questions

1. **Observe:** What did the spoon look like when you removed it from the egg yolk?
2. Egg yolk contains sulfur. What do you think happened when you placed the silver spoon into the egg yolk?
3. How could you remove the tarnish from the silver spoon?

18-6 How are metals plated?

Objective ► Explain how metals are plated.

TechTerms

► **electroplating** (i-LEK-truh-playt-ing): use of an electric current to plate one metal with another metal

► **plating:** coating one metal with another metal

Protecting Metals One way to protect metals from corrosion is through the process of **plating.** In this process, a metal is coated with another metal. For example, a plating of zinc on the surface of iron will prevent the iron from rusting. Iron coated with zinc is called galvanized (GAL-vuh-nyzed) iron. When the zinc corrodes, it forms zinc oxide. The layer of zinc oxide prevents any further corrosion. Garbage cans and snow shovels often are made of galvanized iron.

▶**Define:** What is the process of plating?

coating one metal with a layer of another metal

Figure 1 Galvanized iron

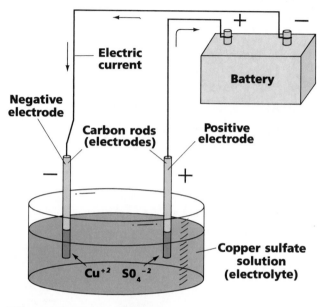

Figure 2 Electroplating

Electroplating Electricity is often used to plate metals onto other metals. This process is called **electroplating** (i-LEK-truh-playt-ing). Nickel, silver, and copper can be used in electroplating. Figure 2 shows how copper can be electroplated onto carbon.

Two carbon rods are placed into a solution of copper sulfate. The copper sulfate is the electrolyte. The carbon rods are attached to the terminals of a battery. The carbon rod attached to the positive terminal becomes the positive electrode. The rod attached to the negative terminal becomes the negative electrode.

Copper ions, Cu^{+2}, move to the negative electrode. At the negative electrode, each ion gains two electrons and becomes a copper atom.

$$Cu^{+2} + 2 \text{ electrons} \longrightarrow Cu$$

The pure copper is then plated onto the carbon rod.

◢*Analyze:* What ion moves to the positive electrode in electroplating with copper?

sulfate, SO_4^{-2}

LESSON SUMMARY

▶ One way to protect metals from corrosion is by the process of plating.

▶ In plating, a protective layer of one metal is placed over another metal.

▶ A type of plating called electroplating uses an electric current.

CHECK *Complete the following.*

1. Iron that has a coating of zinc is called _____ iron. galvanized

2. A layer of zinc will prevent iron from _____ . rusting, or corrosion

3. When a protective layer of zinc corrodes, it forms _____ . zinc oxide

4. Plating a metal by using an electric current is called _____ . electroplating

5. When electroplating with copper, the copper ions will be attracted to the _____ electrode. negative

APPLY *Complete the following.*

▶ 6. **Predict:** A solution of zinc sulfate ($ZnSO_4$) is used as an electrolyte for electroplating.

Which electrode will the zinc ions, Zn^{+2}, be attracted to?

▶ 7. **Hypothesize:** Galvanized iron is used to make garbage cans and snow shovels. What is the advantage of using galvanized iron for these objects?

8. Suppose you want to electroplate an object with silver. To which electrode would you attach the object? Explain.

Designing an Experiment..........

Design an experiment to solve the problem.

PROBLEM: How can you determine if an object is silver plated or is made of pure silver?

Your experiment should:

1. List the materials you would need.

2. Identify safety precautions that should be followed.

3. List a step-by-step procedure.

4. Describe how you would record your data.

ACTIVITY

PROTECTING METALS FROM CORROSION

You will need three iron nails, a beaker of water, vinegar, clear nail polish, and oil or petroleum jelly.

1. Cover one iron nail with clear nail polish. Let the nail polish dry completely.

2. Coat a second nail with oil or petroleum jelly.

3. Do not put anything on the third nail.

4. Put all three nails into a beaker of water. Add a little vinegar to the water to speed up the rusting process.

5. Leave the nails in the water and vinegar overnight.

Questions

1. **Observe:** Which nail shows signs of rust?

2. How do the nail polish and oil or petroleum jelly affect the corrosion of the iron nails?

3. Can you think of other substances that would keep the nails from rusting? What are they?

Water and vinegar

Coated with oil

Coated with clear nail polish

Uncoated

UNIT 18 Challenges

STUDY HINT Before you begin the Unit Challenges, review the TechTerms and Lesson Summary for each lesson in this unit.

TechTerms

alkali metals (334)
alkaline earth metals (334)
alloy (332)

corrosion (336)
electroplating (338)
ore (328)

plating (338)
reduction (330)
roasting (330)

TechTerm Challenges

Matching *Write the TechTerm that matches each description.*

1. rock or mineral from which a useful metal can be removed ore
2. coating of one metal on another metal plating
3. heating an ore in air roasting
4. Group 1 metals alkali metals
5. Group 2 metals alkaline earth metals
6. substance made up of two or more metals alloy
7. process that wears away metal corrosion
8. electrically coating one metal with another metal electroplating
9. removing oxygen from an ore reduction

Fill In *Write the TechTerm that best completes each statement.*

1. Steel is an _____ of iron, chromium, and nickel. alloy
2. In _____ , a copper rod is sometimes used as an electrode. electroplating
3. Sodium is an _____ . alkali metal
4. Another name for the _____ is the calcium family. alkaline earth metals
5. Sulfides and carbonates are converted to oxides by the process of _____ . roasting
6. One of the best examples of _corrosion_ is rust.
7. A layer of zinc placed over iron is an example of _____ . plating
8. Litharge is an _____ of lead. ore
9. The process of _____ is used to separate a metal from an oxide ore. reduction

Content Challenges

Multiple Choice *Write the letter of the term or phrase that best completes each statement.*

1. Corroded metals are
   a. smooth and shiny. **b.** dull and discolored. **c.** silvery. **d.** reddish brown. b

2. Metals are not found in their pure form in nature because they are
   a. chemically active. **b.** rusted. **c.** tarnished. **d.** corroded. a

3. The ores from which it is easiest to remove metals are
   a. sulfides. **b.** carbonates. **c.** oxides. **d.** alloys. c

4. Brass is an alloy of
   a. copper and tin. **b.** iron and gold. **c.** copper and zinc. **d.** copper and chromium. c

5. Corrosion should be prevented because corrosion
   a. wears away metal. **b.** discolors metal. **c.** makes the metal shiny. **d.** plates metal. a

6. All of the following are ways to reduce corrosion *except*
   a. painting. **b.** electroplating. **c.** exposing to air. **d.** using alloys. c

7. Silver is a more active metal than
   a. lithium. **b.** copper. **c.** calcium. **d.** gold. d

8. The ore from which aluminum is removed is
   a. cuprite. **b.** bauxite. **c.** galena. **d.** hematite. b

9. Each of the following is a property of pure metals *except*
   a. ductile. **b.** malleable. **c.** solid. **d.** poor conductor of heat. d

10. The two steps in removing a metal from a sulfide ore are
   a. roasting followed by reduction. **b.** reduction followed by roasting.
   c. reduction followed by oxidation. **d.** oxidation followed by roasting. a

True/False *Write true if the statement is true. If the statement is false, change the underlined term to make the statement true.*

1. An ore is a combination of a <u>metal</u> and another element. true
2. <u>Lead</u> can be removed from bauxite. Aluminum
3. When iron rusts, it forms <u>iron oxide</u>. true
4. Electroplating is one way to prevent <u>corrosion</u>. true
5. Platinum can replace <u>silver</u> in chemical reactions. gold
6. When a metal is oxidized, it <u>gain</u> electrons. loses
7. Two examples of corrosion are rust and <u>tarnish</u>. true
8. Bronze is a combination of copper and <u>iron</u>. tin
9. To remove a metal from an <u>oxide</u> ore, the ore is roasted. true

Understanding the Selections . . . .

Reading Critically *Use the feature reading selections to answer the following. Page numbers for the features are in the parentheses.*

1. What metals is most jewelry made of today? (329)

2. Hypothesize: How do open-pit mines damage the environment? (331)

3. Contrast: What is the difference between low-alloy steel and high-alloy steel? (333)

4. Infer: Why do welders need to take safety precautions? (335)

Concept Challenges . . . .

Critical Thinking *Answer each of the following in complete sentences.*

1. What part of a metal's atomic structure makes it a good conductor of electricity? Explain.

2. Hypothesize: Some periods of human history are known by the metals that were used during that time. The Bronze Age was about 6000 years ago. The Iron Age was about 3000 years ago. What does this tell you about the technology needed to produce bronze and iron?

3. Infer: The Statue of Liberty in New York Harbor is made of copper. Why do you think the statue has a green color? (Hint: You may have seen a similar green coating on old copper pennies.)

4. Lithium is the most active metal in the electromotive series of metals. It is an alkali metal. Why is lithium so active?

Interpreting a Diagram *Use the diagram to answer the following questions.*

1. What does the diagram show?
2. What is the positive electrode made of? The negative electrode?
3. What is the electrolyte?
4. What will happen at the negative electrode?

Finding Out More...

1. Use library references to find out about the structure of the earth. How are metals distributed in the earth's crust? What kinds of metals are found in the deeper layers of the earth? Are metals found in the center of the earth? What are they? Draw a labeled diagram to display your findings to the class.

2. The water from your kitchen faucet may contain calcium carbonate ($CaCO_3$) and magnesium carbonate ($MgCO_3$). This type of water is called "hard water." The name means that it is "hard" to make a lather when soap is added to the water. Test the water in your home for "hardness."

3. Sodium and potassium are two of the most common Group 1 metals. Use library references to learn about the different compounds of sodium and potassium. Make a list of the different compounds and their uses. Share your findings with the class in an oral report.

4. Gold leaf is used to plate objects with gold. Gold leaf has been used for centuries to decorate books, glass, and ceramics. Visit your local museum and ask to see the museum pieces decorated with gold leaf. Find out about the process of making gold leaf.

SOLUTIONS

CONTENTS

STUDY HINT Before beginning each lesson in Unit 19, review the TechTerms for each lesson and read the Lesson Summary.

19-1 What is a solution?

Objective ▶ Describe the characteristics of a solution.

TechTerm

▶ **dissolve:** go into solution

▶ **solution:** mixture in which one substance is evenly mixed with another substance

Salt and Water What would happen if you added some sand to a test tube of water? The sand would settle to the bottom of the test tube. Suppose you then added some salt to another test tube of water. The salt disappears in the water. The salt is still in the water, but you cannot see it. The salt has dissolved (di-ZAHL-ved) in the water. When a substance **dissolves**, it goes into solution. The sand did not dissolve in the water.

Water

Salt and water

Sand

Sand does not dissolve in water.   **Salt dissolves in water.**

▶**Explain:** Why does salt seem to disappear in water? It dissolves in the water.

Solutions A mixture of salt and water is called a solution. A **solution** is a mixture in which one substance is evenly mixed with another substance. In a saltwater solution, particles of salt are evenly mixed with molecules of water.

▶**Define:** What is a solution?
mixture in which one substance is evenly mixed with another substance

Types of Solutions A liquid solution is formed when a solid dissolves in a liquid. Salt water is a liquid solution. A liquid solution may also be formed when a gas dissolves in a liquid. Club soda is a solution of the gas carbon dioxide dissolved in water. One liquid may dissolve in another liquid to form a liquid solution. Water and alcohol form this type of solution.

Liquid solutions are formed when solids, liquids, or gases dissolve in liquids. Solutions can also be formed when different substances dissolve in solids and gases. Table 1 shows some examples of different kinds of solutions.

▶**Analyze:** Why is air called a solution?
It forms when gases dissolve in other gases.

Table 1 Types of Solutions		
SUBSTANCE	**DISSOLVED IN**	**EXAMPLES**
Liquid	Liquid Gas Solid	alcohol in water water vapor in air ether in rubber
Gas	Liquid Gas Solid	club soda (CO_2 in water) air (N_2, O_2, and other gases) hydrogen in palladium
Solid	Liquid Gas Solid	salt in water iodine vapor in air brass (copper and zinc)

LESSON SUMMARY

► When a substance dissolves, it goes into solution.

► A solution is a mixture in which one substance is evenly mixed with another substance.

► Salt water is an example of a solution made from a liquid and a solid.

► Solutions can form when a substance dissolves in a solid or in a gas.

CHECK *Complete the following.*

1. Solutions are formed when substances _____ in other substances. dissolve

2. Salt water is a solution formed when a _____ dissolves in a liquid. solid

3. A mixture in which one substance is evenly mixed with another substance is called a _____ . solution

4. Club soda is an example of a solution formed when a _____ dissolves in a liquid. gas

5. Salt water is a solution formed when a solid dissolves in a _____ . liquid

APPLY *Complete the following.*

6. **Classify:** Which of the following substances are solutions?

   a. sugar   **e.** sea water
   b. mud   **f.** salt and pepper
   c. club soda   **g.** sand
   d. flour   **h.** air

7. For each of the substances you classified as solutions in question 6, identify the type of solution formed.

Skill Builder

△ ***Organizing Information*** When you organize information, you put the information in some kind of order. A table is one way to organize information. Alloys, such as brass, are solid solutions. Make a table with the following headings: "Alloy," "Metals in Solution," and "Uses." Identify three different alloys, the metals they contain, and their common uses. Organize the information in your table.

CAREER IN PHYSICAL SCIENCE

ANALYTICAL CHEMIST

Do you enjoy studying science and mathematics? Can you make careful, precise measurements? Are you determined to find the answer to a problem? If so, you may enjoy a career as an analytical chemist. Analytical chemists analyze the chemical composition of substances. They perform experiments to identify special characteristics of the substances. They test substances to find out what will happen when different substances are combined.

Most analytical chemists work in laboratories. Often, they work as part of a team investigating a certain problem. Their work is very precise and requires careful attention to details. To become an analytical chemist, you need a college degree in chemistry. Usually, analytical chemists have a strong background in mathematics. For some positions, a graduate degree in a special field of interest is necessary. If you would like more information about a career as an analytical chemist, write to the Association of Official Analytical Chemists, 1111 N. 19th St., Suite 210, Arlington, VA 22209.

19-2 What are the parts of a solution?

Objective ▶ Identify the parts of a solution.

TechTerms

▶ **insoluble** (in-SAHL-yoo-bul): not able to dissolve
▶ **soluble** (SAHL-yoo-bul): able to dissolve
▶ **solute** (SAHL-yoot): substance that is dissolved in a solvent
▶ **solvent**: substance in which a solute dissolves

Parts of a Solution All solutions are made when one substance dissolves in another substance. A solution of salt and water forms when salt dissolves in water. The part of a solution that dissolves is called the **solute** (SAHL-yoot). Salt is the solute in a solution of salt and water. The part of the solution in which a solute dissolves is called the **solvent.** Water is the solvent in a saltwater solution.

▶*Contrast:* What is the difference between a solute and a solvent?
The solute dissolves while the solvent does the dissolving.

Soluble Substances Water is the solvent in many types of solutions. Club soda is a solution in which water is the solvent and carbon dioxide is

the solute. Because carbon dioxide dissolves in water, it is said to be **soluble** (SAHL-yoo-bul) in water. A substance is soluble in water if it dissolves in water.

▶*Predict:* What will happen when carbon dioxide gas is mixed with water?
It will dissolve in the water.

Insoluble Substances Many substance do not dissolve in water. A concrete sidewalk does not dissolve in rainwater. Sand does not dissolve in a glass of water. A plastic container does not dissolve when you add water to it. These substances are **insoluble** (in-SAHL-yoo-bul) in water. A substance is called insoluble if it does not dissolve in another substance.

A substance may dissolve in one substance but not in another substance. Sugar dissolves in water. It is soluble in water. Sugar does not dissolve in oil. It is insoluble in oil. The type of solvent determines whether a solute is soluble or insoluble.

Analyze: How can a substance be both soluble and insoluble? It may be soluble in one solvent but insoluble in a different solvent.

LESSON SUMMARY

▶ The substance that dissolves in a solution is called the solute.

▶ The solvent is the substance in which a solute dissolves.

▶ A substance that dissolves in another substance is soluble in that substance.

▶ A substance is insoluble if it does not dissolve in a particular substance.

▶ A substance may be soluble in one solvent but insoluble in a different solvent.

CHECK *Write true if the statement is true. If the statement is false, change the underlined term to make the statement true.*

1. Salt is the <u>solute</u> in a saltwater solution. true

2. Cement is <u>soluble</u> in water. insoluble

3. The substance in which a solute dissolves is called a <u>solvent</u>. true

4. Sugar is <u>insoluble</u> in water. soluble

5. In a solution of sugar and water, water is the <u>solute</u>. solvent

APPLY *Complete the following.*

6. **Classify:** Instant coffee is a solution formed from coffee powder and hot water. Identify the solute and the solvent in this solution.

7. Is wood soluble in water? How do you know?

◆○◆○◆ SCIENCE CONNECTION ◆○◆○◆○◆○◆○◆

CHEMICAL WEATHERING

Weathering is the breaking up of rocks and minerals by natural forces, such as wind and water. In mechanical weathering, rocks are broken down due to the action of wind or water. In chemical weathering, substances in water cause substances in the rock to dissolve. This action weakens the structure of the rock. The rock is then more likely to be broken apart by mechanical weathering.

A common type of chemical weathering takes place when carbon dioxide from the air dissolves in rainwater. A solution called carbonic acid forms in the rainwater. When the rainwater seeps into rocks, the carbonic acid dissolves limestone in the rocks. The dissolved limestone is carried away by the rainwater. As a result, cracks are left in the rocks. The next time the rocks are struck by moving wind or water, they will easily crumble away. Sometimes, the dissolving of soluble minerals in rocks can lead to the formation of caves.

19-3 Why is water a good solvent?

Objective ▶ Explain why water is sometimes called the universal solvent.

TechTerm

▶ **polar molecule:** molecule in which one end has a positive charge and the other end has a negative charge

Water Molecules Water is a type of polar molecule. A **polar molecule** is a molecule in which one end has a positive charge and the other end has a negative charge. A molecule of water is made up of two atoms of hydrogen joined to one atom of oxygen. The hydrogen end of a water molecule has a positive charge. The oxygen end of a water molecule has a negative charge.

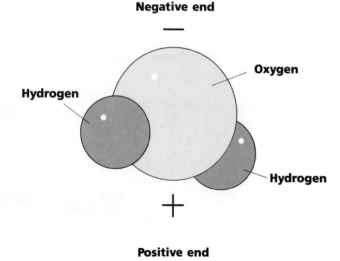

Negative end
Oxygen
Hydrogen
Hydrogen
Positive end

▶*Explain:* Why is a water molecule called a polar molecule? One end has a positive charge and the other end has a negative charge.

The Universal Solvent Water is sometimes called the universal (yoo-nuh-VUR-sul) solvent. This is because many types of substances dissolve in water. The electrical charges in polar water molecules help dissolve different kinds of substances.

A solution forms when a solute mixes evenly with a solvent. The charged ends of a water mole-

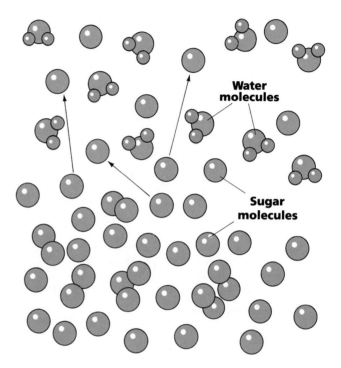

Water molecules

Sugar molecules

cule help spread a solute throughout the water. Suppose you place a sugar cube in a glass of water. The ends of the water molecules attract the molecules in the sugar cube. Each sugar molecule is pulled to a water molecule. As the sugar dissolves, sugar molecules are mixed throughout the water.

▶*Describe:* What happens to the sugar molecules when sugar is placed in water? The sugar molecules are attracted by the water molecules.

Force of Attraction Solutions form when the force of attraction between the solute and solvent is greater than the force of attraction between the particles in the solute. A sugar cube gets its shape from the force of attraction between its molecules. The sugar molecules will break away from the sugar cube only if they are pulled by a greater force of attraction. This is also true of other types of solutes.

▶*Predict:* What will happen if the force of attraction between solute particles is greater than the force of attraction between solute and solvent? The solute will not dissolve.

348

LESSON SUMMARY

▶ A molecule in which one end has a positive charge and the other end has a negative charge is called a polar molecule.

▶ Water is called the universal solvent because it can dissolve many different substances.

▶ Solutions form when the force of attraction between the solute and solvent particles is greater than the force of attraction between the solute particles.

CHECK *Complete the following.*

1. Water is a _____ molecule. polar

2. Water is sometimes called the __universal__ solvent.

3. The hydrogen end of a water molecule has a _____ electrical charge. positive

4. The force of attraction between water molecules and sugar molecules is _____ than the force of attraction between the sugar molecules. greater

5. Solutions form when the force of attraction between the solute and solvent is greater than the force of attraction between particles of the _____. solute

APPLY *Complete the following.*

▶ 6. **Infer:** A substance put in a glass of water does not dissolve. What does this tell you about the force of attraction between particles of the substance?

7. **Hypothesize:** Will a teaspoon of water dissolve in a glass of water? Explain.

Skill Builder

Building Vocabulary A molecule, such as water, with opposite electrical charges is called a polar molecule. Use a dictionary to identify the general meaning of the word "polar." How does this definition relate to the North Pole and South Pole of the earth? How does it apply to the poles of a magnet?

◆◆◆◆ CAREER IN PHYSICAL SCIENCE ◆◆◆◆◆◆◆

WATER PURIFICATION CHEMIST

Water is a very important resource. Cities and towns need large supplies of fresh drinking water. Often, these supplies of water come from rivers, lakes, or reservoirs. The water from these sources must be tested to be sure it is safe to drink.

Water purification chemists analyze water in purification plants. They perform various tests on the water to find out its chemical composition. They analyze the results of their tests. Based on their analysis, they then identify which chemicals should be added to the water to make it suitable for drinking. Most of this work is done in laboratories at the purification plant.

In order to work as a water purification chemist, you need a college degree in chemistry. You also need a good background in mathematics. Some positions require a graduate degree in a specialized field. If you think you would be interested in a career as a water purification chemist, you should write to the American Association for Clinical Chemistry, 1725 K St. NW, Suite 1010, Washington, DC 20006.

How can you change the rate of dissolving?

Objective ▸ Describe three ways to speed up the rate of dissolving.

Stirring Solutions form when a solute dissolves in a solvent. The rate at which a solid solute dissolves can be changed. Certain factors can speed up the rate at which a solute dissolves. Stirring a solution will make the solute dissolve faster. If you put a cube of sugar into a glass of water, it will eventually dissolve. However, stirring the water will cause the sugar to dissolve faster. Stirring the water causes the sugar molecules to leave the cube more rapidly. The sugar dissolves faster.

Particle Size The size of the particles of a solid solute also affects the rate at which it dissolves. The smaller the size of the solute particles, the faster the solute dissolves. A crushed sugar cube dissolves faster in water than a solid sugar cube placed in a equal amount of water at the same temperature. As the size of solute particles decreases, the rate at which the solute dissolves increases.

▸**Infer:** Why does stirring make a sugar cube dissolve faster in water?
Stirring pulls the sugar molecules out of the cube faster.

Temperature The temperature of a liquid solvent affects the rate at which a solid solute dissolves. A cube of sugar dissolves faster in hot water than in an equal amount of cool water. Heat causes the molecules of sugar to leave the cube more rapidly. As the temperature of a liquid solvent increases, the rate at which a solid solute dissolves also increases.

▸**Relate:** What is the relationship between the temperature of a liquid solvent and the rate at which a solid solute dissolves in it?
As the temperature of the liquid increases, the rate at which the solid dissolves also increases.

▸**Predict:** Which would dissolve faster in the same amount of water at the same temperature, a sugar cube or powdered sugar?
powdered sugar

LESSON SUMMARY

► Stirring a solvent increases the rate at which a solute dissolves.

► As the temperature of a solvent increases, the rate at which a solute dissolves also increases.

► The smaller the size of the particles in a solute, the faster the solute dissolves in a solvent.

CHECK *Complete the following.*

1. Stirring a solvent _____ the rate at which a solute dissolves in it. increases

2. Sugar dissolves more slowly in _____ water than in hot water. cold

3. The smaller the size of solute particles, the _____ the rate of dissolving. faster

4. As the _____ of a solvent increases, the rate at which a solute dissolves also increases. temperature

5. Powdered sugar dissolves _____ than a sugar cube placed in equal amounts of water at the same temperature. faster

APPLY *Complete the following.*

► 6. **Infer:** Why are most types of instant coffee made in powdered form?

7. Why is it easier to make tea with boiling water than with cold water?

8. Which of the following will make a solute dissolve faster in a solvent? Explain your answers.
   a. Place the solvent in a freezer.
   b. Grind the solute into small pieces.
   c. Place the solvent in a blender.
   d. Let the solvent stand without stirring.
   e. Heat the solvent.

InfoSearch

Read the passage. Ask two questions that you cannot answer from the information in the passage.

Dissolving Gases Some gases, such as oxygen and carbon dioxide, are soluble in water. Oceans, lakes, and rivers all contain dissolved oxygen. Club soda contains dissolved carbon dioxide. The temperature of a liquid solvent has the opposite effect on gaseous solutes than on solid solutes. As the temperature of the solvent decreases, the dissolving rate of a gaseous solute increases.

SEARCH: Use library references to find answers to your questions.

ACTIVITY

CHANGING THE RATE OF DISSOLVING

You will need four beakers, a graduated cylinder, water, a spoon, and sugar cubes.

1. Put 100 mL of water in each beaker.

2. Place one sugar cube each in the first two beakers. Stir the water in the first beaker with the spoon. Leave the water in the second beaker untouched. Compare the results.

3. Use the spoon to crush a sugar cube. Carefully drop the crushed sugar into the third beaker. Place one sugar cube in the fourth beaker. Compare the results.

Questions

1. What effect did stirring have on the dissolving rate of the sugar cube?

2. What effect did crushing the sugar cube have on the rate of dissolving?

3. **Hypothesize:** What is the fastest way to dissolve a sugar cube in 100 mL of water?

What is the concentration of a solution?

1 g of copper sulfate   **5 g of copper sulfate**   **10 g of copper sulfate**

DILUTE ➡ **CONCENTRATED**

Objective ▶ Differentiate between saturated and unsaturated solutions.

TechTerms

- ▶ **concentrated solution:** strong solution
- ▶ **dilute solution:** weak solution
- ▶ **saturated solution:** solution containing all the solute it can hold at a given temperature
- ▶ **unsaturated solution:** solution containing less solute than it can hold at a given temperature

Dilute and Concentrated A weak solution has a small amount of dissolved solute. Weak solutions are called **dilute solutions.** A strong solution has a large amount of dissolved solute. Strong solutions are called **concentrated solutions.**

▶**Compare:** What is the difference between a dilute solution and a concentrated solution?
A dilute solution has a small amount of solute while a concentrated solution has a large amount of solute.

Unsaturated Solutions There is a limit to the amount of solute a solvent can hold. Some solutions have less solute than they can hold. For example, if you put 1 g of copper sulfate in 100 mL of water, the copper sulfate will dissolve. However, the water could hold an additional amount of copper sulfate. This solution is called an unsaturated solution. An **unsaturated solution** contains less solute than it can hold at a given temperature.

▶**Define:** What is an unsaturated solution?
solution that contains less solute than it can hold at a given temperature

Saturated Solutions Some solutions contain all the solute they can hold at a given temperature. These solutions are called **saturated solutions.** Suppose you put 20 g of copper sulfate in 100 mL of water. Some of the copper sulfate will dissolve in the water. However, some of the copper sulfate will remain on the bottom of the container. It does not dissolve because the water has taken in as much solute as it can hold.

▶**Hypothesize:** Why would some solute remain undissolved in a solution?
The solution is saturated.

1 g of copper sulfate   **20 g of copper sulfate**

Unsaturated solution   **Saturated solution**

Solvent Temperature Heating a saturated solution often causes it to become unsaturated. This is because the temperature of a solvent determines the amount of solute it can hold. As the temperature of a solvent increases, so does the amount of solute it can hold. Heating a saturated solution increases the amount of solute it can dissolve.

▶**Predict:** What will happen if you heat a saturated solution? It will become unsaturated.

352

▶ A weak solution is called a dilute solution, and a strong solution is called a concentrated solution.

▶ An unsaturated solution contains less solute than it can hold at a given temperature.

▶ A saturated solution contains all the solute it can hold at a given temperature.

▶ Heating a saturated solution usually causes it to become unsaturated.

CHECK *Complete the following.*

1. A _____ solution contains a small amount of dissolved solute. dilute

2. An unsaturated solution contains less __solute__ than it can hold at a given temperature.

3. A _____ solution contains a large amount of dissolved solute. concentrated

4. A solution that contains all the solute it can hold at a given temperature is called a _____ solution. saturated

5. When saturated solutions are _____ , they usually become unsaturated. heated

APPLY *Complete the following.*

6. **Hypothesize:** Is a can of frozen juice a concentrated solution or a dilute solution? Explain.

7. **Predict:** Suppose the directions on the can of frozen juice state that it should be mixed with three cans of water. What type of solution would you make if you added five cans of water? Explain.

Ideas in Action ....

IDEA: Many products used in cooking are concentrated solutions that become dilute when water or some other solvent is added.

ACTION: Make a list of five products found in your home that are concentrated solutions. Describe what you must do to dilute them.

ACTIVITY

MAKING A SUPERSATURATED SOLUTION

You will need 6 g of ammonium chloride, a beaker, water, a heat source, a stirring rod, and a thermometer.

1. Put 100 mL of water in the beaker. Add 4 g of ammonium chloride to the beaker and stir.

2. Heat the solution in the beaker to a temperature of 70 °C. Carefully remove the beaker from the heat.

3. Add the remaining 2 g of ammonium chloride to the beaker. Stir the solution. Let it cool to room temperature. Observe the results.

Questions

1. What happened to the solution when it was heated?

2. What happened to the solute when the water was cooled to room temperature?

3. **Analyze:** Supersaturated solutions contain more solute than they can usually hold at a given temperature. What caused the ammonium chloride and water solution to become supersaturated?

70°C

How do solutes affect freezing point?

Objective ▶ Describe how a solute affects the freezing point of a solution.

TechTerms

▶ **freezing point:** temperature at which a liquid changes to a solid

▶ **freezing point depression:** lowering the freezing point of a liquid solvent by adding solute

Freezing Point of Water

The temperature at which a liquid changes to a solid is called its **freezing point.** The freezing point of pure water is 0 °C. When pure liquid water reaches this temperature, it changes to solid ice.

Salt water does not freeze at 0 °C as pure water does. The particles of salt dissolved in the water interfere with the change from a liquid to a solid. Because salt water contains dissolved salt, its freezing point is lower than that of pure water. Salt water freezes at a lower temperature than pure water.

▋▋▶*Describe:* What happens to liquid water at its freezing point? It changes into a solid.

Freezing Point Depression

The amount of solute dissolved in a solvent affects the freezing point of the solvent. This special property of solutions is called **freezing point depression.** Adding solute to a liquid solvent lowers the freezing point of the solvent. The greater the amount of dissolved solute, the lower the freezing point of the solvent.

▶*Predict:* What will happen to the freezing point of water as you add more solute to the water? It will become lower.

Rock Salt

Have you ever thrown rock salt on an icy sidewalk? Salt is sprinkled on icy sidewalks because the salt lowers the freezing point of water. When the ice starts to melt, the salt dissolves in the water and forms salt water. The salt water has a lower freezing point than pure water. The temperature must get lower than 0 °C before the salt water freezes. As a result, ice does not form as fast.

▋▋▶*Explain:* Why is rock salt thrown on an icy sidewalk? to lower the freezing point of water

LESSON SUMMARY

▶ Freezing point is the temperature at which a liquid changes to a solid.

▶ The freezing point of salt water is lower than that of pure water.

▶ Lowering the freezing point of a liquid solvent by adding solute is called freezing point depression.

▶ Rock salt lowers the freezing point of the melted ice on an icy sidewalk.

CHECK *Write true if the statement is true. If the statement is false, change the underlined term to make the statement true.*

1. The <u>freezing point</u> of pure water is 0 °C. true

2. The greater the amount of dissolved solute in a solvent, the <u>higher</u> the freezing point of the solvent. lower

3. The freezing point of salt water is <u>lower</u> than the freezing point of pure water. true

4. Lowering the freezing point of a liquid solvent by adding solute is called <u>freezing point depression</u>. true

5. Putting salt on icy roads <u>raises</u> the freezing point of the melted ice. lowers

APPLY *Complete the following.*

6. **Compare:** Beaker A contains 2 g of sugar dissolved in 100 mL of water. Beaker B contains 10 g of sugar dissolved in 100 mL of water. Which beaker contains the solution with the lower freezing point? Explain your answer.

7. **Analyze:** Beaker 1 and Beaker 2 each contain 5 g of dissolved copper sulfate. Beaker 1 contains 100 mL of water while Beaker 2 contains 250 mL of water. Which beaker contains the solution with the lower freezing point? Explain your answer.

InfoSearch ....

Read the passage. Ask two questions that you cannot answer from the information in the passage.

Hard Water Tap water from your kitchen faucet is not pure water. It does not freeze at exactly 0 °C. It freezes at a lower temperature because it contains dissolved solutes. Some tap water contains dissolved minerals including calcium bicarbonate, magnesium sulfate, calcium chloride, and iron sulfate. Water containing these solutes is called hard water.

SEARCH: Use library references to find answers to your questions.

SCIENCE CONNECTION ◆○◆○◆○◆○◆○◆○◆

FISHES WITH ANTIFREEZE

Fishes that live in the Arctic Ocean must survive freezing and near-freezing water temperatures. Scientists wondered why these fishes did not freeze in such an environment. The scientists studied fishes in the Arctic region of northern Labrador.

Scientists discovered that the blood of the Arctic fishes contained a high concentration of a certain protein. The protein acts like a solute in a solution. The greater the amount of protein in the fishes' blood, the lower the freezing point of the blood.

In order for the fishes' blood to freeze, the water temperature must drop below 0 °C. As a result of this freezing point depression, the fishes can survive in the cold Arctic environment.

How do solutes affect boiling point?

Objective ▶ Describe how a solute affects the boiling point of a solution.

TechTerms

- ▶ **boiling point:** temperature at which a liquid changes to a gas
- ▶ **boiling point elevation:** raising the boiling point of a liquid solvent by adding solute

Boiling Point of Water When water is heated, its temperature rises. When the temperature of the water reaches 100 °C, the liquid water changes to steam. The temperature at which a liquid changes to a gas is called its **boiling point.** The boiling point of water is 100 °C. Adding heat to boiling water does not raise its temperature.

Salt water does not boil at 100 °C as pure water does. This is because salt water contains dissolved salt particles. The particles of salt dissolved in the water interfere with the change from a liquid to a gas. The temperature of salt water must be higher than 100 °C before the water will boil.

▶**Describe:** What happens when the temperature of pure water reaches 100 °C?
The liquid water changes to steam.

101 °C

5 g of salt

102 °C

10 g of salt

104 °C

20 g of salt

Boiling Point Elevation The boiling point of a liquid solvent is raised by adding a solute. This special property of all solutions is called **boiling point elevation.** As the amount of solute in the solvent increases, the boiling point of the solvent also increases. Salt water does not boil at the same temperature as pure water because of boiling point elevation.

▶**Define:** What is boiling point elevation?
raising the boiling point of a liquid solvent by adding solute

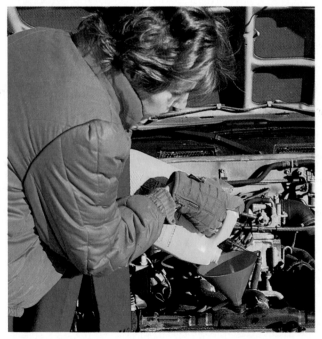

Antifreeze The engine of a car can get very hot. Water is piped around the engine in a cooling system to keep the engine cool. In summer, the temperature of the water in the cooling system can get very high. Drivers add antifreeze to their car's cooling system to prevent overheating. The antifreeze acts like a solute in a solution. Putting antifreeze in the water of the cooling system raises the boiling point of the water. As a result, the chance of the car overheating is reduced.

▶**Compare:** How is antifreeze similar to a solute in a solution?
Both increase the boiling point of a liquid.

LESSON SUMMARY

► The temperature at which a liquid changes to a gas is called its boiling point.

► Raising the boiling point of a liquid solvent by adding solute is called boiling point elevation.

► The greater the amount of solute in a solvent, the higher the boiling point of the solvent.

► Antifreeze is added to a car's cooling system to raise the boiling point of the water.

CHECK *Complete the following.*

1. The _____ of pure water is 100 °C. boiling point

2. The boiling point of salt water is _____ than the boiling point of pure water. higher

3. As the amount of solute in a solution increases, the boiling point of the solution _____. increases

4. Raising the boiling point of a liquid solvent by adding solute is called _____. boiling point elevation

5. Antifreeze added to a car's cooling system _____ the boiling point of water in the cooling system. raises

APPLY *Complete the following.*

6. **Compare:** Solution A contains 5 g of sugar dissolved in 100 mL of water. Solution B contains 20 g of sugar dissolved in 100 mL of water. Which solution has the higher boiling point? Explain your answer.

7. **Analyze:** Two beakers each contain 12 g of salt dissolved in water. Beaker 1 contains 200 mL of water while Beaker 2 contains 100 mL of water. Which solution has the higher boiling point? Explain your answer.

Designing an Experiment............

Design an experiment to solve the problem.

PROBLEM: Does an egg cook faster in tap water or in salt water?

Your experiment should:

1. List the materials you would need.

2. Identify safety precautions that should be followed.

3. List a step-by-step procedure.

4. Describe how you would record your data.

ACTIVITY

OBSERVING BOILING POINT ELEVATION

You will need three beakers, a thermometer, a heat source, a spoon, water, and salt.

1. Put 100 mL of water in each beaker.

2. Add 5 g of salt to the first beaker and stir.

3. Heat the water in the beaker until it begins to boil. Record the temperature.

4. Add 10 g of salt to the second beaker and stir. Repeat Step 3.

5. Add 20 g of salt to the third beaker and stir. Repeat Step 3.

Questions

1. What was the boiling point of the first solution?

2. What was the boiling point of the second solution?

3. What was the boiling point of the third solution?

4. **Analyze:** What is the relationship between the amount of a solute and the boiling point of a solution?

How can a solution be separated?

Objective ▶ Describe two methods for separating the solute from the solvent in a solution.

TechTerms

- **condensation** (kahn-dun-SAY-shun): change of a gas to a liquid
- **distillation** (dis-tuh-LAY-shun): process of evaporating a liquid and then condensing the gas back into a liquid
- **evaporation** (i-vap-uh-RAY-shun): change of a liquid to a gas at the surface of the liquid

Evaporation A solute can be separated from a solution by evaporation. **Evaporation** (i-vap-uh-RAY-shun) is the change of a liquid to a gas at the surface of the liquid. The molecules at the surface of the liquid gain enough energy to break free of the liquid and move into the air as a gas.

Copper sulfate solution

Evaporation of water

Copper sulfate crystals

Suppose you wanted to separate copper sulfate crystals from a solution of copper sulfate and water. You could place the solution in a shallow dish and let it stand. After a few days, all the water would evaporate. Crystals of copper sulfate would remain in the bottom of the dish.

▶ **Infer:** How could salt be separated from a saltwater solution? Let the water evaporate from the solution.

Condensation Have you ever come out of a hot shower to find drops of water on your bathroom mirror? The drops of water are the result of condensation. **Condensation** (kahn-dun-SAY-shun) is the change of a gas to a liquid. Some of the shower water evaporates to form steam. When the steam strikes the mirror, it is cooled. This causes the steam to change back to liquid water.

▶ **Define:** What is condensation? change of a gas to a liquid

Distillation A solution can be separated into its solute and solvent by the process of **distillation** (dis-tuh-LAY-shun). In the process of distillation, a liquid is heated until it evaporates. The gas is then cooled until it condenses back into a liquid.

Evaporation

Condensation

Gas

Solute remains here.

Liquid

Solvent collects here.

When a solution is distilled, both the solvent and the solute can be recovered. The solution to be separated is heated. The solvent evaporates and forms a gas. The gas moves through a tube called a condenser. The condenser cools the gas back to a liquid. The liquid drips into a container. The solute remains in the original container. Both the solute and the solvent are recovered.

▶ **Identify:** What two processes are involved in distillation? evaporation and condensation

LESSON SUMMARY

▶ Evaporation is the process by which a liquid changes to a gas at the surface of the liquid.

▶ Condensation is the process by which a gas changes to a liquid.

▶ Distillation is the process of heating a liquid until it evaporates and then cooling the gas until it condenses back into a liquid.

CHECK *Complete the following.*

1. Evaporation changes a liquid to a ___gas___ .
2. Condensation changes a gas to a ___liquid___ .
3. A solution can be separated into its solute and solvent by _____ . distillation
4. As a liquid is heated, the molecules at the liquid's _____ evaporate first. surface
5. A solvent can be evaporated from a solution to recover the _____ . solute

APPLY *Complete the following.*

▶ **6. Infer:** What causes steam to escape from the spout of a tea kettle?

7. Why do droplets of water form on the top of a pot of boiling water?

TECHNOLOGY AND SOCIETY

FRACTIONAL DISTILLATION OF PETROLEUM

The fossil fuel petroleum contains many important products. Gasoline, kerosene, and heating oil are just some of the products of petroleum. Petroleum is separated into these products by the process of fractional distillation. The petroleum is pumped to the earth's surface from a well deep underground. It is then heated in a fractionating (FRAK-shun-ayt-ing) tower.

The different substances in the petroleum have different boiling points. Each substance evaporates out of the petroleum at a different temperature. The substance with the lowest boiling point evaporates first. Gasoline evaporates when the temperature is between 40 °C and 200 °C. Heating oil evaporates when the temperature is between 250 °C and 400 °C. A condenser cools the substances until they become liquids. The collected liquids are then drained into separate storage tanks.

359

How are crystals formed?

Objectives ► Describe the shape of some crystals. ► Explain how crystals are formed from solutions.

TechTerm

► **supersaturated solution:** solution containing more solute than it can normally hold at a given temperature

Salt Crystals A solution of sodium chloride can be separated by evaporation. The water evaporates from the solution. Crystals of sodium chloride remain in the container. If you were to examine the crystals with a magnifying lens, you would see that salt crystals all have the same shape. Each salt crystal has the shape of a cube. The size of the crystals may be different, but their shape is always a cube.

▁▂▃▶**Describe:** What is the shape of a salt crystal?

a cube

Crystal Shapes The particles that make up a crystal are arranged in a pattern. This pattern gives the crystal a definite shape. All crystals of the same substance have the same shape. Scientists have found that crystals have six basic shapes. The shape of a crystal helps scientists identify unknown substances. Figure 1 shows crystals of copper sulfate.

►*Infer:* Why are all salt crystals shaped like a cube? The particles that make up the salt crystal are arranged in a definite pattern.

Figure 1 Copper sulfate crystals

Growing Crystals One way to grow crystals is to evaporate the solvent from a solution. Another way is to use a supersaturated solution. A **supersaturated solution** contains more solute than it can normally hold at a given temperature. A supersaturated solution can be made from a saturated solution. The saturated solution holds all the solute it can at a given temperature. When a saturated solution is heated, it can hold more solute. When the solution cools, it contains more solute than it would normally hold at room temperature. It is a supersaturated solution.

Figure 2 shows how to grow crystals from a supersaturated solution. First, prepare a supersaturated solution of copper sulfate. Let the solution cool to room temperature. Then add a small copper sulfate crystal to the solution. This extra copper sulfate causes new crystals to form.

▁▂▃▶*Define:* What is a supersaturated solution? solution that contains more solute than it would normally hold at a given temperature

| Saturated | Super saturated | Super saturated | Saturated |

Figure 2 Growing crystals

LESSON SUMMARY

- ► Crystals of the same substance have the same shape.
- ► There are six basic crystal shapes.
- ► A supersaturated solution contains more solute than it would normally hold at a given temperature.
- ► Crystals form when extra solute is added to a supersaturated solution.

CHECK *Complete the following.*

1. When water is evaporated from a saltwater solution, salt _____ remain. crystals

2. All salt crystals are shaped like a ___cube___ .

3. A supersaturated solution contains more _____ than it would normally hold at a given temperature. solute

4. Heating a _____ solution and then adding more solute will form a supersaturated solution. saturated

5. Crystals form when extra _____ is added to a supersaturated solution. solute

APPLY *Complete the following.*

❯ **6. Predict:** What will happen when sugar is added to a supersaturated solution of sugar and water?

7. Suppose you want to identify an unknown substance. You discover that crystals of the substance are shaped like cubes. What could this substance be? How do you know?

Skill Builder ....

▲ *Organizing Information* When you organize information, you put the information in some kind of order. Draw a chart showing the six basic crystal shapes. Use library references to identify a substance whose crystals have each type of shape.

ACTIVITY

GROWING SUGAR CRYSTALS

You will need a glass, boiling water, a spoon, sugar, string, a button, and a pencil.

1. Add some sugar to a half cup of boiling water until no more sugar will dissolve. **Caution:** Be careful not to spill the boiling water on yourself.

2. Pour the sugar solution into a glass.

3. Tie a clean piece of wet string to the pencil. Dip the string in sugar.

4. Tie a button to the end of the string. Place the pencil across the top of the glass so that the string hangs in the water.

5. Let the solution stand for several days.

Questions

👁 **1. Observe:** What formed on the string after several days?

2. Was the sugar solution saturated or supersaturated? Explain.

◣ **3. Analyze:** What caused the result you observed?

STUDY HINT Before you begin the Unit Challenges, review the TechTerms and Lesson Summary for each lesson in this unit.

TechTerms

boiling point (356)
boiling point elevation (356)
concentrated solution (352)
condensation (358)
dilute solution (352)
dissolve (344)
distillation (358)

evaporation (358)
freezing point (354)
freezing point depression (354)
insoluble (346)
polar molecule (348)
saturated solution (352)

soluble (346)
solute (346)
solution (344)
solvent (346)
supersaturated solution (360)
unsaturated solution (352)

TechTerm Challenges

Matching *Write the TechTerm that matches each description.*

1. molecule in which one end has a positive charge and the other end has a negative charge polar molecule
2. temperature at which a liquid changes to a solid freezing point
3. solution containing more solute than it can normally hold at that temperature supersaturated solution
4. change of a gas to a liquid condensation
5. temperature at which a liquid changes to a gas boiling point
6. lowering the freezing point of a liquid solvent by adding solute freezing point depression
7. mixture in which one substance is evenly mixed with another substance solution
8. substance in which a solute dissolves solvent
9. solution containing less solute than it can hold at a given temperature unsaturated solution

Identifying Word Relationships *Explain how the words in each pair are related. Write your answers in complete sentences.* Answers will vary. Accept all logical responses.

1. soluble, insoluble
2. dissolve, solute
3. distillation, evaporation
4. concentrated solution, dilute solution
5. solute, saturated solution
6. boiling point elevation, solute

Content Challenges

True/False *Write true if the statement is true. If the statement is false, change the underlined term to make the statement true.*

1. Stirring a solution <u>speeds up</u> the rate at which a solute dissolves. true
2. The freezing point of salt water is <u>higher</u> than that of pure water. lower
3. Crystals form when extra solute is added to <u>an unsaturated</u> solution. a supersaturated
4. The ends of a water molecule are <u>electrically charged</u>. true
5. The substances in a solution are <u>evenly mixed</u>. true

6. A cement sidewalk is <u>soluble</u> in rainwater. insoluble
7. The <u>solvent</u> in a solution of sugar and water is sugar. solute
8. As the amount of solute in a solution increases, the <u>boiling point</u> of the solution decreases. freezing point
9. A <u>dilute solution</u> contains less solute than solvent. true
10. A saturated solution can be made unsaturated by <u>cooling</u> the solution. heating

Completion *Write the term that best completes each statement.*

1. Extra solute sitting at the bottom of a solution indicates the solution is _____ . saturated
2. As the amount of solute in a solution increases, the _____ of the solution also increases. boiling point
3. In distillation, a liquid _____ and then condenses. evaporates
4. For a solution to form, a _____ must dissolve in a solvent. solute
5. Crystals of the same substance have the same _____ . shape
6. Powdered sugar dissolves _____ than a sugar cube. faster
7. The _____ of pure water is 0 °C. freezing point
8. Adding antifreeze to a car's cooling system raises the _____ of the water it contains. boiling point
9. Placing rock salt on an icy sidewalk lowers the _____ of melted ice. freezing point
10. As water is heated to its boiling point, molecules at the water's _____ evaporate. surface

Understanding the Features ....

Reading Critically *Use the feature reading selections to answer the following. Page numbers for the features are in parentheses.*

1. What do analytical chemists do? (345)
2. How can fishes survive in the cold water of the Arctic Ocean? (355)
3. **Contrast:** How are mechanical weathering and chemical weathering different? (347)
4. **Infer:** Why is the work of water purification chemists important? (349)
5. What is the process by which petroleum is broken down into different products? (359)

Concept Challenges ....

Critical Thinking *Answer each of the following in complete sentences.*

1. The law of conservation of matter states that matter cannot be created or destroyed, but only changed from one form to another. How does the evaporation of water support this law?
2. Compare the effect of an increased amount of solute on both the boiling point and the freezing point of a solution.
3. How could information about crystal shape be used to identify an unknown substance?
4. **Hypothesize:** What effect does evaporation have on the earth's oceans?
5. Club soda contains carbon dioxide gas dissolved in liquid water. Explain why a bottle of club soda goes "flat" when it is left opened at room temperature.

Interpreting a Graph *Use the graph showing the solubility of sodium nitrate in water to answer the following questions.*

1. How many grams of sodium nitrate can be dissolved in 100 g of water at a temperature of 50 °C?
2. At what temperature will 100 g of water dissolve 160 g of sodium nitrate?
3. How much sodium nitrate can be dissolved in 100 g of water at a temperature equal to the boiling point of pure water?
4. How much sodium nitrate can be dissolved in 100 g of water at a temperature equal to the freezing point of pure water?
5. **Analyze:** What is the relationship between the temperature of a solvent and the amount of solvent it can dissolve?

Finding Out More ....

1. Water that contains dissolved minerals is called hard water. Use library references to find out what effect hard water has on pipes and plumbing fixtures. Present your findings to the class.
2. Directions on a box of pasta or spaghetti often say to cook the product in salted, boiling water. Use cookbooks to determine why this is the best way to cook pasta.
3. Honey is an example of a supersaturated solution. Design an experiment by which crystals could be made to grow in honey.
4. Pure drinking water can be made from salt water by the process of distillation. Design an experiment by which salty, ocean water could be distilled to produce pure water and dissolved salts.

SUSPENSIONS

CONTENTS

STUDY HINT Before beginning Unit 20, scan through the lessons in the unit looking for words that you do not know. On a sheet of paper, list these words. Work with a classmate to try to define each word on your list.

20-1 What is a suspension?

Objective ▶ Describe the characteristics of a suspension.

TechTerm

▶ **suspension** (suh-SPEN-shun): cloudy mixture of two or more substances that settle on standing

Suspensions If you add some soil to a jar of water, the water will become cloudy. If you let the mixture stand, you will notice that the soil particles settle to the bottom of the jar. A mixture of soil and water is an example of a **suspension** (suh-SPEN-shun). A suspension is a cloudy mixture of two or more substances that settle on standing. An important thing to remember about suspensions is that they are always temporary. The substances in a suspension may appear to be well mixed at first, but in time they will always separate.

Soil in water

Water

Soil

▐▐▐▶**Define:** What is a suspension? cloudy mixture of two or more substances that settle on standing

Particles in Suspensions The particles in a solution are much too small to be seen, even with the aid of a microscope. That is because the particles in a solution are atoms or molecules. However, the particles in a suspension are much larger than atoms or molecules. You can usually see the particles in a suspension without a microscope.

▐▐▐▶**Compare:** How does the size of particles in a suspension compare with the size of particles in a solution? Particles in a suspension are much larger than particles in a solution.

Properties of Suspensions An important property of suspensions is that the particles of a suspension scatter light. You can see this property if you darken the room and shine a flashlight through a mixture of soil and water. You will see the beam of light as it passes through the cloudy water. One way that you can tell the difference between a solution and a suspension is that the particles of a solution do not scatter light. Table 1 compares some of the properties of solutions and suspensions.

Table 1 Properties of Solutions and Suspensions	
SOLUTION	SUSPENSION
Mixture	Mixture
Clear	Cloudy
Particles evenly mixed	Particles settle on standing
Particles too small to be seen	Particles can be seen

◉**Observe:** What property do suspensions and solutions have in common? Both are mixtures.

Examples of Suspensions A familiar example of a suspension is salad dressing. If you shake a bottle of salad dressing, the contents seem to mix. Once you put the bottle down, however, the ingredients quickly separate. That is why the labels on most bottles of salad dressing state "Shake well before using." Not all suspensions involve liquids. A common suspension of a solid in a gas is dust particles suspended in the air.

▐▐▐▶**List:** What are two common examples of suspensions? salad dressing and dust in the air

366

LESSON SUMMARY

▶ A suspension is a cloudy mixture of two or more substances that settle on standing.

▶ The particles in a suspension are larger than the particles in a solution.

▶ The particles in a suspension scatter light.

▶ Some familiar examples of suspensions include salad dressing and dust in the air.

CHECK *Complete the following.*

1. If a suspension is allowed to stand for a time, the substances will always _____ . separate

2. The appearance of a suspension is cloudy.

3. The _____ in a suspension are larger than atoms or molecules. particles

4. An example of a suspension in a gas is _____ in air. dust

5. Solutions and suspensions are similar in that both are _____ . mixtures

APPLY *Complete the following.*

6. **Contrast:** How are the particles in a suspension different from the particles in a solution?

7. **Classify:** Look at Table 1 on page 366. Are the properties listed physical properties or chemical properties? How do you know?

Ideas in Action ....

IDEA: Many common foods and household items are suspensions.

ACTION: Go through your local supermarket and make a list of the different products you find that are suspensions. How do you know they are suspensions? Note how many of these or similar products are used in your home.

ACTIVITY

READING MEDICINE LABELS

You will need labels or packaging from three different over-the-counter or prescription drugs, paper, and a pencil.

1. Examine one of your drug labels. Write the name of the drug and what it is used for.

2. Note and record whether the drug is an over-the-counter or a prescription drug.

3. Find and record the dosage instructions.

4. Find and record the expiration date.

5. See if there are any special warnings or precautions. Write them down.

6. Repeat Steps 1–5 for each of your drug labels.

Questions

1. What information is given in the dosage instructions?

2. What does the expiration date tell you? Why is this important?

3. How is a prescription drug label different from an over-the-counter drug label?

4. What types of warnings or precautions are given for various drugs? Why are these precautions important?

How can a suspension be separated?

Objective ▶ Describe some ways to separate a suspension.

TechTerms

- ▶ **coagulation** (koh-ag-yoo-LAY-shun): use of chemicals to make the particles in a suspension clump together
- ▶ **filtration:** separation of particles in a suspension by passing it through paper or other substances

Settling Particles in a suspension settle on standing. Large particles settle out quickly. Smaller particles take a longer time to settle. You can see how this works if you mix sand and clay with water, and allow the mixture to stand. The sand will settle to the bottom in a few minutes. The clay will stay in the water much longer. You would have to let the mixture stand overnight in order for the clay to settle.

▶**Explain:** Why does sand settle out much faster than clay when mixed with water? Sand particles are much larger than clay particles.

Filtration One way that a suspension can be separated quickly is by **filtration.** Filtration is the removal of particles in a suspension by passing the suspension through a filter. Filters can be made of paper or other substances. Filters have tiny holes, or pores, through which some substances can pass, but not others. Substances that cannot pass through the filter have particles that are larger than the holes in the filter.

▶*Predict:* What will happen to the particles in a suspension if they are larger than the holes in a filter? The particles will not pass through the filter.

Coagulation Another way to make a suspension separate quickly is to add chemicals to the suspension. The chemicals make the particles of the suspension stick together. The particles form clumps that are larger and heavier than the original particles. As a result, the particles settle out more quickly. This process is called **coagulation** (koh-ag-yoo-LAY-shun). Coagulation takes place when you cut your finger. Chemicals in your blood cause the blood to coagulate and form a clot.

▶*Define:* What is coagulation? use of chemicals to make particles in a suspension clump together

Separation by Centrifuge A third way to separate the substances in a suspension is to spin the mixture at high speeds. The device that is used is called a centrifuge (SEN-truh-fyooj). As the suspension is spun around, the particles in the suspension are pulled down to the bottom of the container. Use of a centrifuge greatly increases the rate at which a suspension settles.

▶*Identify:* What is a centrifuge? device used to spin a suspension at high speeds

Suspension

Particles collect in filter

LESSON SUMMARY

▶ The particles in a suspension settle out on standing.

▶ Filtration is a method of separating a suspension by passing it through a filter.

▶ Coagulation is a process in which chemicals are used to make the particles in a suspension clump together.

▶ A centrifuge is a device that separates a suspension by spinning it at high speeds.

CHECK *Complete the following.*

1. What happens when a suspension is left to stand overnight?

2. What is filtration?

3. What kind of particles will pass through the holes in a filter?

4. What is coagulation?

5. What is one example of coagulation?

6. How is a centrifuge used to separate the particles in a suspension?

7. What is another name for the holes in a piece of filter paper?

APPLY *Complete the following.*

8. **Classify:** Decide whether each statement describes separation of a suspension by filtration, coagulation, or centrifuge.
   a. A solution of ammonium hydroxide and alum is added to a clay-and-water suspension.
   b. A suspension is passed through a piece of linen cloth.
   c. A mixture of clay, sand, and gravel is spun at high speed.
   d. A person is given a drug to make the blood clot more readily.

Skill Builder

▲**Modeling** You can make a model to show how coagulation works. Fill two test tubes about half full of water. Add a small amount of clay to each test tube. To one of the test tubes, add several drops of alum solution. Observe both test tubes for several minutes. You should see the alum solution form a jelly-like material that causes the clay particles to clump together. In which test tube do the clay particles settle faster? Why?

SCIENCE CONNECTION

THE MISSISSIPPI DELTA

As a river flows, it carries particles of clay, sand, and gravel. These particles are called sediments. The sediments are suspended in the water.

A river picks up sediments as it floods and erodes its banks. A fast-moving river carries the most sediments. As the banks of a river widen, the river slows down. When the river slows down, sediments settle out of suspension and are deposited.

The widest part of a river is its mouth, where it empties into a larger body of water. The mouth of the Mississippi River is located at the Gulf of Mexico. Here, the Mississippi moves so slowly that sediments from the river are deposited. Gradually, the sediments form new land. The land that is formed at the mouth of a river is called a delta. The land of the Mississippi delta is good for farming, because new topsoil is always being deposited.

20-3 What is an emulsion?

Objective ▶ Describe and give examples of an emulsion.

TechTerms

▶ **emulsion** (i-MUL-shun): suspension of two liquids

▶ **homogenization** (huh-mahj-uh-ni-ZAY-shun): formation of a permanent emulsion

Emulsions When a liquid is suspended in another liquid, the result in an **emulsion** (i-MUL-shun). An emulsion is a suspension of two liquids. Milk, paint, and many medicines are examples of emulsions.

You can make an emulsion by mixing some cooking oil with water, and then shaking the mixture. This emulsion will not stay mixed for long. If you let the mixture stand, the oil and water will soon separate. An emulsion that does not stay mixed is called a temporary emulsion.

▶**Predict:** What will happen to a temporary emulsion? When the emulsion is left standing, the two liquids will separate.

Permanent Emulsions Many commercial products that are emulsions do not separate on standing. These emulsions are called permanent emulsions. The particles in a permanent emulsion are much smaller than the particles in a temporary emulsion. The particles in a permanent emulsion are small enough to stay in suspension.

A familiar example of a permanent emulsion is homogenized milk. **Homogenization** (huh-mahj-uh-ni-ZAY-shun) is the formation of a permanent emulsion. Fresh milk is a temporary emulsion that quickly separates into milk and cream. Fresh milk is homogenized in a machine that breaks down the cream into very small particles. The small particles of cream remain permanently suspended in the milk.

▶**Contrast:** What is the difference between a temporary emulsion and a permanent emulsion? Temporary emulsions separate on standing; permanent emulsions do not.

Emulsifying Agents Many detergents or other cleaning products contain substances called emulsifying (i-MUL-suh-fy-ing) agents. An emulsifying agent keeps an emulsion from separating. The soap in cleaning products is an emulsifying agent. Soap breaks apart grease or dirt into smaller particles. These particles are small enough to form a permanent emulsion with water. The dirt or grease is washed away in the water. Other emulsifying agents include gelatin and egg yolk. These substances are often used in food to keep ingredients from separating.

▶**Identify:** What is an emulsifying agent? substance that keeps an emulsion from separating

LESSON SUMMARY

▶ An emulsion is a suspension of two liquids.

▶ Temporary emulsions separate on standing, while permanent emulsions do not.

▶ Homogenization is the formation of a permanent emulsion.

▶ Emulsifying agents are substances that prevent an emulsion from separating.

CHECK *Complete the following.*

1. An emulsion is a suspension of a _____ in a liquid. liquid

2. Oil and water separate on standing because they form a _____ emulsion. temporary

3. The particles in a _____ emulsion are small enough to stay in suspension. permanent

4. Soap is an example of an __emulsifying__ agent.

5. Milk and cream form a permanent emulsion through the process of _____ . homogenization

APPLY *Complete the following.*

6. **Hypothesize:** Bile is produced by the liver. It emulsifies the fats that a person eats. Why is this important to the digestive process?

7. **Infer:** When you buy a can of paint, you usually have to stir the paint before you can use it. Why do you think it is necessary to stir the paint?

Skill Builder

▲*Modeling* You can model the effects of an emulsifying agent by performing this activity. Make a mixture of vegetable oil and water. Add equal amounts of the mixture to two test tubes. To one test tube, add some liquid soap. Shake both test tubes thoroughly. Let the mixtures stand. Which mixture does not separate? Can you explain why this happens?

ACTIVITY

MAKING AN EMULSION

You will need vinegar, vegetable oil, an egg, a bowl, a measuring cup, and an eggbeater.

1. Separate the yolk from the white of the egg. Put the yolk in the bowl.

2. Beat the egg yolk until it looks foamy.

3. Add 1/4 cup of vinegar to the egg yolk. Beat the mixture of vinegar and egg yolk.

4. Add 1/8 cup of oil to the mixture one tablespoon at a time. Beat the mixture thoroughly each time you add a tablespoon of oil.

5. Season the mixture with lemon juice, salt and pepper, or other spices.

Questions

1. **Observe:** When did the mixture begin to thicken?

2. If the mixture is allowed to stand, will it separate?

3. What is the emulsifying agent in this mixture?

4. **Predict:** What would happen if you did not add the egg yolk to the mixture?

20-4 What is a colloid?

Objective ▶ Describe and give examples of a colloid.

TechTerm

▶ **colloid** (KAHL-oyd): suspension in which the particles are permanently suspended

Colloids What do whipped cream, fog, mayonnaise, and smoke have in common? All of these substances are colloids (KAHL-oydz). A **colloid** is a suspension in which the particles are permanently suspended. Colloids do not separate on standing.

Colloids can be mixtures of different phases of matter. Table 1 shows some common types of colloids and examples of each type.

Table 1	Types of Colloids	
NAME	PHASE	EXAMPLES
Foam	gas in liquid	shaving cream, whipped cream
Sol	solid in liquid	paint, dyes
Emulsion	liquid in liquid	mayonnaise
Fog	liquid in gas	clouds, fog
Smoke	solid in gas	smoke in air
Gel	liquid in solid	butter, jelly

Define: What is a colloid? suspension in which the particles are permanently suspended

Filter paper

Homogenized milk

Colloid Particle Size The particles in a colloid are not as small as the particles in a solution. However, they are much smaller than the particles in an ordinary suspension. They cannot be seen with a microscope. Because the particles in a colloid are so small, a colloid cannot be separated by filtration. A colloid such as homogenized milk passes right through filter paper. The particles in milk are smaller than the holes in the filter.

Explain: Why can a colloid not be separated by filtration? The particles are small enough to pass through the holes in the filter.

Movement of Colloid Particles Particles in a colloid are kept in suspension because they are being bombarded by the molecules around them. For example, the particles in smoke are always colliding with air molecules. These collisions keep the particles from settling out. These collisions also cause a colloid to scatter light. The cloudy appearance of a colloid such as milk is due to the scattering of light.

Explain: Why does a colloid appear cloudy? The colloid scatters light.

LESSON SUMMARY

▶ A colloid is a suspension in which the particles are permanently suspended.

▶ The particles of a colloid are larger than the particles of a solution, but smaller than those of an ordinary suspension.

▶ Particles of a colloid are kept in suspension because they are always colliding with the molecules around them.

CHECK *Write true if the statement is true. If the statement is false, change the underlined term to make the statement true.*

1. Particles in a colloid are <u>smaller</u> than those in a solution. larger

2. Colloids are <u>permanent</u> suspensions. true

3. Colloids <u>can</u> be separated by filtration. cannot

4. Colloids <u>do not</u> scatter light. do

5. A colloid consisting of a liquid in a solid is called <u>an emulsion</u>. a gel

6. Whipped cream is an example of a <u>foam</u>. true

APPLY *Complete the following.*

▶ 7. **Infer:** Why can the beams of a car's headlights be seen in fog?

8. **Compare:** In what ways are colloids like solutions? In what ways are they different?

📁 9. **Classify:** Identify each of the following colloids as a sol, gel, fog, or foam: **a.** whipped cream **b.** clouds **c.** grape jelly **d.** paint.

....
InfoSearch ....

Read the passage. Ask two questions that you cannot answer from the information in the passage.

Brownian Motion If you look at a colloid through a very powerful microscope, you will see that the particles of a colloid are in continuous random motion. This motion is called Brownian motion. Brownian motion is named for the biologist Robert Brown. Brown first noticed this motion while observing the motion of particles in a suspension of pollen grains in water.

SEARCH: Use library references to find answers to your questions.

◆○◆○◆ SCIENCE CONNECTION ◆○◆○◆○◆○◆○◆○◆○◆○◆○◆○◆○◆○◆○◆○◆○◆◇

HOMOGENIZED MILK

Fresh milk is a suspension that separates on standing into skim milk and cream. Cream is made up mostly of fats, which are less dense than the watery parts of skim milk. As a result, the cream floats on top when skim milk and cream separate.

Homogenization is a process in which skim milk and cream are mixed together to form a permanent suspension, or colloid. In a machine called a homogenizer, the milk-and-cream mixture is forced by a high-pressure pump through small openings in a metal plate. This breaks the fat droplets into smaller pieces. When the fat droplets are small enough, they will remain permanently suspended in the milk.

Unlike fresh milk, homogenized milk is a homogeneous, or uniform, mixture. This means that the particles of cream are spread evenly throughout the milk. According to the U.S. Public Health Service, the fat content of the upper 100 mL of a quart of homogenized milk that has been left standing cannot differ by more than 10% from that of the rest of the milk.

20-5 What are air and water pollution?

Objective ▶ Describe some causes of air and water pollution.

TechTerms

▶ **pollutants** (puh-LOOT-unts): harmful substances in the environment

▶ **potable** (POHT-uh-bul) **water:** water that is safe to drink

Pollution What would happen if you did not have air to breathe or water to drink? You would not be able to live for very long. Every day, these important resources are threatened by pollution. Pollution is the release of harmful substances into the environment. **Pollutants** (puh-LOOT-unts) are the harmful substances that enter the environment as a result of pollution. Pollutants can be gases, liquids, or solids.

▶**Define:** What are pollutants?
harmful substances that enter the environment

Causes of Pollution Most pollution is caused by human activities. Exhaust from cars is the major source of air pollution. Waste products from factories pollute both air and water. Burning fossil fuels such as coal, oil, and natural gas release poisonous gases into the atmosphere.

▶**List:** What are three sources of air and water pollution? car exhaust, waste products from factories,

Reducing Air Pollution Air pollution can be harmful to people in many ways. Particles in the air can cause irritations of the eyes, nose, and throat. They can cause breathing problems and respiratory diseases.

Everyone can help to reduce air pollution. You can ride a bike or walk short distances instead of using a car or bus. You might form a car pool with friends to help reduce the number of cars on the road. By lowering the temperature in your home, you can use less heating oil. Find out what else you can do to help reduce air pollution.

▶**Infer:** How would fewer cars on the road help to reduce air pollution? Pollution from car exhaust would be reduced.

Safe Water Water pollution is caused by dumping sewage and chemical wastes from factories into lakes, river, and streams. These pollutants harm fish and other organisms that live in the water. They also make the water unsafe to drink. Water that is safe to drink is called **potable** (POHT-uh-bul) **water**. Water can be made safe to drink by a series of processes called purification (pyoor-uh-fi-KAY-shun). One way of purifying water is to add chemicals, such as chlorine, that kill germs. Other steps taken to purify water include settling, coagulation, and filtration. These methods all remove pollutant particles that are suspended in water.

▶**Identify:** What is water that is safe to drink?
potable water

LESSON SUMMARY

▶ Pollution is the release of harmful substances into the environment.

▶ Most pollution is caused by human activities.

▶ Air pollution is harmful to people.

▶ Everyone can help to reduce air pollution.

▶ Water that is safe to drink is called potable water.

CHECK *Complete the following.*

1. Pollutants are substances that are _____ to the environment. harmful

2. Car exhaust is the major source of _____ pollution. air

3. Using less fuel in home heating helps to reduce _____ pollution. air

4. Potable water is water that is __safe__ to drink.

5. Chemical wastes and sewage are major causes of _____ pollution. water

APPLY *Complete the following.*

🔲 6. **Hypothesize:** Some trees that are planted on city streets often do not grow well. Can you think of a reason for this?

▶ 7. **Infer:** Why would a water purification technician need to know about suspensions and colloids?

Health and Safety Tip ....

Although the water in a stream or brook may look fresh and clean, you should not drink it. Water pollution often affects water far from the source of the pollution. A mountain stream may be polluted by waste products from a factory many kilometers away. Some harmful chemicals are odorless and tasteless. This makes them very hard to observe in water that looks and tastes pure. Use library references to find out how you can purify water to make it safe for you to drink.

ACTIVITY

OBSERVING POLLUTANTS IN AIR

You will need glass slides, petroleum jelly, and a hand lens.

1. Coat one side of each glass slide with a thin layer of petroleum jelly.

2. Choose several indoor and outdoor spots for testing the air quality.

3. At each spot, place a slide with the coated side up.

4. Record the location and the time when you placed each slide. Leave the slides overnight.

5. Collect the slides the next day. Record the time of collection.

6. Using the hand lens, examine each slide. Record your observations.

Questions

👁 1. **Observe:** What kinds of particles did you see on the slides?

2. Which slide had the most particles?

3. Which slide had the fewest particles?

🔲 4. **Hypothesize:** How can you explain your results?

STUDY HINT Before you begin the Unit Challenges, review the TechTerms and Lesson Summary for each lesson in this unit.

TechTerms . . . .

coagulation (368)	filtration (368)	potable water (374)
colloid (372)	homogenization (370)	suspension (366)
emulsion (370)	pollutants (374)	

TechTerm Challenges . . . .

Matching *Write the TechTerm that matches each description.*

1. water that is safe to drink potable water
2. make the particles is a suspension clump together coagulation
3. formation of a permanent suspension homogenization
4. suspension in which particles are permanently suspended colloid
5. harmful substances in the environment pollutants
6. suspension of two liquids emulsion

Fill In *Write the TechTerm that best completes each statement.*

1. A suspension is a cloudy mixture of two or more substances that settles on standing.
2. Passing a suspension through paper or other substances is called _____ . filtration
3. A suspension of two liquids is called an _____ . emulsion
4. Particles in a suspension are clumped together by the process of _____ . coagulation.

Content Challenges . . . .

Multiple Choice *Write the letter of the term or phrase that best completes each statement.*

1. If a suspension of clay, sand, and gravel is allowed to stand, the particles that settle out first would be
   a. clay. **b.** sand. **c.** gravel. **d.** water. a

2. Colloids cannot be separated by filtration because colloid particles are
   a. round. **b.** too large. **c.** too small. **d.** clumped together. c

3. A process that speeds up the separation of a suspension is
   a. homogenization. **b.** pollution. **c.** emulsification. **d.** coagulation. d

4. An example of an emulsifying agent is
   a. milk. **b.** egg yolk. **c.** fog. **d.** oil and water. b

5. A device that separates a suspension by spinning is called a
   a. centrifuge. **b.** homogenizer. **c.** filter. **d.** coagulator. a

6. Both colloids and suspensions
   a. have large particles. **b.** are clear. **c.** scatter light. **d.** settle on standing. c

7. The particles in a colloid are
   a. smaller than in a solution. b. smaller than in a suspension. c. smaller than molecules.
   d. larger than in a suspension. b

8. An emulsifying agent makes an emulsion that is
   a. temporary. b. liquid. c. permanent. d. soapy. c

9. All of the following are colloids *except*
   a. fog. b. salad dressing. c. smoke. d. whipped cream. b

10. Settling, coagulation, and filtration are examples of
    a. suspensions. b. emulsifying agents. c. homogenization. d. water purification. d

True/False *Write true if the statement is true. If the statement is false, change the underlined term to make the statement true.*

1. The appearance of a suspension is <u>clear</u>. cloudy
2. Light is scattered by particles in a <u>suspension</u>. true
3. Unhomogenized milk is a <u>temporary</u> suspension. true
4. When a suspension is left standing, the <u>larger</u> particles are the last to settle. smaller
5. Solutions, colloids, and suspensions <u>are</u> all mixtures. true
6. An emulsion is a suspension of two <u>gases</u> liquids
7. A <u>permanent</u> emulsion will not settle on standing. true
8. A colloid <u>does not</u> scatter light. does
9. Potable water is <u>unsafe</u> to drink. safe
10. Butter is an example of a <u>gel</u>. true

Understanding the Features ....

Reading Critically *Use the feature reading selections to answer the following. Page numbers for the features are in parentheses.*

1. What is a delta? (369)
2. Why is the land in a delta good for farming? (369)
3. Why does cream float to the top of fresh milk? (373)
4. How does a homogenizer work? (373)
5. **Hypothesize:** Why do you think a fast-moving river can carry more sediment than a slow-moving river? (369)

Concept Challenges ....

Critical Thinking *Answer each of the following in complete sentences.*

1. Explain how an emulsifying agent works.
2. **Contrast:** In what ways do colloids differ from ordinary suspensions?
3. How does the constant motion of colloid particles affect the properties of a colloid?
4. How could you test a mixture to find out if it is a solution, a suspension, or a colloid?

Interpreting a Diagram *The following picture shows several sources of air and water pollution. Use the picture to answer the questions.*

1. What sources of air pollution are shown in the picture?
2. What sources of water pollution are shown?
3. Which pollutants are gases? Which are liquids? Which are solids?
4. How could each source of air pollution be reduced?
5. Would you expect to find healthy fish and plants in the lake? Why or why not?
6. A rural area 20 km away from the area shown in this picture has no factories or other industry. However, the town's water supply is polluted with chemical wastes. How might this be explained?

Finding Out More . . . .

1. Make a chart to show the properties of solutions, suspensions, and colloids. Include in your chart common examples of each type of mixture. You may wish to draw or cut pictures from magazines to illustrate your examples.
2. Certain colloids can be separated by a process called electrophoresis (i-lek-troh-fuh-REE-sis). Use library references to find out what electrophoresis is and how it is used. Write a report of your findings.
3. Find out about the life and work of Thomas Graham, who is known as the "father of colloid chemistry." Present your findings in an oral report to the class.

4. A major pollution problem is acid rain. Find out how acid rain is caused and how it affects the environment. Make a diagram to show the steps in the formation of acid rain. Then find out what is being done to reduce acid rain.
5. Find out how water is purified in your town or city. If possible, visit a water treatment facility and observe the different processes that are being used. Find out what types of pollutants are most likely to affect the water supply in your area.

ACIDS, BASES, AND SALTS

CONTENTS

STUDY HINT Before beginning Unit 21, scan through the lessons in the unit looking for words that you do not know. On a sheet of paper, list these words. Work with a class-mate to try to define each word on your list.

21-1 What is an acid?

Objective ▶ Define and give some examples of acids.

TechTerm

▶ **acid:** substance that reacts with metals to release hydrogen

Acids What do vinegar, lemons, and sour milk have in common? They all contain acids. An **acid** is a substance that reacts with metals to release hydrogen. Acids give vinegar, lemons, and sour milk their sour taste. Remember, however, that you should never taste a substance to find out what it is.

�▶*Identify:* What is it that gives lemons their sour taste? *an acid*

Common Acids Acids are found in many different substances. Citrus fruits, such as lemons and oranges, contain citric acid. Sour milk contains lactic acid. Vinegar contains acetic acid. Table 1 lists some common acids. Notice that the symbol for hydrogen appears first in the chemical formula for an acid.

▶*Analyze:* What is the chemical formula of the acid in vinegar? $HC_2H_3O_2$

Properties of Acids All acids contain hydrogen. The properties of acids are caused by the hydrogen in the acids. When an acid is added to water, the acid forms positive and negative ions. For exam-

Table 1 Common Acids

ACID	CHEMICAL FORMULA	USES
Acetic acid	$HC_2H_3O_2$	vinegar
Boric acid	H_3BO_3	eyewash
Carbonic acid	H_2CO_3	club soda
Citric acid	$H_3C_6H_5O_7$	citrus fruits
Hydrochloric acid	HCl	aids digestion
Nitric acid	HNO_3	fertilizers
Sulfuric acid	H_2SO_4	plastics

ple, hydrochloric acid (HCl) forms positive hydrogen ions and negative chloride ions.

$$HCl \longrightarrow H^+ + Cl^-$$

A hydrogen atom contains one electron and one proton. When the electron is lost to form a positive ion, only the proton is left. Therefore, a hydrogen ion is the same as a proton. For this reason, acids are sometimes called proton donors. The more hydrogen ions an acid releases in water, the stronger the acid.

Acids react with metals to release hydrogen gas. The diagram shows what happens when hydrochloric acid is added to zinc metal.

$$2HCl + Zn \longrightarrow ZnCl_2 + H_2$$

▶*Name:* What is another name for acids?
proton donors

Hydrogen gas

Zinc

Hydrochloric acid

380

LESSON SUMMARY

▸ Acids can be identified by their sour taste.

▸ Acids are found in many common substances.

▸ Acids are also called proton donors.

▸ Acids react with metals to produce hydrogen gas.

CHECK *Write true if the statement is true. If the statement is false change the underlined term to make the statement true.*

1. Sour milk contains <u>acetic</u> acid. lactic

2. The stomach produces <u>hydrochloric</u> acid to aid in digestion. true

3. Acids release <u>hydrogen</u> ions when placed in water. true

4. Acids are <u>electron</u> donors. proton

5. The formula for <u>boric</u> acid is $HC_2H_3O_2$. acetic

6. Lemons contain <u>lactic</u> acid. citric

APPLY *Complete the following.*

7. **Infer:** Why are acids called proton donors?

8. **Classify:** Which of the following chemical formulas are acids? How do you know?
   a. HCl
   b. NaCl
   c. H_3PO_4
   d. $CaSO_4$
   e. $HC_2H_3O_2$
   f. KBr
   g. CO_2
   h. H_2CO_3

Health and Safety Tip

Strong acids, such as sulfuric acid, can be dangerous if spilled on your skin. Use library references to find out how to handle acids safely in the science laboratory.

SCIENCE CONNECTION

ACID RAIN

Acid rain is caused by the release of sulfur dioxide and nitrogen oxides into the atmosphere. Factories that burn fuels with a high sulfur content release these pollutants. The pollutants react with water in the atmosphere to form sulfuric acid and nitric acid. These acids then fall to the earth in rain and snow.

Weather patterns in the Northern Hemisphere move from West to East. For this reason, the northeastern United States and eastern Canada have been most severely affected by acid rain. More power plants and factories all over the earth have made acid rain a worldwide problem.

Acid rain wears away stone and causes damage to metal surfaces. It also kills some types of trees. The acid content of some lakes makes them unsuitable for fish and other aquatic life. Acid rain has caused damage to the environment in many parts of the world. International cooperation is needed to eliminate or reduce the harmful effects of acid rain.

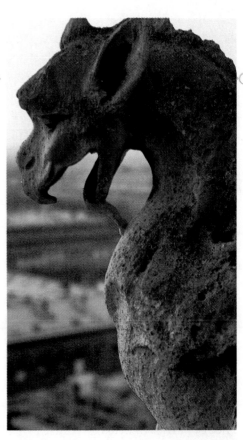

21-2 What is a base?

Objective ► Define and give some examples of a base.

TechTerms

▶ **base:** substance formed when metals react with water

▶ **hydroxyl** (hy-DRAHK-sil) **ion:** negative ion made up of one atom of hydrogen and one atom of oxygen

Bases The early settlers in the United States made their own soap from animal fats and ashes. Soaps are made by chemically combining fats or oils with a type of chemical compound called a **base.** Ashes contain the base potassium hydroxide. A base is a substance that is formed when a metal reacts with water.

||||▶**Describe:** How are soaps made?
by chemically combining fats or oils with a base

Properties of Bases Bases taste bitter and feel slippery. Milk of magnesia, ammonia, and soap are bitter-tasting substances that contain bases. The slippery feel of soap is caused by the base it contains. Like acids, strong bases are dangerous. They can burn the skin.

||||▶**Identify:** What are two properties of bases?
They taste bitter and feel slippery.

Composition of Bases When some metals are placed in water, a chemical reaction takes place. The reaction produces a base plus hydrogen. For example, sodium metal reacts with water to produce sodium hydroxide and hydrogen.

Look at the chemical formulas of the bases listed in Table 1. They all contain a group called a **hydroxyl** (hy-DRAHK-sil) **ion.** The hydroxyl ion, OH^-, is a negative ion made up of one atom of hydrogen and one atom of oxygen. All bases release hydroxyl ions in water.

$$NaOH \longrightarrow Na^+ + OH^-$$

Hydroxyl ions can combine with hydrogen ions to form water. Remember that a hydrogen ion is a proton. Therefore, bases are often called proton acceptors.

||||▶**Explain:** Why are bases often called proton acceptors? The hydroxyl ions in bases combine with hydrogen ions to form water.

Table 1 Common Bases		
BASE	CHEMICAL FORMULA	USES
Potassium hydroxide	KOH	soap
Magnesium hydroxide	$Mg(OH)_2$	milk of magnesia
Calcium hydroxide	$Ca(OH)_2$	mortar
Ammonium hydroxide	NH_4OH	ammonia
Sodium hydroxide	NaOH	soap

LESSON SUMMARY

▶ A base is a chemical compound formed when a metal reacts with water.

▶ Bases taste bitter and have a slippery feel.

▶ All bases contain a negative hydroxyl ion (OH^-).

▶ Bases are often called proton acceptors.

CHECK *Write true if the statement is true. If the statement is false, change the underlined term to make the statement true.*

1. Soaps are made by chemically combining <u>an acid</u> with fats or oils. a base

2. Strong bases are dangerous to handle because they can <u>burn the skin</u>. true

3. Bases form when <u>gases</u> react with water. metals

4. <u>Sodium</u> hydroxide is the base used in milk of magnesia. Magnesium

5. The base used in mortar is <u>potassium</u> hydroxide. calcium

6. The negative ion found in all bases is called a <u>hydrogen</u> ion. hydroxyl

7. Bases are proton <u>acceptors</u>. true

APPLY *Complete the following.*

8. **Hypothesize:** How does the concentration of hydroxyl ions affect the strength of a base?

9. Match each of the following bases with its correct chemical formula.
   a. calcium hydroxide   1. NaOH
   b. potassium hydroxide   2. $Ca(OH)_2$
   c. sodium hydroxide   3. NH_4OH
   d. ammonium hydroxide   4. KOH

Skill Builder ....

Experimenting Bases are used to make soaps and other cleaning agents. You can compare the strengths of different cleaning agents by doing this experiment. Rub some butter or margarine on three plates. Try to clean the first plate with a sponge soaked in plain water. Use a sponge soaked in ammonia water on the second plate, and a sponge soaked in water and liquid soap on the third plate. Which cleaning agent removed the grease most easily?

SCIENCE CONNECTION

SOAP

Soaps are made up of molecules with a split personality. One end of a soap molecule is soluble in water. This end of the soap molecule is attracted to water molecules. The other end is soluble in oil. This is why soap can remove grease and oil from clothes or from your skin.

What happens when you wash dirty clothes in soapy water? Soap molecules gather around a grease spot with their oil-soluble ends pointing in. The water-soluble ends of the molecules point out into the water. The soap-covered grease spot is then washed away by the water.

Soap is made by boiling fats with a solution of sodium hydroxide. Sodium hydroxide is also called lye. The following equation shows the soap-making process:

$$fat + lye \longrightarrow soap + glycerine$$

Sometimes, the glycerine is left in the soap. Glycerine can also be used to make cellophane, printer's ink, cosmetics, and medicines.

21-3 What are indicators?

Objective ► Describe how indicators are used to identify acids and bases.

TechTerm

► **indicator** (IN-duh-kayt-ur): substance that changes color in acids and bases

Litmus Litmus paper can be used to identify an acid or a base. Vinegar is an acid. If you dip one end of a strip of blue litmus paper into vinegar, the blue litmus turns red. If you dip one end of a strip of red litmus paper into soapy water, the red litmus turns blue.

Blue litmus paper

Vinegar

Red litmus paper

Soapy water

👁**Observe:** What color does blue litmus paper turn in an acid? red

Indicators Chemicals that change color in acids or bases are called **indicators** (IN-duh-kayt-urz). Litmus is an indicator that turns red in acids and blue in bases. Phenolphthalein (fee-nohl-THAL-een) is another indicator. Phenolphthalein is colorless in acids and pink to red in bases. The indicator methyl red is red in acids and yellow in bases. Table 1 shows some common indicators.

▮▮▶**Name:** What are three common indicators?
litmus, phenolphthalein, methyl red

Everyday Indicators Many common, everyday substances are indicators. Grape juice is a good

Table 1 Indicators

INDICATOR	COLOR IN ACIDS	COLOR IN BASES
Litmus	red or pink	blue
Phenolphthalein	colorless	pink or red
Methyl red	red	yellow
Congo red	blue	red
Bromthymol blue	yellow	blue

indicator. It is pink or red in acids and green or yellow in bases. Hydrangeas (hy-DRAYN-juhz) have pink flowers in basic soil and blue flowers in acidic soil. Red cabbage, beets, rhubarb, cherries, blueberries, and blackberries all can be used as indicators.

▮▮▶**Explain:** How do hydrangeas act as acid–base indicators?
They have pink flowers in basic soil and blue flowers in acidic soil.

Table 2 Properties of Acids and Bases

ACIDS	BASES
Taste sour	Taste bitter
Contain hydrogen ions and a nonmetallic element	Contain hydroxyl ions and a metal
Turn blue litmus paper red	Turn red litmus paper blue
React with metals to release hydrogen	Feel slippery

LESSON SUMMARY

► Litmus paper can be used to identify acids and bases.

► Indicators are chemical substances that change color in acids and bases.

► Many everyday substances are indicators.

CHECK *Complete the following.*

1. Blue litmus paper turns red in _____ . *acids*
2. Phenolphthalein turns *colorless* in acids.
3. Methyl red indicator turns yellow in ___*bases*___ .
4. Congo red turns _____ in acids. *blue*
5. Grape juice is _____ in bases. *yellow or green*
6. Hydrangeas have __*pink*__ flowers in basic soil.

APPLY *Complete the following.*

7. An unknown solution turns bromthymol blue from yellow to blue. Is the solution an acid or a base? How do you know?

8. **Classify:** Identify each of the following properties as a property of an acid or a base.
   a. Tastes bitter.
   b. Reacts with metals to produce hydrogen.
   c. Turns methyl red from red to yellow.
   d. Feels slippery.
   e. Tastes sour.
   f. Turns phenolphthalein from colorless to pink.

Designing an Experiment..........

Design an experiment to solve the problem.

PROBLEM: If you plant hydrangeas in your garden, will the flowers be pink or blue?

Your experiment should:

1. List the materials you would need.
2. Identify safety precautions that should be followed.
3. List a step-by-step procedure.
4. Describe how you would record your data.

ACTIVITY

MAKING AN INDICATOR

You will need red cabbage leaves, a heat source, water, a beaker, a strainer, samples to be tested, and a graduated cylinder.

1. Boil several red cabbage leaves in water until the mixture turns dark red. **Caution: Be careful when boiling liquids.**
2. Let the mixture cool. Pour the liquid through a strainer into a beaker.
3. Choose several samples to be tested, for example, milk, fruit juice, shampoo, household cleaner, soda, liquid soap. Pour a small amount of each sample into a separate container.
4. Add 2 mL of red cabbage indicator to each of your samples. Observe what happens.

Questions

1. **Observe:** What color did the indicator turn in each sample?
2. What is the color of the indicator in acids?
3. What is the color of the indicator in bases?

21-4 ▸ What is the pH scale?

Objective ▸ Describe how the pH scale is used to measure the strength of acids and bases.

TechTerms

- ▸ **neutral:** neither acidic nor basic
- ▸ **pH scale:** measure of the concentration of hydrogen or hydroxyl ions in a solution

Strength of Acids and Bases Acids and bases can be strong or weak. Sulfuric acid and nitric acid are strong acids that can burn the skin. Carbonic acid and boric acid are weak acids. Boric acid is even used as an eyewash. Sodium hydroxide and potassium hydroxide are strong bases. Ammonium hydroxide is a weak base that is used as a household cleaner. Aluminum hydroxide a weak base that is used as an antacid.

The strength of an acid or a base depends on the number of hydrogen ions (H^+) or hydroxyl ions (OH^-) in the solution. Adding water will reduce the concentration of ions and change the strength of the solution. When water is added to a strong acid or base, the acid or base becomes weaker.

▸*Identify:* What determines the strength of an acid or a base?
the number of hydrogen ions or hydroxyl ions in the solution

The pH Scale Scientists have developed a scale to measure the strength of acids and bases. This scale is called the **pH scale.** The pH scale indicates the concentration of hydrogen ions or hydroxyl ions in solution.

The pH scale is a series of numbers from 0 to 14. A **neutral** solution has a pH of 7. A neutral solution is neither acidic nor basic. Acids have a pH below 7. Bases have a pH above 7. Strong acids have a low pH, while strong bases have a high pH.

▸*Define:* What is a neutral solution?
a solution that is neither acidic nor basic

Indicators and pH Indicators can be used to help find the exact pH of an acid or a base. pH paper is an indicator that changes color depending on the pH of the solution. In strong acids, pH paper is red. It is pink in weak acids and green in weak bases. In strong bases, pH paper is blue.

▸*Predict:* What color will sulfuric acid turn pH paper? red

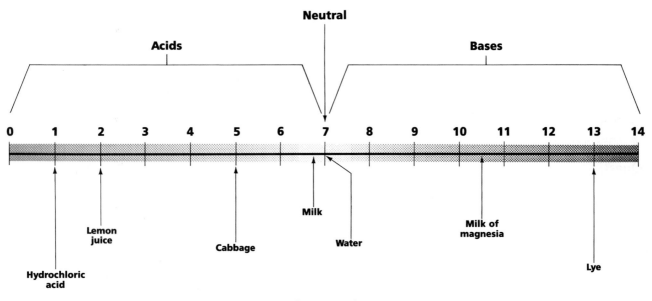

The pH Scale

LESSON SUMMARY

▶ Acids and bases can be strong or weak.

▶ The strength of an acid or a base depends on the number of hydrogen or hydroxyl ions in the solution.

▶ The pH scale is a measure of the strength of acids and bases.

▶ Indicators can be used to help determine the strength of an acid or a base.

CHECK *Write true if the statement is true. If the statement is false, change the underlined term to make the statement true.*

1. Sulfuric acid is a <u>strong</u> acid. true
2. Nitric acid is a <u>weak</u> acid. strong
3. Ammonium hydroxide is a <u>strong</u> base. weak
4. Sodium hydroxide is a <u>weak</u> base. strong
5. The strength of a base depends on the number of <u>hydroxyl</u> ions in a solution. true
6. Acids have a pH <u>above</u> 7. below
7. A solution with a pH of 12 is <u>an acid</u>. a base

APPLY *Use the pH scale on page 386 to answer the following questions.*

8. What is the pH of the following substances?
   a. milk
   b. water
   c. lye
   d. hydrochloric acid
   e. lemon juice
   f. milk of magnesia
   g. cabbage

9. a. Which of the substances shown on the pH scale are acids? b. Which are bases?
   c. Which substance is neutral?

Ideas in Action ....

IDEA: Many common household products are acids or bases.

ACTION: Make a list of products around your house that are acids and a list of those that are bases. Use library references to try to identify the acid or base present in each product.

ACTIVITY

TESTING THE ACIDITY OF FOODS

You will need samples of several foods, test tubes, water, a graduated cylinder, and pink and blue litmus paper.

1. Place a small amount of each food sample to be tested in a separate test tube. Solid foods should be broken into small pieces.
2. Add 5 mL of water to the solid food samples. Gently shake the test tubes to mix the water and solids.
3. Test each sample with pink and blue litmus paper.
4. Observe and record your results.

Questions

1. Which foods were acids?
2. Which foods were bases?
3. Were any of the foods neutral? How do you know?

What is neutralization?

Objective ▶ Describe what happens when an acid reacts with a base.

TechTerms

- ▶ **neutralization** (noo-truh-li-ZAY-shun): reaction between an acid and a base to produce a salt and water
- ▶ **salt**: substance formed from the negative ion of an acid and the positive ion of a base

Mixing Acids and Bases Some acids and bases are dangerous, corrosive compounds. However, different compounds are formed when acids and bases are mixed together. Hydrochloric acid and sodium hydroxide can each cause burns if they are spilled on the skin. When they are mixed together, they form sodium chloride and water. Sodium chloride is a harmless substance known as table salt. The chemical reaction between hydrochloric acid and sodium hydroxide produces two new compounds with different properties.

▶**Describe:** What happens when hydrochloric acid is mixed with sodium hydroxide?

Sodium chloride and water are formed.

Neutralization The reaction between an acid and a base forms a neutral substance. When red litmus paper is placed in a solution of table salt and water, the paper does not change color. When blue litmus is placed in a solution of table salt and water, the paper does not change color. Salt water is neither acidic nor basic. It is a neutral substance.

When acids and bases are mixed, a **neutralization** (noo-truh-li-ZAY-shun) reaction takes place. Neutralization reactions are double-replacement reactions. Water and a salt are always formed in a neutralization reaction.

▶**Define:** What is a neutralization reaction?

reaction between an acid and a base

Salts Sodium chloride is called table salt. It is just one of many different salts. A **salt** is a substance produced when an acid reacts with a base. A salt is made up of the negative ion from the acid and the positive ion from the base. Mixing sulfuric acid with sodium hydroxide produces the salt sodium sulfate and water.

▶**Identify:** When is a salt produced?

when an acid reacts with a base

Acid	+	Base	⟶	Salt	+	Water
H₂SO₄ sulfuric acid	+	NaOH sodium hydroxide	⟶	Na₂SO₄ sodium sulfate	+	2H₂O water
HCl hydrochloric acid	+	KOH potassium hydroxide	⟶	KCl potassium chloride	+	H₂O water
HCl hydrochloric acid	+	NaOH sodium hydroxide	⟶	NaCl sodium chloride	+	H₂O water

LESSON SUMMARY

▶ When acids and bases are mixed, new compounds with different properties are formed.

▶ The reaction between an acid and a base forms a neutral substance.

▶ Water and a salt are always formed in a neutralization reaction.

▶ A salt is made up of the negative ion from an acid and the positive ion from a base.

CHECK *Find the sentence in the lesson that answers each question. Then, write the sentence.*

1. What is a neutral substance?
2. Why is sodium hydroxide dangerous?
3. What happens when hydrochloric acid and sodium hydroxide are mixed together?
4. What is sodium chloride?
5. What is a neutralization reaction?
6. What is a salt?
7. What happens when red litmus paper and blue litmus paper are placed in a solution of salt and water?

APPLY *Match the salts in the second column with the acids and bases in the first column from which they were formed.*

Acid + Base ⟶ Salt

8. $HCl + NaOH$	$CaSO_4$
9. $H_2SO_4 + Ca(OH)_2$	$NaCl$
10. $HNO_3 + NH_4OH$	K_3PO_4
11. $H_3PO_4 + KOH$	NH_4NO_3

Skill Builder

Experimenting You can compare the strengths of acids by neutralizing them with a base. Try this experiment. Place 10 mL of an acid into a beaker. Add a few drops of phenolphthalein. Add a base one drop at a time to the acid and indicator. Stop when the phenolphthalein turns light pink. How many drops of base did you add? Repeat this experiment with a different acid. The stronger the acid, the more base you will have to add to neutralize it.

▾▾▾ LOOKING BACK IN SCIENCE ▾▾▾▾▾▾▾▾▾▾▾▾▾▾▾▾▾▾▾▾▾▾▾▾

HISTORY OF SALT

The saying that people are "worth their salt" comes from a time when salt had great value. Being described as "the salt of the earth" was a great compliment. Spilling salt was considered bad luck because it was too valuable to be wasted. Salt was so valuable at one time that Roman soldiers were paid with it. The part of their wages that was paid in salt was called their "salary."

When salt is added to water a substance called brine is produced. Brine is used to pickle and preserve many different types of foods. Before canning and refrigeration were available, pickling was used to keep meat and vegetables from spoiling.

Today, salt is used to make ice cream, dyes, rubber, soap, leather, and many other items. Salt can be readily obtained by the evaporation of sea water, but it is usually mined like other minerals. Deposits of salt are found where prehistoric inland seas have dried up.

21-6 What is an electrolyte?

Objective ▶ Explain how ions in solution conduct electricity.

TechTerms

- ▶ **electrolyte** (i-LEK-truh-lyt): substance that conducts an electric current when it is dissolved in water
- ▶ **nonelectrolyte:** substance that will not conduct an electric current when it is dissolved in water
- ▶ **ionization** (y-ahn-uh-ZAY-shun): formation of ions

Electrolytes A substance that conducts an electric current when it is dissolved in water is called an **electrolyte** (i-LEK-truh-lyt). Pure, or distilled, water is not a good conductor of electricity. Figure 1 shows what happens when electrodes connected to a battery are placed in distilled water. Electricity does not flow through the circuit. The bulb does not light. Figure 2 shows what happens when acetic acid is added to the distilled water. The bulb lights. A solution of acetic acid and water is a good conductor of electricity.

▶*Infer:* Is a solution of acetic acid and water an electrolyte? Why or why not?

Yes, because it conducts an electric current.

Ionization Electrolytes can conduct electricity because they form ions when they dissolve in water. A solution that contains ions can conduct an electric current. The ions move and carry the current through the solution.

The formation of ions is called **ionization** (y-ahn-uh-ZAY-shun). When sodium chloride dissolves in water, it forms Na^+ and Cl^- ions. When hydrochloric acid is added to water, it separates into H^+ and Cl^- ions. The base potassium hydroxide forms K^+ and Cl^- ions in solution.

▶*Define:* What is ionization? formation of ions

Nonelectrolytes A substance that does not conduct an electric current when it is dissolved in water is called a **nonelectrolyte.** A sugar solution does not conduct electricity. Sugar is a nonelectrolyte. Nonelectrolytes do not conduct an electric current because they do not form ions in solution.

▶*Classify:* Is sugar an example of an electrolyte or a nonelectrolyte? nonelectrolyte

Figure 1

Figure 2

LESSON SUMMARY

▶ An electrolyte is a substance that conducts electricity when it is dissolved in water.

▶ Electrolytes conduct electricity because they form ions when they dissolve in water.

▶ The formation of ions is called ionization.

▶ Nonelectrolytes do not conduct electricity because they do not form ions when they dissolve in water.

CHECK *Complete the following.*

1. Pure water is called _____ water. distilled

2. A solution of acetic acid and water is an example of an _____ . electrolyte

3. Acids, bases, and salts are all good _____ . electrolytes

4. A substance that forms _____ in solution is a good electrolyte. ions

5. Sugar is an example of a _____ . nonelectrolyte

6. The separation of a salt into ions is called _____ . ionization

APPLY *Complete the following.*

7. What ions will be formed in solutions of the following acids, bases, and salts?
   a. KCl
   b. Ca(OH)$_2$
   c. NH$_4$OH
   d. HNO$_3$
   e. CaCl$_2$
   f. NaBr
   g. HI
   h. MgSO$_4$

Designing an Experiment

Design an experiment to solve the problem.

PROBLEM: How can you identify a substance as an electrolyte or a nonelectrolyte?

Your experiment should:

1. List the materials you would need.

2. Identify safety precautions that should be followed.

3. List a step-by-step procedure.

4. Describe how you would record your data.

ACTIVITY

CLASSIFYING ACIDS AND BASES

You will need red and blue litmus paper, distilled water, a spoon, a dropper, and several different samples to be tested.

1. Choose several samples to be tested, for example, table salt, sugar, vinegar, ammonia, lemon juice, and baking soda.

2. To test solids, wet a piece of litmus paper with a drop of distilled water. Place the wet litmus paper into a small amount of the solid.

3. To test liquids, add a few drops of the liquid to the end of the litmus paper.

4. Test each of your samples. Observe and record your results in a table.

Questions

1. Which of your samples are acids? How do you know?

2. Which of your samples are bases? How do you know?

3. Are any of your samples neutral substances? How do you know?

STUDY HINT Before you begin the Unit Challenges, review the TechTerms and Lesson Summary for each lesson in this unit.

TechTerms . . . .

acid (380)
base (382)
electrolyte (390)
hydroxyl ion (382)

indicator (384)
ionization (390)
neutral (386)
neutralization (388)

nonelectrolyte (390)
pH scale (386)
salt (388)

TechTerm Challenges . . . .

Matching *Write the TechTerm that matches each description.*

1. formed when metals react with water base
2. neither acidic nor basic neutral
3. changes color in acids and bases indicator
4. reaction between an acid and a base neutralization
5. formation of ions ionization
6. conducts an electric current in water electrolyte
7. negative ion made up of a hydrogen atom and an oxygen atom hydroxyl ion

Fill In *Write the TechTerm that best completes each statement.*

1. A _____ is always formed in a neutralization reaction. salt
2. Sugar is a _____ because it does not conduct an electric current when it is dissolved in water. nonelectrolyte
3. Scientists use the _____ to measure the strength of acids and bases. pH scale
4. Vinegar is an example of an _____ . acid
5. A substance with a pH below 7 is an _____ . acid

Content Challenges . . . .

Multiple Choice *Write the letter of the term that best completes each statement.*

1. The acid in lemons and oranges is
   a. lactic acid. **b.** citric acid. **c.** acetic acid. **d.** hydrochloric acid. b

2. Acids are sometimes called
   a. proton donors. **b.** proton acceptors. **c.** electron donors. **d.** electron acceptors. a

3. When they are placed in water, all bases release
   a. chloride ions. **b.** ammonium ions. **c.** hydrogen ions. **d.** hydroxyl ions. d

4. Bases taste
   a. sweet. **b.** sour. **c.** bitter. **d.** salty. c

5. Phenolphthalein and bromthymol blue are
   a. acids. **b.** bases. **c.** electrolytes. **d.** indicators. d

6. If you put a strip of blue litmus paper into a cup of vinegar, the litmus paper will turn
   a. red. **b.** yellow. **c.** green. **d.** colorless. a

7. The pH of a neutral solution is
   a. 0. **b.** 7. **c.** 10. **d.** 14. b

8. A solution with a pH of 2 is a
   a. strong base. **b.** weak base. **c.** strong acid. **d.** weak acid. c

9. A neutralization reaction is a reaction between
   a. an acid and a base. b. a salt and water. c. an acid and water. d. a base and water. a

10. A reaction between sulfuric acid and sodium hydroxide produces water and
    a. sodium chloride. b. sulfur hydroxide. c. hydrochloric acid. d. sodium sulfate. d

11. Electrolytes can conduct electricity when they are dissolved in water because they form
    a. molecules. b. compounds. c. ions. d. salts. c

12. Of the following substances, the one that is a nonelectrolyte is
    a. sodium chloride. b. acetic acid. c. potassium hydroxide. d. sugar. d

True/False *Write true if the statement is true. If the statement is false, change the underlined term to make the statement true.*

1. An acid that aids in the digestion of food is <u>lactic</u> acid. hydrochloric
2. All acids have a <u>sour</u> taste. true
3. Acids react with metals to release <u>oxygen</u> gas. hydrogen
4. Soap and ammonia both contain <u>acids</u>. bases
5. All bases release <u>hydrogen</u> ions in water. hydroxyl
6. Chemicals, such as phenolphthalein, that change color in acids and bases are called <u>indicators</u>. true
7. Hydrangeas have <u>pink</u> flowers in acidic soil. blue
8. <u>Carbonic</u> acid is a weak acid that is used as an eyewash. Boric
9. A neutral solution has a pH of <u>7</u>. true
10. A neutralization reaction is a <u>double</u>-replacement reaction. true
11. A salt contains <u>positive</u> ions from an acid. negative
12. Distilled water is <u>polluted</u> water. pure
13. A solution that contains ions <u>can</u> conduct an electric current. true

Understanding the Features . . . .

Reading Critically *Use the feature reading selections to answer the following. Page numbers for the features are shown in parentheses.*

1. What are the two acids that cause acid rain? (380)
2. What is the equation for the soap-making process? (382)
3. Why was spilling salt thought to be bad luck? (388)
4. **Hypothesize:** Why do you think fish cannot live in lakes that have a high acid content? (380)

Concept Challenges . . . .

Critical Thinking *Answer each of the following in complete sentences.*

1. **Hypothesize:** Why should you never taste an unknown substance to identify it as an acid or a base?
2. **Classify:** An unknown solution is found to have a pH of 3. Congo red turns blue in this solution, and bromthymol blue turns yellow. Is the unknown solution an acid or a base?
3. **Predict:** Will the unknown solution in question 2 conduct an electric current? How do you know?
4. Use specific examples to explain the differences between an acid and a base.

Interpreting a Table *Use the table to answer each of the following questions.*

Table 1 Common Acids

ACID	CHEMICAL FORMULA	USES
Acetic acid	$HC_2H_3O_2$	vinegar
Boric acid	H_3BO_3	eyewash
Carbonic acid	H_2CO_3	club soda
Citric acid	$H_3C_6H_5O_7$	citrus fruits
Hydrochloric acid	HCl	aids digestion
Nitric acid	HNO_3	fertilizers
Sulfuric acid	H_2SO_4	plastics

1. What element do all of the acids listed have in common?
2. What is the chemical formula for acetic acid?
3. Which acid is used to make fertilizers?
4. What is the chemical formula of the acid found in citrus fruits? What is the name of this acid?
5. **Hypothesize:** Why do you think boric acid can be used as an eyewash?

Finding Out More . . . .

1. Use library reference to find out how salt is obtained from sea water. Describe this process to the class.
2. Look up the word "saponification" in a dictionary. What does this word mean? How is saponification related to the soap-making process? Write a report of your findings.
3. Visit a local nursery or garden center. Ask how gardeners determine the pH of soil before planting flowers or shrubs. How can the pH of soil be changed?
4. Hydrochloric acid in the stomach is important to the process of chemical digestion. However, too much acid in the stomach can be harmful. Use library references to find out how too much stomach acid can cause ulcers or other medical problems. How do antacids work to reduce the amount of acid in the stomach?

Appendices

Appendix A

THE METRIC SYSTEM AND SI UNITS

The metric system is an international system of measurement based on units of 10. More than 90 percent of the nations of the world use the metric system. In the United States, both the English or Imperial Measurement System and the metric system are used.

Systeme International, or SI, has been used as the international measurement system since 1960. SI is a modernized version of the metric system. Like the metric system, SI is a decimal system based on units of 10.

In both SI and the metric system, prefixes are added to base units to form larger or smaller units. Each unit is 10 times larger than the next smaller unit, and 10 times smaller than the next larger unit. For example, the meter is the basic unit of length. The next larger unit is a dekameter. A dekameter is 10 times larger than a meter. The next smaller unit is a decimeter. A decimeter is 10 times smaller than a meter. Ten decimeters is equal to one meter. How many meters equal one dekameter? 10

When you want to change from one unit in the metric system to another unit, you multiply or divide by a multiple of 10.

• When you change from a smaller unit to a larger unit, you divide.

• When you change from a larger unit to a smaller unit, you multiply.

SI UNITS	
The basic unit is printed in capital letters.	
Length	*Symbol*
kilometer	km
METER	m
centimeter	cm
millimeter	mm
Area	*Symbol*
square kilometer	km²
SQUARE METER	m²
square millimeter	mm²
Volume	*Symbol*
CUBIC METER	m³
cubic millimeter	mm³
liter	L
milliliter	mL
Mass	*Symbol*
KILOGRAM	kg
gram	g
tonne	t
Temperature	*Symbol*
KELVIN	K
degree Celsius	°C

SOME COMMON METRIC PREFIXES		
Prefix		*Meaning*
micro-	=	0.000001, or 1/1,000,000
milli-	=	0.001, or 1/1000
centi-	=	0.01, or 1/100
deci-	=	0.1, or 1/10
deka-	=	10
hecto-	=	100
kilo-	=	1000
mega-	=	1,000,000

SOME METRIC RELATIONSHIPS	
Unit	*Relationship*
kilometer	1 km = 1000 m
meter	1 m = 100 cm
centimeter	1 cm = 10 mm
millimeter	1 mm = 0.1 cm
liter	1 L = 1000 mL
milliliter	1 mL = 0.001 L
tonne	1t = 1000 kg
kilogram	1 kg = 1000 g
gram	1 g = 1000 mg
centigram	1 cg = 10 mg
milligram	1 mg = 0.001 g

SI–ENGLISH EQUIVALENTS

	SI to English	*English to SI*
Length	1 kilometer = 0.621 mile (mi)	1 mi = 1.61 km
	1 meter = 0.914 yards (yd)	1 yd = 1.09 m
	1 meter = 3.28 feet (ft)	1 ft = 0.305 m
	1 centimeter = 0.394 inch (in)	1 in = 2.54 cm
	1 millimeter = 0.039 inch	1 in = 25.4 mm
Area	1 square kilometer = 0.3861 square mile	$1\ mi^2 = 2.590\ km^2$
	1 square meter = 1.1960 square yards	$1\ yd^2 = 0.8361\ m^2$
	1 square meter = 10.763 square feet	$1\ ft^2 = 0.0929\ m^2$
	1 square centimeter = 0.155 square inch	$1\ in^2 = 6.452\ cm^2$
Volume	1 cubic meter = 1.3080 cubic yards	$1\ yd^3 = 0.7646\ m^3$
	1 cubic meter = 35.315 cubic feet	$1\ ft^3 = 0.0283\ m^3$
	1 cubic centimeter = 0.0610 cubic inches	$1\ in^3 = 16.39\ cm^3$
	1 liter = .2642 gallon (gal)	1 gal = 3.79 L
	1 liter = 1.06 quart (qt)	1 qt = 0.94 L
	1 liter = 2.11 pint (pt)	1 pt = 0.47 L
	1 milliliter = 0.034 fluid ounce (fl oz)	1 fl oz = 29.57 mL
Mass	1 tonn = .984 ton	1 ton = 1.016 t
	1 kilogram = 2.205 pound (lb)	1 lb = 0.4536 kg
	1 gram = 0.0353 ounce (oz)	1 oz = 28.35 g
Temperature	Celsius = 5/9 (°F −32)	Fahrenheit = 9/5°C + 32
	0°C = 32°F (Freezing point of water)	72°F = 22°C (Room temperature)
	100°C = 212°F (Boiling point of water)	98.6=F = 37°C
		(Human body temperature)

Appendix B

SAFETY IN THE SCIENCE CLASSROOM

Safety is very important in the science classroom. Science classrooms and laboratories have equipment and chemicals that can be dangerous if not handled properly. To avoid accidents in the science laboratory, always follow proper safety rules. Listen carefully when your teacher explains precautions and safety rules that must be followed. You should never perform an activity without your teacher's direction. By following safety rules you can help insure the safety of yourself and your classmates. Safety rules that should be followed are listed below. Read over these safety rules carefully. Always look for caution statements before you perform an activity.

Clothing Protection • Wear your laboratory apron. • Confine loose clothing.

Eye Safety • Wear safety goggles in the laboratory. • If anything gets in your eyes, flush them with plenty of water. • Be sure you know how to use the emergency wash system in the laboratory.

Heat and Fire Safety • Be careful when handling hot objects. • Use proper procedures when lighting Bunsen burners. • Turn off all heat sources when they are not in use. • Tie back long hair when working near an open flame. • Confine loose clothing. • Turn off gas valves when not in use.

Electrical Safety • Keep all work areas clean and dry. • Never handle electrical equipment with wet hands. • Do not overload an electric circuit. • Do not use wires that are frayed.

Glassware Safety • Never use chipped or cracked glassware. • Never pick up broken glass with your bare hands. • Allow heated glass to cool before touching it. • Never force glassware into a rubber stopper.

Chemical Safety • Never taste chemicals as a means of identification. • Never transfer liquids with a mouth pipette. Use a suction bulb. • Be very careful when working with acids or bases. Both can cause serious burns. • Never pour water into an acid or base. Always pour an acid or base into water. • Inform your teacher immediately if you spill chemicals or get any chemicals on your skin. • Use a waving motion of your hand to observe the odor of a chemical. • Never put your nose near a chemical. • Never eat or drink in the laboratory.

Sharp Objects • Use knives, scissors, and other sharp instruments with care. • Cut in the direction away from your body.

Cleanup • Clean up your work area before leaving the laboratory. • Follow your teacher's instructions for disposal of materials. • Wash your hands after an activity.

Appendix C

USING DECIMALS

As you study physical science, you may find it necessary to work with decimal numbers. This Appendix is intended to help you solve problems involving decimals. You may wish to review this material before working on problems in your text.

Adding and Subtracting Decimals   When adding or subtracting decimal numbers, always be sure to line up the decimal points correctly. Examples:

1. Add 3.4 km, 20.95 km, and 153.6 km.

$$
\begin{array}{r}
3.4 \\
20.95 \\
+153.6 \\
\hline
177.95 \text{ km}
\end{array}
$$

2. Subtract 13.5 mL from 35.75 mL.

$$
\begin{array}{r}
35.75 \\
-13.5 \\
\hline
22.25 \text{ mL}
\end{array}
$$

Multiplying and Dividing Decimals   When multiplying or dividing decimal numbers, it is not necessary to line up the decimal points. Examples:

1. Multiply 0.5 N by 11.25 m to find the amount of work done in joules.

$W = F \times d$
$W = 0.5 \text{ N} \times 11.25 \text{ m}$
$W = 5.625 \text{ J}$

Notice that the number of places to the right of the decimal point in the answer is equal to the sum of the places to the right of the decimal point in the numbers being multiplied.

2. Divide 4.05 m by 0.5 m to find the mechanical advantage of a lever.

MA = effort arm length/resistance arm length
MA = 4.05 m/0.5 m
MA = 8.1

When dividing a decimal by another decimal, you must first change the divisor to a whole number. For example, change 0.5 to 5 by moving the decimal point one place to the right. You must also move the decimal point in 4.05 one place to the right. The result is 40.5/5 = 8.1.

Changing a Decimal to a Percent   To change a decimal number to a percent, multiply the decimal by 100%.
Example:
Find the efficiency of a machine if the work output is 5 J and the work input is 10 J.

efficiency = work output/work input × 100%
efficiency = 5 J/10 J × 100%
efficiency = 0.5 × 100%
efficiency = 50%

Notice that when you multiply 0.5 by 100%, the decimal point moves two places to the right.

Appendix D
PREFIXES AND SUFFIXES

Prefixes and suffixes are word parts that can be helpful in determining the meaning of an unfamiliar term. Prefixes are found at the beginning of words. Suffixes are found at the end of words.

Both prefixes and suffixes have meanings that mainly come from Latin and Greek words. Some meanings of prefixes and suffixes commonly used in physical science are listed below.

Prefix	Meaning	Example
bar, baro-	weight, pressure	barometer
bi-	two	binary
carbo-	containing carbon	carbonate
co-	with, together	coagulation
de-	remove from	decomposition
di-	twice, two	diatomic
electro-	electricity	electrolyte
hydro-	water, containing hydrogen	hydrometer, hydrocarbon
in-	not	insoluble
magneto-	magnetism	magnetosphere
non-	not	nonmetal
photo-	light	photoelectric
poly-	many	polyatomic
re-	again, back	reflection
sub-	under, beneath	subscript
super-	above, more than	supersonic
therm-, thermo-	heat	thermometer
trans-	across, beyond	transparent
ultra-	beyond	ultraviolet
un-	not	unsaturated

Suffix	Meaning	Example
-ate	salt of an acid	nitrate
-graph	write	thermograph
-ide	binary compound	sulfide
-logy	study of	cosmology
-lysis	decomposition	electrolysis
-meter	measuring device	manometer
-sonic	sound	supersonic
-sphere	ball, globe	magnetosphere

Glossary

Pronunciation and syllabication have been derived from *Webster's New World Dictionary*, Second College Edition, Revised School Printing (Prentice Hall, 1985). Syllables printed in capital letters are given primary stress. (Numbers in parentheses indicate the page number, or page numbers, on which the term is defined.)

PRONUNCIATION KEY

Symbol	Example	Respelling	Symbol	Example	Respelling
ah	velocity	(vuh-LAHS-uh-tee)	oo	amplitude	(AM-pluh-tood)
ay	radiation	(ray-dee-AY-shun)	oy	colloid	(KAHL-oyd)
e	convection	(kuhn-VEK-shun)	s	solute	(SAHL-yoot)
ee	decomposition	(dee-kahm-puh-ZISH-un)	sh	suspension	(suh-SPEN-shun)
f	coefficient	(koh-uh-FISH-unt)	uh	barometer	(buh-RAHM-uh-tur)
ih	specialization	(SPESH-uh-lih-zay-shun)	y	binary	(BY-nur-ee)
j	homogenization	(huh-mahj-uh-ni-ZAY-shun)	yoo	insoluble	(in-SAHL-yoo-bul)
k	calorie	(KAL-uh-ree)	z	ionization	(y-ahn-uh-ZAY-shun)
oh	coagulation	(koh-ag-yoo-LAY-shun)			

A

absolute zero: lowest possible temperature; temperature at which particles of matter almost stop moving **(96)**

acceleration (uk-sel-uh-RAY-shun): change in speed or direction (62)

acid: substance that reacts with metals to release hydrogen (380)

action force: force acting in one direction (70)

air resistance: force that opposes the downward motion of a falling object

alkali metals: metals in Group 1 of the periodic table (334)

alkaline earth metals: metals in Group 2 of the periodic table (334)

alloy: substance made up of two or more metals (332)

alternating current: current in which electrons change direction at a regular rate (180)

amino acids: building blocks of proteins (288)

ampere: unit used to measure electric current (188)

amplitude (AM-pluh-tood): height of a wave (118)

atom: smallest part of an element that can be identified as that element (258)

atomic mass: total mass of the protons and neutrons in an atom, measured in atomic mass units (264)

atomic number: number of protons in the nucleus of an atom (262)

B

balanced forces: forces that are equal in size but opposite in direction (64)

barometer (buh-RAHM-uh-tur): instrument used to measure air pressure (34)

base: substance formed when metals react with water (382)

battery: series of electrochemical cells connected together (182)

Bernoulli's principle: principle which states that as the speed of a fluid increases, its pressure decreases (36)

binary (BY-nur-ee) **compound:** compound containing two elements (298)

boiling point: temperature at which a liquid changes to a gas (100, 356)

boiling point elevation: raising the boiling point of a liquid solvent by adding solute (356)

buoyancy (BOI-uhn-see): upward force exerted by a gas or liquid (250)

C

calorie (KAL-uh-ree): unit of heat; amount of heat needed to raise the temperature of 1 g of water 1 °C (94)

Calorie: 1000 calories, or 1 kilocalorie (94)

carbohydrates (kahr-buh-HY-drayts): sugars and starches (288)

chemical change: change that produces new substances (226)

chemical equation: statement in which chemical formulas are used to describe a chemical reaction (312)

chemical formula: way of writing the name of a compound using chemical symbols (294)

chemical reaction: process in which new substances with new chemical and physical properties are formed (310)

chemical symbols: shorthand way of writing the names of the elements (230)

chemistry (KEM-is-tree): study of matter and its reactions (220)

chlorophyll (KLOWR-uh-fil): green substance in plants (160)

circuit breaker: switch that opens a circuit if too much current is flowing (192)

coagulation (koh-ag-yoo-LAY-shun): use of chemicals to make the particles in a suspension clump together (368)

coefficient (koh-uh-FISH-unt): number that shows how many molecules of a substance are involved in a chemical reaction (312)

colloid (KAHL-oyd): suspension in which the particles are permanently suspended (372)

compound: substance made up of two or more elements chemically combined (274)

compound machine: machine that combines two or more simple machines (86)

compression (kum-PRESH-un): part of a medium where the particles are close together (116)

concave lens: lens that curves inward (164)

concentrated solution: strong solution (352)

condensation (kahn-dun-SAY-shun): change of a gas to a liquid (224, 358)

conduction (kun-DUK-shun): heat transfer in solids (102)

conductors: materials that allow electric charges to flow through them easily (102, 176); substances that conduct heat easily (178)

convection (kuhn-VEK-shun): heat transfer in gases and liquids (104)

convection currents: up and down movements of gases or liquids caused by heat transfer (104)

convex lens: lens that curves outward (164)

corrosion: wearing away of a metal (336)

covalent bond: bond formed when atoms share electrons (284)

crest: high point of a wave (116)

crystal lattice (LAT-is): ions arranged in a regular pattern (282)

D

data (DAY-tuh): information (6)

decibel: unit used to measure the intensity or loudness of a sound (138)

decomposition (dee-kahm-puh-ZISH-un) **reaction:** reaction in which a complex substance is broken down into simpler substances (318)

degree Celsius (SEL-see-us): metric unit of temperature (14)

density (DEN-suh-tee): mass per unit volume (242)

diatomic molecule: molecule made up of two atoms of the same element (302)

diffuse reflection: reflection that forms a fuzzy image (162)

dilute solution: weak solution (352)

direct current: current in which electrons always flow in the same direction (180)

displacement (dis-PLAYS-muhnt): amount of water an object replaces (248)

dissolve: go into solution (344)

distillation (dis-tuh-LAY-shun): process of evaporating a liquid and then condensing the gas back into a liquid (358)

Doppler effect: apparent change in the frequency of waves (124)

double-replacement reaction: reaction in which elements from two different compounds replace each other, forming two new compounds (322)

ductile (DUK-tul): able to be drawn into thin wires (236)

E

ear: sense organ that detects sound (146)

echo: reflected sound waves (136)

efficiency (uh-FISH-un-see): comparison of work output to work input (78)

effort force: force applied to a machine (76)

electric circuit: path that an electric current follows (184)

electric current: flow of electrons through a conductor (180)

electric motor: device that changes electrical energy into mechanical energy (212)

electricity (i-lek-TRIS-uh-tee): form of energy caused by moving electrons (176)

electrochemical cell: device that changes chemical energy into electrical energy (182)

electrode: positive or negative pole of an electrochemical cell (182)

electrolysis (i-lek-TRAHL-uh-sis): process by which water is decomposed using an electric current (318)

electrolyte (i-LEK-truh-lyt): substance that conducts an electric current when it is dissolved in water (182); substance that is an electrical conductor (390)

electromagnet: temporary magnet made by wrapping a current-carrying wire around an iron core (208)

electromagnetic induction: process by which an electric current is produced by moving a wire in a magnetic field (206)

electromagnetic spectrum: range of electromagnetic waves (166)

electromagnetism: relationship between electricity and magnetism (206)

electron: atomic particle with a negative electric charge (176, 260)

electroplating (i-LEK-truh-playt-ing): use of an electric current to plate one metal with another metal (338)

elements (EL-uh-munts): simple substances that cannot be broken down into simpler substances (228)

emulsion (i-MUL-shun): suspension of two liquids (370)

energy: ability to do work (42)

energy level: place in an atom where an electron is most likely to be found (268)

evaporation (i-vap-uh-RAY-shun): change from a liquid to a gas at the surface of the liquid (100, 224, 358)

eye: sense organ that detects light (156)

F

filtration: separation of particles in a suspension by passing it through paper or other substances (368)

fluid pressure: pressure in gases and liquids (32)

force: push or pull (20, 64)

formula mass: sum of the mass numbers of all the atoms in a molecule (304)

freezing: change from a liquid to a solid (224)

freezing point: temperature at which a liquid changes to a solid (98, 354)

freezing point depression: lowering the freezing point of a liquid solvent by adding solute (354)

frequency (FREE-kwun-see): number of complete waves passing a point in a given time (118)

friction: force that opposes the motion of an object (28)

fulcrum (FUL-krum): point at which a lever is supported (80)

fundamental (fun-duh-MEN-tul) **tone:** lowest-pitched sound produced when a whole string vibrates (142)

fuse: wire that melts and breaks a circuit if too much current is flowing (192)

G

gas: phase of matter that has no definite shape or volume (222)

generator: device that changes mechanical energy into electrical energy (214)

gravity: force of attraction between all objects in the universe (22)

group: vertical column of elements in the periodic table (232)

H

hearing: one of the five human senses (146)

heat: form of energy in moving particles of matter (92)

hertz (HURTS): unit used to measure the frequency of a wave (118)

homogenization (huh-mahj-uh-ni-ZAY-shun): formation of a permanent emulsion (370)

hydrometer (hy-DRAHM-uh-tuhr): device used to measure specific gravity (246)

hydroxyl (hy-DRAHK-sil) **ion:** negative ion made up of one atom of hydrogen and one atom of oxygen (382)

hypothesis (hy-PAHTH-uh-sis): suggested solution to a problem (4)

I

illuminated (i-LOO-muh-nayt-ed) **objects:** objects that reflect light (154)

image: picture formed by the eye (156)

incident wave: wave that strikes a barrier (120)

inclined plane: slanted surface, or ramp (84)

indicator (IN-duh-kayt-ur): substance that changes color in acids and bases (384)

inert (in-URT) **gases:** six elements that make up the last group in the periodic table of the elements (238)

inertia (in-UR-shuh): tendency of an object to stay at rest or in motion (66)

insoluble (in-SAHL-yoo-bul): not able to dissolve (346)

insulators: substances that do not conduct heat easily (102); materials that prevent electric charges from flowing through them easily (178)

intensity: amount of energy in a sound wave (138)

ion (Y-un): charged particle (282)

ionic bond: bond formed when atoms gain and lose electrons (282)

ionization (y-ahn-uh-ZAY-shun): formation of ions (390)

isotope (Y-suh-tohp): atom of an element with the same number of protons but a different number of neutrons (266)

J

joule (JOOL): metric unit of work; equal to 1 N-m (52)

K

kilogram (KIL-ih-gram): basic unit of mass (10)

kinetic (ki-NET-ik) **energy:** energy of motion (42)

L

laser: device that produces a powerful beam of light (168)

law of conservation of energy: law which states that energy cannot be made or destroyed, but only changed in form (48)

lens: transparent material that bends light (164)

lever (LEV-ur): bar that is free to turn about a fixed point (80)

light: form of electromagnetic (i-lek-troh-mag-NET-ik) energy made up of streams of photons (150)

lipids: fats and oils (288)

liquid: phase of matter with a definite volume, but no definite shape (222)

liter (LEE-tur): basic metric unit of volume (12)

longitudinal (lahn-juh-TOOD-un-ul) **wave:** wave in which the particles of the medium move back and forth in the direction of the wave motion (116, 132)

lubricants (LOO-bruh-kunts): substances that reduce friction (30)

luminous (LOO-muh-nus) **objects:** objects that give off their own light (154)

luster (LUS-tur): shine (236)

M

magnetic field: area around a magnet where magnetic forces can act (200)

magnetic induction: process by which a material can be made into a magnet (202)

magnetic lines of force: lines that show the shape of a magnetic field (200)

magnetism: force of attraction or repulsion (198)

magnetosphere (mag-NEET-uh-sfir): region of the earth's magnetic field (204)

malleable (MAL-ee-uh-bul): able to be hammered into different shapes (236)

manometer (muh-NAHM-uh-tur): instrument used to measure pressure in a liquid (34)

mass: amount of matter in an object (10)

mass number: number of protons and neutrons in the nucleus of an atom (264)

matter: anything that has mass and takes up space (220)

mechanical advantage: number of times a machine multiplies the effort force (76)

medium: substance through which waves can travel (114, 132)

melting: change from a solid to a liquid (224)

melting point: temperature at which a solid changes to a liquid (98)

meniscus (mi-NIS-kus): curved surface of a liquid in a graduated cylinder (12)

metals (MET-uls): elements that have the properties of luster, ductility, and malleability (236)

meter (MEE-tur): basic SI and metric unit of length (8)

mirror: smooth surface that reflects light and forms images (162)

mixture: two or more substances that have been combined, but not chemically changed (278)

molecule: smallest part of a compound that has all the properties of the compound (276)

motion: change in position (60)

music: sounds combining a pleasing quality, melody, harmony, and rhythm (144)

N

neutral: having neither a positive nor a negative electric charge (260); neither acidic nor basic (386)

neutralization (noo-truh-li-ZAY-shun): reaction between an acid and a base to produce a salt and water (388)

neutron: atomic particle with neither a negative nor a positive electric charge (176, 260)

newton: metric unit of force equal to one kilogram-meter per second per second (24, 68)

noise: unpleasant sounds, with irregular patterns of vibration (144)

nonelectrolyte: substance that will not conduct an electric current when it is dissolved in water (390)

nonmetals: elements that have none of the properties of metals (236)

normal: line at right angles to a barrier (120)

nucleus: center, or core, of an atom (260)

O

ohm: unit used to measure resistance (188)

Ohm's law: law which states that the current in a wire is equal to the voltage divided by the resistance (190)

opaque (oh-PAYK): material that blocks light (154)

ore: rock or mineral from which a useful metal can be removed (328)

organic chemistry: study of organic compounds (286)

organic compound: compound containing carbon (286)

overtones: high-pitched sounds produced when parts of a string vibrate (142)

oxidation (ahk-suh-DAY-shun): chemical reaction in which electrons are lost (314)

oxidation number: number of electrons an atom gains, loses, or shares when it forms a chemical bond (296)

P

parallel circuit: circuit in which electric current can follow more than one path (186)

period: horizontal row of elements in the periodic table (232)

periodic (pir-ee-AHD-ik): repeating pattern (232)

pH scale: measure of the concentration of hydrogen ions or hydroxyl ions in a solution (386)

phase (FAYZ): form of matter (222)

photon (FOH-tahn): tiny bundle of energy (150)

photosynthesis (foht-uh-SIN-thuh-sis): process by which plants use energy from the sun to make food (160)

physical change: change that does not produce new substances (226)

pitch: how high or low a sound is (140)

plating: coating one metal with another metal (338)

polar molecule: molecule in which one end has a positive charge and the other end has a negative charge (348)

poles: two ends of a magnet (198)

pollutants (puh-LOOT-unts): harmful substances in the environment (374)

polyatomic (PAHL-i-uh-tahm-ik) **ion:** group of atoms that acts as a single atom when combining with other atoms (300)

potable (POHT-uh-bul) **water:** water that is safe to drink (374)

potential (puh-TEN-shul) **energy:** stored energy (42)

power: amount of work done per unit of time (54)

precipitate (pri-SIP-uh-tayt): solid that settles to the bottom of a mixture (322)

pressure: force per unit area (32)

prism (PRIZ-um): triangular piece of glass that breaks up white light into a band of colors (158)

product: substance that is formed in a chemical reaction (310)

properties (PROP-ur-tees): characteristics used to describe a substance (220)

proteins: compounds needed to build and repair the body (288)

proton: atomic particle with a positive electric charge (176, 260)

pulley: rope wrapped around a wheel (82)

R

radiation (ray-dee-AY-shun): transfer of heat through space (106)

rarefaction (rer-FAK-shun): part of a medium where the particles are far apart (116)

ray: straight line that shows the direction of light (150)

reactant: substance that is changed in a chemical reaction (310)

reaction force: force acting in the opposite direction from an action force (70)

real image: image that can be projected onto a screen (164)

reduction (ri-DUK-shun): chemical reaction in which electrons are gained (314); process of removing oxygen from an ore (330)

reflected wave: wave that bounces back from a barrier (120)

reflection: bouncing back of a wave after striking a barrier (120)

refraction: bending of a wave as it moves from one medium to another (122)

regular reflection: reflection that forms a clear image (162)

resistance: opposition to the flow of electric current (188)

resistance force: force that opposes the effort force (76)

roasting: process in which an ore is heated in air to produce an oxide (330)

S

salt: substance formed from the negative ion of an acid and the positive ion of a base (388)

saturated solution: solution containing all the solute it can hold at a given temperature (352)

scientific method: model or guide used to gather information and solve problems (6, 48)

scientific theory: idea supported by evidence over a period of time

series circuit: circuit in which electric current follows only one path (184)

sight: one of the five human senses (156)

single-replacement reaction: reaction in which one element replaces another element in a compound. (320)

solid: phase of matter with a definite shape and volume (222)

soluble (SAHL-yoo-bul): able to dissolve (346)

solute (SAHL-yoot): substance that is dissolved (346)

solution: mixture in which one substance is evenly mixed with another substance (344)

solvent: substance in which a solute dissolves (346)

sound: form of energy that travels as waves (130)

specialization (SPESH-uh-lih-zay-shun): studying or working in only one part of a subject. (2)

specific (spi-SIF-ik) **gravity:** density of a substance compared with the density of water (246)

speed: distance traveled per unit of time (60)

structural formula: molecular model of an organic compound (286)

subscript: number written to the lower right of a chemical symbol (294)

supersaturated solution: solution containing more solute than it can normally hold at a given temperature (360)

supersonic: faster than the speed of sound (134)

suspension (suh-SPEN-shun): cloudy mixture of two or more substances that settle on standing (366)

synthesis (SIN-thuh-sis) **reaction:** reaction in which substances combine to form a more complex substance (316)

T

temperature: measure of how hot or cold something is (14); measure of the average kinetic energy of the particles of a substance (96)

terminal velocity: speed at which air resistance and gravity acting on a falling object are equal (26)

thermal (THUR-mul) **expansion:** expansion of a substance caused by heating (108)

thermal pollution: damage that occurs when waste heat enters the environment (46)

timbre (TAM-bur): sound quality (142)

transformer: device in which alternating current in one coil of wire induces a current in a second coil (210)

translucent (trans-LOO-sunt): material that transmits some light (154)

transparent (trans-PER-unt): material that transmits light easily (154)

transverse (trans-VURS) **wave:** wave in which the particles of the medium move up and down at right angles to the direction of the wave motion (116, 152)

trough (TROWF): low point of a wave (116)

U

ultrasonic (ul-truh-SAHN-ik): above 20,000 Hz frequency (140)

unbalanced forces: forces that cause a change in the motion of an object (64)

unit (YOU-nit): amount used to measure something (8)

unsaturated solution: solution containing less solute than it can hold at a given temperature (352)

V

vacuum: empty space (26, 104)

valence electrons: electrons in an atom's outer energy level (296)

velocity (vuh-LAHS-uh-tee): speed and direction of motion (60)

vibration: rapid back-and-forth movement (130)

visible spectrum: seven colors that make up white light (158)

volt: unit used to measure voltage (188)

voltage: force that makes electrons move (188)

volume: amount of space something takes up (12)

W

watt: metric unit of power; equal to 1 J/sec (54)

wavelength: distance between two neighboring crests or troughs (118)

waves: disturbances that transfer energy from place to place (114)

weight: measure of the pull of gravity on an object (10)

work: force times distance (50)

work input: work done on a machine (78)

work output: work done by a machine (78)

INDEX

Photo Credits